ELECTRICITY
Principles and Applications

ELECTRICITY
Principles and Applications

RICHARD J. FOWLER
WESTERN WASHINGTON UNIVERSITY

McGraw-Hill Book Company
Gregg Division

New York
St. Louis
Dallas
San Francisco
Auckland
Bogotá
Düsseldorf
Johannesburg
London
Madrid
Mexico
Montreal
New Delhi
Panama
Paris
São Paulo
Singapore
Sydney
Tokyo
Toronto

Library of Congress Cataloging in Publication Data

Fowler, Richard J.
Electricity, principles and applications.

(Basic skills in electricity and electronics)
1. Electricity. I. Title. II. Series.
QC523.F75 537 78-26824
ISBN 0-07-021704-1

Acknowledgments
Students, teachers, school administrators, and industrial trainers have contributed to the development of the *Basic Skills in Electricity and Electronics* series. Classroom testing of preliminary editions has been conducted at the following sites:

Burr D. Coe Vocational Technical High School (East Brunswick, New Jersey)
Chantilly Secondary School (Chantilly, Virginia)
Nashoba Valley Technical High School (Westford, Massachusetts)
Platt Regional Vocational Technical High School (Milford, Connecticut)
United States Steel Corporation: Edgar Thomson, Irvin Works (Dravosburg, Pennsylvania)

The publisher gratefully acknowledges the helpful comments and suggestions received from these participants.

Electricity: Principles and Applications

234567890 VHVH 786543210

The editors for this book were Gordon Rockmaker and Mark Haas. The designer was Tracy A. Glasner. The art supervisor was George T. Resch. The production supervisor was Kathleen Morrissey. Cover photography by Martin Bough/Studio Inc. It was set in Electra by Trufont Typographers, Inc.
Printed and bound by Von Hoffman Press, Inc.

Contents

Editor's Foreword

The Gregg/McGraw-Hill *Basic Skills in Electricity and Electronics* series has been designed to provide entry-level competencies in a wide range of occupations in the electrical and electronics fields. The series consists of instructional materials geared especially for the career-oriented student. Each major subject area covered in the series is supported by a textbook, an activities manual, and a teacher's manual. All the materials focus on the theory, applications, and experiences required for those just beginning their chosen vocations.

There are two basic considerations in the preparation of educational materials for such a series: the needs of the learner and the needs of the employer. The materials in the series have been designed to meet these needs. They are based on many years of experience in the classroom and with electricity and electronics. In addition, these books reflect the needs of industry and commerce as developed through questionnaires, surveys, interviews with employers, government occupational trend reports, and various field studies.

Further refinements both in pedagogy and technical content resulted from actual classroom experience with the materials. Preliminary editions of selected texts and manuals were field-tested in schools and in-plant training programs throughout the country. The knowledge from this testing has enhanced the effectiveness and the validity of the materials.

Teachers will find the materials in each of the subject areas well coordinated and structured around a framework of modern objectives. Students will find the concepts clearly presented with many practical references and applications. In all, every effort has been made to prepare and refine the most effective learning tools possible.

The publisher and editor welcome comments from teachers and students using this book.

Charles A. Schuler
Project Editor

BASIC SKILLS IN ELECTRICITY AND ELECTRONICS

Charles A. Schuler, Project Editor

Books in this series

Introduction to Television Servicing by Wayne C. Brandenburg
Electricity: Principles and Applications by Richard J. Fowler
Instruments and Measurements by Charles M. Gilmore
Microprocessors by Charles M. Gilmore (*in preparation*)
Motors and Generators by Russell L. Heiserman and Jack D. Burson
Small Appliance Repair by Phyllis Palmore and Nevin E. André
Residential Wiring by Gordon Rockmaker
Electronics: Principles and Applications by Charles A. Schuler
Your Future in Electricity and Electronics by William A. Stanton
Digital Electronics by Roger L. Tokheim

Preface

This book is written for students just beginning their study of electricity. No previous formal training in the subject is assumed. Only arithmetic and basic algebra (solving equations with one unknown quantity) are used in explaining and solving electrical problems. A few topics in alternating current require the student to use trigonometry. In those specific cases, the elements of trigonometry that relate to electric circuits are covered as needed.

The materials in this textbook are arranged in a logical, sequential order. The terms and concepts that are developed and discussed in Chapter 1 are needed to fully understand Chapter 2. The concepts of Chapter 2 are used in Chapter 3, and so on. The first half of this text is devoted, in general, to the fundamentals of direct current, while the second half focuses on subjects usually associated with alternating currents. The transition from dc to ac is distinct enough to allow use of the material in a traditional dc/ac sequence. However, all of the material is structured to provide a unified introduction to the broad subject area called *electricity*.

Although no prior formal training in electricity is assumed, it is highly unlikely that a student will not have had an extensive exposure to the world of electricity and electronics. For this reason, certain words associated with electricity are expected to have already entered the students' vocabulary (for example, switch, plug, motor, television, etc.). Nevertheless, strict technical definitions are used throughout (which may help some students rid themselves of misconceptions learned earlier!).

Since concepts build from chapter to chapter, and within a chapter from section to section, students are provided with a number of stopping points for review questions and problems. These reviews serve two purposes: They reveal areas of weakness that require re-study, and they serve as positive reinforcement for material students know. Each chapter concludes with a summary of the key concepts covered in the chapter. The summaries will also help identify areas of weakness and reinforce areas of strength. Students should not skim through this material lightly. Finally, a unique color strip along the margin of each page is used to highlight key terms and phrases on each page.

This truly is the age of electricity and electronics. The solid foundation in electricity provided by this text is likely to prove useful, whether a student intends to pursue a career in electricity and electronics or simply needs the material as background information for other pursuits. The content and organization of this book reflects the experience and feedback of years of classroom testing. The author welcomes comments and suggestions from students and teachers alike.

Richard J. Fowler

Safety

Electric devices and circuits can be dangerous. Safe practices are necessary to prevent electrical shock, fires, explosions, mechanical damage, and injuries resulting from the improper use of tools.

Perhaps the greatest of these hazards is electrical shock. A current through the human body in excess of 10 milliamperes can paralyze the victim and make it impossible to let go of a "live" conductor. Ten milliamperes is a small amount of electrical flow: It is *ten one-thousandths* of an ampere. An ordinary flashlight uses more than 100 times that amount of current! If a shock victim is exposed to currents over 100 milliamperes, the shock is often *fatal*. This is still far less current than the flashlight uses.

A flashlight cell can deliver more than enough current to kill a human being. Yet it is safe to handle a flashlight cell because the resistance of human skin normally will be high enough to greatly limit the flow of electric current. Human skin usually has a resistance of several hundred thousand ohms. In low-voltage systems, a high resistance restricts current flow to very low values. Thus, there is little danger of an electrical shock.

High voltage, on the other hand, can force enough current through the skin to produce a shock. The danger of harmful shock increases as the voltage increases. Those who work on very high-voltage circuits must use special equipment and procedures for protection.

When human skin is moist or cut, its resistance can drop to several hundred ohms. Much less voltage is then required to produce a shock. Potentials as low as 40 volts can produce a fatal shock if the skin is broken! Although most technicians and electrical workers refer to 40 volts as a *low voltage*, it does not necessarily mean *safe voltage*. Obviously, you should, therefore, be very cautious even when working with so-called low voltages.

Safety is an attitude; safety is knowledge. Safe workers are not fooled by terms such as *low voltage*. They do not assume protective devices are working. They do not assume a circuit is off even though the switch is in the OFF position. They know that the switch could be defective.

As your knowledge of electricity and electronics grows, you will learn many specific safety rules and practices. In the meantime:

1. Investigate before you act
2. Follow procedures
3. When in doubt, *do not act:* ask your instructor

GENERAL SAFETY RULES FOR ELECTRICITY AND ELECTRONICS

Safe practices will protect you and those around you. Study the following rules. Discuss them with others. Ask your instructor about any that you do not understand.

1. Do not work when you are tired or taking medicine that makes you drowsy.
2. Do not work in poor light.
3. Do not work in damp areas.
4. Use approved tools, equipment, and protective devices.
5. Do not work if you or your clothing is wet.
6. Remove all rings, bracelets and similar metal items.
7. Never assume that a circuit is off. Check it with a device or piece of equipment that you are sure is operating properly.
8. Do not tamper with safety devices. *Never* defeat an interlock switch. Verify that all interlocks operate properly.

9. Keep your tools and equipment in good condition. Use the correct tool for the job.
10. Verify that capacitors have discharged. Some capacitors may store a lethal charge for a long time.
11. Do not remove equipment grounds. Verify that all grounds are intact.
12. Do not use adaptors that defeat ground connections.
13. Use only an approved fire extinguisher. Water can conduct electrical current and increase the hazards and damage. Carbon dioxide (CO_2) and certain halogenated extinguishers are preferred for most electrical fires. Foam types may also be used in some cases.
14. Follow directions when using solvents and other chemicals. They may explode, ignite, or damage electrical circuits.
15. Certain electronic components affect the safe performance of the equipment. Always use the correct replacement parts.
16. Use protective clothing and safety glasses when handling high-vacuum devices such as television picture tubes.
17. Do not attempt to work on complex equipment or circuits before you are ready. There may be hidden dangers.
18. Some of the best safety information for electrical and electronic equipment is in the literature prepared by the manufacturer. Find it and use it!

Any of the above rules could be expanded. As your study progresses, you will learn many of the details concerning proper procedure. Learn them well, because they are the most important information available.

Remember, always practice safety; your life depends on it.

Basic Concepts

- **Electricity is a form of energy. When properly controlled, it can do much of the work required to keep our society going. However, uncontrolled electric energy, such as lightning, can be very destructive.**

  When you have finished this first chapter, you will know the kinds of energy into which electric energy can be converted. You will be able to specify exact amounts of energy. You will know about atoms and be able to sketch and label the parts of an atom. Finally, you will be able to explain some important uses of static charges and static electricity.

  The study of electricity is concerned primarily with learning how to control electric energy. Electric energy is so much a part of our daily lives that we tend to take it for granted. Yet without it our lives would be quite different, and much harder. Electric energy lights our homes and industries, operates our radios and television sets, and turns the many motors used in clocks, washing machines, clothes dryers, vacuum cleaners, etc.

1-1 WORK AND ENERGY

Energy is the ability, or capacity, to do *work*. For example, it requires energy to pull a rowboat out of the water onto the shore or beach. Of course, work is also done in pulling the boat out of the water. It takes energy to do work. The symbol for work or energy is W. In the case of pulling the boat, the energy comes from the human body. The work consists of a force moving through a distance. The force is required to overcome friction and gravity. When the boat is pulled out of the water, the force overcomes the friction of the boat on the sand and the gravitational pull on the boat. Gravity is involved because the boat is raised from water level to above water level.

1-2 UNIT OF ENERGY

Base units are the terms used to indicate the amount of something. The *joule* is the base unit of energy. The symbol for the joule is J. Specifying energy in joules is the same as specifying butter in pounds or money in dollars. All are base units used to specify amount. Base units are important because nearly all relationships in electricity are expressed in base units.

A joule of energy (or work) is very small compared with the amount of energy you use each day. For example, an electric toaster uses approximately 100,000 joules of energy to make two slices of toast. It requires 360,000 joules to operate a small (100-watt) table lamp for 1 hour.

The work or energy involved in a mechanical system (such as pulling a boat) can be determined by the following relationship:

$$\text{Work (W)} = \text{force} \times \text{distance}$$

In the metric system, the base unit for force is the *newton* (approximately 0.2248 pounds). The base unit for distance is the *meter* (approximately 39.4 inches), and the base unit for work (energy) is the joule. The joule is equal to a newton-meter, which is a convenient unit for mechanical energy.

Let us return to the problem of pulling a boat out of the water and calculate how much work is done in performing the task.

From page 1:
Energy

Work

Base units

Joule

Newton

Meter

On this page:
Converting energy

Example 1-1

It requires a force of 200 newtons (45 lb) to pull the boat, and the boat is moved 5 meters (16.4 ft). How much work is done in moving the boat?

Given:	Force = 200 newtons distance = 5 meters
Find:	Work (W)
Known:	Work (W) = force × distance 1 newton-meter = 1 joule
Solution:	W = 200 newtons × 5 meters = 1000 newton-meters = 1000 joules
Answer:	Work = 1000 joules W = 1000 J

Note the procedure used in solving the problem given in the above example. First, the information (values) *given* in the problem is listed. Next, the information you are required to *find* is recorded. Finally, the relationship (formula) between the two is written. For simple problems, this formal procedure may seem unnecessary. But solving more complex problems will be easier if you establish the habit of using this formal procedure.

So far, the calculation of specific amounts of work or energy has been limited to mechanical examples. Once we learn some new terms like voltage, current, and power, we will be able to solve problems involving electric energy.

In Example 1-1, we found that the work done in moving the boat was 1000 joules. The amount of energy required to move the boat was also 1000 joules. Work and energy have the same base unit. They are basically the same thing. Work is the use of energy to perform some task. For example, a car battery has energy stored in it. When the car engine is started, the work of cranking the engine is done by using energy from the battery. The "work done" and the "energy used" are two ways of saying the same thing.

Review Questions

1. What is energy?
2. What is a base unit?
3. What is the base unit of energy?

Review Problem

1. How much work is required to push a car 40 meters, if you push with a steady force of 410 newtons?

1-3 ENERGY CONVERSION

One of the fundamental laws of physics states that energy can be neither created nor destroyed. There is only a specific amount of energy in the universe. It exists in various forms such as heat energy, light energy, and electric energy. We cannot add to or subtract from this amount. When we say we "use" electric energy, we do not mean that we have destroyed, or lost, the energy. We mean we have *converted* that electric energy into a more useful form of energy. For example, when we operate an electric lamp, we are converting electric energy into light energy and heat energy (Fig. 1–1). We have "used" the electric energy in the sense that it no longer exists as electric energy. But, we have not "used up" the energy. It still exists as heat energy and light energy.

The study of electricity deals with the study of converting energy from one form to another form. Electric energy itself is obtained by converting other forms of energy to electric energy. Batteries convert chemical energy to electric energy. Solar cells convert light energy to electric energy. And generators convert mechanical (rotational) energy into electric energy.

We seldom use energy directly in the form of electric energy. Yet the electric form of

Fig. 1-1 Converting electric energy to light energy and heat energy.

energy is very desirable because it can be easily moved from one location to another. Electric energy produced at an electric power plant many miles from your home can be easily transferred from the plant to your home. Once the electric energy is delivered to your home, it can be converted to a more useful form.

We have previously discussed how a light bulb converts electric energy into light energy and heat energy. Another familiar example is the electric stove, which converts electric energy to heat energy. Changing electric energy to mechanical (rotational) energy with an electric motor is also a common conversion.

Although the process is very involved and complex, radio receivers convert electric energy into sound energy. The electric energy comes from electric signals sent through the air. The sound energy radiates from the speaker. In a similar fashion, a television receiver converts electric energy into sound energy and light energy.

1-4 EFFICIENCY

Efficiency

No conversion process is 100 percent efficient. That is, not all the energy put into a device, or a system, is converted into the form of energy we desire. When 1000 joules of electric energy is put into a light bulb, only about 200 joules of light energy is produced. The other 800 joules is converted into heat energy. We could say that the efficiency of the light bulb is low.

The efficiency of a system is usually expressed as a percentage. It is calculated by the formula

$$\%\ \text{efficiency} = \frac{\text{useful energy out}}{\text{total energy in}} \times 100$$

By abbreviating efficiency to eff. and using the symbol W for energy, we can write this formula as

$$\%\ \text{eff.} = \frac{W_{out}}{W_{in}} \times 100$$

Let us determine the efficiency of the light bulb mentioned above.

Example 1-2

What is the efficiency of a light bulb which uses 1000 joules of electric energy to produce 200 joules of light energy?

Given: Energy in = 1000 joules
energy out = 200 joules

Find: % eff.

Known: $\%\ \text{eff.} = \frac{W_{out}}{W_{in}} \times 100$

Solution: $\%\ \text{eff.} = \frac{200\ \text{joules}}{1000\ \text{joules}} \times 100$
$= 0.2 \times 100$
$= 20$

Answer: Efficiency = 20 %

Notice that in the efficiency formula both the denominator and the numerator have base units of joules. The base units, therefore, cancel, and the answer is a pure number (it has no units). We could reword the answer to read "the efficiency of the light bulb is 20 percent" or, "the light bulb is 20 percent efficient."

Not all electrical devices have such a low efficiency as the light bulb. Electric motors like those used in washing machines, clothes dryers, and refrigerators have efficiencies of 50 to 75 percent. This means that 50 to 75 percent of the electric energy put into the motor is converted to mechanical (rotational) energy. The other 25 to 50 percent is converted into heat energy.

So far, we have illustrated the efficiency of converting electric energy into other desired forms of energy. Of course, we are just as interested in the efficiency of converting other forms of energy into electric energy.

Example 1-3

What is the efficiency of an electric generator that produces 5000 joules of electric energy from the 7000 joules of mechanical energy used to rotate the generator?

Given: Energy in = 7000 joules
energy out = 5000 joules

Known: $\%\ \text{eff.} = \frac{W_{out}}{W_{in}} \times 100$

Solution: $\%\ \text{eff.} = \frac{5000\ \text{joules}}{7000\ \text{joules}} \times 100$
$= 0.714 \times 100$
$= 71.4$

Answer: Efficiency = 71.4%

Atom

Element

Compound

Electron

Proton

Neutron

Nucleus

Review Questions

4. List the forms of energy into which the electric energy from a car battery is converted.
5. What is the undesirable form of energy produced by both light bulbs and electric motors?
6. What causes the temperature of an electric battery to increase when the battery is discharged?

Review Problems

2. An electric motor produces 1320 joules of mechanical energy for every 1890 joules of electric energy supplied to it. What is the efficiency of the motor?
3. A flashlight battery uses 120 joules of chemical energy to supply 100 joules of electric energy to the flashlight bulb. What is the efficiency of the battery?

1-5 STRUCTURE OF MATTER

All matter, materials, and objects are composed of *atoms*. Atoms are the basic building blocks of nature. Regardless of their physical characteristics, glass, chalk, rock, and wood are all made from atoms. Rock is different from wood because of the type of atoms of which it is composed.

There are more than 100 different types of atoms. Each different type of atom is called an *element*. Thus, there must also be over 100 different elements. Some common elements are copper, silver, and gold.

There are thousands and thousands of different materials in the world. Obviously, most materials must be composed of more than one element. When different kinds of atoms combine chemically they form materials called *compounds*. An example of a simple compound is water, which is composed of the elements oxygen and hydrogen. Many of the materials used in electronic circuits are composed of compounds.

It has been shown that all matter is composed of atoms. But to really understand electricity, we must "break the atom down" into still smaller particles. To understand electricity we need to be familiar with the three major particles of the atom. These are the *electron*, *proton*, and *neutron*. A pictorial representation of a helium atom showing its three major particles is illustrated in Fig. 1-2. The center of the atom is called the *nucleus*. It contains the protons and neutrons. The electrons revolve around the nucleus in elliptical paths. The electron is much larger (nearly 2000 times larger) than either the proton or the neutron. Although larger, the electron is much lighter than either the proton or neutron (about $\frac{1}{2000}$ as heavy). Thus, the center (nucleus) of the atom contains most of the weight, while the electrons make up most of the volume. It should also be noted that the distance between the nucleus and the electron is very great compared with the size of the electron. In fact, this distance is approximately 60,000 times greater than the diameter of the electron.

An analogy may help us to further visualize the relative sizes of the atomic particles and the spaces between them. Let the nucleus of the hydrogen atom be represented by a common marble. The electron could then be represented by a 31-meter (100-ft) sphere located 1610 kilometers (1000 miles) from the marble. Although the distance between the nucleus and the electron is very great relative to the size of either, we must remember that these sizes and distances are submicroscopic. For example, the diameter of an electron is only 4×10^{-13} centimeters (0.0000000000004 cm).

Electrons rotate or orbit around the nucleus of the atom in much the same manner as the earth rotates around the sun (Fig. 1-2). In the more complex atoms which contain more than one electron, each individual electron has its own orbit. With proper coordination of the orbiting electrons, it is possible for two or

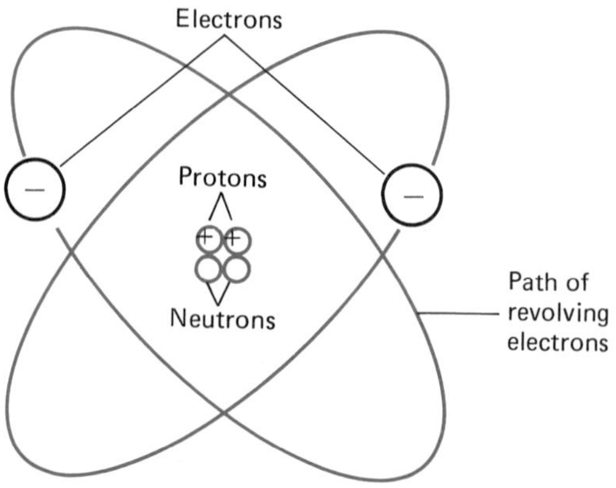

Fig. 1-2 Structure of an atom of helium.

more atoms to share common space. Indeed, in many materials, neighboring atoms share electrons as well as space.

Figure 1-3 represents the aluminum atom in a two-dimensional form. Remember that each electron is actually orbiting around the nucleus in its own elliptical path. The two electrons closest to the nucleus will not actually follow the same orbital path. Their orbital paths are merely the same average distance from the nucleus. The two electrons closest to the nucleus are said to occupy the first *shell* or *orbit* of the atom. This first shell of the atom can accommodate only two electrons. Atoms which have more than two electrons, such as the aluminum atom, must have a second shell or orbit.

The second shell of the aluminum atom contains eight electrons. This is the maximum number of electrons that the second shell of any atom can contain. The third shell can contain a maximum of 18 electrons, and the fourth shell a maximum of 32 electrons. Since the aluminum atom (Fig. 1-3) has only 13 electrons, its third shell has 3 electrons.

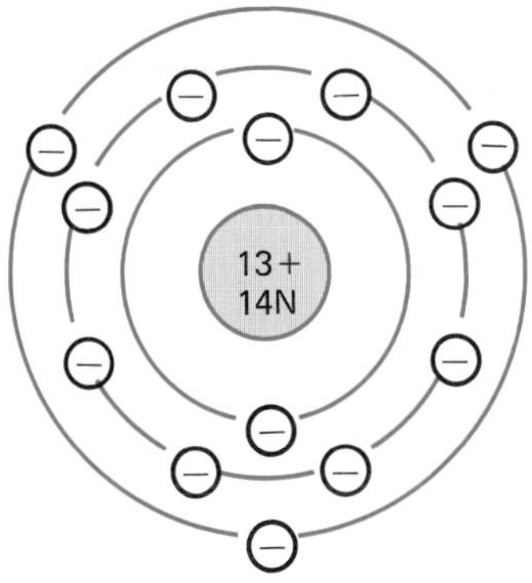

Fig. 1-3 Simplified presentation of an aluminum atom showing its 13 electrons (−), 13 protons (+), and 14 neutrons (N).

Review Questions

7. What are the major particles of an atom?
8. Which part of the atom contains most of its weight?
9. Which part of the atom contains most of its volume?
10. Which is greater, the diameter of an electron or the distance from an electron to a proton?

1-6 ELECTRIC CHARGE

Orbit

Polarity

Both electrons and protons possess electric charges, but these charges are of opposite *polarity*. Polarity refers to the type (negative or positive) of charge. The electron possesses a negative (−) charge, while the proton possesses a positive (+) charge. These electric charges create electric fields of force which behave much like magnetic fields of force. In Fig. 1-4, the lines with arrows on them represent the electric fields. Two positive charges or two negative charges will repel each other. Two opposite electric charges will attract each other (Fig. 1-4). The force of attraction between the positive proton and the negative electron aids in keeping the electron in orbit around the nucleus. The neutron in the nucleus of the atom has no electric charge. The neutron can be ignored when considering the electric charge of the atom.

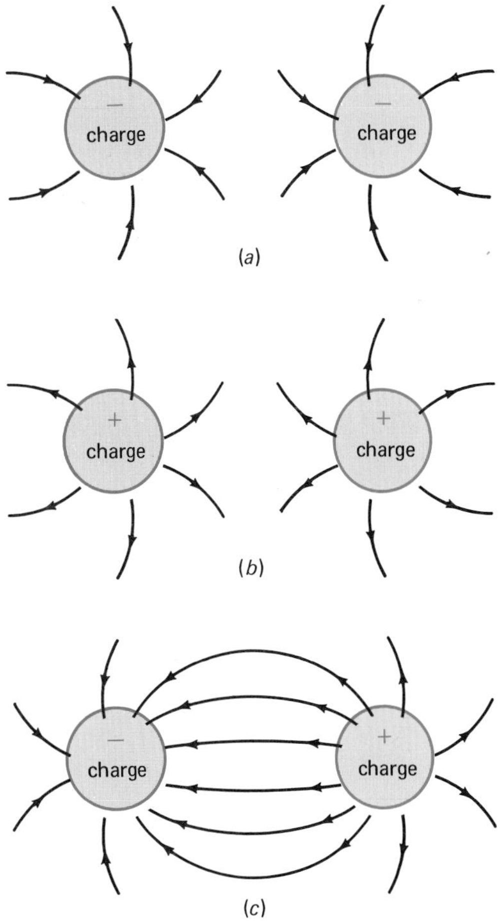

Fig. 1-4 Electric fields between charges: (a) and (b) like charges repel each other, (c) unlike charges attract each other.

An atom in its natural state always has a net electric charge of zero, that is, it always has as many electrons as it has protons. For example, look at the aluminum atom illustrated (in simplified form) in Fig. 1-3. It has 13 electrons orbiting around the nucleus, and the nucleus contains 13 protons in addition to the 14 neutrons. We can say the aluminum atom is electrically neutral, even though individual electrons and protons are electrically charged.

Valence electrons

Free electrons

Positive ion

1-7 VALENCE ELECTRONS

The electrons in the outermost shell of an atom are called *valence electrons*. The valence electrons are the atomic particles that are involved in chemical reactions and electric currents.

One of the forces that helps to hold electrons in orbit is the force of attraction between unlike charges. The closer together electric charges are, the greater the electrical attraction between them. Therefore, the attraction between the proton of the nucleus and the electron decreases as the electron gets farther from the nucleus. Thus, the valence electrons are held to the nucleus with less attraction than are the electrons in the inner shells. They can be more easily removed from the parent atom than can the electrons in the other shells.

All electrons possess energy. They possess energy because they have weight and they are moving. Thus, they are capable of doing work. The valence electron possesses more energy than the electrons in the other shells. In general, the farther the electron is from the nucleus of the atom, the more energy it possesses. When a valence electron is separated from the parent atom, its energy level increases. As a free electron, the electron possesses more energy than it did as a valence electron.

1-8 FREE ELECTRONS

Free electrons are those valence electrons that have been temporarily separated from an atom. They are free to wander about in the free space around the atom. They are unattached to any particular atom. Only the valence electrons are capable of becoming free electrons. Electrons in the inner shells are very tightly held to the nucleus. They cannot be separated from the parent atom. Since free electrons contain more energy than valence electrons, they can be created by giving additional energy to the atomic structure. The additional energy allows the valence electron to escape the force of attraction between the electron and the nucleus. One way to provide the additional energy needed to free an electron is to heat the atom. Another way is to subject the atom to an electric field.

1-9 IONS

When a valence electron leaves an atom to become a free electron, it takes with it one negative electric charge. This absence of one negative electric charge from the parent atom leaves that parent atom with a net positive charge. In the case of the aluminum atom, there would be 13 protons (positive charges) but only 12 electrons (negative charges). Atoms which have more than, or less than, their normal complement of electrons are called *ions*. When an atom loses electrons, it becomes a positive ion. Conversely, atoms with an excess of electrons contain a net negative charge and become negative ions. The amount of energy required to free a valence electron and create an ion varies from element to element.

The energy required to create a free electron is related to the number of valence electrons contained in the atom. In general, the fewer the electrons in the valence shell, the smaller the amount of additional energy

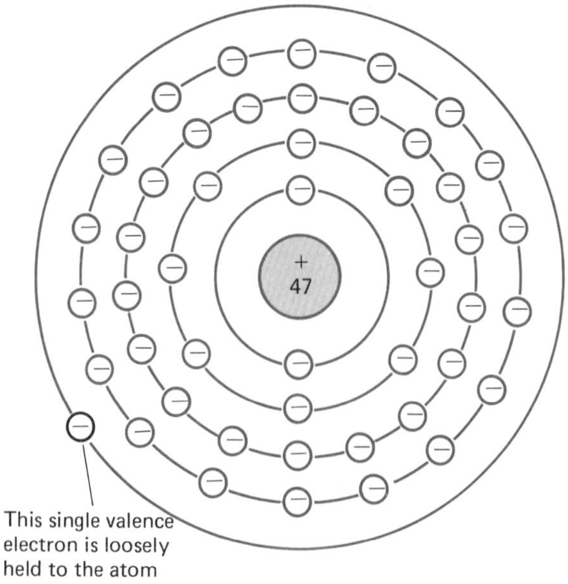

Fig. 1-5 Simplified silver atom.

needed to free an electron. The silver atom, illustrated in Fig. 1-5, with its single valence electron requires relatively little energy to free the valence electron. Carbon, with four valence electrons, requires much more energy to free an electron. Elements with five or more electrons in the outer shell do not readily release valence electrons to become free electrons.

Negative ions are created when an atom with five or more electrons in its outer shell accepts additional electrons. For example, when table salt (sodium chloride) is added to water, the chlorine atom receives an extra electron and becomes a negative ion. At the same time, the sodium atom, which gave up an electron, becomes a positive ion (Fig. 1-6). The concept of ions is important in understanding electric circuits involving batteries and gas-filled devices.

Review Questions

11. What polarity is the charge of an electron? A proton?
12. Is an atom electrically charged? Explain.
13. What is an atom called when it has lost a valence electron?
14. Does a free electron or a valence electron possess the most energy?
15. Is it easier to obtain free electrons from an element (A) whose atoms have one valence electron, or (B) whose atoms have five valence electrons?

1-10 STATIC CHARGE AND STATIC ELECTRICITY

Negative ion

Positive static charge

Negative static charge

Static electricity is a common phenomenon that all of us have observed. Probably the most dramatic example of static electricity is lightning. Static electricity is responsible for the shock you may receive when reaching for a metal doorknob after walking across a thick rug. It is also responsible for hair clinging to a comb, and some synthetic clothes clinging to themselves.

All the above phenomena have one thing in common. They all involve the transfer of electrons from one object to another object or from one material to another material.

A *positive static charge* is created when a transfer of electrons leaves an object with a deficiency of electrons. A *negative static charge* results when an object is left with an excess of electrons. Static charges can be created by rubbing a glass rod with a piece of silk cloth. The cloth is vigorously rubbed against the glass rod. Some of the valence electrons from the glass rod become free electrons and are transferred to the cloth. The cloth takes on a negative charge, while the glass rod takes on a positive charge. The charges on the glass and the cloth tend to remain stationary, thus the name "static electricity."

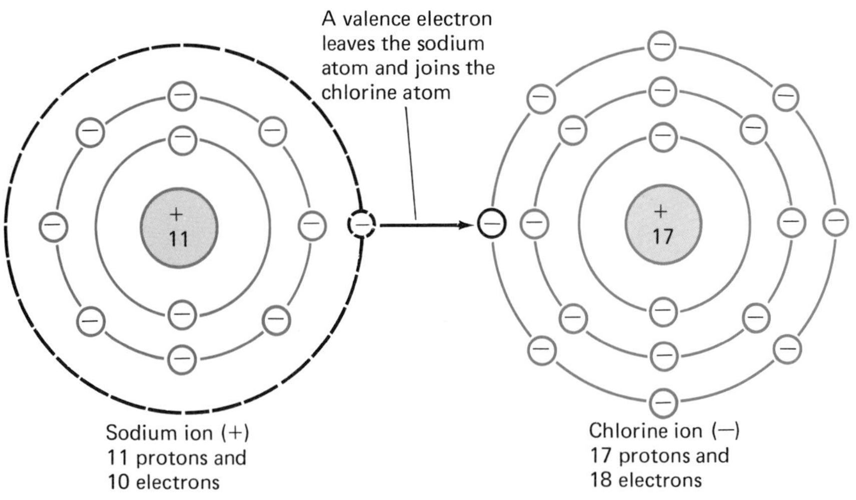

Fig. 1-6 Creation of positive (sodium) ions and negative (chlorine) ions.

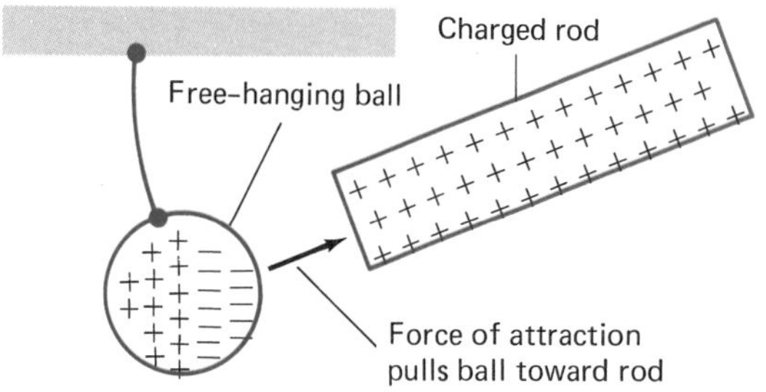

Fig. 1-7 Inducing a static charge. When the ball touches the rod, electrons are transferred to the rod. The ball is left with a positive charge.

Induced charge

Static discharge

Electrostatic precipitator

An object which possesses a static charge can attract other objects that are not charged. This happens because the charged object can induce a charge on the surface of the uncharged object (Fig. 1-7). The induced charge, being of the opposite polarity to the charged object, is then attracted to the charged object. If the two objects (the ball and the rod in Fig. 1-7) are allowed to touch, part of the positive charge of the rod will be transferred to the ball. Then both objects will be positively charged and a force of repulsion will result. The ball will then be pushed away from the rod.

Let us clarify how a positive charge is transferred from the positive rod to the ball (Fig. 1-7). When the rod and ball touch, protons *do not* travel from the rod to the ball. Remember, the electron is the only particle of the atom that can be easily detached and moved about. When the rod and ball touch, electrons travel from the ball to the rod. This leaves the ball with a shortage of negative charge (electrons). The ball still has its normal number of positive charges (protons). Since the ball now has more protons than electrons, it possesses a net positive charge.

1-11 STATIC DISCHARGE

Static discharge occurs when the electric force field (Fig. 1-4) between a positive charge and a negative charge becomes too strong. Then electrons are pulled off the negatively charged object and travel through the air to the positively charged object. The observable spark is the result of the air path between the objects becoming ionized and deionized. When the air is ionized, electrons are raised to a higher energy level. When the air deionizes, the electrons return to a lower energy level. The difference in the two energy levels is given off as light energy when the electron returns to the lower energy level. Lightning is an ionized column of air caused by electrons traveling from a negatively charged cloud to a positively charged portion of land.

1-12 USES OF STATIC ELECTRICITY

Most of the useful applications of static electricity do not rely on discharges that ionize the air. Rather, they make use of the force of attraction between unlike charges or the force of repulsion between like charges. These forces are used to move charged particles to desired locations. For example, dust particles can be removed from air by this method. In Fig. 1-8, the air is forced between negatively charged rods and positively charged plates. A negative charge is transferred from the rods to the dust particles. When the air stream goes by the positively charged plate, the negatively charged dust particles are pulled out. Electrical devices of this type are often called *electrostatic precipitators*. An electrostatic precipitator is shown in Fig. 1-9. The cutaway section shows the charged rods and plates.

Static charges are also used in some spray-painting operations. The paint mist leaving the spray gun nozzle is negatively charged and the object to be painted is positively charged. This process tends to give a uniform coat of paint on an object that has an irregular surface. As paint builds up on one part of the object, the charge on that part of the object is canceled. The force of attraction disappears. If excess paint builds up on part of the object, then that part becomes negatively charged and repels additional paint. The repelled paint is attracted to those parts of the object that are still positively charged.

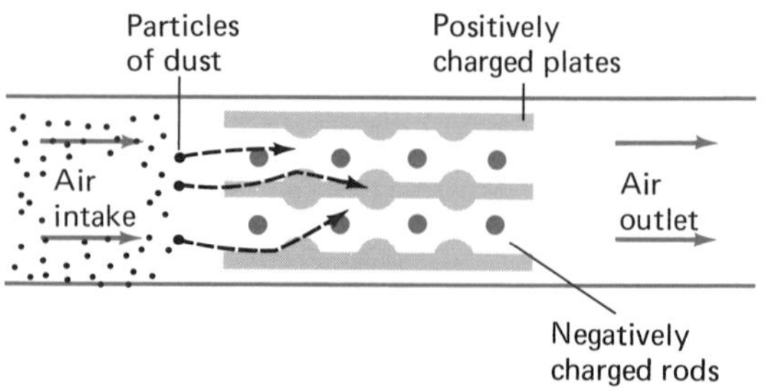

Fig. 1-8 Principle of a dust precipitator. Dust particles are given a negative charge so they are attracted to positively charged plates.

Fig. 1-9 Cutaway view of an electrostatic precipitator. (Courtesy of American Air Filter.)

Several properties of static electricity can be used to advantage in manufacturing abrasive paper (sandpaper). The backing paper is coated with an adhesive (glue) and given a static charge. The abrasive particles are given the opposite charge. As the paper is passed over the abrasive particles, the particles are attracted to the paper and adhere to it (Fig. 1-10). Once the abrasive particles are distributed on the adhesive backing paper, both the paper and the abrasive particles are given like charges. The like charges repel each other and try to push the abrasive particles from the paper. However, the adhesive is strong enough to hold the particles to the paper. The abrasive particles "stand up" on the paper so that the sharpest points of the particles are exposed as the cutting surface. The adhesive material is then heated and hardened to hold the abrasive particles in place. The reason the abrasive particles stand up is because static charges concentrate on the sharpest point of an object. Since like charges repel each other, they will push away from each other as far as possible. The sharpest point of the abrasive particle will be the most highly charged part of the particle. Therefore, it will be pushed farthest from the surface of the backing paper.

Review Questions

16. What is a static charge?
17. What atomic particle is transferred from one object to another when a static discharge occurs?
18. List two industrial applications of static charges.

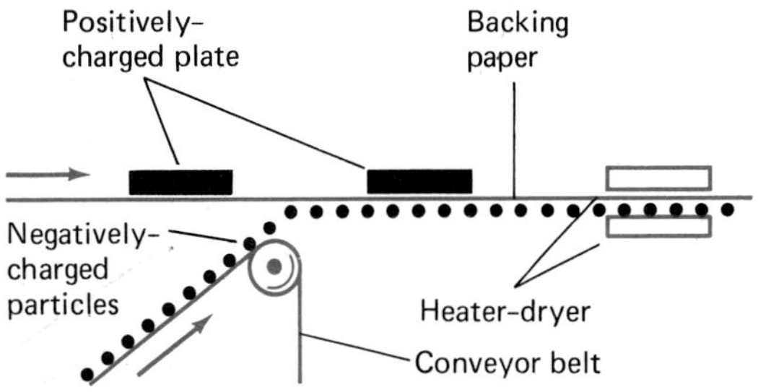

Fig. 1-10 Principle of making abrasive paper. Static charges position the abrasive particles so that their sharpest points are exposed.

Summary

1. Energy is the ability to do work.
2. The symbol for energy is W.
3. Electricity is a form of energy.
4. Work equals force times distance.
5. Joule is the base unit of energy.
6. Joule is the base unit of work.
7. The symbol for joule is J.
8. Energy is neither created nor destroyed.
9. Percent efficiency $= \frac{W_{out}}{W_{in}} \times 100$
10. All matter is made up of atoms.
11. Atoms contain electrons, protons, and neutrons.
12. Protons and neutrons are found in the nucleus of the atom.
13. Electrons have a negative electric charge.
14. Protons have a positive electric charge.
15. Neutrons do not have an electric charge.
16. Atoms have an equal number of protons and electrons.
17. Valence electrons are found in the outermost shell of the atom.
18. Valence electrons can become free electrons when their energy level is raised.
19. Negative ions are atoms that have attracted and captured an extra (free) electron.

20. Positive ions are atoms that have given up an electron.
21. Static charges are the result of an object possessing either more or fewer electrons than protons.
22. A negative charge means an excess of electrons.
23. Like charges repel each other.
24. Unlike charges attract each other.

Chapter Review Questions

Here is a chance to see what you have learned. For questions 1-1 to 1-7, determine whether the statement is true or false.

1-1. Unlike electric charges repel each other.

1-2. Valence electrons are at a higher energy level than are free electrons.

1-3. Energy and work have the same base unit.

1-4. Atoms have twice as many electrons as protons.

1-5. An object which has a static charge can possess either more or fewer electrons than protons.

1-6. The second shell or orbit of an atom can contain any number of electrons.

1-7. If an electric device is less than 100 percent efficient, part of the energy provided to it is destroyed.

For questions 1-8 to 1-11, choose the letter that best completes the sentence.

1-8. The symbol for energy is
(A) *t* (B) W (C) *D* (D) J

1-9. The base unit of energy is the
(A) Ion (B) Proton (C) Joule (D) Pound

1-10. The abbreviation for the base unit of energy is
(A) *t* (B) W (C) *D* (D) J

1-11. A positive ion is an atom which has
(A) Captured one, or more, electrons (B) Given up one, or more, electrons (C) Captured one, or more, protons (D) Given up one, or more, protons

Answer the following question.

1-12. What is the efficiency of a motor which requires 780 base units of electric energy to produce 585 joules of mechanical energy?

Answers to Review Questions

1. Energy is the ability to do work.
2. A base unit is a term used to specify the amount of something.
3. Joule (J)
4. Heat (due to inefficiencies of lights

and motors), light, and mechanical
5. Heat energy
6. Inefficiency—part of the available energy is converted to heat energy within the battery.
7. Proton, electron, and neutron
8. Nucleus (or protons and neutrons)
9. Electron
10. The distance from an electron to a proton
11. Electron is negative and proton is positive.
12. No, the negative charge of the electrons cancels the positive charge of the protons, leaving the atom neutral.
13. Negative ion
14. Free electron
15. (*A*)
16. A static charge is the excess or deficiency of electrons on an object.
17. Electron
18. Removal of dust particles from air, spray painting, and manufacturing of abrasive paper

Answers to Review Problems

1. *Given:* Distance = 40 meters
   force = 410 newtons
   Find: Work (W)
   Known: Work (W) = force × distance
   Solution: W = 410 newtons × 40 meters
   = 16,400 joules
   Answer: W = 16,400 joules

2. *Given:* W_{out} = 1320 joules
   W_{in} = 1890 joules
   Find: % eff.
   Known: $\% \text{ eff.} = \frac{W_{out}}{W_{in}} \times 100$
   Solution: $\% \text{ eff.} = \frac{1320 \text{ joules}}{1890 \text{ joules}} \times 100 = 69.8$
   Answer: Efficiency = 69.8%

3. *Given:* W_{out} = 100 joules
   W_{in} = 120 joules
   Find: % eff.
   Known: $\% \text{ eff.} = \frac{W_{out}}{W_{in}} \times 100$
   Solution: $\% \text{ eff.} = \frac{100 \text{ joules}}{120 \text{ joules}} \times 100 = 83.3$
   Answer: Efficiency = 83.3%

Electrical Quantities and Units

▪ If you were describing a shopping trip to the grocery store, you would probably use many quantities and units in your description. You might start out by telling the distance (a quantity) in blocks or miles (units) to the store. In describing what you purchased, you could use many quantities and units such as pounds (a unit) of sugar (a quantity), heads of lettuce, cans of soup, quarts of milk, dozens of eggs, or bars of soap.

In describing electric circuits you will also need to use quantities and units. This chapter will tell you about many of the basic electrical quantities and units. After completing this chapter, you will be able to describe current, voltage, resistance, and power. You will be able to list and define the unit for each of these quantities. Finally, you will be able to convert from large to small units or from small to large electrical units. Converting from a small electrical unit to a large unit is like converting from cups to quarts of milk.

2-1 CHARGE

In Chap. 1 we defined an *electric charge* as the electrical property possessed by electrons and protons. The proton has a positive charge and the electron a negative charge. However, we never did specify exact amounts of charge. That is, we never did define the base unit of charge so that we could specify exact amounts of charge.

2-2 UNIT OF CHARGE

The base unit of charge is the *coulomb*. A coulomb of charge is the amount of charge possessed by 6.25×10^{18} (6,250,000,000,000,000,000) electrons. We do not use the charge on a single electron as the base unit because it is a very small charge—too small for most practical applications. The coulomb, the base unit of charge, is also used in defining the base units of other electrical quantities such as current and voltage. The coulomb is named after Charles Augustin Coulomb, a French physicist.

In electricity, we use many symbols for electrical quantities and units. The symbol for charge is Q. The abbreviation for coulomb is C. Use of symbols for electricity allows us to condense ideas and statements. For example, instead of writing "the charge is 5 coulombs," we can just write "$Q = 5$ C."

2-3 CURRENT AND CURRENT CARRIERS

Electric current is the movement of charged particles in a specified direction. The charged particle may be an electron, a positive ion, or a negative ion. The charged particle is often referred to as a *current carrier*. The movement of the charged particle may be through a solid, a gas, a liquid, or a vacuum. In a solid, such as copper wire, the charged particle (current carrier) is the electron. The ions in a copper wire, and other solids, are rigidly held in place by the atomic (crystalline) structure of the material. Thus, ions cannot be current carriers in solid materials. However, in both liquids and gases, the ions are free to move about and become current carriers.

The symbol for current is I. The symbol I was chosen because early scientists talked about the intensity of the electricity in a wire.

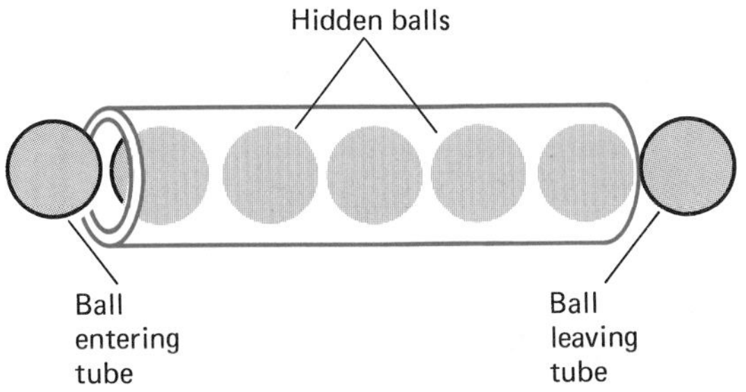

Fig. 2-1 Apparent speed illustrated. A ball exits the instant another ball enters the tube.

2-4 CURRENT IN SOLIDS

When thinking about current, keep two points in mind. First, the effect of current is almost instantaneous. Current in a wire travels at nearly the speed of light: 186,000 miles per second or 3×10^8 meters per second. Second, the individual electron moves much more slowly than the effect of the current. It may take minutes for an individual electron to travel a few feet in the wire.

The idea of the "effect of the current being instantaneous" and the "individual electron moving much slower" is illustrated in Fig. 2-1. Suppose you had a very long cardboard tube with a diameter just large enough to pass a tennis ball. You lay the tube on the floor and fill it full of tennis balls. When you push an extra ball into one end of the tube, another ball immediately comes out the other end of the tube (Fig. 2-1). If you did not know that the tube was full of tennis balls, you might think that the tennis ball you pushed in one end of the tube traveled very quickly down the tube and out the other end. The effect (the movement of tennis balls from one end of the tube to the other end) is very fast. Yet each tennis ball moves only a short distance. Suppose you stacked up six tubes filled with balls (Fig. 2-2) and alternately pushed a ball into one tube and then another. Now you could have a steady stream of balls appearing at one end of the tubes, yet only the balls in one tube would be moving at any one time. Even within that tube each individual ball would move only a short distance. This is comparable to the way in which current carriers (electrons) move through a wire when current is flowing in the wire.

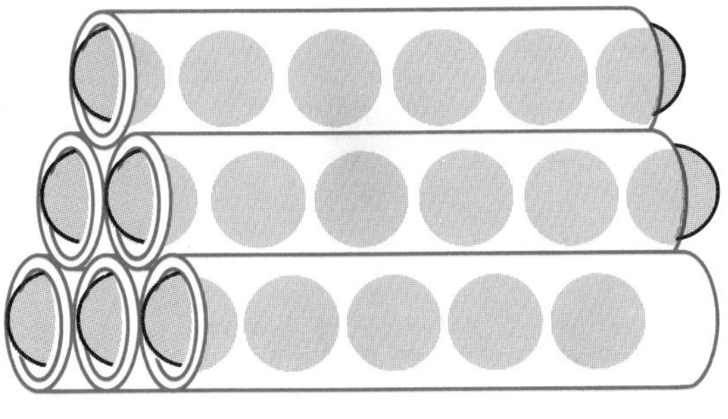

Fig. 2-2 Electron movement illustrated.

Assume you could look inside an aluminum wire and view the individual atoms and their particles (Fig. 2-3). The aluminum atom actually has 13 electrons and 13 protons, but, for simplicity, only 3 of each are shown. Now, suppose the ends of the wire are connected to a flashlight cell. The cell will provide an electric field through the wire. The electric field will free some of the valence electrons, as shown in Fig. 2-3, by giving them additional energy. At the moment that an individual electron is freed, it may be traveling in a direction opposite to that of the main current. However, in the presence of the electric field, it will soon change its direction of movement. For every electron that is freed, a positive ion will be created. This positive ion will have an attraction for an electron. Eventually, one of the free electrons will migrate close to the positive ion. The electron will be captured by that positive ion, which then, of course, becomes a neutral atom.

Notice that a free electron does not remain as a free electron and travel the full length of the conductor. Rather it travels a short distance down the wire and is captured by one of the positive ions. At some later time this particular electron may again gain enough energy to free itself of its new parent atom. It then travels farther down the wire as a free electron. We can think of the electrons as hopping down the conductor from atom to atom to atom. As long as there is a new free electron created every time a free electron is captured, the net number of free electrons moving down the wire will remain constant. Current will continue to flow. It will continue to flow in the same direction through the conductor. Current that travels in the same direction all the time is called *direct current*, which is abbreviated dc. It is the type of current you get from flashlight cells and batteries.

If you remember that the individual electron travels rather slowly from ion to ion, you will be able to understand alternating current. *Alternating current* (abbreviated ac) is the type of current you have in your home and

From page 12:
Electric charge
Coulomb
Electric current
Current carrier
On this page:
Speed of light
Direct current
Alternating current

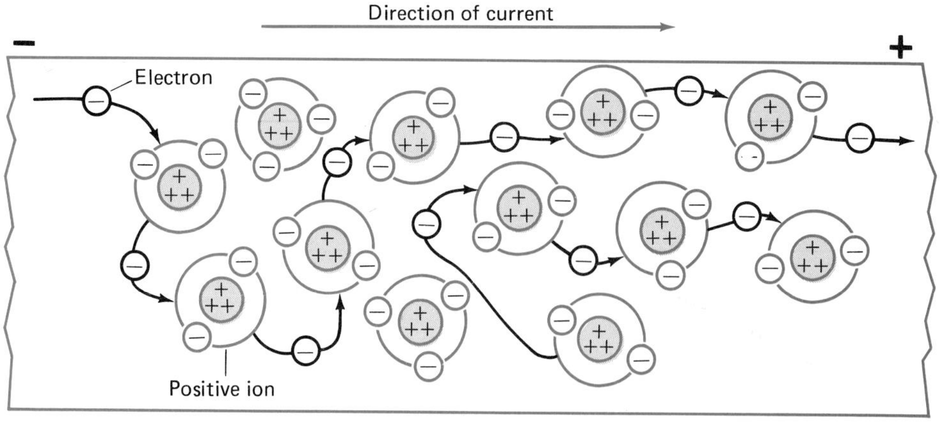

Fig. 2-3 Current in a solid. A free electron travels only a short distance before it combines with a positive ion.

your school. It is the type of current that periodically reverses the direction in which it is moving. The current in all the electric wires in your home reverses its direction every $\frac{1}{120}$ of a second. Some of the current in your television receiver reverses every $\frac{1}{67{,}000{,}000}$ of a second. Currents that reverse direction this often are easier to visualize if you think of the individual electrons as "swinging" back and forth between several ions.

Ionize

2-5 CURRENT IN LIQUIDS AND GASES

In gases and liquids, all three types of carriers (electrons, positive ions, and negative ions) can be involved in current flow. When a gas, such as neon, is subjected to a strong electric field, the gas ionizes. Once ionized, the neon gas allows current to flow. Figure 2-4 illustrates current flow in ionized neon gas. The negative and positive signs indicate that the neon bulb is connected to an electric force field such as a battery. A neon atom has eight electrons in its outermost (valence) shell. When the atom is ionized, one electron is freed. The resulting positive ion travels toward the negative plate (Fig. 2-4). The resulting free electron travels toward the positive plate. Once the positive ion arrives at the negative plate, it receives an electron and becomes a neutral atom. It will drift around in the glass enclosure until it is again ionized. The free electron is received by the positive plate and travels out the connecting wire. The system of Fig. 2-4 requires that electrons be supplied to the negative plate and removed from the positive plate. That job is done by the power supply, which also supplies the electric force field.

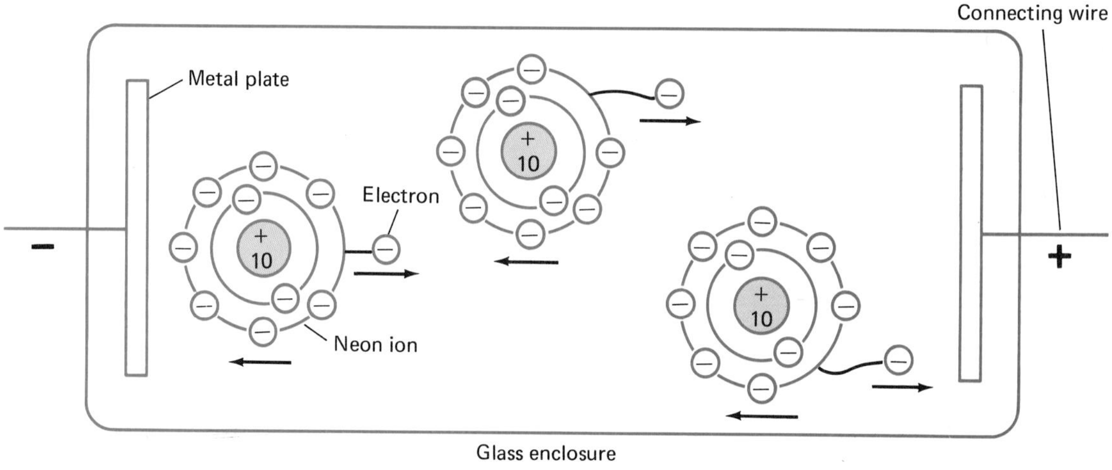

Fig. 2-4 Current in a gas. Both electrons and ions serve as current carriers.

Current flow in a liquid consists of both negative and positive ions moving through the liquid. A simplified diagram of current flow in a sodium chloride (salt) solution is shown in Fig. 2-5. In the diagram Na is the chemical symbol for sodium and the plus sign (+) in the circle means a positive ion. Likewise, Cl is the chemical symbol for chlorine and the minus sign (−) in the circle means a negative ion. When an electric field is created between the plates, the positive sodium ions move to the negative plate. The negative chlorine ions move to the positive plate. Notice that the current in the liquid is composed entirely of ions. However, the current in the wires and plates connected to the liquid is composed of electrons in motion. The changeover from electron-charge carriers to ion-charge carriers occurs at the surface where the plates and liquid meet. This change of carrier is actually more complex than implied by Fig. 2-5. The change also involves some ions created by the water in which the salt is dissolved. However, whether or not we account for the water ions, the end results are the same. That is, negative ions give up electrons at the positive plate and positive ions pick up electrons from the negative plate. A liquid solution which is capable of carrying current is called an *electrolyte*. A saltwater solution is an electrolyte; it is an ionized solution.

Electrolyte

One industrial use of current flow in electrolytes is electroplating. Electroplating is a process by which a thin layer of one type of metal can be plated (surface covered) over another material. The other material may be another metal or a piece of plastic coated with a conductive material. Figure 2-6 illustrates electroplating copper onto iron. The electrolyte is copper sulfate, which ionizes into a copper ion (Cu^{++}) with two positive charges and a sulfate ion (SO_4^{--}) with two negative charges. The copper ions are attracted to the iron plate, where they pick up two electrons and adhere (stick to) to the iron plate as copper atoms. The sulfate ions move to the copper plate, where they chemically react with the copper to create more copper sulfate. The copper sulfate goes back into the solution. The reaction that created the copper sulfate leaves two electrons on the copper plate. These two electrons move out through the wire connected to the copper plate (Fig. 2-6). Notice that this electroplating system is always in balance. That is, for every two electrons that enter the negative iron plate, two more electrons leave the positive copper plate. This type of balance is present in all electric devices which carry current.

Electroplating

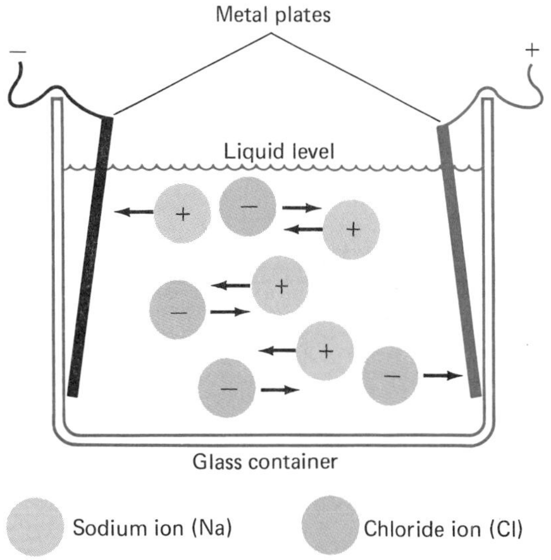

Fig. 2-5 Current in a liquid. Both positive and negative ions serve as current carriers.

2-6 UNIT OF CURRENT—THE AMPERE

We have now developed a concept of electric current. Our next problem is to develop a logical unit for keeping track of the amount of current. The simplest method might be to keep track of the number of electrons, or

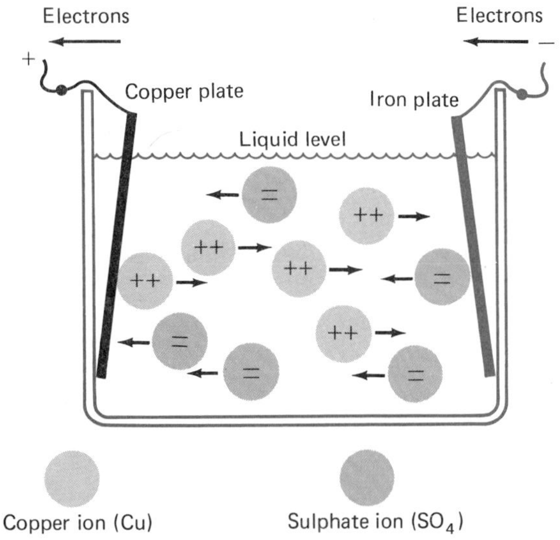

Fig. 2-6 Electroplating copper onto an iron plate. Current in the liquid consists of ions in motion.

coulombs, that move down the conductor. However, this method would leave much to be desired. It would not take into account the amount of time required to move the charge. This would be somewhat comparable to tabulating automobile traffic without considering the time involved. For example, on a freeway, 1000 cars may pass a given point in 1 hour of time. On a two-lane country road it may take 20 hours for 1000 cars to pass a given point. Certainly the traffic is greater on the freeway. A more meaningful way to compare the traffic on these two roads would be to talk in terms of the number of cars per hour. Thus, the traffic on the freeway would be 1000 cars per hour, while the traffic on the two-lane road would be 50 cars per hour.

Coulombs per second

Ampere

$I = \frac{Q}{t}$

In electricity, the amount of current is specified in terms of the charge and the time required to move the charge by a given point. The amount of electric current is therefore specified in *coulombs per second*. But, since coulombs per second is a rather long term, the base unit of current has been named the *ampere*. Ampere is the base unit of current. An ampere is equal to one coulomb per second. Ampere was chosen as the base unit of current in honor of André Marie Ampere, a French scientist who did some early work in the field of electricity.

The abbreviation for ampere is A. To indicate that the current in the wire is 10 amperes, we would write $I = 10$ A.

Notice that our definition of ampere involves time. In electricity, time is abbreviated with the symbol t. The base unit of time is the second, which we abbreviate with s. The relationship between time, charge, and current is

$$\text{Current } (I) = \frac{\text{charge } (Q)}{\text{time } (t)}$$

$$\text{or} \quad I = \frac{Q}{t}$$

Review Questions

1. What is an electric charge?
2. What is the symbol for charge?
3. What is a coulomb?
4. What is the abbreviation for coulomb?
5. What is a current carrier?
6. What are the two current carriers in an ionized gas?
7. What is an electric current?
8. What is the symbol for current?
9. What is the current carrier in a copper wire?
10. Describe the way in which an electron travels through a copper wire.
11. How is alternating current different from direct current?
12. What is the base unit of current?
13. Define an ampere in terms of charge and time.
14. What is the abbreviation for ampere?
15. Rewrite each of the statements below by using the correct symbols for the electrical quantities and units:
    a. The current is 8 amperes.
    b. The charge is 6 coulombs.
16. If 20 coulombs of charge pass a specified point in 1 second, how much current is flowing?

2-7 VOLTAGE

Electromotive force

Potential difference

Kinetic energy

Voltage is the electric pressure that causes current to flow. Voltage is also known as *electromotive force* (abbreviated emf), or potential difference. All these terms refer to the same thing, that is, the force that sets charges in motion. Potential difference is the most descriptive term because a voltage is actually a potential energy difference that exists between two points. The symbol for voltage is V. To really understand voltage, one must first understand what is meant by potential energy and potential energy difference. We must, therefore, extend the discussion of energy which we started in Chap. 1.

There are two types of energy. One is called potential energy and the other kinetic energy. *Kinetic energy* refers to energy in motion, or energy doing work, or energy being converted into another form. When you swing a baseball bat, the baseball bat has kinetic energy. It does work when it hits the ball; that is, it exerts a force on the ball which moves the ball through a distance. Anything that has mass (weight) and is in motion possesses kinetic energy.

Potential energy is energy at rest. It is energy that can be stored for long periods of time in its present form. It is capable of doing work when we provide the right condition to convert it from its stored form into another form. Of course, when it is being converted, it is changed from potential to kinetic energy. Water stored in a lake behind a dam of a hydroelectric plant possesses potential energy due to gravitational forces. The potential energy of the water can be stored for long periods of time. When the energy is needed, it is converted to kinetic energy by letting the water flow through the hydroelectric plant.

An electric charge possesses potential energy. When you scuff around on the carpet, you collect an electric charge on your body. That charge (static electricity) is potential energy. Then, when you touch some object in the room, a spark occurs. The potential energy becomes kinetic energy as the electric charge is converted to the light and heat energy of the spark.

An object, such as this book, resting on a table, also possesses potential energy. The book is capable of doing work when it moves from the table to the floor. Thus, the book has potential energy with respect to the floor. When the book falls from the table to the floor, its potential energy is converted into mechanical energy and heat energy. Potential energy is dependent on mass. If we replace the book with an object that weighs more, then the potential energy will increase. The potential energy *difference* is independent of the amount of weight. It is a function of the distance between the two surfaces and gravitational force. Knowing the potential energy difference that exists between two points, we can easily figure the potential energy possessed by an object of any given weight.

Example 2-1

What is the potential energy (with reference to the floor) of a 5.5-kilogram block resting on a table top if the potential energy difference between the table top and the floor is 8 joules per kilogram?

Given: Weight = 5.5 kilograms, potential energy difference = 8 joules per kilogram

Find: Potential energy (PE)

Known: Potential energy = potential energy difference × weight

Solution: PE = 8 joules per kilogram × 5.5 kilograms = 44 joules

Answer: Potential energy = 44 joules

Potential energy

Potential energy difference

In the above discussions, we have always started with the weight on the table top and considered the energy the weight possessed with reference to the floor. We can just as well reverse the situation and consider the energy required to move the weight from the floor to the table top. The potential energy difference is the same in either case. In the one case you remove energy from the system; in the other you must put energy into the system. That is, you do work in lifting the book from the floor to the table top.

So far in our discussion of potential energy, we have been using mechanical examples. In these examples, the potential energy of the block and the potential energy difference between the floor and the table top are due to weight and gravitational forces. In electricity, the potential energy and the potential energy differences are due to electric fields and electric charges.

Voltage is a potential energy difference similar to the mechanical case discussed above. Instead of moving weights with the force of gravity, electric charges are moved with the force of an electric field. In Fig. 2-7, the electron charge "loses" energy as it moves from a negatively charged point to a positively charged point. (This is the same as the book "losing" energy as it moved from the table top to the floor.) The "lost" energy of the charge could be converted to heat and light as the charge moves through a lamp. There is a voltage (potential energy difference) between the

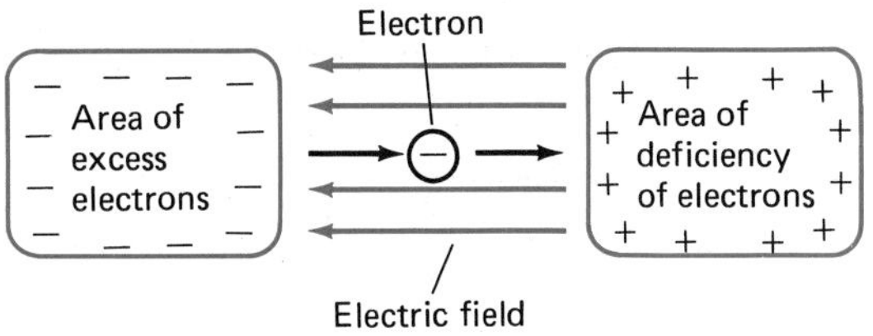

Fig. 2-7 Charge moving through electric field. Part of the electron's energy is converted to another form of energy.

Joule per coulomb

Volt

$W = V \times Q$

Polarize

negative and positive areas shown in Fig. 2-7. These negative and positive areas of Fig. 2-7 could represent the terminals of an electric battery. A lead-acid storage battery, such as the one shown in Fig. 2-8, is a common source of voltage. A potential energy difference (voltage) exists between the negative and the positive terminals of a battery. This voltage is the result of an excess of electrons at the negative terminal and a deficiency of electrons at the positive terminal. When electrons move from the negative terminal to the positive terminal of the battery, work is done. Energy is taken from the battery and converted to another form of energy.

Like the mechanical system, the electric system can also be reversed. That is, the electron charge can gain energy when it is moved from the positively charged point to the negatively charged point. This is what happens when a battery is charged. The battery charger forces electrons back through the battery in the reverse direction.

2-8 UNIT OF VOLTAGE—THE VOLT

We need a unit to indicate the potential energy difference (voltage) between two points such as the terminals of a battery. This unit must specify the energy available when a given charge is transported from a negative to a positive point. We already have the joule as the base unit of energy and the coulomb as the base unit of charge. Therefore, the logical unit of voltage is the joule per coulomb. The joule per coulomb is called the *volt*. The volt is the base unit of voltage. It is abbreviated V. The battery shown in Fig. 2-8 is a 48-volt battery. In symbolic form we would indicate the voltage of this battery as V = 48 V. A potential difference (voltage) of 48 V means that each coulomb of charge will provide 48 joules of energy. For example, 1 coulomb flowing through a lamp will convert 48 joules of the battery's energy into heat and light energy.

Fig. 2-8 Lead-acid battery. (Courtesy of Exide Power Systems.)

The relationship between charge, energy, and voltage can be expressed as

$$\text{Voltage (V)} = \frac{\text{energy }(W)}{\text{charge }(Q)}$$

or by rearranging:

$$W = V \times Q$$

This relationship can be used to determine electric energy in the same way as mechanical energy was found in Example 2-1.

Example 2-2

Determine the potential energy (*W*) of a 6-volt (6-V) battery that has 3000 coulombs (3000 C) of charge (*Q*) stored in it.

Given:	V = 6 V
	Q = 3000 C
Find:	W
Known:	W = V Q
Solution:	W = 6 V × 3000 C
	= 18,000 joules (J)
Answer:	Energy = 18,000 joules

Notice in Example 2-2 that multiplying volts by coulombs yields joules. This is because a volt is a joule per coulomb and the coulombs cancel:

$$\frac{\text{joule}}{\text{coulomb}} \times \frac{\text{coulomb}}{1} = \text{joule}$$

2-9 POLARITY

Polarity is a term that is used in several different ways. We can say that the polarity of a charge is negative, or we can say the polarity of a terminal is positive. We can also use the term to indicate how to connect the negative and positive terminals of electric devices. For example, when putting new cells in a transistor radio, we must install the cells with correct polarity (Fig. 2-9). The positive terminal of each cell connects to the positive terminal of the radio and the negative to the negative.

Electric devices which have negative and positive identifications on their terminals are said to be *polarized*. When connecting such devices to a source of voltage (such as a battery), we must observe the polarity marking. Again, the negative terminal of the device is connected to the negative terminal of the source and the positive to the positive. If polarity is not observed (that is, a positive connected to a negative), the device will not function and it may be damaged or ruined.

In electric-arc welding, the welder can weld with either *straight polarity* or *reverse polarity*. The straight or reverse refers to whether the negative or the positive terminal of the voltage source is connected to the welding rod.

2-10 SOURCES OF VOLTAGE

Voltage can be created by a number of techniques. All involve the conversion of some other form of energy into electric energy. All of them create a voltage by producing an excess of electrons at one terminal and a deficiency of electrons at another terminal.

The most common way of producing a voltage is by an electric generator. Generators convert mechanical energy into electric energy. Large generators like the one shown in Fig. 2-10 produce the voltage (and energy) that is provided to our homes, schools, and industries. Most of these generators are turned by mechanical devices such as steam turbines.

The electric cell is the next most common

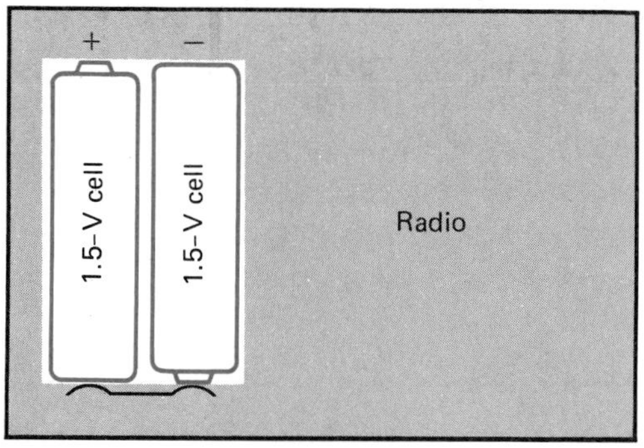

Fig. 2-9 When installing cells in a radio, the correct polarity must be observed.

Fig. 2-10 Electric generator shown being shipped on a railroad flatcar. (Courtesy of General Electric.)

source of voltage. Cells convert chemical energy into electric energy. Several cells can be connected together to form a battery. A heavy-duty battery is shown in Fig. 2-8. It is the type that would be used to operate the electric system of a large truck. A wide variety of batteries and cells are manufactured. They range from dry cells which weigh a few grams to industrial batteries which weigh hundreds of kilograms.

Other devices (Fig. 2-11) which produce voltage are thermocouples, crystals, and solar cells. Static electricity, which also produces voltage, has been discussed in Chap. 1.

The thermocouple converts heat energy into electric energy. This process is known as the *Seebeck effect*. Thermocouples are used extensively for measuring temperatures, especially high temperatures.

Crystals produce voltage by the *piezoelectric effect*. A voltage is produced by applying a varying pressure to the surface of the crystal. Crystals are used in such devices as phonograph pickup cartridges and microphones. In a microphone, the sound energy of the voice is first converted to mechanical energy by a diaphragm which applies pressure to the crystal. Thus, crystals convert mechanical energy into electric energy.

Solar cells are semiconductor devices. They convert light energy into electric energy. *Photovoltaic* is the term used to describe this conversion process. Solar cells can be used to provide the voltage needed to operate exposure meters in photography or a satellite communications system.

Reverse polarity

Electric generator

Chemical cells

Thermocouple

Crystal

Solar cell

Photovoltaic

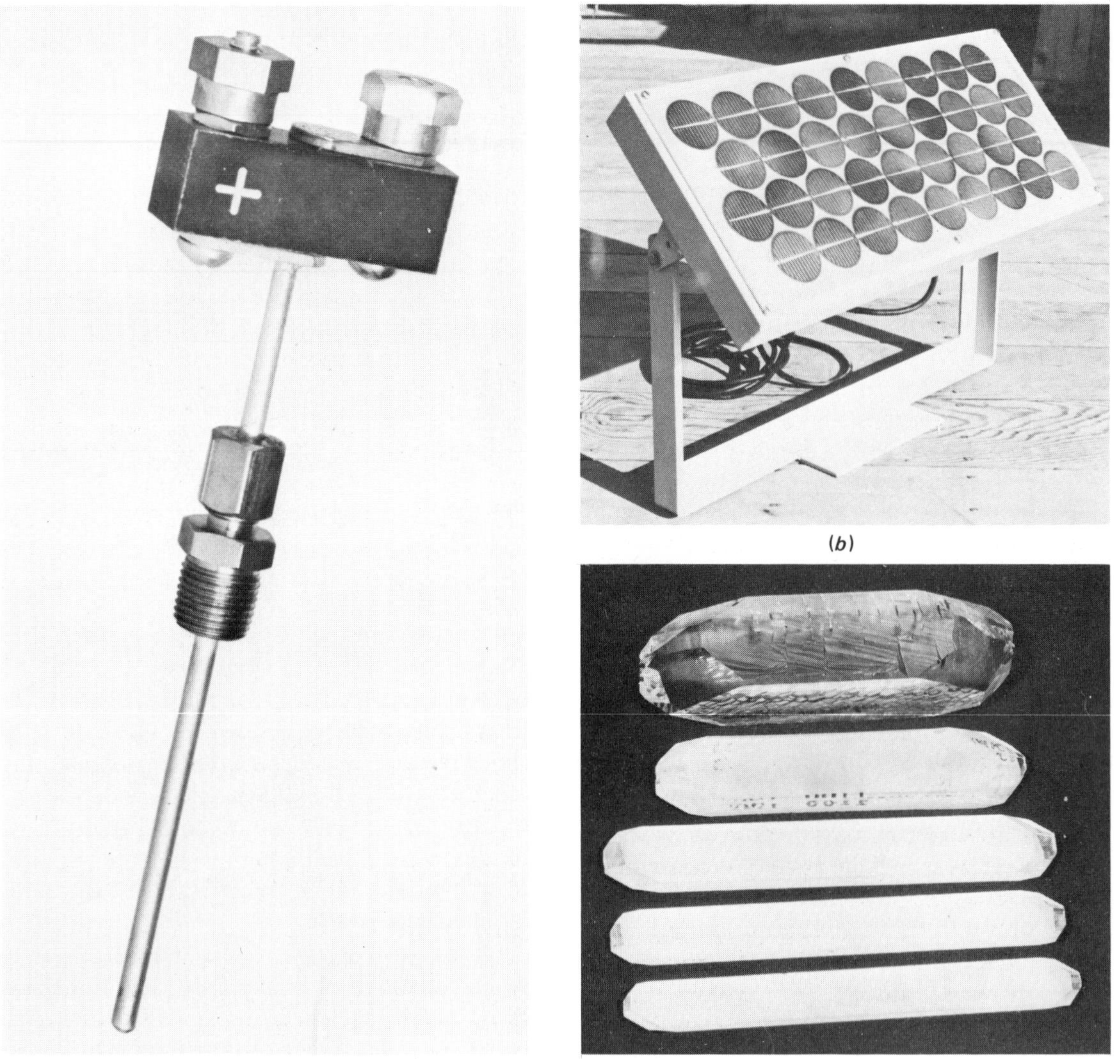

(*a*) (*b*) (*c*)

Fig. 2-11 Voltage-producing devices: (a) thermocouple. (Courtesy of Foxboro Company.) (b) Solar cells. (Courtesy of Solar Power Corp.) (c) Uncut quartz crystals. (Courtesy of CTS Knights, Inc.)

Review Questions

17. What is voltage?
18. What is the symbol for voltage?
19. Explain the difference between kinetic energy and potential energy.
20. Does a potential energy difference exist between the left-hand half and the right-hand half of your desk top? Explain.
21. Is it meaningful to say, "The potential energy difference of the desk top is 7 joules per kilogram?" Explain.
22. Is it meaningful to say, "The voltage at point A is 12 volts?" Explain.
23. What happens when an electron is allowed to move from the negative terminal of a battery through a lamp back to the positive terminal of the battery?
24. What is the base unit of voltage?
25. What is the abbreviation for the base unit of voltage?
26. Define the base unit of voltage in terms of units of energy and charge.
27. Define *polarized*.

28. List five devices that can produce a voltage. Also specify what type of energy they convert into electric energy.

2-11 RESISTANCE

The opposition a material offers to current is called *resistance*. The symbol for resistance is R. All materials offer some resistance to current. However, there is extreme variation in the amount of resistance offered by various materials. It is harder to obtain free electrons (current carriers) from some materials than others. It requires more energy to free an electron in high-resistance materials than in low-resistance materials. Resistance converts electric energy into heat energy when current is forced through a material.

2-12 CONDUCTORS

Those materials that offer very little resistance (opposition) to current are called *conductors*. Copper, aluminum, and silver are good conductors. They have very low resistance. In general those elements which have three or fewer electrons in the valence shell can be classified as conductors. However, even within those elements classified as conductors there is wide variation in the ability to conduct current. For example, iron has nearly six times the resistance of copper, though they are both considered conductors. Although silver is a slightly better conductor than copper, it is too expensive for common use. Aluminum is not as good a conductor as copper, but it is cheaper and lighter. Large aluminum conductors are used to bring electric energy into homes.

2-13 INSULATORS

Materials which offer a high resistance to current are called *insulators*. Even the best of insulators do release an occasional free electron to serve as a current carrier. However, for most practical purposes we can consider an insulator as a material that allows no current to flow through it. Common insulator materials used in electric devices are paper, wood, plastics, rubber, glass, and mica. Notice that common insulators are not pure elements. They are materials in which two or more elements are joined together to form a new substance. In the process of joining together, elements share their valence electrons. This sharing of valence electrons is called *covalent bonding*. It takes a lot of additional energy to break an electron free of a covalent bond.

A variety of insulators are shown in Fig. 2-12. These are the types of insulators used in high-temperature devices.

2-14 SEMICONDUCTORS

In between the extremes of conductors and insulators is a group of elements known as *semiconductors*. Semiconductor elements have four valence electrons. Two of the best known semiconductors are silicon and germanium. Semiconductors are neither good conductors nor good insulators. They allow some current to flow, yet they have a considerable amount of resistance. Semiconductors

Resistance

Conductor

Insulator

Covalent bonding

Semiconductor

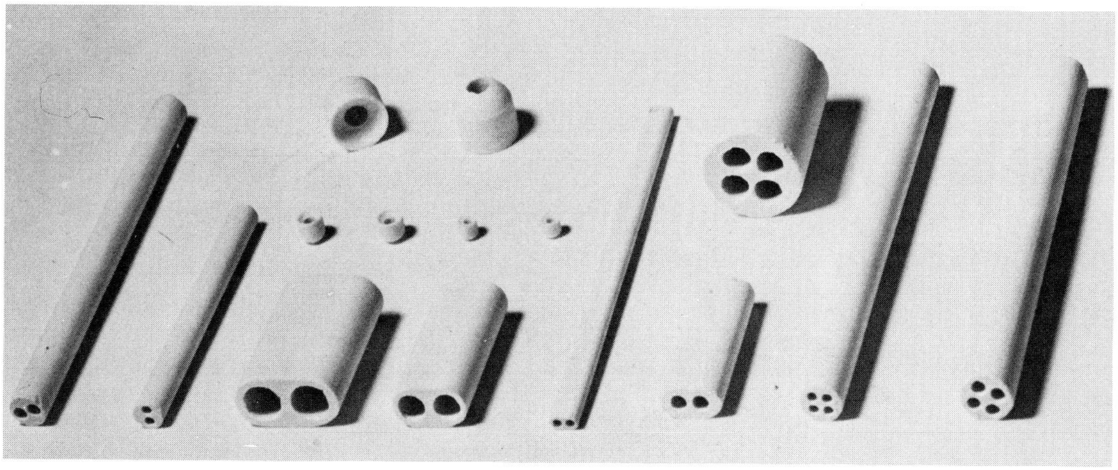

Fig. 2-12 Insulators. (Courtesy of Foxboro Co.)

are extremely important industrial materials. They are the materials from which electronic devices such as transistors, integrated circuits (IC), and solar cells are manufactured.

2-15 UNIT OF RESISTANCE—THE OHM

Ohm

Temperature coefficient of resistance

So far we have discussed the amount of resistance in terms of *low resistance* and *high resistance*. In order to work with electric circuits, we must be able to state the amount of resistance more specifically. The unit used to specify the amount of resistance is the *ohm*. Ohm is the base unit of resistance. The symbol used as an abbreviation for ohm is Ω (the capital Greek letter omega). The ohm is named in honor of Georg Ohm who worked out the relationship between current, voltage, and resistance. An ohm can be defined in several different ways. First, it is the amount of resistance that allows 1 ampere of current to flow when the voltage is 1 volt. Second, it is the amount of resistance of a column of mercury 106.3 centimeters in length, 1 millimeter square, and at a temperature of 0° Celsius (centigrade). From this last definition of an ohm, you can see that the amount of resistance of an object is determined by four factors: (1) the type of material from which the object is made, (2) the length of the object, (3) the cross-sectional area (height × width) of the object, and (4) the temperature of the object. The amount of resistance of an object is directly proportional to its length and inversely proportional to its cross-sectional area. For example, if the length of a piece of copper wire is doubled, then its resistance will also double (Fig. 2-13). If the cross-sectional area of the copper wire is made twice as great, then its resistance will be one-half its former value. The shaded ends of the conductors in Fig. 2-13 are the cross-sectional areas of the conductors.

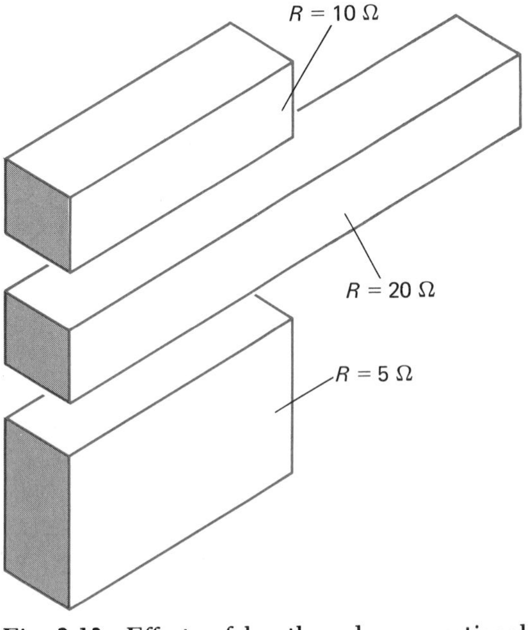

Fig. 2-13 Effects of length and cross-sectional area on resistance.

No simple relationships exist between resistance and temperature. The resistance of most materials increases as the temperature increases. However, with some materials, such as carbon, the resistance decreases when temperature increases.

2-16 TEMPERATURE COEFFICIENT

The change in resistance corresponding to a change in temperature is known as the *temperature coefficient of resistance*. Each material has its own temperature coefficient. Carbon has a negative temperature coefficient, while most metals have a positive temperature coefficient. A temperature coefficient is expressed as the number of ohms of change per million ohms per degree Celsius. The expression is abbreviated ppm/°C. For example, carbon has a negative temperature coefficient of 500 ppm/°C at 20°C. This means that a piece of carbon which had 1,000,000 ohms of resistance at 20°C would have 1,000,500 ohms at 19°C. At 18°C it would have 1,001,000 ohms. In many electric and electronic devices, changes in resistance due to changes in temperature are so small that they can be ignored. In those devices where small changes in resistance are important, such as electric meters, special low-temperature-coefficient materials are used. One of these materials is constantan (a mixture of copper and nickel). Constantan has a positive temperature coefficient of 18 ppm/°C at 20°C. Notice that the temperature coefficient is defined at a specific temperature. This means that the temperature coefficient itself changes with temperature. However, these changes are extremely small over the range of temperature in which most electric devices operate. A table of temperature coefficients of resistance for some common materials is given in Appendix 5.

2-17 RESISTIVITY

The characteristic resistance of a material is given by its *resistivity* or its *specific resistance*. These terms mean the same thing. The resistivity of a material is really just the resistance (in ohms) of a 1-meter cube of material. Resistivity ratings allow us to directly compare the abilities of various materials to conduct current. Figure 2-14 illustrates the way in which resistivity is determined. Resistivity is specified in ohm-centimeters. For example, annealed copper at 20°C has a resistivity of 1.72×10^{-6} (0.00000172) ohm-centimeter. This means that a cube of copper 1 centimeter on each side would have 1.72×10^{-6} ohm of resistance between any two opposite faces of the cube. A table of resistivity (in ohm-centimeters) is shown in Appendix 4. The lower the resistivity of a material, the better conductor it is.

The relationship of resistance to length, cross-sectional area, and resistivity is given by the formula

$$\text{Resistance } (R) = \frac{\text{resistivity } (K) \times \text{length } (L)}{\text{area } (A)}$$

Or, using only symbols,

$$R = \frac{KL}{A}$$

Resistance will be in ohms if all other quantities are also in base units, that is, if resistivity is in ohm-centimeters, length in centimeters, and area in square centimeters.

A common unit used in electrical wiring is the circular mil. Many wire tables give the diameter of wire in circular mils. The resistivity formula can be used with this unit if K is given in ohm-circular mils per foot, L is given in feet, and A is given in circular mils.

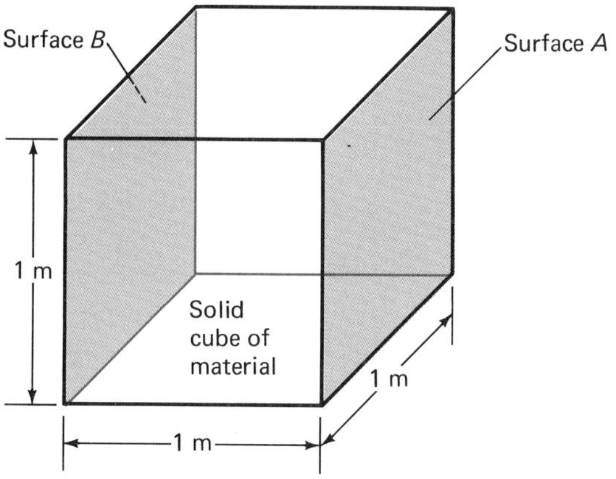

Fig. 2-14 Resistivity equals the resistance between surface A and surface B.

Example 2-3

What is the resistance, at 20°C, of an electric motor winding that uses 200 meters of copper conductor which is 0.26 square centimeter (approximately 3/16 in by 3/16 in)? The resistivity of copper at 20°C is 0.00000172 ohm-centimeter.

Given: K = 0.00000172 ohm-centimeter
L = 20,000 centimeters
A = 0.26 square centimeter

Find: R

Known: $R = \frac{KL}{A}$

Solution: $R = \frac{0.00000172\ \Omega\text{-cm} \times 20{,}000\ \text{cm}}{0.26\ \text{sq cm}}$

$= 0.132$ ohm

Answer: R = 0.132 ohm (Ω)

2-18 RESISTORS

In many electric and electronic devices, the resistance of the conductors is so small compared with other parts of the device that it can be ignored. In most cases the resistance of the conductors is not calculated or figured into the design of the device. However, in certain devices, such as radio and television receivers, large amounts of resistance are needed. When we need resistance to control current, we use resistors. *Resistors* are the physical devices which can be used to control the amount of current in an electric system. They are manufactured in a wide variety of shapes and sizes. They are available in resistance values that range from less than 1 ohm to many millions of ohms.

Resistivity

Specific resistance

$R = \frac{KL}{A}$

Resistor

Review Questions

29. What is resistance?
30. What form of energy is electric energy converted into when current flows through resistance?
31. List two ways in which conductors are different from insulators.

32. What devices are manufactured from semiconductor material?

33. What is the base unit of resistance?

34. What symbol is used as the abbreviation for the base unit of resistance?

35. Define the base unit of resistance in terms of the base units of voltage and current.

36. List the four factors that determine the resistance of an object.

37. Explain what is meant by a statement such as "this material has a positive temperature coefficient of 380 ppm/°C at 20°C."

38. Classify the following materials as either insulator or conductor:

Iron	Glass
Rubber	Carbon
Paper	Copper
Silver	Mica
Aluminum	

39. Which of the materials that you classified as conductors in question 38 have a positive temperature coefficient?

40. Define resistivity.

41. What is another name for resistivity?

42. What is a resistor?

2-19 POWER AND ENERGY

Power

Rate of using energy

Power refers to how rapidly energy is used or converted to another form of energy. Since energy is the ability to do work, we may also say power is concerned with how fast work is done. We will combine these two ideas in our definition of power. *Power* is the rate of using energy or doing work. The symbol for power is *P*.

The power needed to do a specified amount of work depends on how much time it takes to do the work. Suppose two workers are going to load 100 bricks onto a truck. They agree to each load one-half of the bricks. The first worker carries five bricks at a time. He finishes his half of the work in 10 trips to the truck. He is finished in 40 minutes (4 minutes per trip). The second worker carries two bricks at a time, and she finishes in 25 trips. She requires 100 minutes (4 minutes per trip). Both workers have done the same amount of work (loaded 50 bricks). However, carrying five bricks requires more power than carrying two bricks; so the first worker used more power than the second worker. He, the first worker, has done more work per unit of time.

The companies from which we buy our electricity are very often named "power companies." However, when we pay our electric bill, we are paying for electric energy, not electric power. The electric power company is less concerned with the power we use (how rapidly we consume electric energy) than with the amount of energy we use.

2-20 UNIT OF POWER

Joule per second

Watt

$W = Pt$

In electricity, we use the joule as our base unit of energy and the second as our base unit of time. Therefore, the logical unit for power is the *joule per second* (J/s). In honor of a Scottish scientist and inventor, James Watt, the joule per second has been named the *watt*. The base unit of power, therefore, is the watt. One watt is equal to one joule per second. The abbreviation for the watt is W.

The relationship between power, energy, and time is:

$$\text{Power } (P) = \frac{\text{energy } (W)}{\text{time } (t)}$$

By rearranging terms, we can state the relationship as:

$$W = Pt$$

Let us use these relationships to solve some electrical problems. As with other symbols in electricity, the W is used to denote two entirely different things. Don't confuse W meaning energy for W meaning watts.

Example 2-4

What is the power rating of an electric device that converts 940 joules of energy in 10 seconds?

Given: $W = 940$ joules (J)
$t = 10$ seconds (s)

Find: P

Known: $P = \frac{W}{t}$

Solution: $P = \dfrac{940\ \text{J}}{10\ \text{s}} = 94$ joules per second
$= 94$ watts (W)

Answer: The power rating is 94 watts.

Example 2-5

How much energy is required to operate a 60-watt lamp for 30 minutes?

Given: Power (P) = 60 watts (W)
time (t) = 30 minutes

Find: Energy (W)

Known: $W = Pt$

Solution: Since there are 60 seconds in a minute, 30 minutes equals

$30 \times 60 = 1800$ seconds (s)

$W = 60\ \text{W} \times 1800\ \text{s}$
$= 108{,}000$ watt-seconds
$= 108{,}000$ joules (J)

Answer: Energy required is 108,000 joules.

Notice in Example 2-5 we first converted time from minutes to the base unit of seconds. By doing this, the answer came out in joules—the base unit of energy. When first learning electrical relationships, it is best if you work with all quantities in base units, whenever possible. Also notice in Example 2-5 that watt × second = watt-second. One watt-second is equal to one joule. This can be shown by substituting joule per second for watt:

Watt-second

$$\text{Watt-second} = \frac{\text{joule}}{\text{second}} \times \frac{\text{second}}{1} = \text{joule}$$

The term watt-second is often used to express the amount of energy.

Since electric power companies charge their customers for energy use, electric bills must be based on a unit of energy. The joule is much too small to be practical. Imagine getting a bill for several hundred million joules each month! Instead, power companies use the *kilowatt-hour* which is equal to 3.6 million joules.

Review Problems

1. An 800-watt toaster requires 3 minutes to make a slice of toast. How much energy is required to make a slice of toast? Give your answer in joules and in kilowatt-hours.
2. A transistor receiver requires 6240 joules to operate for 40 minutes. What is the power rating of the receiver?

2-21 EFFICIENCY

In Sec. 1-4 of Chap. 1, we dealt with the concept of efficiency. In that section we viewed efficiency in terms of "useful energy out" and "total energy in." Efficiency can also be viewed in terms of power. The formula is the same except power is substituted for energy.

$$\%\ \text{eff.} = \frac{P_{out}}{P_{in}} \times 100$$

Example 2-6

What is the efficiency of a radio receiver that requires 4 watts of electric power input to deliver 0.5 watt of power output?

Given: $P_{in} = 4$ watts (W)
$P_{out} = 0.5$ watts (W)

Find: % eff.

Known: $\%\ \text{eff.} = \dfrac{P_{out}}{P_{in}} \times 100$

Solution: $\%\ \text{eff.} = \dfrac{0.5\ \text{W}}{4\ \text{W}} \times 100 = 0.125 \times 100$
$= 12.5$

Answer: Efficiency = 12.5%

The efficiency formula can also be rearranged to allow us to determine the amount of input power:

$$P_{in} = \frac{P_{out}}{\%\ \text{eff.}} \times 100$$

Example 2-7

How much power input is required to operate a stereo amplifier producing 50 watts of output power that is 30 percent efficient?

Given: $P_{out} = 50$ watts (W)
% eff. = 30

Find: P_{in}

Known: $P_{in} = \dfrac{P_{out}}{\%\ \text{eff.}} \times 100$

Solution: $P_{in} = \frac{50\text{ W}}{30} \times 100 = 1.7\text{ W} \times 100$
$= 170\text{ W}$

Answer: The power input would be 170 watts

Review Questions

Powers of 10

Exponent

43. Define power.
44. What is the base unit of power?
45. Define the base unit of power in terms of the base units of other quantities.
46. How many base units of energy are there in a watt-second?
47. What is the relationship between energy, power, and time?
48. Efficiency can be calculated using either of two quantities. What are they?

Review Problem

3. What is the efficiency of an electric motor that requires 1000 watts of electric power to deliver 680 watts of mechanical power?

2-22 POWERS OF 10

By now you may have noticed that some very large numbers are used in electricity. These numbers are easier to manage if they are expressed in powers of 10. *Powers of 10* refer to numbers that are exponents of the base 10. *Base 10* refers to the number system we use in our everyday life. The exponent (or power) refers to the number of times the digit 1 will be multiplied by 10. For example, 10^2 means that 1 will be multiplied by 10 twice:

$$1 \times 10 \times 10 = 100$$

Ten to the first power (10^1) would be $1 \times 10 = 10$; $10^0 = 1$. Powers of 10 and their base-10 equivalents which are commonly used in electricity are listed in Table 2-1. Notice that the power (exponent) of 10 tells you how many places to move the decimal place. For example, 10^4 can be changed to a number without an exponent by writing a 1 and moving the decimal point four places to the right. Thus, 10^4 is equal to 10,000. Also, a number can be converted to powers of 10 by making the exponent (power) equal to the number of places you move the decimal point. Thus, 2100 can be expressed as 2.1×10^3, or 105,000 can be written 1.05×10^5. Notice that it is conventional to move the decimal so that there is only one digit to the left of it. The number multiplied by the power of 10 is called the *coefficient*. In the above examples, 1.05 and 2.1 are the coefficients.

Numbers smaller than 1 are expressed in powers of 10 with negative exponents. The negative exponent tells you how many times 1 is to be divided by 10. Thus, 10^{-2} is the same as $1 \div 10 \div 10 = 0.01$. Notice again that the easy way to handle exponents is by moving the decimal point as many places as the exponent. When the exponent (power) is negative, move the decimal to the left. To convert 10^{-3}, just write a 1 and move the decimal three places to the left. This procedure yields 0.001, which is equal to 10^{-3}. To convert to a negative power

Table 2-1 Powers of 10 and Base-10 Equivalents

Power of 10	Number (Base-10 Equivalent)
10^{12}	1,000,000,000,000
10^{11}	100,000,000,000
10^{10}	10,000,000,000
10^{9}	1,000,000,000
10^{8}	100,000,000
10^{7}	10,000,000
10^{6}	1,000,000
10^{5}	100,000
10^{4}	10,000
10^{3}	1000
10^{2}	100
10^{1}	10
10^{0}	1
10^{-1}	0.1
10^{-2}	0.01
10^{-3}	0.001
10^{-4}	0.0001
10^{-5}	0.00001
10^{-6}	0.000001
10^{-7}	0.0000001
10^{-8}	0.00000001
10^{-9}	0.000000001
10^{-10}	0.0000000001
10^{-11}	0.00000000001
10^{-12}	0.000000000001

of 10, just move the decimal to the right of the first digit larger than zero. The negative exponent equals the number of places you moved the decimal place. For example, 0.000000054 is equal to 5.4×10^{-8} and 0.03816 is equal to 3.816×10^{-2}.

The big advantage to expressing numbers in powers of 10 is that it simplifies arithmetic involving large numbers. To multiply two numbers expressed in powers of 10 just multiply the coefficients and then *add* the powers. The product is the new coefficient multiplied by the new power of 10. Some examples are

$$10^4 \times 10^2 = 10^6$$
$$10^{-2} \times 10^4 = 10^2$$
$$10^{-5} \times 10^3 = 10^{-2}$$
$$1.4 \times 10^2 \times 1.2 \times 10^6 = 1.4 \times 1.2 \times 10^2 \times 10^6 = 1.68 \times 10^8$$
$$6.3 \times 10^3 \times 8.4 \times 10^4 = 6.3 \times 8.4 \times 10^3 \times 10^4 = 52.92 \times 10^7 = 5.292 \times 10^8$$

To divide with powers of 10, first divide the coefficients. Then subtract the exponent in the divisor from the exponent in the dividend. This procedure is illustrated below:

$$1 \times 10^4 \div 1 \times 10^2 = 1 \times 10^2$$

$$4 \times 10^3 \div 2 \times 10^{-2} = (4 \div 2)\,(10^3 \div 10^{-2}) = 2 \times 10^{3-(-2)} = 2 \times 10^5$$

$$6 \times 10^{-10} \div 4 \times 10^{-8} = (6 \div 4)\,(10^{-10} \div 10^{-8}) = 1.5 \times 10^{-2}$$

To either add or subtract numbers expressed as powers of 10, both numbers must have the same exponent. The exponent then remains the same. For example:

$$(2.4 \times 10^6) + (3.5 \times 10^6) = 5.9 \times 10^6$$

$$(2.4 \times 10^4) + (3.5 \times 10^5) = (0.24 \times 10^5) + (3.5 \times 10^5) = 3.74 \times 10^5$$

$$(3.8 \times 10^3) - (1.6 \times 10^3) = 2.2 \times 10^3$$

Review Problems

4. Express the following numbers in powers of 10:
   a. 320
   b. 400,000
   c. 1,000,000
5. Convert the following powers of 10 to ordinary numbers:
   a. 6.8×10^3
   b. 10^5
   c. 3.64×10^2
6. Convert the following powers of 10 to ordinary numbers:
   a. 10^{-4}
   b. 2.81×10^{-3}
   c. 6.3×10^{-4}
   d. 6.3×10^2
7. Convert the following numbers to powers of 10:
   a. 0.0000001
   b. 0.028
   c. 0.0072
   d. 1000
8. Solve the following problems:
   a. $2 \times 10^8 \times 4 \times 10^3 = ?$
   b. $1.4 \times 10^2 \times 2.8 \times 10^{-3} = ?$
   c. $\dfrac{6.6 \times 10^4}{3 \times 10^{-2}} = ?$
   d. $\dfrac{4 \times 10^{-3}}{2 \times 10^2} = ?$
   e. $(4 \times 10^3) + (6 \times 10^4) = ?$

2-23 MULTIPLE AND SUBMULTIPLE UNITS

Prefix

Multiple and submultiple units

For some applications of electricity, the base unit of a quantity may seem to be very large. For other applications, the same base unit may seem rather small. For example, in solid-state devices, we work with currents of less than 0.0000001 ampere (A). In an aluminum reduction plant, we are working with currents greater than 110,000 amperes (A). Although these numbers could be shortened by expressing them in powers of 10, they would still be long expressions. Also, they would be long when spoken. For example, 1.1×10^5 A would be spoken as "one point one times 10 to the fifth amperes." To avoid such long expressions, scientists use prefixes to indicate units that are smaller and larger than the base unit.

The prefixes and their symbols commonly used in electricity are shown in Table 2-2. Also shown in Table 2-2 are the relationships (both powers of 10 and base 10) of the prefixes to the base unit. Notice that adjacent prefixes are related by factors of 1000. Adjacent prefixes

Table 2-2 Prefixes and Symbols

Prefix	Symbol	Number (Base 10)	Powers of 10
Mega	M	1,000,000	10^6
Kilo	k	1000	10^3
Base unit		1	10^0
Milli	m	0.001	10^{-3}
Micro	μ	0.000001	10^{-6}
Nano	n	0.000000001	10^{-9}
Pico	p	0.000000000001	10^{-12}

Multiples of three

Kilowatt-hour

Horsepower

are either 1000 times larger or $\frac{1}{1000}$ as large as their neighbor.

Multiple and submultiple units are designated by adding the appropriate prefix to the base unit. Now we can specify the 110,000 amperes (A) used in an aluminum plant as 0.11 megampere (MA) or 110 kiloamperes (kA). The 0.0000001 ampere (A) used in a solid-state device can be written as 0.1 microampere (μA). Some other examples of conversions between units are

2200 ohms (Ω) = 2.2 kilohms (kΩ)
0.083 watt (W) = 83 milliwatts (mW)
450,000 volts (V) = 450 kilovolts (kV)
2.7×10^6 ohms (Ω) = 2.7 megohms (MΩ)
3700 microamperes (μA) = 3.7 milliamperes (mA)
4700 kilohms (kΩ) = 4.7 megohms (MΩ)
0.27 megohm (MΩ) = 270 kilohms (kΩ)
6.2×10^{-2} ampere (A) = 62 milliamperes (mA)
6,800,000 ohms (Ω) = 6.8×10^6 ohms (Ω) = 6.8 megohms (MΩ)

Notice that in all the above examples the conversion is made by moving the decimal point either three places or multiples of three places. If you are converting from a small unit to a larger unit, the decimal point is moved to the left. Remember that going from micro to milli is going from a small to a large unit. When converting from a large unit to a small unit, you shift the decimal point to the right.

It is important that you become very familiar with converting from one unit to another. A repair manual for an electrical device may list a resistor as 2.2 kΩ. When the electrician goes to order a replacement resistor, the parts manufacturer may list the resistor as 2200 ohms. The electrician must make the conversion.

Review Problems

9. Complete the conversions listed below.
   a. 30 millivolts = _?_ volt
   b. 3800 ohms = _?_ kilohms
   c. 490 microamperes = _?_ amperes
   d. 5.6×10^5 ohms = _?_ megohm
   e. 6000 millicoulombs = _?_ coulombs
   f. 0.048 watt = _?_ milliwatts
   g. 9,600,000 joules = _?_ megajoules
   h. 9.8 milliamperes = _?_ $\times 10^{-3}$ amperes

2-24 SPECIAL UNITS AND CONVERSIONS

In this book, standard metric units are used whenever practical. However, there are some areas of electrical work in which nonmetric units are so common that we must consider them.

As noted earlier, electric power companies sell their electric energy by the kilowatt-hour. A *kilowatt-hour* = 1000 watt-hours. An hour = 3600 seconds. Therefore, a watt-hour = 3600 watt-seconds. And finally, a kilowatt-hour = 3,600,000 watt-seconds or joules. As you can see, the joule would be a small unit of energy for a power company to use. Consider, for instance, that many homes use more than 1000 kilowatt-hours of energy every month. Remember that the watt-second and watt-hour are perfectly usable units of energy, even though the base unit is the joule.

Example 2-8

How much energy is used by a 1200-watt heater in 4 hours of continuous operation?

Given: Power (P) = 1200 watts (W)
time (t) = 4 hours (h)

Find: Energy (W)

Known: $W = Pt$

Solution: $W = 1200\ \text{W} \times 4\ \text{h} = 4800\ \text{Wh}$

Answer: The energy used is 4800 Wh or 4.8 kWh.

The output power of an electric motor (and many mechanical devices) is specified in horsepower (hp) rather than in watts. One horsepower is equal to 746 watts. When figuring the efficiency of a motor, you must first convert the output power to watts.

Example 2-9

What is the efficiency of a $\frac{3}{4}$-horsepower (hp) motor that requires an input of 1000 watts (W) of electric power?

Given: Power input (P_{in}) = 1000 W
power out (P_{out}) = $\frac{3}{4}$ hp

Find: Efficiency (eff.)

Known: $\% \text{ eff.} = \frac{P_{out}}{P_{in}} \times 100,$
1 hp = 746 W

Solution: $\frac{3}{4}$ hp = 0.75 hp

$$P_{out} = 0.75 \text{ hp} \times 746 \frac{\text{W}}{\text{hp}} = 559.5 \text{ W}$$

$$\% \text{ eff.} = \frac{559.5 \text{ W}}{1000 \text{ W}} \times 100 = 55.95$$

Answer: The efficiency of the $\frac{3}{4}$-horsepower motor is about 56 percent.

Review Questions

49. List two ways by which you can condense a large number such as 6,800,000 ohms.
50. What is horsepower?

Summary

1. Charge is the electrical property of electrons and protons.
2. One coulomb is the charge possessed by 6.25×10^{18} electrons.
3. Current is the movement of charge in a specified direction.
4. Current can flow in solids, gases, liquids, and a vacuum.
5. In solids the current carrier is the electron.
6. In gases the current carriers are both electrons and ions.
7. In liquids the current carriers are ions—both positive and negative.
8. Current travels at approximately the speed of light.
9. Individual current carriers (electrons) travel much slower than the speed of light.
10. Direct current (dc) never reverses direction.
11. Alternating current (ac) periodically reverses direction.
12. An ionized liquid is an electrolyte.
13. An ampere is a coulomb per second.
14. Voltage is a potential energy difference between two points.
15. Energy = voltage × charge
16. Polarity indicates whether a point is negative or positive.
17. Sources of voltage include generators, batteries, thermocouples, solar cells, and crystals.
18. Resistance is opposition to current.
19. Resistance converts electric energy to heat energy.
20. Conductors are materials that have low resistance.
21. Silver, coppper, and aluminum (in that order) are the best conductors.
22. Insulators do not allow any current to flow (for practical purposes).
23. An ohm is a volt per ampere. It is the resistance of a specified column of mercury at a specified temperature.
24. The resistance of an object is determined by the resistivity of the material, the length, the cross-sectional area, and the temperature.
25. Resistance is directly proportional to the length.
26. Resistance is inversely proportional to the cross-sectional area.
27. Most conductors have a positive temperature coefficient.
28. A temperature coefficient specifies the amount of change in resistance per million ohms per degree Celsius change.
29. Resistivity (specific resistance) of a material is the resistance of a specified cube of the material.
30. Resistors are devices used to control current.
31. Power is the rate of doing work or converting energy.

32. A watt is a joule per second.
33. Energy = power × time
34. $\% \text{ eff.} = \frac{\text{power output}}{\text{power input}} \times 100$
35. A kilowatt-hour is a unit of energy.
36. Horsepower is a unit of power.
37. 1 horsepower = 746 watts
38. Table 2-3 is a summary of the units and symbols used throughout this book.

Table 2-3 Symbols and Abbreviations

Quantity	Symbol	Base Unit	Abbreviation
Charge	Q	Coulomb	C
Current	I	Ampere	A
Time	t	Second	s
Voltage	V	Volt	V
Energy	W	Joule	J
Resistance	R	Ohm	Ω
Power	P	Watt	W

Chapter Review Questions

Here is a chance to see what you have learned. For questions 2-1 to 2-10, determine whether the statement is true or false.

2-1. A proton can act as a current carrier in an electric circuit.

2-2. A joule is equal to a watt-second.

2-3. 5 microamperes = 0.005 ampere

2-4. Semiconductors generally have five valence electrons.

2-5. The liquid in a battery is called an electrolyte.

2-6. A potential energy difference between two points is called power.

2-7. Aluminum is a better conductor than silver.

2-8. $1.4 \times 10^{-2} = 0.014$

2-9. $6 \times 10^{8} \times 3 \times 10^{-6} = 1.8 \times 10^{3}$

2-10. 0.038 watts = 38 milliwatts

For questions 2-11 to 2-17, choose the letter that best completes the sentence.

2-11. A joule per second defines a
(A) Volt (B) Ampere (C) Watt (D) Ohm

2-12. A coulomb per second defines a
(A) Volt (B) Ampere (C) Watt (D) Ohm

2-13. The base unit of charge is the
(A) Ampere (B) Watt (C) Proton (D) Coulomb

2-14. Electric energy is converted into heat energy by
(A) Voltage (B) Current (C) Power (D) Resistance

2-15. The base unit of voltage is the
(A) Volt (B) Ampere (C) Watt (D) Ohm

2-16. The base unit of resistance is the
(A) Volt (B) Ampere (C) Watt (D) Ohm

2-17. The base unit of current is the
(A) Volt (B) Ampere (C) Watt (D) Ohm

Answer the following questions.

2-18. What are the four factors that determine the resistance of an object?

2-19. Do most conductors have a positive or a negative temperature coefficient?

2-20. What is the symbol for each of the following terms?
a. Volt
b. Ampere
c. Resistance
d. Ohm
e. Watt
f. Charge
g. Milliampere

2-21. An electric motor requires 980 watts to operate it. It operates for 4 hours. How much energy does it use?

2-22. A 1-horsepower motor requires 1100 watts to operate. What is its percent efficiency?

2-23. A television receiver uses 216,000 joules of energy to operate for 1 hour. What is the power rating of the receiver?

Answers to Review Questions

1. An electric charge is the electrical property exhibited by protons and electrons.
2. *Q*
3. A coulomb is the base unit of charge. It is equal to the charge possessed by 6.25×10^{18} electrons.
4. C
5. A current carrier is a charged particle, i.e., an electron or an ion.
6. Electrons and positive ions
7. An electric current is the movement of electric charges, in a specified direction, through some material such as a solid, a gas, or a liquid.
8. *I*
9. The electron
10. The electron travels from ion to ion. The individual electron moves down the wire slowly.
11. Alternating current periodically reverses its direction, while direct current flows continuously in the same direction.
12. Ampere
13. An ampere is equal to one coulomb per second.
14. A
15. *a*. $I = 8$ A
    b. $Q = 6$ C
16. 20 A
17. Voltage is potential energy difference; or: voltage is an electric pressure.
18. *V*
19. Kinetic energy is energy being converted or energy in use. Potential energy is energy at rest or stored energy.
20. No, because the two halves are at the same height; or: no, because no useful work or energy can be gained by having an object move from one side to the other.
21. No, because potential energy differences can exist only between two points. The reference point (floor) must be specified.
22. No, because voltage is a potential energy difference. This means that voltage always exists between two points.
23. Electric energy is converted to heat and light energy.
24. Volt
25. V
26. The volt is equal to a joule per coulomb.
27. Polarized means that an electric device has a positive and a negative terminal.
28. Battery—chemical energy
    Crystal (microphone, cartridge)—mechanical energy
    Generator—mechanical energy
    Thermocouple—heat energy
    Solar cell—light energy
29. Resistance is opposition to current.
30. Heat energy

31. Conductors have very low resistance, while insulators have extremely high resistance. Conductors are usually elements with no more than three valence electrons, while insulators are usually compounds with covalent bonding.
32. Transistors and integrated circuits
33. Ohm
34. Ω
35. One ohm equals one volt per ampere.
36. Resistivity of the material from which the object is made, length of the object, cross-sectional area of the object, and, temperature of the object.
37. The statement means that the resistance of the material will change 380 ohms for every million ohms for each degree Celsius that the temperature changes above or below 20°C.
38. Conductors: iron, silver, carbon, copper, and aluminum; insulators: rubber, paper, glass, and mica
39. Iron, silver, copper, and aluminum
40. Resistivity is the characteristic resistance of a material. It is the resistance of a specified cube of the material. The cube is often defined as being either 1 cm, 1 m, or 1 ft on each side.
41. Specific resistance
42. A resistor is a physical device designed to provide a controlled amount of resistance.
43. Power is the rate of doing work or using energy.
44. Watt
45. A watt is a joule per second.
46. 1 (1 joule = 1 watt-second)
47. Energy = power × time
48. Power and energy
49. By using powers of 10 (6.8×10^6 ohms), or by using multiple units (6.8 megohms)
50. Horsepower is a nonmetric unit of power. It is equal to 746 watts.

Answers to Review Problems

1. *Given:* $P = 800$ W, $t = 3$ minutes
   Find: W
   Known: $W = Pt$
   60 seconds = 1 minute
   Solution: $t = 3 \text{ min} \times 60 \dfrac{\text{s}}{\text{min}} = 180 \text{ s}$
   $W = 800 \times 180 = 144{,}000 \text{ J}$
   Answer: Energy = 144,000 joules

2. *Given:* $W = 6240$ joules, t = 40 minutes
   Find: P
   Known: $P = \dfrac{W}{t}$
   Solution: $T = 40 \text{ min} \times 60 \dfrac{\text{s}}{\text{min}} = 2400 \text{ s}$
   $P = \dfrac{6240}{2400} = 2.6 \text{ W}$
   Answer: Power = 2.6 watts

3. *Given:* $P_{in} = 1000$ W, $P_{out} = 680$ W
   Find: % eff.
   Known: $\% \text{ eff.} = \dfrac{P_{out}}{P_{in}} \times 100$
   Solution: $\% \text{ eff.} = \dfrac{680}{1000} \times 100 = 68$
   Answer: % eff. = 68%

4. *a*. 3.2×10^2
   b. 4×10^5
   c. 1×10^6 or 10^6
5. *a*. 6800
   b. 100,000
   c. 364
6. *a*. 0.0001
   b. 0.00281
   c. 0.00063
   d. 630
7. *a*. 1×10^{-7} or 10^{-7}
   b. 2.8×10^{-2}
   c. 7.2×10^{-3}
   d. 1×10^3 or 10^3
8. *a*. 8×10^{11}
   b. 3.92×10^{-1}
   c. 2.2×10^6
   d. 2×10^{-5}
   e. 6.4×10^4
9. *a*. 0.03 V
   b. 3.8 kΩ
   c. 0.00049 A
   d. 0.56 MΩ
   e. 6 C
   f. 48 mW
   g. 9.6 MJ
   h. 9.8×10^{-3} A

Basic Circuits, Laws, and Measurements

You are now familiar with electrical quantities and units. Now you are ready to explore the circuits, laws, and devices used to control and measure these quantities and units.

When you have finished this chapter, you will be able to calculate the expected current, voltage, and resistance in a device. You will also be able to measure these same quantities. Comparing measured values with calculated, or expected, values is one way of determining if a device is operating properly.

Electric components (parts) such as resistors and batteries are represented by symbols in electrical drawings. In this chapter you will learn about components and their symbols. Electric components are connected together to form electric circuits. You will learn about circuits and how to represent circuits with electrical symbols.

3-1 CIRCUIT ESSENTIALS

Most complete electric circuits contain six parts. These are:

1. An *energy source* to provide the voltage needed to force current (electrons) through the circuit
2. *Conductors* through which the current can travel
3. *Insulators* to confine the current to the desired paths (conductors, resistors, etc.)
4. A *load* to control the amount of current and convert the electric energy taken from the energy source
5. A *control device*, often a switch, to start and stop the flow of current
6. A *protection device* to interrupt the circuit in case of a circuit malfunction

The first four of the above six parts are essential parts. All complete circuits use them. The control device (item 5) is occasionally omitted. Protection devices are often omitted from circuits. A complete electric circuit has an uninterrupted path for current (electrons) to flow from the negative terminal of the energy source through the load and control device to the positive terminal of the energy source.

The simplest electric circuit contains only one load, one voltage source, and one control device. It is sometimes referred to as a *simple circuit* to distinguish it from the more complex circuits. A single-cell flashlight is an example of a simple circuit. Figure 3-1(*a*) shows (in cross section) the construction details of such a flashlight. Electron flow (current) in the flashlight circuit can be traced by referring to Fig. 3-1. Electrons leave the negative end of the cell, travel through the spring, the metal case, the leaf spring of the switch, the metal reflector, the light bulb, and back to the positive end of the cell. Note that the spring, case, and reflector are conductors for the flashlight circuit. It is quite common in electric devices for structural parts of the device to also serve as conductors. For example, both the frame of an automobile and the chassis of many radio receivers serve as circuit conductors.

3-2 CIRCUIT SYMBOLS AND DIAGRAMS

In describing an electric circuit, you will find it more convenient to use symbols to represent electric components than to draw pic-

From page 33:
Simple circuit

On this page:
Schematic diagram

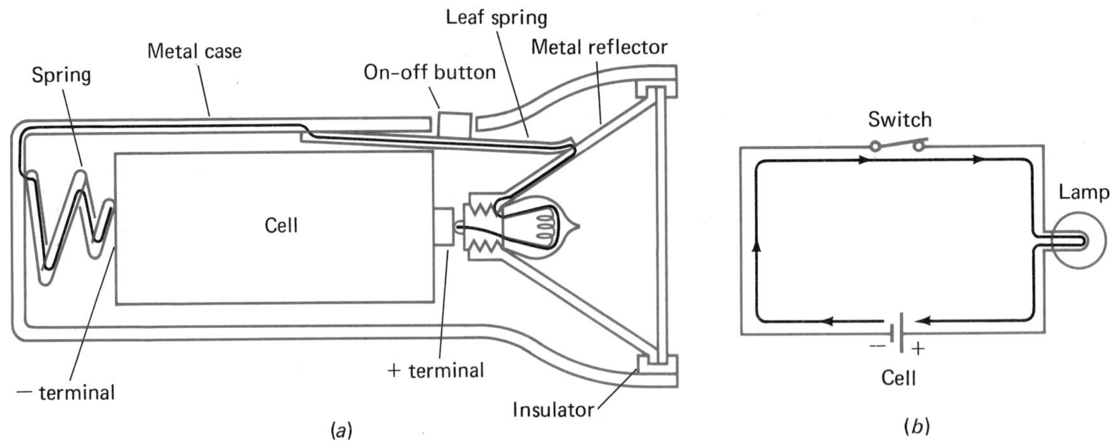

Fig. 3-1 Flashlight: (a) cross-sectional drawing, (b) schematic diagram.

tures of the components. A resistor and the symbol used to represent it are shown in Fig. 3-2. This same symbol is used for all fixed resistors regardless of the material from which the resistor is made. Other common electric components and their symbols are shown in Fig. 3-3. It should be noticed that there is no symbol to distinguish insulated from noninsulated conductors. Insulation is assumed to be present wherever it is needed to keep components and conductors from making electrical contact. In constructing an electric circuit which is described by electrical symbols, the circuit builder must determine where insulation is needed.

A drawing which uses only symbols to show how components are connected together is called a *schematic diagram*. A schematic diagram shows only how the parts are electrically interconnected. The physical size and the mechanical arrangements of the parts are in no way indicated. Also, accessories such as battery or lamp holders are not indicated. Unless the schematic diagram is accompanied by some form of pictorial drawing, the physical arrangement of the electric components is left to the discretion of the circuit builder.

A schematic diagram of the flashlight illustrated in Fig. 3-1(*a*) is shown in Fig. 3-1(*b*). Current can also be traced in the schematic diagram. The line with the arrowheads indicates the direction of current flow. The electrons (current) flow from the negative end of the cell through the closed switch and the lamp

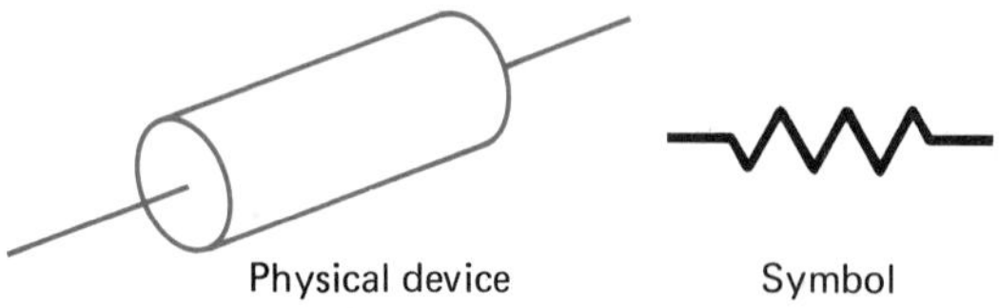

Fig. 3-2 Resistor and its symbol.

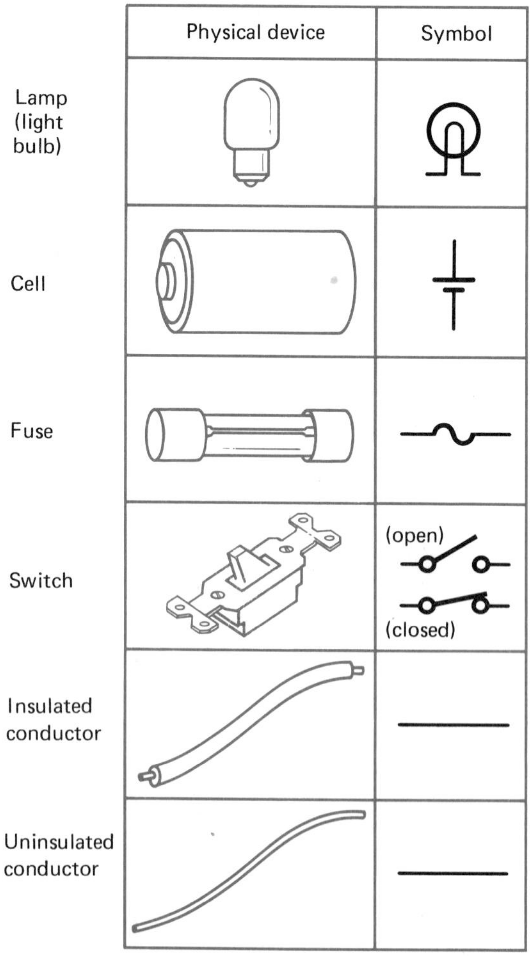

Fig. 3-3 Electric components and their symbols.

and back to the positive end of the cell. The conductor between the negative terminal of the cell and the switch in Fig. 3-1(*b*) represents the spring and the metal case in Fig. 3-1(*a*). The contact between the positive end of the cell and the light bulb in Fig. 3-1(*a*) is represented by the long line in the schematic diagram. Notice that lines in a schematic do not necessarily indicate a wire, but they do indicate a path for current to flow through.

The electrical values of the components used in the circuit can also be included on the schematic diagram. This is done in one of two ways. In the first method, the values of the components are printed beside the symbol for the components (Fig. 3-4). In the second method, shown in Fig. 3-5, an identifying letter or symbol is printed beside the component, and the value of the components is given in an accompanying parts list.

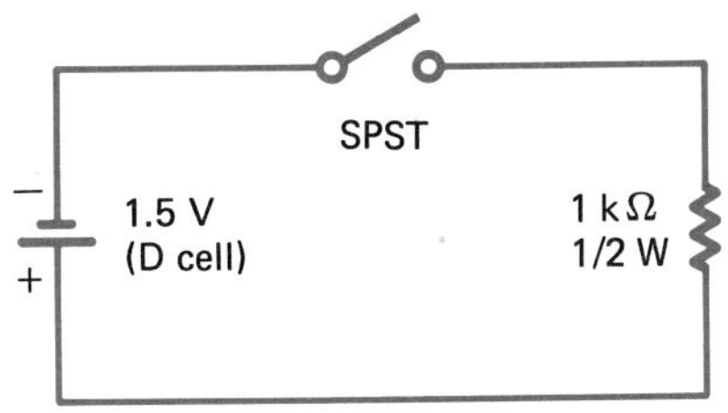

Fig. 3-4 Specifying component values on the schematic diagram.

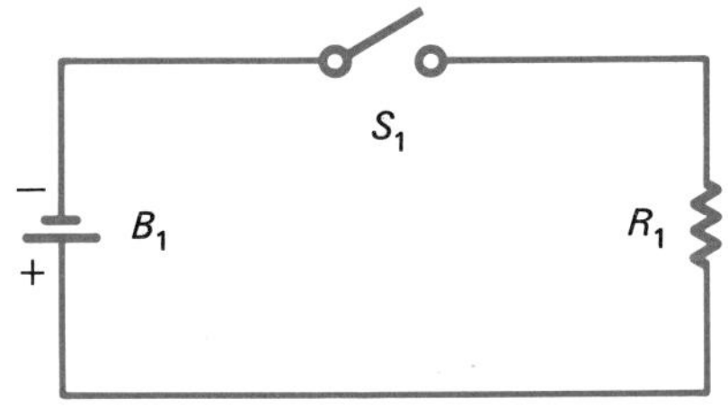

Parts list

R_1 Resistor, 1 kΩ, 1/2 W
S_1 Switch, SPST
B_1 Cell, 1.5 V, size D

Fig. 3-5 Specifying component values in a parts list.

Review Questions

1. List the six parts of a complete circuit.
2. Which two of the parts in question 1 are sometimes omitted?
3. Are conductors in an electric circuit always wire conductors?
4. Draw the symbol for a lamp, a resistor, and a conductor.
5. What is a schematic diagram?

3-3 CALCULATING ELECTRICAL QUANTITIES

In Chaps. 1 and 2 you learned how to use relationships between electrical quantities to calculate other electrical quantities. In this section we will work with more relationships among electrical quantities. These new relationships will emphasize quantities which can be easily measured or are commonly specified by manufacturers of electrical products.

Ohm's Law

The relationship between current (I), voltage (V), and resistance (R) was discovered by a German scientist named Georg Ohm. This relationship is named Ohm's law in his honor. Ohm found that the current in a circuit varies directly with the voltage when the resistance is kept constant. While keeping the resistance constant, Ohm varied the voltage across the resistance and measured the current through it. In each case, when he divided the voltage by the current, the result was the same. In short, this is Ohm's law. Ohm's law can be stated as, "The current is directly proportional to the voltage and inversely proportional to the resistance."

Written as a mathematical expression, Ohm's law is

$$\text{Resistance } (R) = \frac{\text{voltage (V)}}{\text{current } (I)} \quad \text{or} \quad R = \frac{V}{I}$$

The above equation allows you to determine the value of the resistance when the voltage and current are known.

Of course, Ohm's law can be rearranged to solve for either resistance or voltage. The rearranged relationships are

$$\text{Current } (I) = \frac{\text{voltage (V)}}{\text{resistance } (R)} \quad \text{or} \quad I = \frac{V}{R}$$

and

$$\text{Voltage (V)} = \text{current } (I) \times \text{resistance } (R)$$

or $V = IR$

Electrical ratings

$R = \frac{V}{I}$

$I = \frac{V}{R}$

$V = IR$

An aid to remembering the Ohm's law relationships is shown in the divided circle of Fig. 3-6. To use the aid, just cover the quantity you want to find and perform the multiplication or division indicated. Cover the V of Fig. 3-6 and the remainder of the circle indicates I times R. Thus, voltage (V) equals current (I) times resistance (R). Cover the R and the remainder of the circle shows voltage (V) divided by current (I). Finally, if the I is covered, the indicated operation is to divide voltage (V) by resistance (R).

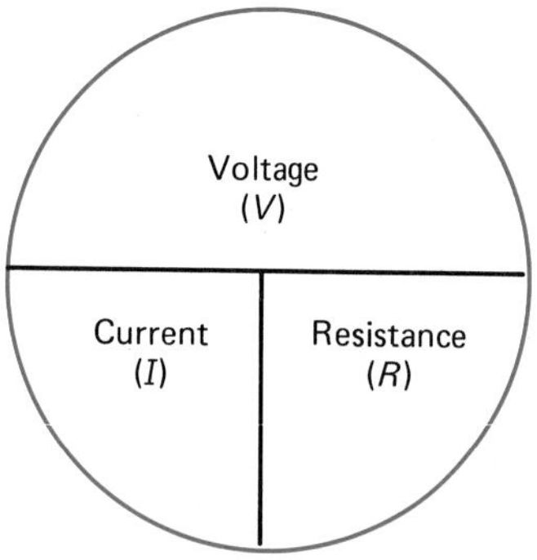

Fig. 3-6 Ohm's law circle.

Example 3-1

How much current (I) flows in the circuit shown in Fig. 3-7?

Given: Voltage (V) = 2.2 volts (V)
Resistance (R) = 1 kilohm (1 kΩ)

Find: Current (I)

Known: $I = \frac{V}{R}$, 1 kΩ = 1000 Ω

Solution: $I = \frac{2.2\ \text{V}}{1000\ \Omega} = 0.0022$ ampere (A)

Answer: Current in the circuit is 0.0022 A.

Example 3-2

A lamp has a resistance of 90 ohms. How much current flows through the lamp when it is connected to 120 volts?

Given: R = 90 Ω
V = 120 V

Find: I

Known: $I = \frac{V}{R}$

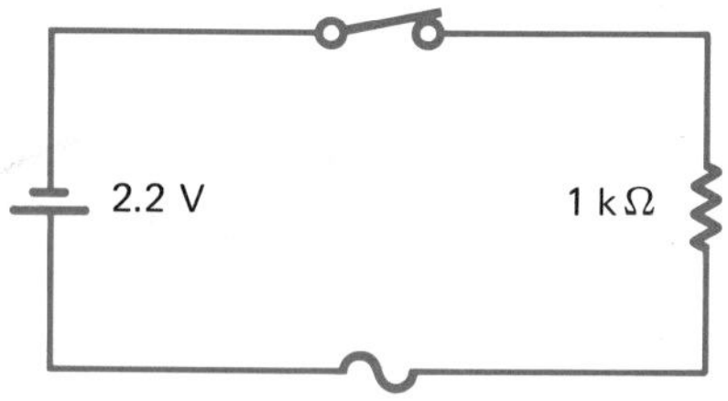

Fig. 3-7 Circuit diagram for Example 3-1.

Solution: $I = \frac{120\ \text{V}}{90\ \Omega} = 1.3$ A

Answer: Current through the lamp equals 1.3 A.

Notice in the above examples that the answers for current are in their base units. This is because both voltage and resistance are also in base units. Remember that an ohm was defined as a volt per ampere. Therefore, Ohm's law expressed in base units is

$$1 \text{ ampere} = \frac{1 \text{ volt}}{1 \text{ volt/ampere}}$$

This expression reduces to 1 ampere = 1 ampere, which shows that proper units were used.

Example 3-3

The manufacturer specifies a certain lamp will allow 0.8 amperes of current when 120 volts are applied to it. What is the resistance of the lamp?

Given: Current (I) = 0.8 ampere (A)
Voltage (V) = 120 volts (V)

Find: Resistance (R)

Known: $R = \frac{V}{I}$

Solution: $R = \frac{120\ \text{V}}{0.8\ \text{A}} = 150$ ohms (Ω)

Answer: Resistance of the lamp is 150 Ω.

Example 3-4

An electric device has a resistance of 50 ohms. How much voltage must be applied to the device to cause 1.6 amperes of current to flow?

Given: R = 50 Ω
I = 1.6 A

Find: V

Known: $V = IR$

Solution: $V = 1.6\text{ A} \times 50\ \Omega = 80\text{ V}$

Answer: Voltage applied to the device must be 80 V.

Example 3-5

The current flowing through a 10-kΩ resistor is 350 mA. What is the potential energy difference (voltage) across the resistor?

Given: $R = 10\text{ k}\Omega = 10{,}000\ \Omega$
$I = 350\text{ mA} = 0.35\text{ A}$

Find: V

Known: $V = IR$

Solution: $V = 0.35\text{ A} \times 10{,}000\ \Omega = 3500\text{ V}$

Answer: The voltage across the resistor is 3500 V.

Calculating Power

In the previous chapter we learned how to find the amount of power when energy and time were known. Now we are going to work with the relationship between current, voltage, and power. Since current and voltage are easily measured quantities, you will be using this relationship quite often in electricity.

Power is equal to current times voltage. Expressed as a formula, we have

$$\text{Power } (P) = \text{current } (I) \times \text{voltage } (V)$$

or

$$P = IV$$

Therefore, power is equal to current times voltage. Power will be in its base unit of watts when voltage is in volts and current is in amperes.

Example 3-6

What is the power input to an electric heater that draws 6 amperes from a 120-volt outlet?

Given: Current (I) = 6 amperes (A)
Voltage (V) = 120 volts (V)

Find: Power (P)

Known: $P = IV$

Solution: $P = 6\text{ A} \times 120\text{ V} = 720\text{ watts (W)}$

Answer: Power input to the electric heater is 720 W.

By rearranging the power formula, we can solve for current when power and voltage are known. The formula is

$$\text{Current } (I) = \frac{\text{power } (P)}{\text{voltage } (V)} \quad \text{or} \quad I = \frac{P}{V}$$

This formula is used to find the current a conductor must carry to serve a load of specified power.

$P = IV$

$I = \frac{P}{V}$

Example 3-7

How much current flows through a 120-volt, 150-watt lamp?

Given: Voltage (V) = 120 volts (V)
Power (P) = 150 watts (W)

Find: Current (I)

Known: $I = \frac{P}{V}$

Solution: $I = \frac{150\text{ W}}{120\text{ V}} = 1.25\text{ A}$

Answer: Current flowing through the lamp is 1.25 A.

Example 3-8

The heating element in a clothes dryer is rated at 4 kilowatts (kW) and 240 volts (V). How much current does it draw?

Given: $P = 4\text{ kW} = 4000\text{ W}$
$V = 240\text{ V}$

Find: I

Known: $I = \frac{P}{V}$

Solution: $I = \frac{4000\text{ W}}{240\text{ V}} = 16.7\text{ A}$

Answer: Current drawn by the clothes dryer is 16.7 A.

Using Ohm's law and the power formula, we can find the power when resistance and current or voltage are known.

Example 3-9

Find the power used by the resistor in the circuit shown in Fig. 3-8.

$P = \frac{V^2}{R}$

$P = I^2R$

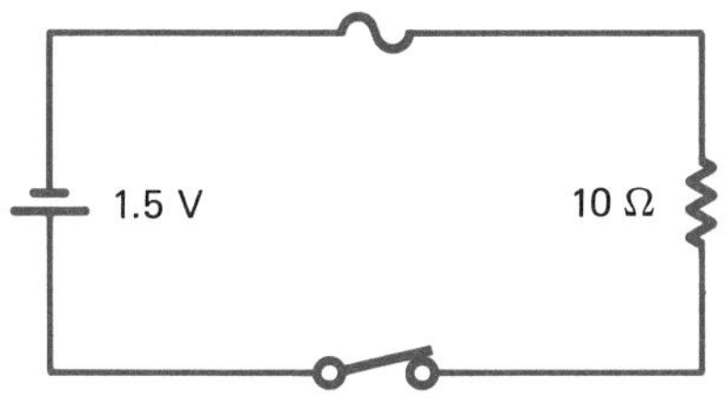

Fig. 3-8 Circuit diagram for Example 3-9, Example 3-10, and Example 3-13.

Given: $V = 1.5\ V$
$R = 10\ \Omega$

Find: P

Known: $P = IV,\ I = \frac{V}{R}$

Solution: $I = \frac{1.5}{10\ \Omega} = 0.15\ A$

$P = 0.15\ A \times 1.5\ V = 0.225\ W$

Answer: Power used (dissipated) by the resistor is 0.225 W.

The procedure used in Example 3-9 is a two-step process. If Ohm's law and the power formula are combined, the procedure can be shortened to one step. The result of combining Ohm's law and the power formula yields two formulas.

The basic formula for power is

$$P = I \times V$$

From Ohm's law we know

$$I = \frac{V}{R}$$

Substituting this in the basic formula

$$P = \frac{V}{R} \times V$$

$$P = \frac{V^2}{R}$$

Thus we can solve for the power if we know the voltage and the resistance.

From Ohm's law we also know

$$V = I \times R$$

Again, substituting this in the basic formula

$$P = I \times I \times R$$

$$P = I^2R$$

Thus we can solve for the power given the current and the resistance.

Example 3-10 will be solved using one of the above formulas.

Example 3-10

Find the power dissipated (used) by the resistor in Fig. 3-8 using the appropriate power formula.

Given: $V = 1.5\ V$
$R = 10\ \Omega$

Find: P

Known: $P = \frac{V^2}{R}$

Solution: $P = \frac{(1.5)^2}{10} = \frac{1.5 \times 1.5}{10} = 0.225\ W$

Answer: Power dissipated by the resistor is 0.225 W.

Example 3-11

How much power does a 100-ohm resistor use when 0.1 ampere of current flows through it?

Given: $R = 100\ \Omega$
$I = 0.1\ A$

Find: P

Known: $P = I^2R$

Solution: $P = (0.1)^2 \times 100 = 0.1 \times 0.1 \times 100$
$= 0.01 \times 100$
$= 1\ W$

Answer: The resistor uses 1 W of power.

Calculating Energy

You already know from the previous chapter that energy is equal to power times time. You have just learned that power is equal to current times voltage. Thus, energy can also be determined by knowing current, voltage, and time.

Example 3-12

How much energy is converted by a device that draws 1.5 amperes from a 24-volt battery for 2 hours?

Given: $I = 1.5\ A$
$V = 24\ V$
$t = 2$ hours (h)

Find: Energy (W)
Known: $W = Pt$, $P = IV$
Solution: $P = 1.5\ \text{A} \times 24\ \text{V} = 36\ \text{W}$
$W = 36\ \text{W} \times 2\ \text{h} = 72\ \text{Wh}$
Answer: The energy is 72 Wh.

Energy can also be determined if resistance and voltage or current are known. Of course, it is always necessary to know the amount of time the power is being used.

Example 3-13

In Fig. 3-8, how much energy is taken from the battery by the resistor if the switch remains closed for 30 minutes?

Given: $V = 1.5\ \text{V}$
$R = 10\ \Omega$
$t = 30\ \text{min}$
Find: W
Known: $W = Pt$
$P = \frac{V^2}{R}$
1 joule = 1 watt-second
1 min = 60 s
Solution: $P = \frac{(1.5)^2}{10} = \frac{2.25}{10}$
$= 0.225\ \text{W}$

$W = 0.225\ \text{W} \times 1800\ \text{s}$
$= 405\ \text{Ws}$
$= 405$ joules (J)
Answer: Energy from the battery is 405 J.

Calculating Cost

The cost of electric energy can be determined from the amount of energy used and the cost rate. The cost rate is usually specified in cents per kilowatt-hour. Specifying the cost rate of energy in ¢ per kWh is like specifying the cost rate of gasoline in cents per gallon. The cost of anything is equal to the total quantity times the cost rate. If potatoes cost 10¢ per pound (cost rate) and you buy 10 pounds, they cost \$1. That is, the cost is equal to the rate times the amount of the quantity. The cost of electric energy is

$$\text{Cost} = \text{rate} \times \text{energy}$$
$$= (\text{¢ per kWh}) \times \text{kWh}$$

Cents per kilowatt-hour

Cost = rate × energy

Example 3-14

What is the cost of 200 kWh of energy if the rate is 4¢ per kWh?

Given: W = 200 kWh
rate = 4¢ per kWh
Find: Cost
Known: Cost = rate × energy
Solution: Cost = 4¢ per kWh × 200 kWh
= 800¢ = \$8
Answer: Cost is \$8.

Example 3-15

What is the cost of operating a 100-watt lamp for 3 hours if the rate is 3¢ per kWh?

Given: $P = 100\ \text{W}$
$t = 3\ \text{h}$
rate = 3¢ per kWh
Find: Cost
Known: Cost = rate × energy, $W = Pt$
Solution: $W = 100\ \text{W} \times 3\ \text{h}$
$= 300\ \text{Wh} = 0.3\ \text{kWh}$
Cost = 3¢ per kWh × 0.3 kWh
= 0.9¢
Answer: It cost 0.9¢ to operate the lamp for 3 hours.

Example 3-16

An electric iron operates from a 120-volt outlet and draws 8 amperes of current. How much does it cost to operate the iron for 2 hours if the rate is 5¢ per kWh?

Given: $V = 120\ \text{V}$
$I = 8\ \text{A}$
$t = 2\ \text{h}$
rate = 5¢ per kWh
Find: Cost
Known: Cost = rate × energy,
$W = Pt$, $P = IV$
Solution: $P = 8\ \text{A} \times 120\ \text{V} = 960\ \text{W}$
$W = 960\ \text{W} \times 2\ \text{h} = 1920\ \text{Wh}$
$= 1.92\ \text{kWh}$

Cost = 5¢ per kWh × 1.92 kWh
= 9.6¢
Answer: Cost is 9.6¢.

Voltmeter

Ammeter

Ohmmeter

Wattmeter

Panel meter

Notice the order in which the "Known" in Example 3-16 is listed. The procedure outlined below was used in developing the order in which the "Known" formulas are listed.

1. Write the formula needed to solve for the quantity to be found. Look at the quantities to the right of the equals sign in this formula. If one of these quantities is not listed in the "Given" row, write the formula needed to find it ($W = Pt$).
2. Look at the right-hand half of the formula just written ($W = Pt$). If there is a quantity there that is not listed in the "Given," then write the formula needed to find it ($P = IV$).
3. Look at the right-hand half of $P = IV$. The "Given" lists both I and V. The problem can now be solved.

The above procedure should always be used to reduce complex problems into a series of simple steps.

Review Problems

1. Refer to Fig. 3-4. When the switch is closed, how much current will flow?
2. What is the resistance of a semiconductor device that allows 200 milliamperes of current when 600 millivolts of voltage is connected to it?
3. What is the power of a hair dryer that operates from 120 volts and draws 2.5 amperes of current?
4. How much current is required by a 1000-watt toaster that operates from a 120-volt outlet?
5. A tape deck in an automobile requires 2 amperes from a 12.6-volt battery. What is its power rating?
6. A miniature lamp has 99 ohms of resistance. It is designed to operate from a 12.6-volt battery. What is the power rating of the lamp?
7. An electric heater draws 8 amperes from a 240-volt source. How much energy does it convert in 6 hours?
8. The current through a 100-ohm resistor is 200 milliamperes. How much energy does the resistor convert to heat in 10 minutes?
9. An electric water heater has a 3000-watt heating element. The element is on for 3 hours. What is the cost if the rate is 4¢ per kilowatthour?
10. A string of Christmas tree lights draws 0.5 amperes from a 120-volt outlet. At 4¢ per kilowatthour, how much does it cost to operate the lights for 50 hours?

3-4 MEASURING ELECTRICAL QUANTITIES

Most electrical quantities are measured with a device called a *meter*. Voltage is measured with a *voltmeter*, current with an *ammeter*, resistance with an *ohmmeter*, and power with a *wattmeter*.

Panel Meters

A meter that measures only one of the above quantities is called a *panel meter*. A panel meter is shown in Fig. 3-9. This panel meter can measure up to 50 microamperes. The amount of current flowing through the meter is read directly from the 0 to 50 scale. Before reading a meter scale, you must figure out the value of each division of the scale. Look at the scale on the meter in Fig. 3-9. Notice there is a heavy line halfway between 0 and 10. This

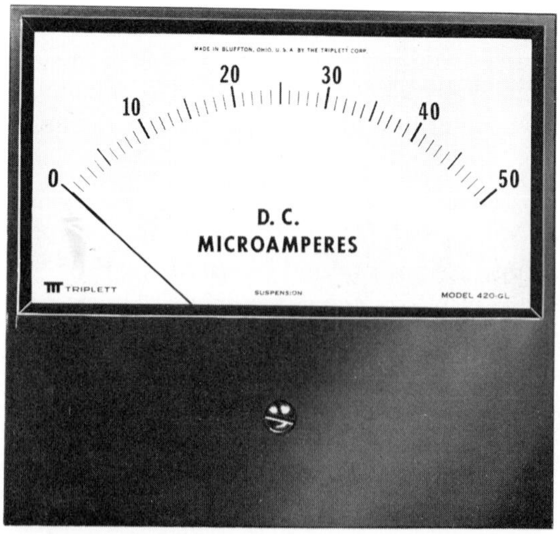

Fig. 3-9 Panel meter. (Courtesy of Triplett Corporation.)

heavy line represents 5 units. Now count the number of divisions (called minor divisions) between 0 and 5. Since there are 5 minor divisions between 0 and 5, each minor division has a value of 1 unit. Therefore, each minor division on the scale represents 1 microampere. Suppose the needle (pointer) on the meter in Fig. 3-9 was pointing to the second mark to the right of the 30 mark. The meter would be indicating 32 microamperes. If the needle pointed 3 minor divisions to the left of the 20 mark, it would indicate 17 microamperes.

The meter shown in Fig. 3-9 is an analog meter. The needle can point anywhere on the scale. It can be any place between two marks. When an analog meter is read, the reading is generally taken to the nearest minor-division mark. If the pointer is halfway between marks, it is read as a half division.

The digital panel meter (DPM) shown in Fig. 3-10 eliminates the need to decide which mark is closest to the pointer. It always provides a reading in digits. There is no guesswork in trying to decide if the meter reading is 1.999 or 1.998. Digital meters are usually specified by the number of digits in their readout (although many include ½ in their specifications). Thus, a digital panel meter may be called a 2-digit DPM or a 3½-digit DPM.

Panel meters are often permanently connected (wired) into a circuit to provide continuous monitoring of an electrical quantity. The meters may be either analog or digital.

Multimeters

Often a single meter is used to measure two or more electrical quantities. That is, a single meter serves as a voltmeter, an ammeter, and an ohmmeter. Meters which are capable of measuring two or more electrical quantities are known as multimeters (Fig. 3-11). Multimeters use the same basic mechanism to indicate the amount of a quantity as panel meters. However, the multimeter also includes some circuitry (switches, resistors, etc.) inside its housing or case. The analog multimeter also has more than one scale printed on the face of the meter.

There are a number of different types of analog multimeters. Some of the more common types are:

1. The volt-ohm-milliammeter (usually called the VOM). This meter uses no transistors or integrated circuits.
2. The solid-state VOM, also called the FET (field-effect transistor) VOM, or the transistorized VOM.
3. The vacuum tube voltmeter (VTVM). These meters are rapidly being replaced by the solid-state types listed above.

Fig. 3-10 Digital panel meter. (Courtesy of Data Technology Corporation.)

Reading Multimeters

While the various analog multimeters may look different, all of them have functions, ranges, and scales. *Function* refers to the *quantity being measured* (voltage, current, or resistance). Saying, "The meter is on the dc-voltage function" means the meter is all set to measure dc voltage. *Range* refers to the *maximum amount* of the quantity that can be measured. Saying, "The meter is on the 50-V range" means the meter is all set to measure *up to* 50 volts. Whether the 50 volts is to be ac voltage or dc voltage depends on the function to which the meter is set. Which *scale* of the meter is used depends on both the function and the range to which the meter is set. Proper use of the multimeter involves selection of the correct function, range, and scale. Once you understand the relationships between (1) function and scale and (2) range and scale, using any multimeter will be possible.

Look at the VOM (multimeter) in Fig. 3-11. Notice the four circular brackets around the large switch (selector switch) in the lower center of the meter. The bracket labeled ACV marks the ac-voltage function of the meter. Within the ACV function are six ranges. The highest range is 1000 volts and the lowest is 3

Volt-ohm-milliammeter

Solid-state VOM

Vacuum tube voltmeter

Function

Range

Scale

ACV function

Fig. 3-11 Multimeter (VOM). (Courtesy of Triplett Corporation.)

DCV function

Ohm function

volts. If the large center switch were turned clockwise three positions, the white pointer would be pointing to 100. The meter would then be set on the ac-voltage function and the 100-volt range. It would be all set to measure any ac voltage from 0 to 100 volts.

Again referring to Fig. 3-11, find the DCV (direct-current-voltage) function. The ranges for ACV and DCV are nearly identical. The only difference is that the DCV function has a 1-V range.

Next, find the DCMA (direct-current-milliampere) function. Notice the ranges within this function cover from 0.1 milliampere (100 microamperes) to 1000 milliamperes (1 ampere).

The last function on the meter in Fig. 3-11 is the Ω (ohm) function. This function is used to measure resistance. Look carefully at the ranges within the ohm function. They are quite different from the ranges of the other functions. The ranges within the ohm function are just multipliers. They indicate the amount by which the ohm scale is to be multiplied. The ohm scale is the top scale on the meter. Suppose the selector switch were turned to the ×100 range and the needle pointed to the 30 mark on the ohm scale. The meter would be reading 3000 ohms (30 × 100 = 3000). If the meter were on the ×10-k range and the needle pointed to 6, the reading would be 60,000 ohms or 60 kΩ (6 × 10 k = 60 k).

The ohm scale on the meter in Fig. 3-11 is different from most voltage and current scales in three ways. First, it is reverse reading. Second, it is nonlinear. Third, the number of minor divisions between the heavy lines (and the numbered lines) is not the same throughout the scale. Therefore, the value of a minor division varies across the scale. Between 0 and 1, there are 5 minor divisions; each minor division is, therefore, equal to 0.2 units. Between 6 and 10 there are 4 major divisions (longer lines); therefore, the major division is worth 1 unit. Between each major division there are 2 minor divisions. These minor divisions (between 6 and 10) represent 0.5 units. Now move to the left half of the scale. Between 50 and 100 there are 5 minor divisions; each minor division is equal to 10 units (50 ÷ 5 = 10). Finally, look between 100 and 500. Here there are 4 major divisions, each represents 100 units. The first major division (between the 100 mark and the first long line) is divided into two minor divisions. Thus, these two minor divisions are equal to 50 units each.

Example 3-17

The meter in Fig. 3-11 is set to the ohm function, ×1-k range. The needle points to the first mark left of the 30 mark. How much resistance is the meter indicating?

Solution: Between 30 and 50 there are two major divisions; each division is worth 10 units (50 − 30 = 20; 20 ÷ 2 = 10). Two minor divisions are located between 30 and 40. Therefore, each minor division has a 5-unit value (40 − 30 = 10; 10 ÷ 2 = 5). The scale indicates 35 units (30 + 5). Since the meter is set on the 1-k range

$$R = 35 \times 1\ \text{k}\Omega$$
$$= 35\ \text{k}\Omega$$

Answer: The meter indicates 35,000 ohms or 35 kΩ.

Now we need to learn how to select and use the scales for the other functions. Let us start with the scales for the DCV function. The second and third scales (Fig. 3-11) from the top are marked AC-DC. All dc voltages and currents are read from one of these two scales. Look at the end of these scales. One ends at 10, the other at 30 (actually, the scale goes to 31, but the last numbered mark is 30). Notice the relationship between the ends of these scales and the ranges. All the ranges are either (1) equal to or (2) related by a factor of 10 to the end of one of the scales. When the scale and the range are equal, the scale reads directly in volts. When the scale does not equal the range, the scale must be either multiplied or divided by a factor of 10. This will make the scale equal to the range. For example, if the meter is on the 1-volt range, the 0–10 scale is read. The reading is then *divided* by 10. Dividing by 10 makes the scale equal to the range (10 ÷ 10 = 1). If the range were set to 300 volts, then the 0–30 scale would be used. The scale reading would have to be *multiplied* by 10 to make the scale equal to the range.

Example 3-18

Refer to Fig. 3-11. Assume the selector switch is set on the 1000 range and the DCV function. The needle points to the first division to the right of the 7 on the 0–10 scale. What is the value of the indicated voltage?

Solution: There are 5 minor divisions between 7 and 8. Therefore, each division represents 0.2 units. The scale reading is therefore 7.2. The 0–10 scale must be multiplied by 100 to make it equal to the 1000 range. Thus the voltage would be 720 volts (7.2 × 100 = 720).

Answer: The meter reads 720 volts dc.

Example 3-19

Assume the needle location and the function are the same as in Example 3-18. However, the meter is now on the 30-volt range. How much voltage would the meter indicate?

Solution: Read the 0–30 scale. Between 20 and 25 there are 5 minor divisions. Each division has a value of 1 unit. To the nearest division, the scale reading would be 23. Since the end of the scale equals the range, the voltage is also 23 volts.

Answer: The voltage is 23 volts.

All the ranges in the DCMA function are related by factors of 10 to the 0–10 scale. Therefore, all dc values measured by this

Polarity switch

Ohms-adjust control

Voltage measurement

multimeter are read on the 0-10 scale (Fig. 3-11).

Except for the 3-V range, all ranges in the ACV function are read on the two AC-DC scales. These are the same scales as used for all ranges of the DCV function. The procedure used is exactly the same as that used for reading dc voltage. On the 3-volt range, the fourth scale from the top is used. This scale is marked AC AMPS, 3 ACV ONLY. Since it is a 0–30 scale, its reading must be divided by 10 when on the 3-volt range. Having a separate scale for the lowest ac voltage range is very common for multimeters. Many multimeters, however, do not have the ac current function.

Using the Multimeter

All electrical quantities to be measured are applied to the meter through test leads. A typical test lead is illustrated in Fig. 3-12. The test leads plug into the two jacks in the lower right-hand corner of the meter shown in Fig. 3-11. A black lead goes in the COM jack, and a red lead in the V · Ω · MA jack. The black lead will be negative and the red lead positive when the polarity switch is in the + position. The polarity switch is located to the left of the range/function selector switch. When the polarity switch is in the minus (–) position, the black lead (COM jack) becomes positive. Of course, the red lead also reverses polarity. The polarity switch should be left in the positive position except when special measurements are being made. In this way the red lead will be positive. Red is the color often used to indicate positive in electric circuits.

Measuring Resistance

When resistance is measured, the multimeter must first be *ohms-adjusted*. The ohms-adjust control is marked Ω in Fig. 3-11. This control is rotated until the meter indicates zero on the ohms scale *when the test-lead tips are touching each other*. The ohms-adjust control must be adjusted for each range of the ohms function. Thus, every time the range is changed, the leads must be touched together and the meter adjusted to zero.

The ohmmeter function of the multimeter uses a cell or battery inside the meter housing. That is, it has its own source of energy. Therefore, any other energy source must be disconnected from any circuit in which resistance is to be measured. Never measure the resistance of a load when power (the energy source) is connected to the circuit. To do so damages the ohmmeter. Figure 3-13(*a*) illustrates the correct technique for measuring the resistance of a lamp. Notice in Fig. 3-13(*b*) the symbols used for the ohmmeter and the test-lead connection.

The procedure used in measuring resistance is summarized as follows:

1. Remove power from the circuit.
2. Select an appropriate range within the ohm function. The appropriate range is the one that gives an indication in the right-hand portion of the scale.
3. Short (touch) the test leads. Turn the ohms adjust control until the pointer reads zero ohms.
4. Connect, or touch, the test leads to the terminals of the device whose resistance is to be measured. Except for some electronic components, the polarity of the ohmmeter leads is unimportant.

When measuring resistance, do not touch the metal parts of the test leads with your hands. If you do, you will be measuring your body's resistance as well as the circuit's resistance. This will not harm you, but you will not obtain the correct resistance reading for the circuit.

Measuring Voltage

Voltage measurements are the easiest electrical measurements to make. They are also the most common. Voltage measurements are made with power connected to the circuit. The following procedure is used to make voltage measurements:

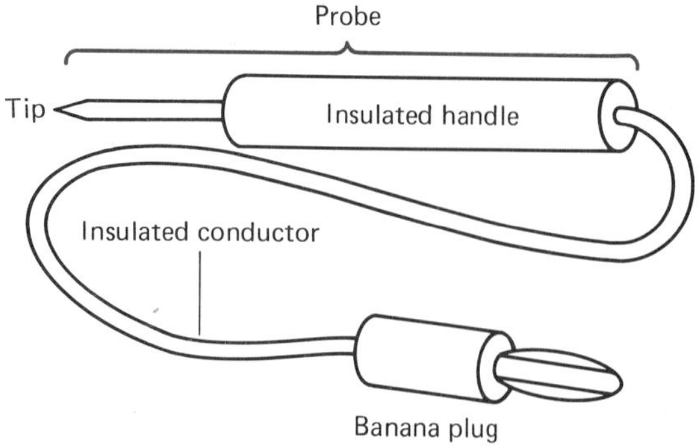

Fig. 3-12 Test lead.

Current measurement

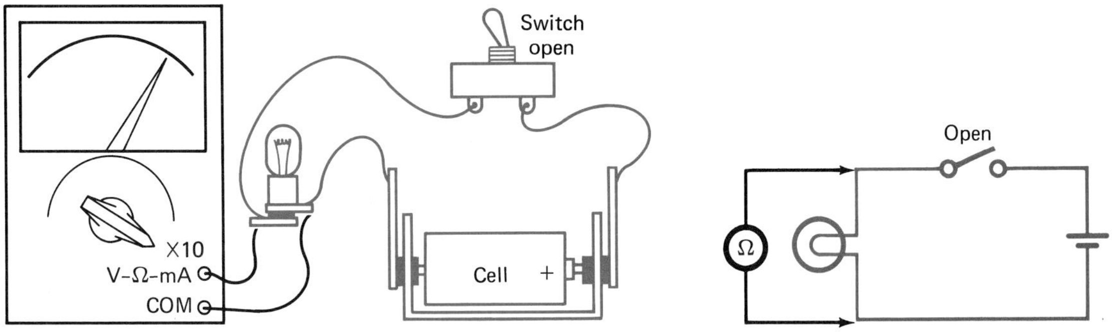

(*a*) Pictorial presentation (*b*) Schematic diagram

Fig. 3-13 Measuring resistance. Notice that the power source is disconnected from the load by the open switch.

1. Select the correct voltage function (ac or dc) for the type of voltage used in the circuit.
2. Select a range that is greater than the expected voltage.
3. Determine the polarity of the voltage to be measured by looking at the schematic diagram or at the battery terminals. This step is omitted when measuring ac because the polarity reverses every fraction of a second.
4. Connect the negative (black) lead of the multimeter to the negative end of the voltage to be measured. Touch (or connect) the positive (red) lead of the meter to the positive end of the voltage. In other words, observe polarity when measuring voltage with a multimeter or voltmeter. If you do not, the meter pointer may bend when it tries to rotate counterclockwise. Again, with ac voltage, polarity does not have to be observed.

Figure 3-14 shows the correct connections for measuring dc voltage. In this figure, the meter is measuring the voltage across the lamp. Notice that the switch is in the closed position. If the switch were open, there would be no voltage across the lamp and the meter would indicate 0 volts. A load has a voltage across it only when current is flowing through it. If the meter were connected across the cell, as in Fig. 3-15, the meter would indicate the cell's voltage. The meter would indicate the cell's voltage regardless of whether the switch were open or closed. As long as the switch is closed, the meter reading is the same in Figs. 3-14 and 3-15. That is, the voltage output of the cell appears across the lamp when the switch is closed.

Measuring Current

Current measurements are made much less frequently than either resistance or voltage measurements. This is because the circuit has to be physically interrupted to insert the meter in most cases. In Fig. 3-16, the circuit has been physically interrupted by disconnecting one end of the lead between the cell and the lamp. The meter is then connected between the end of the wire and the lamp. As shown in

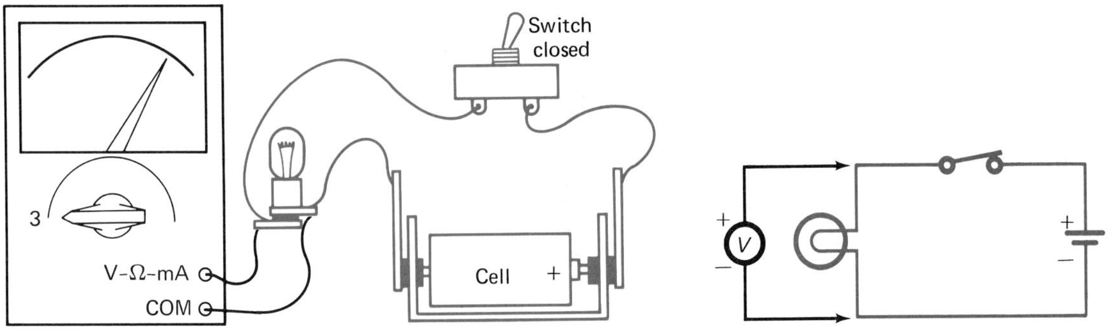

(*a*) Pictorial presentation (*b*) Schematic diagram

Fig. 3-14 Lamp voltage being measured. The switch must be closed.

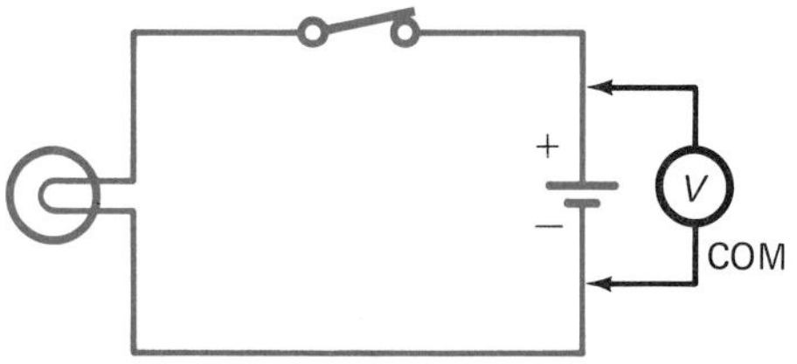

Fig. 3-15 Cell voltage being measured. The switch may be either open or closed.

Zero-adjust control

Fig. 3-17, the meter could just as well have been connected on either side of the switch. All three meter locations in Fig. 3-17 are correct and all three locations will yield the same results. Remember, the meter must be inserted into the circuit so that the circuit current flows through the meter as well as the load.

In using the current function of a multimeter or an ammeter, follow these steps:

1. Select the current function.
2. Select a range that is greater than the expected current.
3. Physically interrupt the circuit.
4. Observing polarity, connect the multimeter or ammeter between the points created by the interruption. Correct polarity can be determined by tracing current (electron flow). As indicated in Fig. 3-17, current enters the negative terminal and leaves the positive terminal of the meter.

A meter can be easily damaged, so it should be used with care. Reverse polarity connections may bend the pointer. However, most damage to multimeters or ammeters occurs by incorrectly connecting them to the circuit. If you connect an ammeter or multimeter into a circuit in the same way that you would a voltmeter, you may destroy it. If the meter is not completely destroyed, its accuracy is greatly decreased.

Variations in Multimeters

In all the discussion and examples to this point, we have used a nonelectronic VOM. Now that we have mastered one VOM, let us look at some other types and styles. The meter shown in Fig. 3-18 is a solid-state VOM. Solid-state meters have a zero-adjust control that compensates for temperature changes as

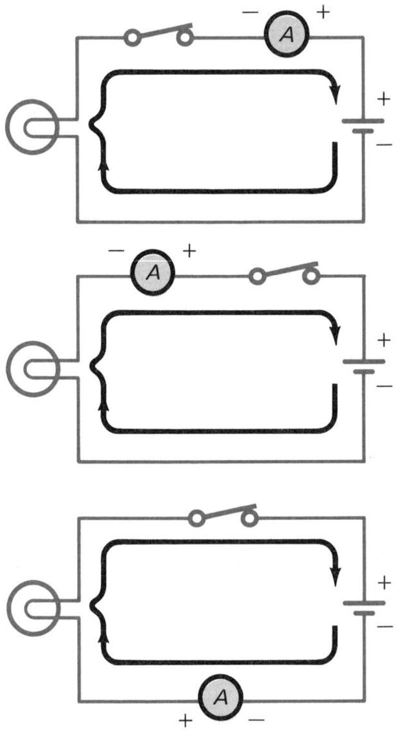

Fig. 3-17 Ammeter location. The location of the ammeter with respect to the lamp or the switch does not change the amount of current.

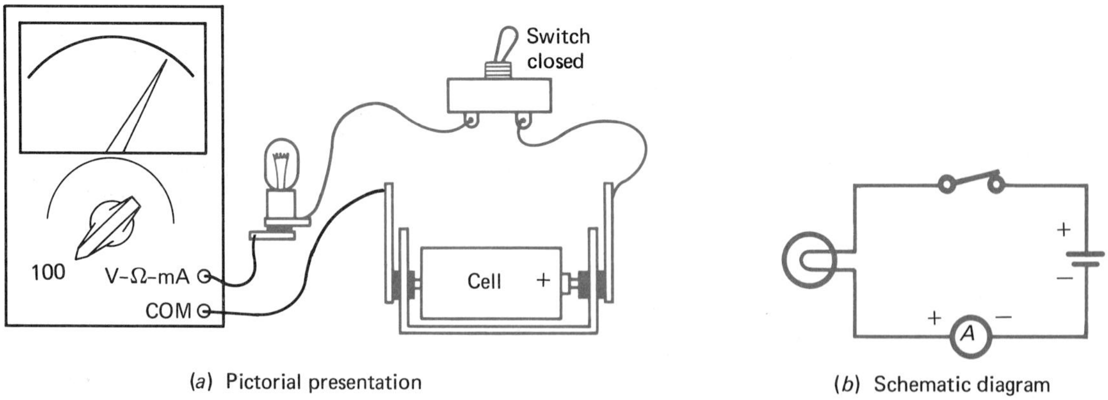

Fig. 3-16 Current being measured. Current must flow through both the meter and the load.

the meter warms up. This control is used to adjust the needle (pointer) to read zero when the two test leads are shorted together. The zero setting needs to be checked (and adjusted if necessary) every time the function or range is changed. Also, the zero-adjust needs to be checked periodically, even when it is used on the same function and range. This is especially necessary when the meter is first turned on and is still warming up.

The ohms-adjust on this type of meter (Fig. 3-18) is adjusted when the leads are open (not touching). It is adjusted until the needle indicates infinity (∞) on the ohm scale. Notice this is just the opposite of the meter in Fig. 3-11.

Many multimeters have separate switches for function and range. The meter in Fig. 3-18 combines the polarity reversing switch with the function switch. Notice that this meter can measure ac as well as dc. It also has a *low power* ohms function. This function is used to measure the resistance of devices that cannot handle the current of the regular ohms function.

Look at the range switch on the meter in Fig. 3-18. Many of the settings of this switch have three range markings. For example, in the same position we find 30 V, 1 mA and $R \times 1$ k. This is the 30-V range for the voltage setting of the function switch. For the current setting of the function switch, it is the 1-mA range. And, for the resistance function, the ohm-scale reading would be multiplied by 1 k or 1000.

The negative and positive leads for the meter of Fig. 3-18 are contained in a single cable. The cable connects to the jack in the lower center of the meter. On the other end of the cable the alligator clip is the negative lead and the bare tip of the probe is the positive lead.

Other variations in VOMs can be seen in Fig. 3-19. One of the range switch positions has two voltage ranges (300-V and 1000-V) indicated. If the test leads are in the normal jacks, common (–) and +, the meter is on the lower (300-V) range. To change to the 1000-V range, the red (positive) lead in the + jack is moved to one of the 1000-V jacks. It is moved to 1000 V DC if the function switch is on DC and to 1000 V AC if the switch is on AC. The multiple-range setting for current (30 mA, 1 A, and 10 A) is handled in the same way as the voltage setting is.

Zero-center

Notice the function switch on the meter in Fig. 3-19. The "zero-center" setting means the needle will be in the center of the scale when no voltage (or current) is applied to the meter. The bottom two scales are used with this function. This function is very useful for adjusting certain types of electronic circuits. Also notice that the function switch has a plus (+) ohm setting and a minus (−) ohm setting. Switching from minus ohm to plus ohm reverses the polarity of the test leads.

Digital multimeter

Figure 3-20 shows a digital multimeter. Digital multimeters are easy to use. You do not have to multiply or divide the readout to find

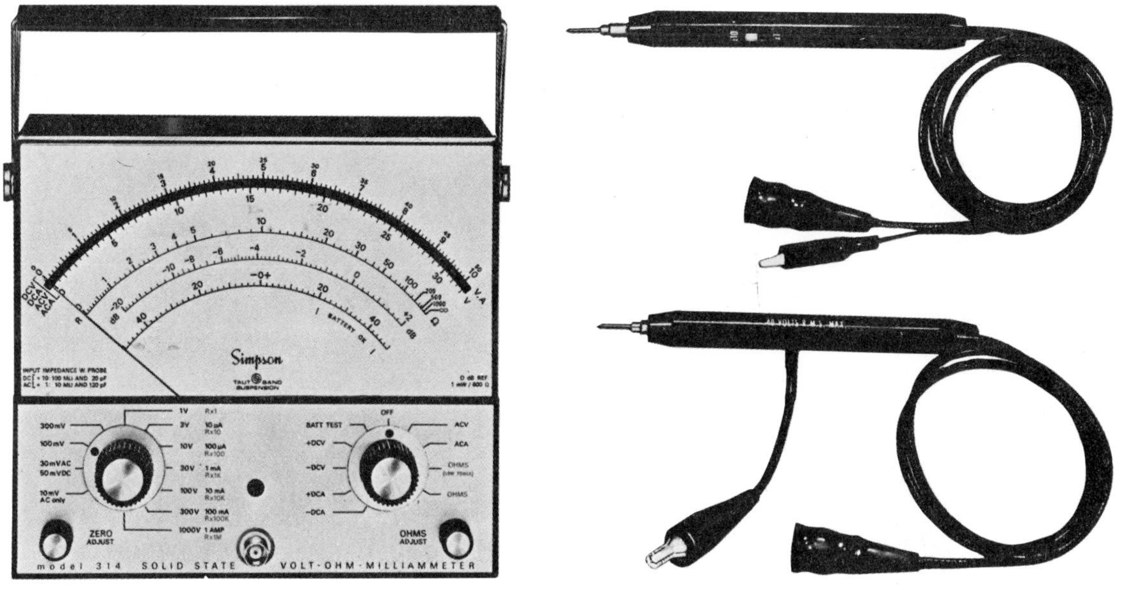

Fig. 3-18 Solid-state VOM. (Courtesy of Simpson Electric Company.)

Fig. 3-19 VOM with multiple jacks. (Courtesy of Simpson Electric Company.)

the actual meter reading. The readout is direct. Whenever the range is changed, the decimal point automatically shifts. For example, if you change from the 0.5-V range to the 50-V range, the decimal point in the display would move two places to the right. Also notice that the meter in Fig. 3-20 has no polarity markings on the input jacks. This meter, like most digital multimeters, internally compensates for the polarity of the input voltage.

The meter shown in Fig. 3-21 has no range switch. Internal circuitry automatically switches ranges to the correct range for the amount of input voltage. The quantity to be measured (function) is selected by pushing one of the buttons in the row of function buttons. Meters of this type are said to be *autoranging*.

Review Questions

6. Refer to Fig. 3-9. If the needle points 4 divisions to the left of the 40 mark, how much current is the meter indicating?
7. Refer to Fig. 3-9. Suppose the needle points to the heavy mark (line) between the 40 mark and the 50 mark. How much current is indicated?
8. Refer to Fig. 3-11. The needle points one mark to the right of 15. The range is set on ×100. What is the resistance?
9. Refer to Fig. 3-11. The needle points three marks to the left of 6. The range is set on ×10 k. What is the resistance?
10. Refer to Fig. 3-11. Assume the meter is on the DCMA function and the 100 range. The needle is pointing to 20 on the third scale from the top. How much current is indicated by the meter?

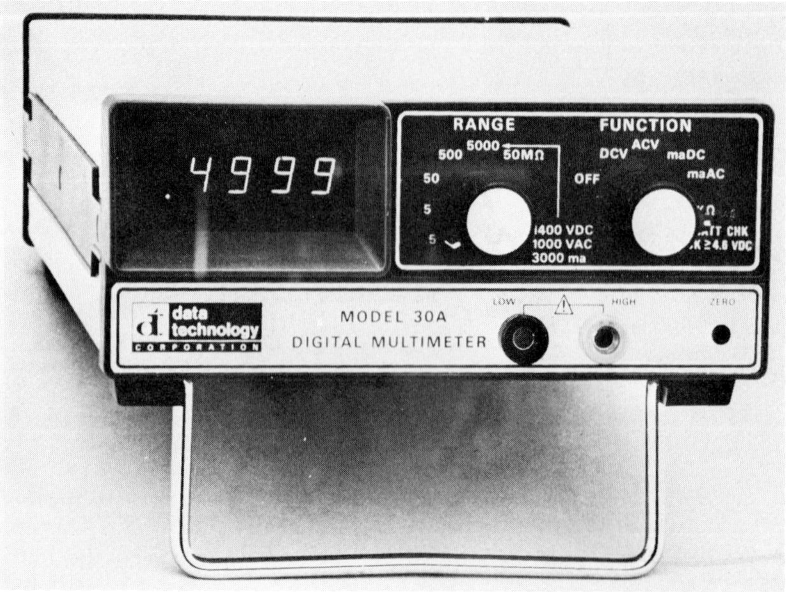

Fig. 3-20 Digital multimeter. (Courtesy of Data Technology Corporation.)

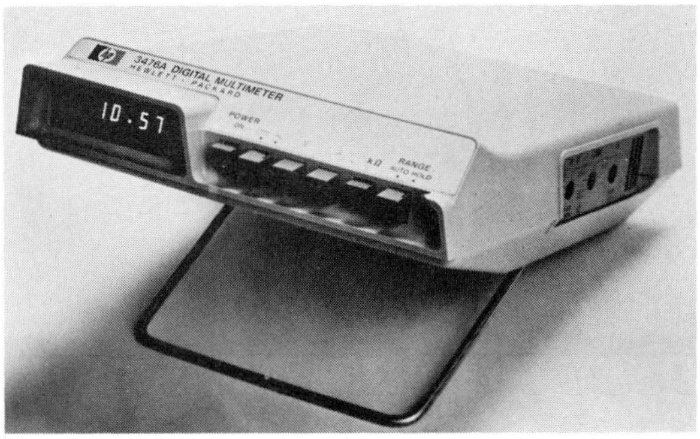

Fig. 3-21 Digital multimeter with autorange. (Courtesy of Hewlett-Packard.)

11. Refer to Fig. 3-11. The meter is set on the 3 range and the ACV function. The needle is pointing to 20 on the third scale from the top. What is the meter indicating?
12. Summarize the procedure for measuring resistance.
13. Summarize the procedure for measuring voltage.
14. Summarize the procedure for measuring current.
15. What does the polarity switch on a VOM do?
16. How does ohms-adjusting an electronic VOM differ from ohms-adjusting a nonelectronic VOM?

Summary

1. Electric components are represented by symbols in schematic diagrams.
2. Sometimes the physical structure of a device also serves as a conductor in a circuit.
3. The current is directly proportional to the voltage and inversely proportional to the resistance.
4. The Ohm's law formulas are:

$$V = IR$$
$$I = \frac{V}{R}$$
$$R = \frac{V}{I}$$

5. The power formulas are:

$$P = IV$$
$$P = I^2R$$
$$P = \frac{V^2}{R}$$

6. Energy equals power multiplied by time.
7. Cost of energy equals energy times rate.
8. A panel meter measures only one quantity.
9. Multimeters measure several quantities.
10. Multimeters have functions, ranges, and scales.
11. Ohmmeters must be adjusted every time the range is changed.
12. Polarity must be observed when measuring dc voltage.
13. Polarity must be observed when measuring direct current.
14. Ammeters are easily damaged by incorrect connection to a circuit.
15. Electronic VOMs need to be zero-adjusted for all functions and ranges.
16. Multimeters may have separate switches for selecting function and range.
17. With some meters, ranges are changed by changing the location of the positive test lead.
18. Many digital meters have automatic range and polarity switching.

Chapter Review Questions

Here is a chance to see what you have learned. For questions 3-1 to 3-12, determine whether the statement is true or false.

3-1. The metal chassis of an electric device is often used as a conductor.

3-2. Current is directly proportional to voltage and inversely proportional to resistance.

3-3. A watt is equal to a volt divided by an ampere.

3-4. A 60-watt, 120-volt lamp requires more current than a 100-watt, 120-volt lamp.

3-5. A 60-watt, 120-volt lamp has more resistance than a 100-watt, 120-volt lamp.

3-6. An electric circuit must be physically interrupted to measure voltage in the circuit.

3-7. An ohmmeter contains its own energy source.

3-8. Always observe polarity when measuring resistance.

3-9. Correct polarity for an ammeter requires that current (electrons) enter the positive terminal of the meter.

3-10. Ohmmeters must be checked for ohm-adjustment each time the range switch is changed.

3-11. Multirange ohmmeters have only a single ohmmeter scale.

3-12. The multimeter jack with the V·Ω·mA label will be the negative terminal if the polarity switch is the negative position.

Answer the following questions.

3-13. What are the six parts of an electric circuit?

3-14. Draw the schematic symbols for the following components:
a. Lamp
b. Conductor
c. Resistor
d. Cell
e. Open switch

3-15. A lamp draws 240 mA when connected to a 12-V battery. What is the resistance of the lamp?

3-16. How much current does a 200-W lamp draw from a 120-V source?

3-17. How much power does a heater require if it draws 10 amperes from a 230-volt circuit?

3-18. A toaster draws 8 amperes from a 120-volt outlet. How much energy does it use in 2 hours?

3-19. How much would it cost to operate the toaster in the above question if energy costs 10¢ per kWh?

3-20. How much voltage is required to force 20 milliamperes of current through a 1-kilohm resistor?

Answers to Review Questions

1. Source of energy, conductors, insulators, load, control device, and protection device
2. The control device and the protection device
3. No, the conductor can be any conductive material of any size or shape.
4.

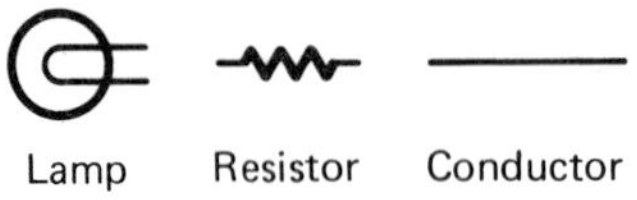

5. A schematic diagram is a drawing which uses symbols for electric components. It indicates the connections between the components.
6. 36 microamperes
7. 45 microamperes
8. 1400 ohms (14 × 100)
9. 54,000 Ω (5.4 × 10 k)
10. 64 milliamperes
11. 19.5 volts ac
12. Remove power from circuit, select ohm

function, select correct range, ohms-adjust, and connect test leads to device or component being measured. If an electronic VOM is used, the meter must also be zero-adjusted.

13. Select correct voltage function, select correct range, determine polarity of voltage, and observing polarity, connect meter to circuit. Again, zero-adjust after selecting the range if using an electronic VOM.

14. Select current function, select correct range, zero-adjust if using an electronic VOM, interrupt the circuit, and observing polarity, connect meter between interrupted points.

15. The polarity switch reverses the polarity of the meter jacks to which the test leads are connected.

16. Ohms-adjusting an electronic VOM involves adjusting for infinity with the leads open. Ohms adjusting the nonelectronic VOM involves adjusting for zero with the leads shorted.

Answers to Review Problems

1. *Given:* $V = 1.5\ \text{V}, R = 1\ \text{k}\Omega = 1000\ \Omega$
   Find: I
   Known: $I = \dfrac{V}{R}$
   Solution: $I = \dfrac{1.5\ \text{V}}{1000\ \Omega} = 0.0015\ \text{A}$
   Answer: Current is 0.0015 A, or 1.5 mA.

2. *Given:* $I = 200\ \text{mA} = 0.2\ \text{A}$, $V = 600\ \text{mV} = 0.6\ \text{V}$
   Find: R
   Known: $R = \dfrac{V}{I}$
   Solution: $R = \dfrac{0.6\ \text{V}}{0.2\ \text{A}} = 3\ \Omega$
   Answer: Resistance is 3 Ω.

3. *Given:* $V = 120\ \text{V}, I = 2.5\ \text{A}$
   Find: P
   Known: $P = IV$
   Solution: $P = 2.5\ \text{A} \times 120\ \text{V} = 300\ \text{W}$
   Answer: The dryer is a 300-W dryer.

4. *Given:* $P = 1000\ \text{W}, V = 120\ \text{V}$
   Find: I
   Known: $I = \dfrac{P}{V}$
   Solution: $I = \dfrac{1000\ \text{W}}{120\ \text{V}} = 8.3\ \text{A}$
   Answer: Current for the 1000-W toaster is 8.3 A.

5. *Given:* $I = 2\ \text{A}, V = 12.6\ \text{V}$
   Find: P
   Known: $P = IV$
   Solution: $P = 2\ \text{A} \times 12.6\ \text{V} = 25.2\ \text{W}$
   Answer: The tape deck requires 25.2 W.

6. *Given:* $R = 99\ \Omega, V = 12.6\ \text{V}$
   Find: P
   Known: $P = \dfrac{V^2}{R}$
   Solution: $P = \dfrac{(12.6)^2}{99} = 1.6\ \text{W}$
   Answer: Power rating of the lamp is 1.6 W.

7. *Given:* $I = 8\ \text{A}, V = 240\ \text{V}, t = 6\ \text{h}$
   Find: W
   Known: $W = Pt, P = IV$
   Solution: $P = 8\ \text{A} \times 240\ \text{V} = 1920\ \text{W}$
   $W = 1920\ \text{W} \times 6\ \text{h}$
   $= 11{,}520\ \text{Wh} = 11.52\ \text{kWh}$
   Answer: Energy converted is 11.52 kWh.

8. *Given:* $R = 100\ \Omega, I = 200\ \text{mA} = 0.2\ \text{A}$, $t = 10\ \text{min} = 600\ \text{s}$
   Find: W
   Known: $W = Pt, P = I^2R, \text{Ws} = \text{J}$
   Solution: $P = (0.2)^2 \times 100 = 4\ \text{W}$
   $W = 4\ \text{W} \times 600\ \text{s}$
   $= 2400\ \text{Ws} = 2400\ \text{J}$
   Answer: The resistor converts 2400 J of electric energy to heat energy.

9. *Given:* $P = 3000\ \text{W}, t = 3\ \text{h}$, rate = 4¢ per kWh
   Find: Cost
   Known: Cost = rate × W, $W = Pt$
   Solution: $W = 3000\ \text{W} \times 3\ \text{h} = 9000\ \text{Wh}$
   $= 9\ \text{kWh}$
   Cost = 4¢ per kWh × 9 kWh
   = 36¢
   Answer: The cost is 36¢.

10. *Given:* $I = 0.5$ A, $V = 120$ V, rate = 4¢ per kWh, $t = 50$ h
Find: Cost
Known: Cost = rate × W, $W = Pt$, $P = IV$
Solution: $P = 0.5 \text{ A} \times 120 \text{ V} = 60 \text{ W}$
$W = 60 \text{ W} \times 50 \text{ h} = 3000 \text{ Wh} = 3 \text{ kWh}$
Cost = 4¢ per kWh × 3 kWh = 12¢
Answer: Cost is 12¢.

Circuit Components

- In the previous chapter you learned about the requirements for a complete electric circuit. In this chapter you will learn about some of the components (parts) used to construct electric circuits. When you have completed this chapter, you will be able to identify and use basic components such as cells and batteries, lamps, resistors, switches, conductors, and circuit protectors.

4-1 BATTERIES AND CELLS

Batteries and cells are chemical devices that provide dc voltage. Since voltage is a potential energy difference, batteries and cells are also known as sources of electric energy. And, since power is the rate of using energy, these devices are also called power sources. They are used to power electric devices.

Terminology and Symbols

A *cell* is an electrochemical device consisting of two electrodes made of different materials and an electrolyte. The chemical reaction between the electrodes and the electrolyte produces a voltage.

A *battery* consists of two or more cells electrically connected together and packaged as a single unit or device. Although technically a battery has two or more cells, the term "battery" is often used to indicate either a single cell or a group of cells.

The schematic symbols for a cell and a battery are shown in Fig. 4-1. The voltage rating of the cell or battery is specified beside the symbol. The short line in the symbol is always the negative terminal. Polarity indicators (negative and positive signs) are not always shown with the symbol.

Cells and batteries are classified as either primary or secondary. *Primary cells* are those cells that are not rechargeable. That is, the chemical reaction that occurs during discharge is not easily reversed. When the chemicals used in the reactions are all converted, the cell is fully discharged. It must then be replaced by a new cell. Included in the primary cell category are the following types: carbon-zinc, zinc chloride, most alkaline, silver oxide, and mercury cells.

Secondary cells may be discharged and recharged many times. The number of discharge-charge cycles a cell can withstand depends on the type and size of the cell and the operating conditions. The number of cycles will vary from less than one hundred to many thousands. Secondary cells include the following types: rechargeable alkaline, lead-acid, nickel-cadmium, and nickel-iron cells.

Cells and batteries are also classified as either dry or wet. Historically, a *dry cell* was one that had a paste or gel electrolyte. It was semisealed and could be used in any position. With newer designs and manufacturing techniques, it is possible to completely (hermetically) seal a cell. With complete seals and chemical control of gas buildup, it is possible to use liquid electrolyte in dry cells. Today the term "dry cell" refers to a cell that can be operated in any position without electrolyte leakage.

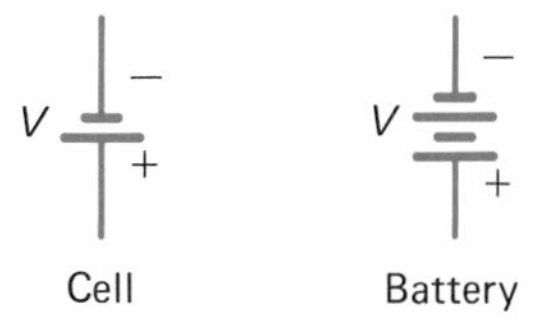

Fig. 4-1 Cell and battery symbols.

From page 53:
Cell

Electrode

Electrolyte

Battery

Primary cell

Secondary cell

Dry cell

On this page:
Wet cell

Cell capacity

Ampere-hour

Internal resistance

Depolarizing agent

Wet cells are cells that must be operated in an upright position. These cells have vents to allow gases generated during charge or discharge to escape. The most common wet cell is the lead-acid cell.

Ratings of Cells and Batteries

Cell (or battery) capacity refers to the amount of energy the cell can provide under specified conditions. The conditions specified are temperature, current drain, discharge schedule (minutes per hour), and final output voltage when discharge is complete. The amount of energy available varies widely as these specified conditions change. For example, a typical flashlight cell provides approximately twice as much energy when the discharge rate is changed from 300 to 50 mA. For the same cell, the capacity decreases about 30 percent when the temperature is decreased from 21 to 5°C. In general, a cell (or battery) provides the most energy when the temperature is high and the discharge schedule, current drain, and final voltage are low.

The capacity of a cell (or battery) is expressed in *ampere-hours* (Ah). When the voltage of the cell is specified, ampere-hour becomes a unit of energy. The energy of a cell is

$$W = Pt$$

Since $P = IV = VI$ we have

$$W = VIt$$

If t is given in hours and I is in amperes, then the energy of a cell is expressed in watt-hours.

Internal Resistance

The output voltage from a cell varies as the load on the cell changes. *Load on a cell* refers to the amount of current drawn from the cell. As the load increases, the voltage output drops, and vice versa. The change in output voltage is caused by the *internal resistance* of the cell. Since materials from which the cells are made are not perfect conductors, they have resistance. Current flowing through the external circuit also flows through the internal resistance of the cell. According to Ohm's law, a current flowing through a resistance (either external or internal) results in a voltage drop ($V = IR$). Any voltage developed across the internal resistance is not available at the terminals of the cell. The voltage at the terminals is the voltage due to the chemical reactions *minus the voltage dropped across the internal resistance*. The terminal voltage of a cell, therefore, depends on both the internal resistance of the cell and the amount of load current.

In some applications, the changes in cell terminal voltage are so small they make no practical difference. In other applications, the changes are very noticeable. For example, when starting an automobile engine, the battery output voltage changes from about 12.6 to 8 V.

As a cell is discharged, its internal resistance increases. Therefore, its output voltage decreases for a given value of load current.

Polarization

When a primary cell discharges, gas ions (such as hydrogen) are created around the positive electrode. The gas ions polarize the cell and reduce the cell's terminal voltage. If the gas ions are allowed to build up, the polarization can become so great that the cell is useless. Therefore, cells include a depolarizing agent in their chemical composition. The depolarizing agent is a chemical compound that will react with the polarizing gas to remove it. For example, the depolarizing compound may provide oxygen to react with hydrogen gas and create water. Often the depolarizing agent is part of the positive electrode.

Review Questions

1. What is a cell?
2. What is a battery?
3. Describe the basic structure of a cell.
4. Define the following terms:
   a. Primary cell
   b. Secondary cell
   c. Dry cell
   d. Wet cell
   e. Ampere-hour
5. List three secondary-type cells.
6. List four primary-type cells.
7. What causes a cell's voltage to decrease when a load is connected to the cell?
8. What is polarization?

4-2 LEAD-ACID CELLS

The lead-acid cell is a very common secondary cell. It is the power source for the electric system of most cars, trucks, boats, and tractors. It can provide the large currents (hundreds of amperes) needed to crank internal-combustion engines.

A lead-acid cell produces about 2.1 V. Higher voltages are obtained by connecting cells together to form batteries. A 12-V automobile battery contains six cells.

The structure of a lead-acid cell is shown in Fig. 4-2. Notice that the cell terminals connect to a group of plates. The groups are tied together (both electrically and mechanically) at the top. The meshing together of the positive and negative groups (Fig. 4-2), effectively provides one large negative and one large positive plate.

Chemical Action

Cell terminals

When a load is connected to the cell of Fig. 4-3(*a*), current flows. Electrons leave the negative terminal (plate), flow through the load, and return to the positive terminal of the cell. At the surface of the positive plate in Fig. 4-3(*a*), a molecule of PbO_2 goes into solution as three ions. Notice that each ion of oxygen (O) has two excess electrons, while the ion of lead (Pb) has a deficiency of only two electrons. Thus, the plate is left with a deficiency of two electrons; that is, it has a positive charge. In the electrolyte, two molecules of H_2SO_4 provide four hydrogen (H) ions and two sulfate (SO_4) ions. The four hydrogen ions combine with the two oxygen ions to form two molecules of water (H_2O). The water molecules do not ionize, as shown in Fig. 4-3(*b*). One of the sulfate ions combines

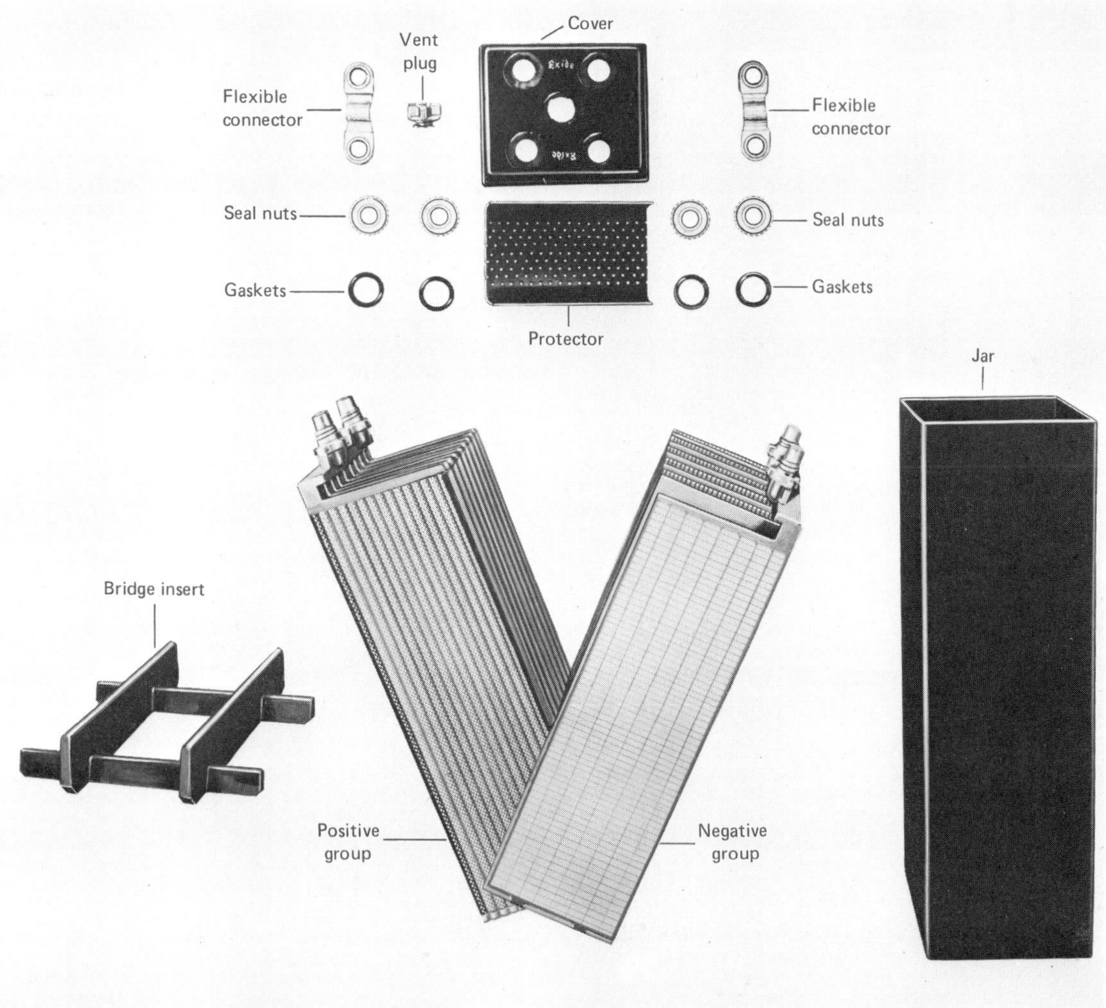

Fig. 4-2 Structure of lead-acid cell. (Courtesy of Exide Power Systems.)

Sulfuric acid

Lead

Lead peroxide

Rate of charge

Charging

with the lead ion at the positive plate to form a molecule of lead sulfate ($PbSO_4$). The lead sulfate settles out of the solution.

At the negative plate [Fig. 4-3(*a*)], an atom of lead goes into the solution as a positive ion. In becoming a positive ion, the lead atom left two electrons on the plate. Thus, the plate has a negative charge. This lead ion combines with a sulfate ion from the electrolyte and forms a molecule of lead sulfate. The lead sulfate again drops out of the solution [Fig. 4-3(*b*)].

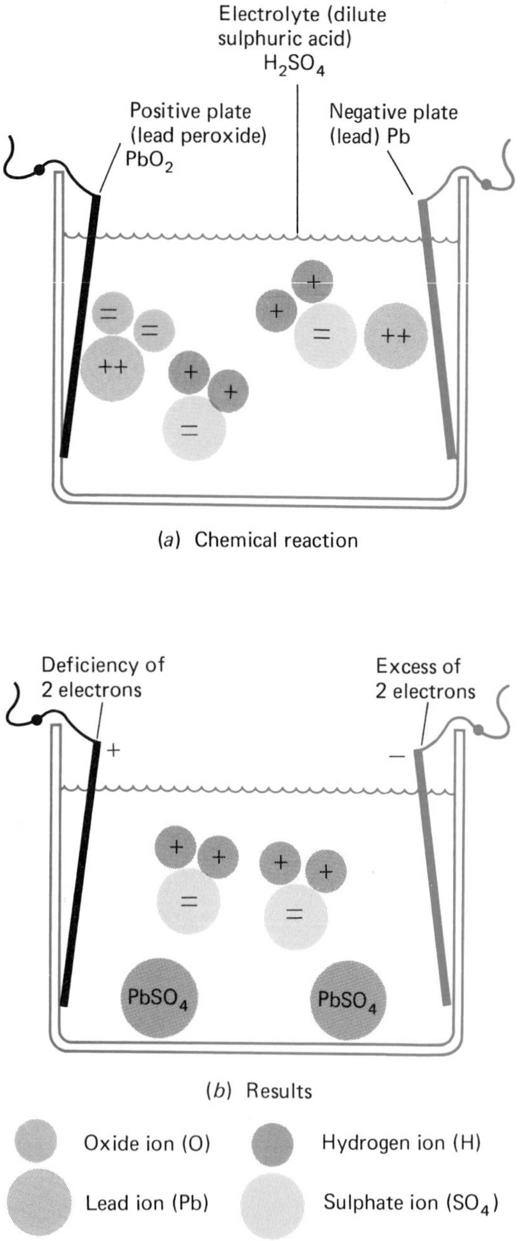

Fig. 4-3 Lead-acid cell. (*a*) Chemical reaction and (*b*) results.

The above-described chemical reaction continues until one of two things happens:

1. The load is removed; the chemical reaction stops. At this time, the charges on the plates and the charges of the electrolyte ions are in a state of equilibrium (balance), or
2. All the sulfuric acid solution has been converted to water and/or all the lead and lead peroxide have been converted to lead sulfate. In this case, the battery is fully discharged.

Recharging

A lead-acid cell is recharged by forcing a reverse current through the cell. That is, electrons enter the negative plate and leave the positive plate. When a cell is being charged, all the chemical reactions described above are reversed. The water and lead sulfate are converted back into sulfuric acid, lead, and lead peroxide. Cells are charged by connecting them to a voltage source that is greater than the cell's voltage. In other words, the cell becomes the load rather than the energy source. As the load, the cell is converting electric energy (from the other voltage source) into chemical energy.

The six-cell (12-V) battery in Fig. 4-4 is being recharged. Notice that the negative terminal of the charger is connected to the negative terminal of the battery. Since the charger's voltage is greater than the battery's voltage, current will flow in the direction indicated in Fig. 4-4. The exact voltage of the charger will depend on the condition of the battery and the rate of charge desired. Generally, the voltage will be between 13 and 15 V

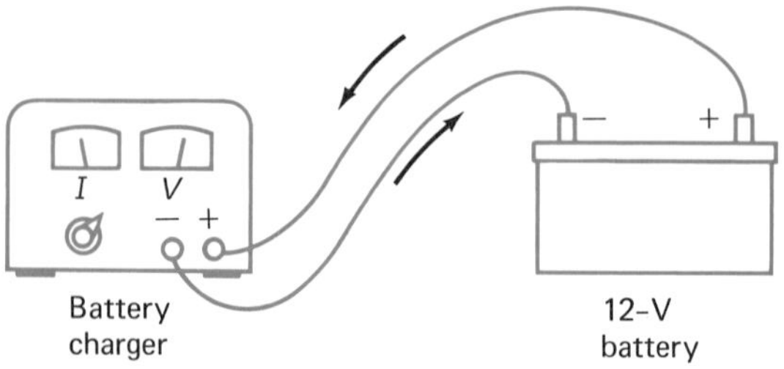

Fig. 4-4 Charging a battery. The charger voltage must exceed the battery voltage.

for a 12-V battery. The charger voltage is adjusted to provide the desired charge current. The manufacturer's recommendation should be followed in determining the rate of charge for a specific type of battery or cell. With an automobile battery, charging currents of 25 to 50 A are used when the battery is fully discharged. As the battery charges, its charging rate is reduced to a few amperes.

Too fast a charge rate must be avoided, as it will overheat the battery. The battery temperature should never exceed 49°C (120°F). Higher temperatures greatly reduce the life expectancy of the battery.

A battery should not be overcharged. Overcharging can weaken the plate structure of the battery. When a battery is overcharged, water from the electrolyte is converted into hydrogen gas and oxygen gas. Having to add more than normal amounts of water to a battery is an indication that it is being overcharged.

Specific Gravity

The charge of a battery can be determined by measuring the specific gravity of the electrolyte. The specific gravity of a substance is the ratio of its weight compared with the weight of water. If a substance has a specific gravity of 1.251, it is 1.251 times as heavy as water. Sulfuric acid is heavier than water. Therefore, the more sulfuric acid in the electrolyte, the higher will be the specific gravity of the electrolyte. Since the sulfuric acid increases as the battery is charged, specific gravity indicates the state of charge of the battery. The specific gravity of a fully charged cell is adjusted at the factory to suit the structure and intended purpose of the cell. For fully charged lead-acid cells, the specific gravity ranges from about 1.21 to 1.28. The typical automotive battery is fully charged at a specific gravity of about 1.26.

A lead-acid cell is considered to be completely discharged when the specific gravity drops to 1.12. It should not be left in this discharged state for extended periods. To do so shortens the life of the battery. Discharged cells and batteries need to be protected from low temperature. A completely discharged cell will freeze at about −9°C (16°F). At 50 percent discharge it freezes at −24°C (−11°F). When a cell freezes, the electrolyte expands and can break the jar (case) of the cell.

Hydrometer

Specific gravity is measured with a hydrometer like the one shown in Fig. 4-5. The hydrometer is used as follows:

1. Squeeze the rubber bulb. Insert the flexible tube into the electrolyte of the cell. Slowly release the rubber bulb, drawing the electrolyte into the hydrometer. When the float in the hydrometer lifts free, remove the flexible tube from the electrolyte and finish releasing the rubber bulb.
2. Read the specific gravity from the float at the top surface of the electrolyte.
3. Reinsert the flexible tube into the cell and slowly squeeze the bulb to return the electrolyte.
4. When finished testing, flush the hydrometer (both inside and out) with clean water.

Overcharged

State of charge

Hydrometer

Specific gravity

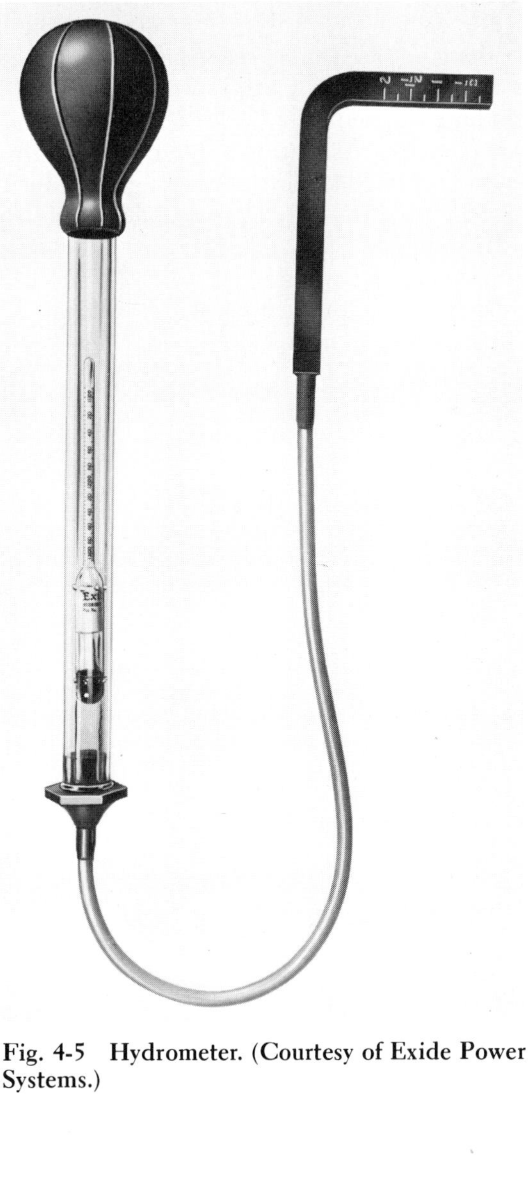

Fig. 4-5 Hydrometer. (Courtesy of Exide Power Systems.)

A hydrometer test is useful in determining either the percentage of charge or the condition of a cell. For a new cell, the specific gravity decreases approximately 0.01 for each 7 percent of the charge taken from the cell. For example, a cell with a specific gravity of 1.25 when fully charged will have approximately 23 percent of its charge left when its specific gravity has fallen to 1.15. When the specific gravity cannot be raised by charging to within 0.05 of the manufacturer's specification, it is a questionable cell. It will probably completely fail in the near future.

Percentage of charge

Battery safety

Nickel-cadmium cells

Battery Safety

If not properly handled, lead-acid cells and batteries can be dangerous. The acid used in the electrolyte can cause skin burns and burn holes in clothing. It is extremely harmful to the eyes. Always wear safety glasses when working with lead-acid cells and batteries. If acid does come in contact with the skin or clothing, immediately flush with clean water. Then wash with soap and water, except for the eyes. If the acid is in the eyes, get medical attention immediately after repeated flushing with water. Wash hands in soap and water after handling batteries.

The gases released from charging batteries are explosive. Charge batteries in a ventilated area where there are no sparks and open flames.

Review Questions

9. What happens in a lead-acid cell as it is discharged?
10. How is a lead-acid battery recharged?
11. List two precautions to follow when charging a battery to prevent battery damage.
12. What is specific gravity?
13. How is specific gravity measured?
14. Should a lead-acid battery be left in a discharged state? Why?
15. Is sulfuric acid heavier than or lighter than water?
16. What safety rules should be followed in handling or charging lead-acid batteries?

4-3 NICKEL-CADMIUM CELLS

The nickel-cadmium cell is a secondary cell which uses a solution of potassium hydroxide as an electrolyte. The positive electrode is nickelic hydroxide and the negative electrode is cadmium. Nickel-cadmium cells are dependable, rugged units. As vented wet cells, they have been in limited use for many years in many types of service. When technological advances made it possible to produce a nickel-cadmium dry cell, the popularity of the cell greatly increased. The dry cell is hermetically sealed and can be operated in any position. Figure 4-6 illustrates the structure of a typical sealed nickel-cadmium cell. These cells are most commonly available in button or cylindrical shape and in capacities ranging from a few milliampere-hours to 5 or 6 Ah.

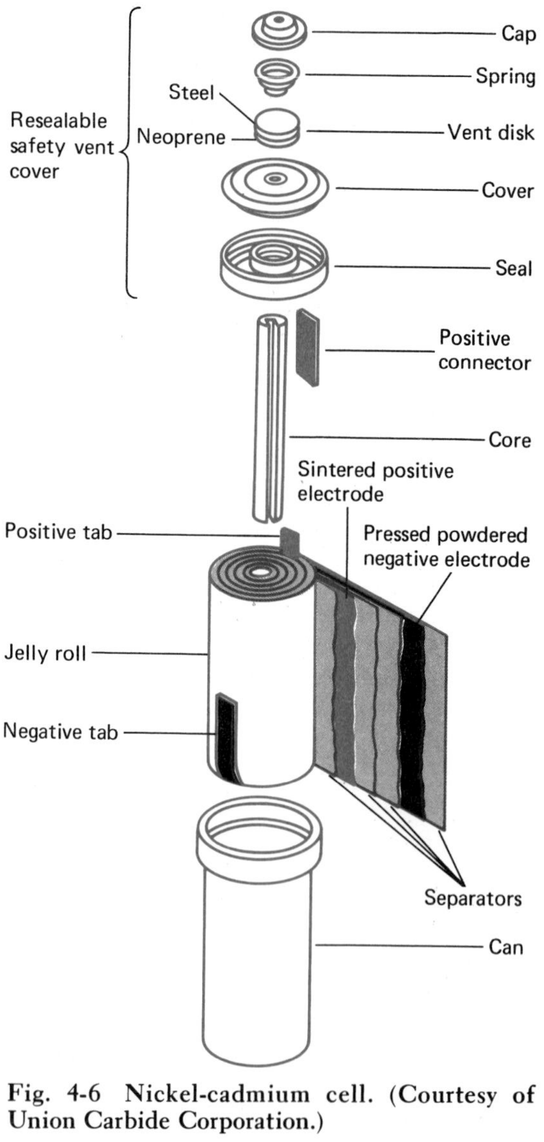

Fig. 4-6 Nickel-cadmium cell. (Courtesy of Union Carbide Corporation.)

Sealed nickel-cadmium cells have a life ranging from several hundred to several thousand charge/discharge cycles. They have extremely long shelf life (storage life) and can be stored either charged or discharged. They can be stored at temperatures ranging from −40 to 60°C (−40 to 140°F).

Other notable characteristics of the nickel-cadmium cell are:

1. Good low-temperature operating characteristics.
2. High initial cost but low operating cost.
3. Very low internal resistance. Thus, they can provide high currents with little drop in output voltage.
4. Nearly constant output voltage (approximately 1.2 V) as the cell is discharged by a moderate load. See Fig. 4-7 for a general comparison with other types of cells.

4-4 CARBON-ZINC AND ZINC CHLORIDE CELLS

The carbon-zinc cell is the most common of the dry cells. It is also the least expensive of the primary cells.

The composition of a carbon-zinc cell is shown in Fig. 4-8. The zinc can is the negative electrode and the manganese dioxide is the positive electrode. The carbon rod (Fig. 4-8) makes electric contact with the manganese dioxide and conducts current to the positive terminal. However, the carbon rod is not involved in the chemical reaction that produces the voltage. Electrolyte for the chemical system is a solution of ammonium chloride and zinc chloride.

Although extensively used because it is cheap, the carbon-zinc cell has many weaknesses. Some of these weaknesses are:

1. Poor low-temperature operating characteristics.
2. Gradual decrease in output voltage as cell discharges (Fig. 4-7).
3. Low energy-to-weight ratio and low energy-to-volume ratio. Other types of primary cells have ratios two to three times as great as the carbon-zinc cell.
4. Poor efficiency under heavy loads due to high internal resistance.

Carbon-zinc cells and batteries are available in a wide range of sizes and shapes. Batteries with capacities of 30 Ah are available.

Zinc chloride cells are constructed much like the carbon-zinc cell, but they use a modified electrolyte system. The electrolyte is a solution of zinc chloride. Using only zinc chloride in the electrolyte provides several advantages over the standard carbon-zinc cell. First, the zinc chloride cell is more efficient at higher current drain. Second, the output voltage does not decrease as rapidly as it does in the carbon-zinc cell.

4-5 ALKALINE-MANGANESE DIOXIDE CELLS

Alkaline cells use the same electrodes as carbon-zinc cells (zinc and manganese

Charge/discharge cycle

Carbon-zinc cell

Zinc chloride cell

Shelf life

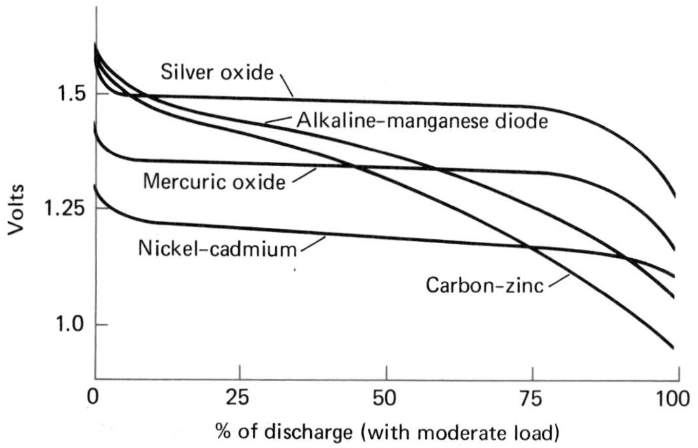

Fig. 4-7 Comparison of various cell voltages as the cells are discharged.

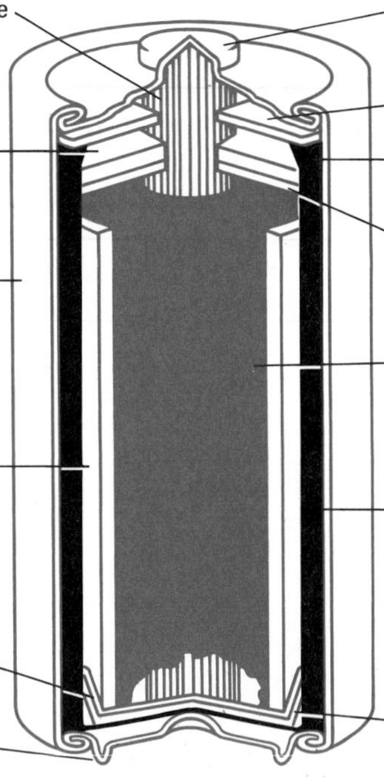

Fig. 4-8 Cutaway view of a carbon-zinc cell. (Courtesy of Union Carbide Corporation.)

Primary alkaline cell

Secondary alkaline cells

Mercuric oxide cell

Charge retention

Cycle life

dioxide). However, a solution of potassium hydroxide is used as the electrolyte. There are two types of alkaline cells—primary and secondary (rechargeable). The construction of a primary cell is shown in Fig. 4-9.

Alkaline cells produce about 1.5 V. The cell voltage decreases gradually as the cell is discharged. However, the change is not as great as it is with carbon-zinc cells (Fig. 4-7).

The primary alkaline cell is more expensive than the carbon-zinc cell. However, it has several advantages. It can be discharged at high current drains and still maintain reasonable efficiency and cell voltage. It can operate effectively at lower temperatures (−30°C versus −7°C for carbon-zinc cells). It has a lower internal resistance and a higher energy-to-weight ratio. An alkaline cell stores at least 50 percent more energy than the carbon-zinc cell of the same size.

Secondary alkaline cells are much cheaper than nickel-cadmium cells. They have better charge retention (ability to remain charged when stored) and a wider operating temperature range than do nickel-cadmium cells. The internal resistance of the two types of cells is comparable. However, the alkaline cell has several shortcomings compared with the nickel-cadmium cell:

1. The cycle life (number of discharge/recharge cycles) is less than 75.
2. The cell voltage is not as constant (Fig. 4-7).
3. Cycle life is more dependent on proper discharge and recharge of the cell. Discharging beyond rated capacity and extended overcharging both shorten the cycle life.
4. Charging circuits for optimum cycle life are more complex.

4-6 MERCURIC OXIDE CELLS

Mercuric oxide cells (commonly called mercury cells or batteries) have some distinct advantages over the primary cells previously discussed. These advantages are:

1. More uniform cell voltage as the cell discharges (Fig. 4-7).
2. Capacity is less dependent on discharge rate.
3. An energy-to-volume ratio 2 to 4 times as great as either the alkaline or zinc-carbon cell.
4. A higher energy-to-weight ratio.

The mercury cell is also mechanically rugged, has low internal resistance, and can op-

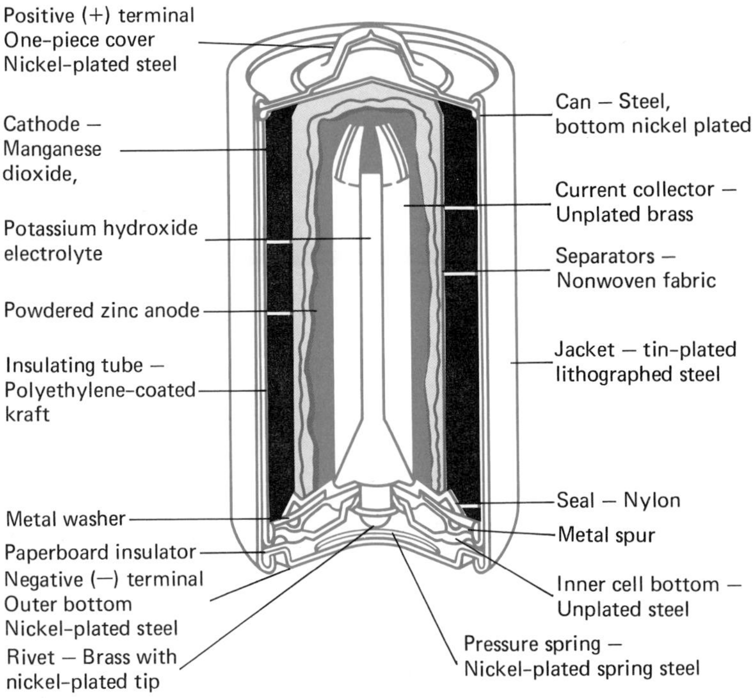

Fig. 4-9 Cutaway view of an alkaline cell. (Courtesy of Union Carbide Corporation.)

Voltage standard

Light load

erate at high temperatures (54°C). Since the output voltage (1.35 V) is so constant over most of its discharge, the mercury cell is used as a voltage standard. (A standard is used to check the accuracy of other voltage sources and measuring devices.)

Some of the disadvantages of the mercury cell are its poor low-temperature (0°C) performance and its relatively high cost. Both the initial cost and the operating cost are high compared with costs of carbon-zinc and alkaline cells.

Figure 4-10 shows the internal structure of the mercury cell. Again the negative electrode is zinc. The positive electrode is mercuric oxide, and the electrolyte is a solution of either potassium or sodium hydroxide. Mercury cells and batteries are available in a wide range of capacities (up to 28 Ah) and shapes.

4-7 SILVER OXIDE CELLS

Silver oxide cells are much like the mercury cell. However, they provide a higher voltage (1.5 V) and they are made for light loads. The loads can be continuous such as those encountered in hearing aids and electronic watches. Like the mercury cell, the silver oxide cell has good energy-to-weight and energy-to-volume ratios, poor low-temperature response, and flat output voltage characteristics. The structures of the mercuric and silver oxide cells are very similar. The main difference is that the positive electrode of the silver cell is silver oxide instead of mercuric oxide. Figure 4-7 compares the voltage characteristics of the silver oxide cell with those of other cells.

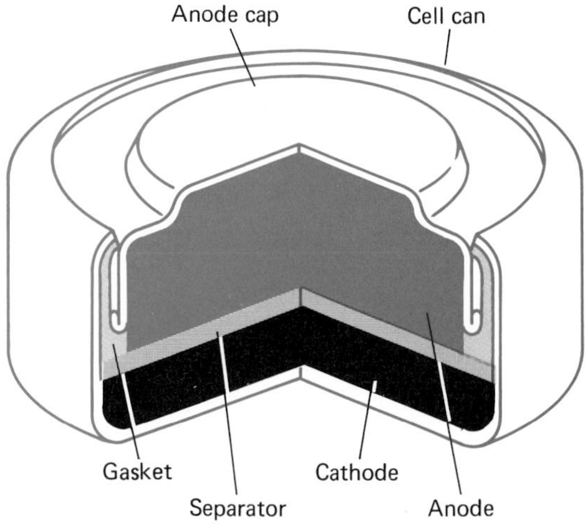

Fig. 4-10 Cutaway view of a mercuric oxide cell. (Courtesy of Union Carbide Corporation.)

Review Questions

17. Are nickel-cadmium cells damaged by leaving them in a discharged condition?
18. What is the cycle life of a nickel-cadmium cell?
19. Compare the nickel-cadmium and rechargeable alkaline cell in terms of:
    a. Voltage during discharge
    b. Cycle life
    c. Cost
20. What is the major advantage of the carbon-zinc cell compared with other dry cells?
21. Compare the alkaline and carbon-zinc cells in terms of:
    a. Efficiency at high current drains
    b. Voltage at high current drains
    c. Energy storage
22. What are the advantages of both the mercury and silver oxide cells over other primary cells?

4-8 MINIATURE LAMPS

Neon glow lamp

Incandescent lamp

Tungsten filament

Wedge base

Miniature lamps are used extensively in electric and electronic circuits. They are used for indicator lights and instrument illumination in automobiles, aircraft, home appliances, coin-operated machines, and all kinds of electronic instruments. Miniature lamps are also used in flashlights, lanterns, and signal lights. Both *neon glow lamps* and *incandescent lamps* are used as indicators.

Incandescent Lamps

The heart of the incandescent lamp is the tungsten filament (wire) which emits light when heated by an electric current. When the tungsten wire is long, thin, and coiled, as in Fig. 4-11, it is often supported in the center.

The tungsten filament is enclosed in a glass bulb from which the air is evacuated. Often the air is replaced by an inert gas. As the last of the air is removed from the bulb, the bulb is sealed. It is sealed by melting the glass around the evacuation hole. The seal on the lamp in Fig. 4-11 was made on the small tip of glass at the bottom of the bulb.

The heavy wire leads connecting to the filament (Fig. 4-11) carry current to the filament. For this lamp, these leads are also the terminals which make electric contact with the lamp holder. This configuration of terminal connections is called a *wedge base*. It provides for a simple, compact lamp holder.

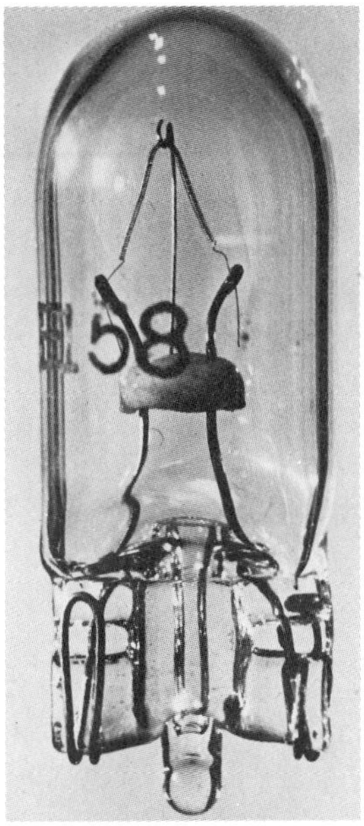

Fig. 4-11 Wedge-base lamp. (Courtesy of General Electric.)

The two principal bases used in miniature incandescent lamps are the *screw-type base* and the *bayonet base*. The three lamps in the top row, left, of Fig. 4-12 are examples of screw-type and bayonet-type miniature lamps. The lamp on the extreme left has a screw-type base. The screw is stamped on the portion of the base called the *shell*. The next two lamps have bayonet bases. Again, the straight portion of the base is the shell. The screw-type lamp is screwed into a lamp holder in the same way as the common household incandescent lamp. The bayonet base lamp usually has one or two small projections on the shell base. These projections slide into grooves in the bayonet lamp holder. When the lamp is in place, it is given a slight twist locking it in the holder. Lamp bases are usually made of brass for good electrical conductivity and long wear.

Figure 4-12 shows a variety of other base and bulb styles used with miniature lamps. All the bases provide two contacts which make connections to the filament. Each base was designed to best fulfill a specific need. For example, the base on the lamp in the lower right-hand corner of Fig. 4-12 provides for exact location of the lamp. This is important when the light is to be focused. Some lamps, like the one in the center of Fig. 4-12, are designed to be soldered into the circuit. Some bases are designed to withstand high temperature. Others are made to withstand severe vibration.

Miniature lamps are available in a wide range of voltage and current ratings. Voltages range from about 1 to 70 V and currents from a few milliamperes to 7 A. The voltage rating of a lamp is the voltage for which the manufacturer has designed the lamp. At rated voltage, a lamp draws its rated current, provides its specified light, and operates for a specified time, usually called the *life* of the lamp.

Operating above rated voltage increases the current draw and light output (specified in units called *lumens*). However, operating with overvoltage decreases the life expectancy so greatly that it is not recommended. Operation below rated voltage is permissible. Under this condition, current decreases, light output decreases, and expected life greatly increases. The efficiency in terms of light output per watt input is reduced when operating below rated voltage.

Sometimes the manufacturer specifies the voltage and power rating of a lamp. In this case, the current can be determined by using the power relationship $I = P/V$.

Example 4-1

A no. 1385 lamp is rated at 28 V and 20 W. How much current does it draw?

Given: Voltage (V) = 28 V, power (*P*) = 20 W

Find: Current (*I*)

Known: $I = \frac{P}{V}$

Solution: $I = \frac{20\text{ W}}{28\text{ V}} = 0.714\text{ A}$

Answer: The lamp draws 0.714 A or 714 mA.

A few miniature lamps are made with built-in flasher units. Flasher lamps are used as

Current rating

Voltage rating

Flasher lamp

Lamp life

Lumen

Fig. 4-12 Miniature lamps. Sizes may be judged by comparing the lamps to the 6-in scale. (Courtesy of General Electric.)

Bimetallic strip

Hot resistance

Neon glow lamp

Two metal electrodes

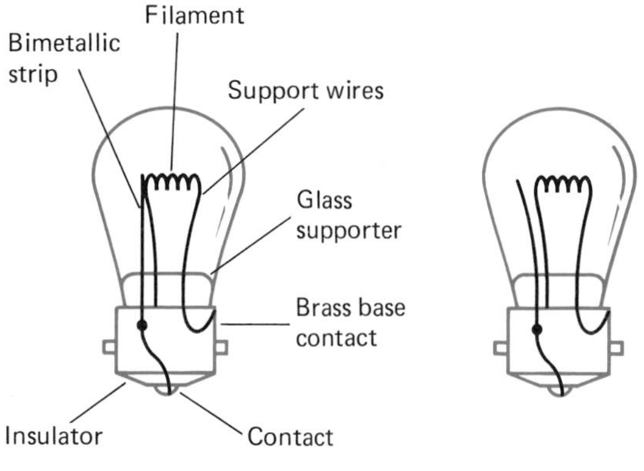

Fig. 4-13 Flasher lamp. The bimetallic strip turns the lamp off and on as it warms up and cools down. (*a*) Lamp on, (*b*) lamp off.

warning indicators, such as seat belt and emergency brake warnings in automobiles. The flasher action is caused by a *bimetallic strip* located next to the filament. This bimetallic strip is one of the conductors carrying current to the filament, as shown in Fig. 4-13(*a*). When the filament heats the bimetallic strip, the strip bends out [Fig. 4-13(*b*)] and opens the circuit, turning off the lamp. When the lamp is off, the strip cools and straightens out. When it returns to its normal shape, it makes contact with the filament again and turns on the lamp. This cycle repeats as long as the lamp is connected to a power source.

Figure 4-14 illustrates the structure of a bimetallic strip which is composed of two different metals. The two metals are rigidly joined where they meet. The metals have different temperature coefficients of expansion. That is, one of the metals increases its length more rapidly than the other as the temperature increases. This unequal expansion causes the strip to curve when heated, as shown in Fig. 4-14(*b*).

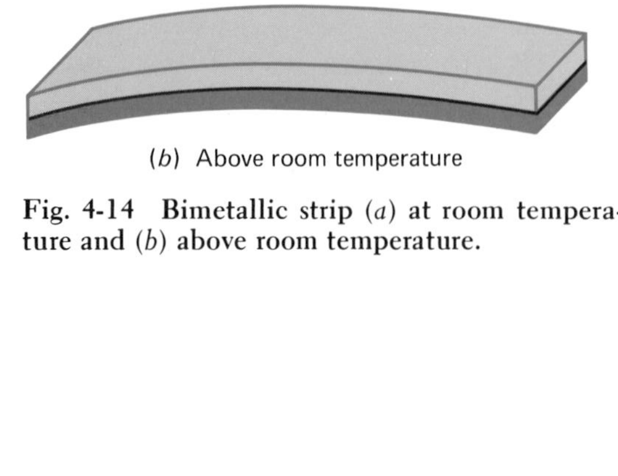

Fig. 4-14 Bimetallic strip (*a*) at room temperature and (*b*) above room temperature.

The resistance of a lamp filament when cool is only a fraction of the resistance of the filament when it is hot. Therefore, a lamp draws a large current when it is first turned on. However, the filament comes up to operating temperature in a few milliseconds, and so this large current does not last long. The reason for the change in resistance of the lamp is the temperature coefficient of resistance of the tungsten filament. Although the coefficient of tungsten is not much higher than that of copper, the temperature change is tremendous.

Measuring the resistance of a lamp with an ohmmeter only indicates whether or not the lamp is burned out. The indicated resistance will be far less than the actual hot resistance of the lamp. This is because the current from the meter is too small to bring the lamp up to operating temperature.

Neon Glow Lamps

The center lamp in Fig. 4-15 is a neon glow lamp. The glass bulb is filled with neon gas. When neon gas is not ionized, it is an insulator. Once it ionizes, it is a fair conductor.

Ionization of the neon in the lamp in Fig. 4-15 occurs between the two metal electrodes inside the bulb. If the ionization voltage is dc, only one electrode glows. If it is ac, both electrodes glow. It requires between 70 and 95 V to ionize a neon bulb. After ionization occurs, it requires 10 to 15 percent less voltage, or from 60 to 80 V, to maintain ionization.

Because the internal resistance of an ionized lamp drops to a very low value, a resistor must be connected in series with the lamp. The resistor (Fig. 4-16) limits the current through the lamp to a safe value. In Fig. 4-16, the neon bulb and the resistor are represented by their schematic symbols. Some neon lamps are manufactured with a resistor in the base. In this case no external resistor is needed.

Neon lamps are made in a number of different bulb and base styles. Most of the styles are the same as for the incandescent lamps. Neon lamps require very little power to operate. Most are less than ½ W and require no more than 2 mA.

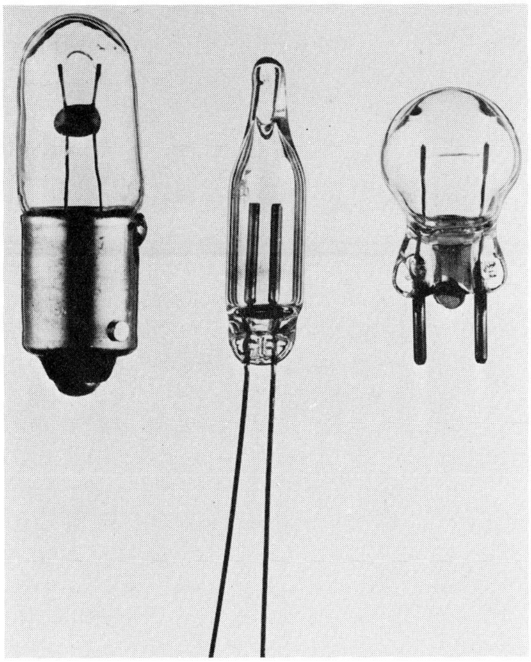

Fig. 4-15 Neon lamp (center). The lamps on either side are miniature incandescent lamps. Note the absence of a filament in the neon lamp. (Courtesy of General Electric.)

Light-Emitting Diodes

The light-emitting diode (LED) is often used as an indicator lamp. It is also used for numeric readouts on calculators and other digital devices. LED lamps are small, low-powered, and reliable. They can operate for years without failure. They are available in either a wire-terminal or socket-type base.

Although most LEDs are red, they are also available in green and yellow. The color is determined by the combination of materials used in the manufacturing process. The materials used include various compounds of gallium, arsenic, and phosphorus.

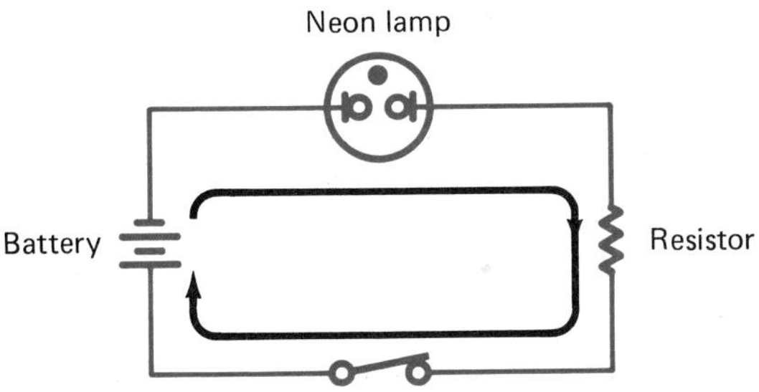

Fig. 4-16 Neon lamp and current-limiting resistor. Without the resistor the neon lamp would immediately burn out.

Like neon lamps, LEDs need current-limiting resistors to control the current once the lamp starts conducting. However, the typical LED needs only about 2 V before it starts conducting. Most of these lamps can operate on 3 to 40 mA of current. Typically, they are operated at about 20 mA.

The LED lamp operates only on direct current. It is a polarized lamp. The polarity of its leads needs to be determined before it is connected to a voltage source.

Light-emitting diode

Polarized lamp

Working with Miniature Lamps

Lamp safety

Current-limiting resistor

Some miniature lamps operate at very high temperatures. The quartz lamp in the lower left corner of Fig. 4-12 can operate with a bulb temperature of 350°C. Wedge-based quartz lamps can operate as high as 450°C. At these temperatures, dirt on the surface of the glass can be quite harmful. Needless to say, severe burns can result from touching lamps operating at such high temperatures.

When connecting wire-terminal lamps, like the neon lamp in Fig. 4-15, you must be careful in bending the wire leads. Do not bend the leads within 3 mm (⅛ in) of the glass bulb. Also, do not solder within 3 mm of the glass. Either bending or soldering too close to the bulb can damage the metal-to-glass seal.

Lamp Safety

Following the procedures listed below will reduce the probability of accidents when working with lamps:

1. Turn off the power to a lamp before changing the lamp.
2. Allow a lamp to cool before changing it.
3. Be certain a replacement lamp has the correct voltage. If the voltage of the power source greatly exceeds the voltage rating of the lamp, the lamp may explode.
4. When replacing lamps that fit tightly in their sockets, wear a glove (or wrap the bulb in a heavy cloth) in case the bulb breaks.

Review Questions

23. What material is used in the filament of a miniature lamp?

24. List two types of miniature lamps used as indicators.
25. What makes a flasher lamp blink off and on?
26. How much current does a 12-V, 3-W lamp draw?
27. Does a hot or a cold lamp have the most resistance?
28. Does tungsten have a negative or positive temperature coefficient?
29. Does ac voltage cause one or both electrodes of neon lamps to glow?
30. What is the purpose of the resistor in a neon-lamp circuit?
31. What precaution should be observed when using wire-terminal lamps?
32. Why is it dangerous to put a 1.5-V lamp in a 28-V circuit?

4-9 RESISTORS

One of the most common and most reliable electrical-electronic components is the resistor. The resistor is used as a load, or part of the load, in most electronic circuits. Its major purposes are to control current and divide voltage. Some types of resistors can operate at temperatures as high as 300°C. Resistors are manufactured in a wide range of resistance values. Resistances of less than 1 Ω to over 100 MΩ are available.

Classification and Symbols

Resistors can be classified into four broad categories: fixed, variable, adjustable, and tapped. Sketches and symbols for these four categories are shown in Fig. 4-17. Notice that the variable and the adjustable resistor use the same symbol.

As shown in Fig. 4-17, there are two types of variable resistors. The potentiometer has three terminals. Rotation of the shaft changes the resistance between the middle terminal and the two end terminals.

Potentiometer

Most potentiometers are *linear*. That is, one degree of shaft rotation results in the same change of resistance regardless of the shaft location. Other potentiometers have *nonlinear tapers*. This means that the rate of change of resistance varies as the shaft is rotated. A variety of tapers are available. Tapered potentiometers are sometimes used as volume and tone controls in stereo amplifiers.

Nonlinear taper

Ganged potentiometers

Potentiometers are often ganged together so that the resistance in several circuits can be changed simultaneously. The top row of Fig. 4-18 shows some multiple ganged potentiometers. The bottom left potentiometer in Fig. 4-18 is a dual-shaft unit. One shaft controls one of the ganged potentiometers and the other shaft controls the other two. The other dual-shaft potentiometer in Fig. 4-18 has

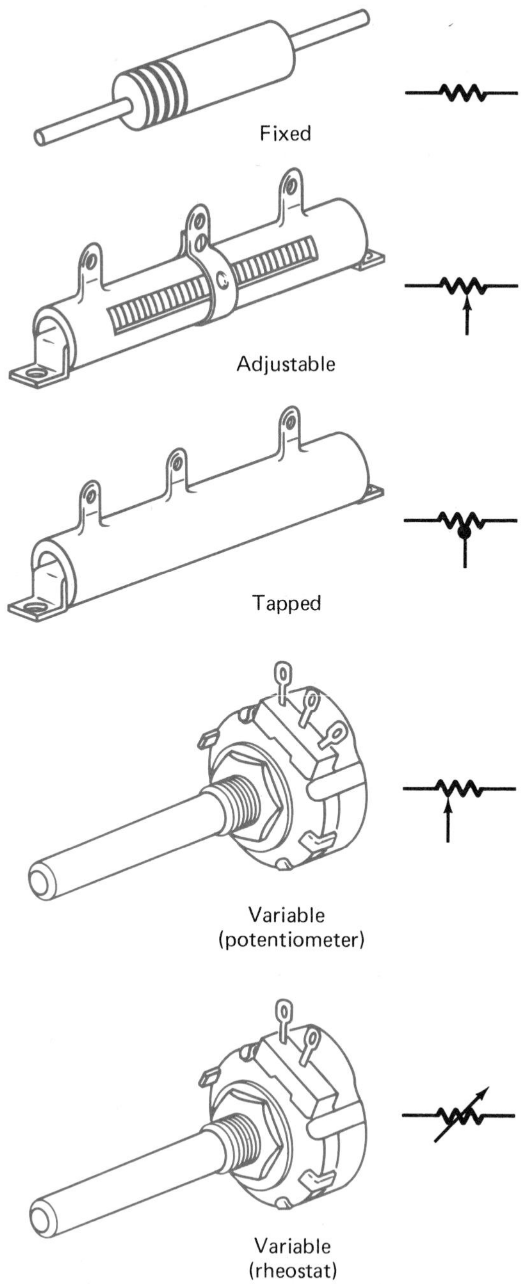

Fig. 4-17 Classification and symbols of resistors.

a switch ganged to a potentiometer. One shaft operates the switch, and the other shaft controls the potentiometer. Often the same shaft will operate both the switch and the potentiometer when the two are ganged together.

Rheostats (Fig. 4-17) have only two terminals. Turning the shaft changes the resistance between these two terminals. Rheostats are used to adjust the current in a circuit to a specified value. Very often potentiometers are used as rheostats by just using two of the three terminals.

Adjustable resistors serve about the same functions as potentiometers and rheostats. However, adjustable resistors are used in higher-power circuits. They are used only when infrequent changes in resistance are required. Unlike rheostats and potentiometers, adjustable resistors are not usually adjusted while the circuit is in operation.

A single-tap resistor is illustrated in Fig. 4-17. Multiple-tap resistors are also available. Tapped resistors, like adjustable resistors, are mostly used in high-power circuits (greater than 2 W).

Power Ratings

Resistors have power ratings as well as resistance ratings. The power rating of a resistor indicates the amount of power the resistor can safely dissipate without being destroyed. As current passes through the resistor, heat is produced. The resistor is converting electric energy into heat energy. If the current were allowed to increase, the heat would burn up the resistor. Thus, some safe level of heat must be specified and this is what the power rating does.

There is no relationship between the resistance rating and the power rating of a resistor. The same resistance value can be obtained in resistors ranging in power ratings from less than a watt to many watts. The power rating of a resistor is primarily determined by the physical size of the resistor and the type of materials used. Figure 4-19 shows five fixed resistors with different power ratings. All the resistors have one resistance rating. The largest resistor is a 2-W resistor. In descending order, the others are 1, $\frac{1}{2}$, $\frac{1}{4}$, and $\frac{1}{8}$ W.

The power rating of a resistor is assigned by the manufacturer under specified conditions. These conditions include free air circulation around the resistor and resistor leads soldered to sizable terminals. Often these conditions are not fully satisfied in the devices which use resistors. Therefore, a resistor with a power rating higher than the value calculated for the circuit is used. As a rule of thumb many circuit designers specify a power rating twice as

Rheostat

Adjustable resistor

Tapped resistor

Physical size

Power rating

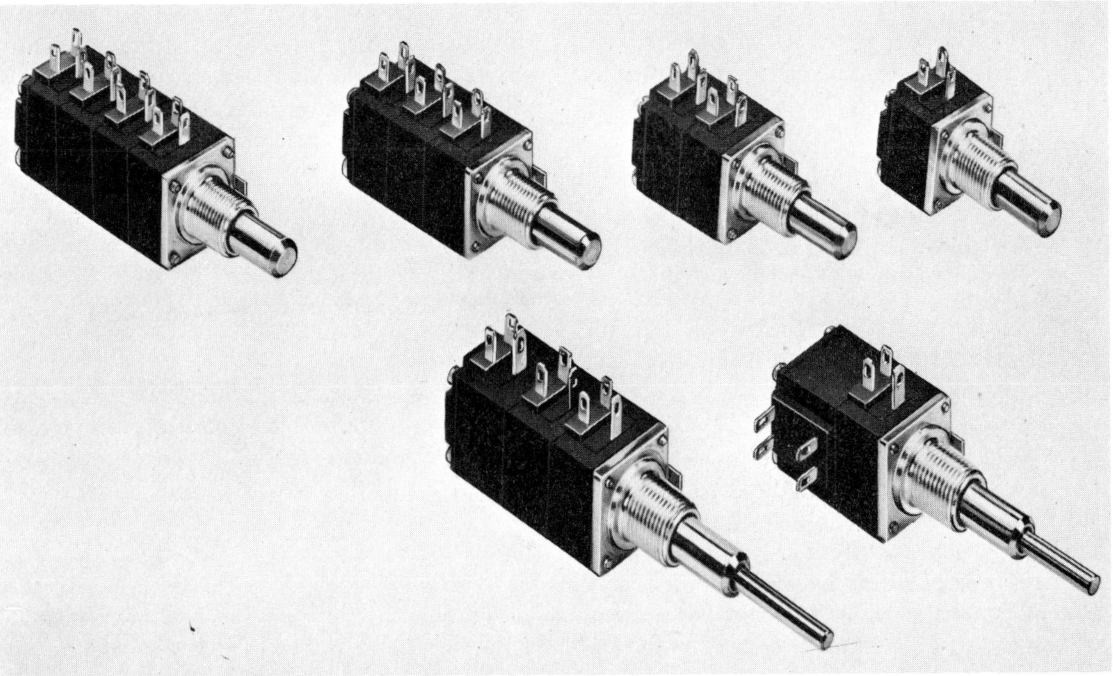

Fig. 4-18 Potentiometers. The two potentiometers in the bottom row have concentric shafts. (Courtesy of Allen-Bradley.)

Safety factor

Tolerance

Carbon composition

Wirewound

Power resistor

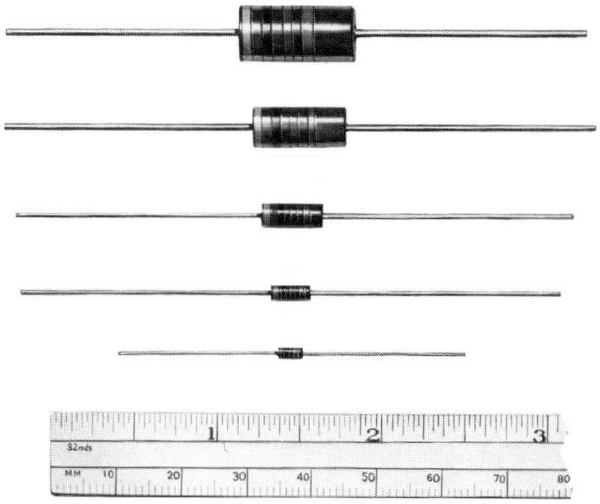

Fig. 4-19 Relative size and power rating of resistors. The physical size of a resistor determines its power rating rather than its resistance. (Courtesy of Allen-Bradley.)

great as the calculated value. In most applications this is a reasonable safety factor.

Example 4-2

How much power is used (dissipated) by a 1000-Ω resistor when 100 V is connected to it? What power rating would be specified for the resistor?

Given: Resistance (R) = 1000 Ω
voltage (V) = 100 V

Find: Power (P)

Known: $P = \frac{V^2}{R}$

Solution: $P = \frac{(100\text{ V})(100\text{ V})}{1000\ \Omega} = \frac{10{,}000}{1000}$
$= 10\text{ W}$

Answer: Calculated power is 10 W. Using the rule of thumb, the specified power rating should be twice the calculated value, or

Resistor rating = 10 W × 2 = 20 W
Therefore, use a 20-W resistor.

Resistor Tolerance

It is very difficult to mass produce resistors that have exactly the same resistance value. Therefore, manufacturers specify tolerances for their resistors. *Tolerance* is specified as a percentage of the nominal (stated) resistance. For example, a 10% 1000-Ω resistor can have a resistance anywhere between 900 Ω (1000 − 10% of 1000) and 1100 Ω (1000 + 10% of 1000). A 2700-Ω resistor with a ± 5% tolerance could have a resistance between 2565 and 2835 Ω.

Common tolerances are 10 percent and 5 percent for general electronic circuits. Meters and other electronic instruments use resistors with 2 percent, 1 percent, or less tolerance. Resistors are available with tolerances of less than 0.01 percent. Of course, low-tolerance resistors are more expensive than high-tolerance resistors.

Types of Resistors

Resistors are grouped according to the type of material or process used to make the resistive element. The major types and their characteristics are listed below:

Carbon Composition The resistance element is made of finely powdered carbon held together with an inert binding material. The resistance of the element is determined by the ratio of carbon to binding material. The physical size determines the power rating, not the resistance. Carbon composition is used for both fixed and variable resistors. A fixed, carbon-composition resistor is shown in Fig. 4-20. These resistors are reliable and relatively inexpensive. They are available in a wide range of resistances and in power ratings up to 2 W.

Wirewound Wirewound resistors are used for all classifications of resistors—variable, fixed, adjustable, and tapped. In fact, nearly all tapped and adjustable resistors are wirewound and are classed as power resistors (above 2 W). The wire for wirewound resistors is made of alloys such as nickel-chromium or copper-nickel. The wire is wound (evenly spaced) on an insulator form and the ends are connected to solder-coated copper leads. For

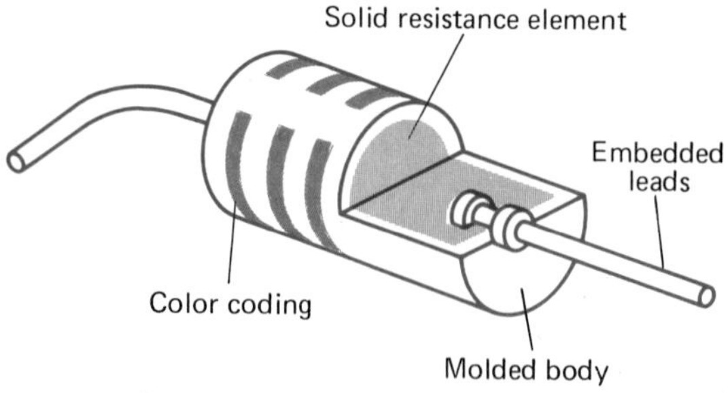

Fig. 4-20 Carbon composition resistor. (Courtesy Allen-Bradley.)

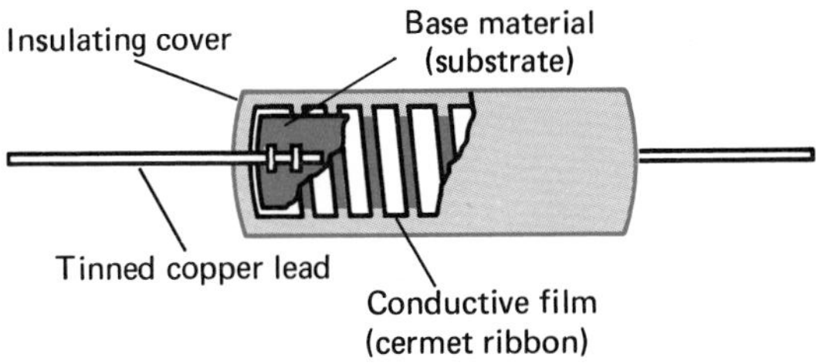

Fig. 4-21 Cermet or film resistor. This resistor is similar to the deposited film resistor.

fixed resistors, the entire assembly, except for the terminals, is coated with insulating material.

Fixed wirewound resistors come in many sizes and shapes. Some look just like carbon-composition resistors. Wirewound resistors are made with power ratings in excess of 1000 W.

Wirewound resistors have low temperature coefficients, good stability, and can be made to close tolerances.

Cermet Cermet is a mixture of fine particles of glass (or ceramic) and finely powdered metal (or oxides) such as silver, platinum, or gold. This mixture, in a paste form, is applied to a base material made of ceramic, glass, or alumina. Leads are attached to the form and the cermet mixture, and the whole device is fired (baked) in a kiln. The firing fuses everything together. The cermet is put on the substrate in a spiral ribbon. This effectively makes a long, thin resistive element for the resistor. The structure of a cermet (or cermet film as it is sometimes called) resistor is like that shown in Fig. 4-21.

Cermet is used in variable as well as fixed resistors. It has a low temperature coefficient, and it is very stable and rugged. As a potentiometer, cermet offers fine adjustment and long life. As a fixed resistor, it is compact and durable.

Deposited Film Deposited-film resistors are made like and look like the resistor in Fig. 4-21. The film is applied to the substrate by vaporizing (in a vacuum) the film material. The material will form a very thin, uniform film on the substrate. The film is deposited in the form of a spiral ribbon.

Either carbon or metal may be deposited as a film. One of the nickel-chromium alloys is often used as the metal to be deposited. Deposited-film resistors are often referred to as either carbon-film or metal-film resistors. These film resistors have characteristics much like the cermet resistors. They operate well at high frequencies.

Conductive Plastic Conductive plastics are used as the resistive element in some potentiometers. They are made by combining carbon powder with a plastic resin such as polyester or epoxy. The carbon-resin mixture is applied to a substrate such as ceramic. Conductive-plastic potentiometers are relatively cheap. They wear very well in applications that require much adjusting of the potentiometer.

Cermet

Deposited film

Carbon film

Metal film

Conductive plastic

Color code

Color Code

Many fixed resistors use a color code system to indicate their resistance and their tolerance. The resistors are marked with color bands. The color of the band and its position relative to the other bands on the body of the resistor indicate the value of resistance and tolerance. The color code is shown in Fig. 4-22. The first band is always the band closest to one end of the resistor. The first two bands indicate the first two digits of the value. The third band indicates the number by which the first two digits are multiplied. Examples of the use of the color code are given in Table 4-1.

Table 4-1 Color Code Examples

Color				Resistance	Tolerance
1st Band	2nd Band	3rd Band	4th Band	(Ω)	(%)
Brown (1)	Black (0)	Black (× 1)	None	10	20
Yellow (4)	Violet (7)	Orange (× 1000)	Silver	47,000	10
Red (2)	Red (2)	Red (× 100)	Gold	2200	5
Brown (1)	Black (0)	Gold (× 0.1)	Gold	1	5
Orange (3)	White (9)	Silver (× 0.01)	Silver	0.39	10

Color	Digit	Multiplier	
Black	0	1	or 10^0
Brown	1	10	or 10^1
Red	2	100	or 10^2
Orange	3	1000	or 10^3
Yellow	4	10,000	or 10^4
Green	5	100,000	or 10^5
Blue	6	1,000,000	or 10^6
Violet	7	10,000,000	or 10^7
Gray	8	100,000,000	or 10^8
White	9	1,000,000,000	or 10^9
Gold		0.1	or 10^{-1}
Silver		0.01	or 10^{-2}

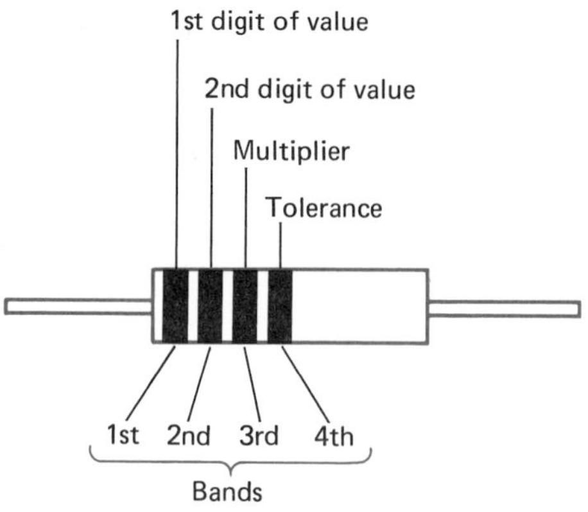

Tolerance Band	
Gold	± 5%
Silver	±10%
None	±20%

Fig. 4-22 Resistor color code.

Reliability

Sometimes a resistor will have a fifth color band. This fifth band indicates the reliability of the resistor. The reliability of a resistor refers to what percentage of the resistors fail within 1000 h.

Networks

Resistor Networks

Resistor networks are composed of many film resistors, such as cermet, on a single substrate. The resistors and substrate are usually housed in a dual-in-line package like those in Fig. 4-23. These resistor networks are used primarily with digital electronic circuits on printed-circuit boards.

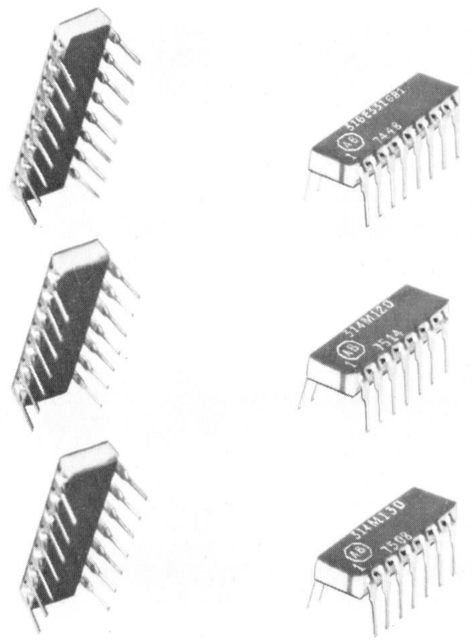

Fig. 4-23 Resistor networks for printed circuits. (Courtesy of Allen-Bradley.)

Special Resistors

A number of special resistors are manufactured. Three of them and their typical applications are listed below:

Fuse resistor

Fuse Resistor Fuse resistors have low resistance (less than 200 Ω) and limited current-carrying capacity. They are wirewound resistors. These resistors are used in circuits that require high initial currents but low operating currents. If the current remains high, the resistor burns out (opens) and protects the circuit.

Thermal Resistor

Thermal Resistor These resistors are commonly called *thermistors*. They are resistors which have very high temperature coefficients. The large changes in resistance for small changes in temperature make the thermistor useful for measuring temperature. The thermistor is also used to compensate for other temperature-induced changes in electronic circuits.

Voltage-sensitive resistor

Voltage-Sensitive or Voltage-Dependent Resistors (VDR) Also called *varistors*, these devices are used to protect circuits from very sudden increases in source voltage. If the sudden increase only lasts for a fraction of a second, the varistor will limit the voltage to a safe value.

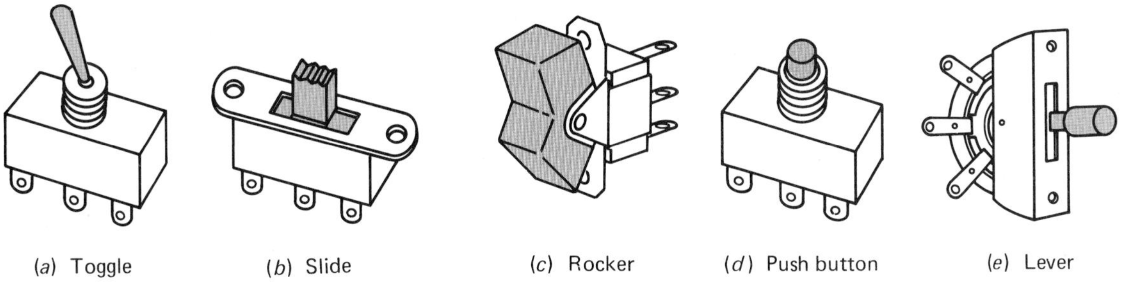

Fig. 4-24 Some common types of switches.

Switches

SPST

SPDT

Review Questions

33. List four categories of resistors.
34. What is a potentiometer?
35. What is a rheostat?
36. What factors determine the power rating of a resistor?
37. What are the resistance and tolerance of a resistor with the following color-band combinations?
    a. Brown, black, orange, silver
    b. Red, red, silver, gold
    c. Blue, gray, green
    d. Yellow, violet, black, silver
38. What is the minimum and maximum permissible resistance of a 220-Ω, 10% resistor?
39. List four common types of resistors.
40. What is the power dissipated by a 200-Ω resistor with 6 V across it?
41. What type of resistor is generally used for high-power applications?

4-10 SWITCHES

There are many types of switches available for electric circuits. They all perform the same basic function of opening or closing circuits. The type used in a given application is often a matter of style and/or convenience of operation. When the switching requirements are complex, the choice narrows to the rotary switch.

Types, Symbols, and Uses

Some of the more common types of small switches are shown in Fig. 4-24. These types are fairly simple switches which usually control only one or two circuit paths. Figure 4-25 shows the schematic symbols of these switches. The number of poles indicates the number of separate circuit connecting points; the number of throws indicates the number of different contact points each pole can connect to. A typical on-off switch has one circuit connection (single pole) and one contact point (single throw). The dashed line in Fig. 4-25(*c*) and (*d*) means that the two poles are mechanically connected but not electrically connected. Figure 4-26 shows the internal structure of a toggle switch similar to those used as wall switches in the home.

Controlling a lamp from two different locations is illustrated in Fig. 4-27. The circuit uses two SPDT switches. As shown, the lamp would be off. Changing the position of either switch would put the lamp on. The lamp can be turned on or off by either switch regardless of the condition of the other switch.

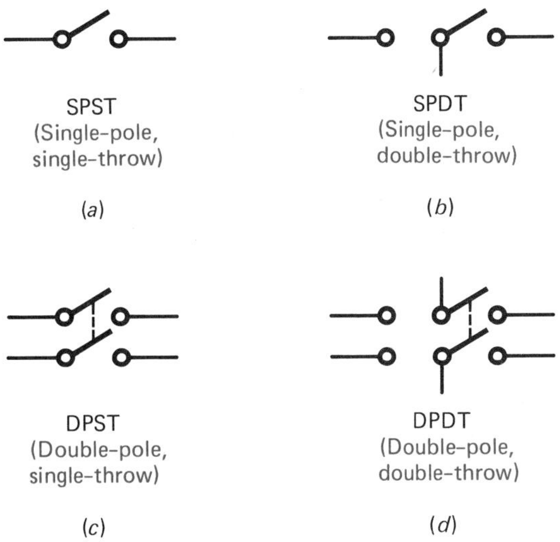

Fig. 4-25 Switch symbols and names.

Normally open

Normally closed

Spring-loaded

Rotary switches

Shorting

Nonshorting

Make-before-break

Arcing

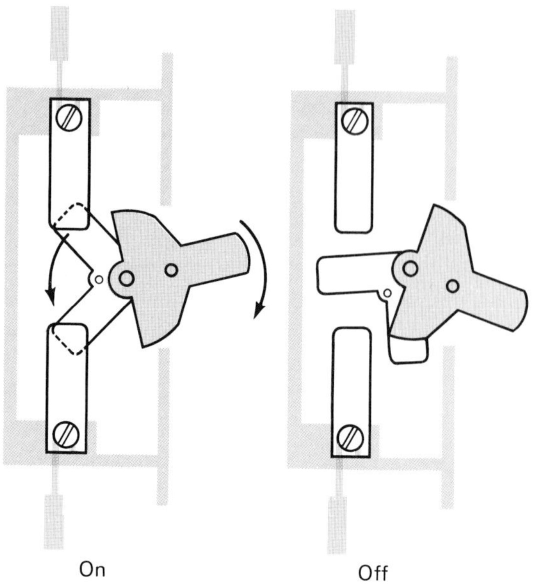

Fig. 4-26 Internal structure of a switch.

Some switches are constructed so that they always return to the same condition when released by the operator. These are called *momentary contact* switches. They can be either *normally open (NO)* or *normally closed (NC)*. Schematic symbols for such switches are shown in Fig. 4-28. This type of switch is sometimes called *spring-loaded* because a spring returns it to its normal position.

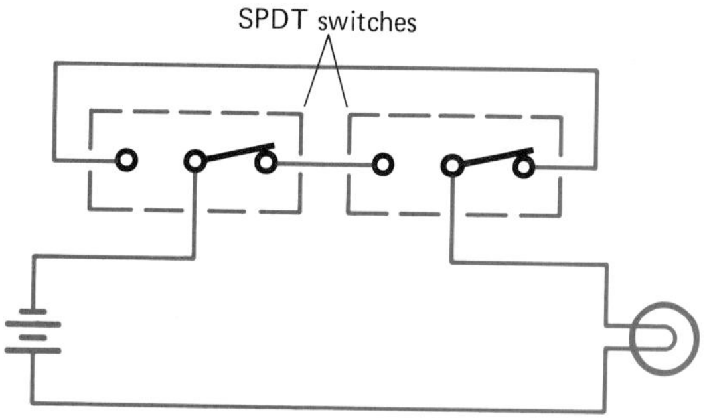

Fig. 4-27 Controlling a lamp from two different locations using SPDT switches. The lamp can be turned off or on with either switch.

Rotary switches are illustrated in Fig. 4-29. They range in complexity from a simple on-off switch to a multideck switch. Figure 4-30 shows the symbols and terminology used for rotary switches.

Rotary switches come in either *shorting* or *nonshorting* configurations (Fig. 4-31). The shorting configuration is sometimes referred to as a *make-before-break switch*. This means the pole makes contact with the new position before it breaks contact with the old position. The wiper in Fig. 4-31 will contact position 2 before it loses contact with position 1.

An example of the use of a rotary switch is shown in Fig. 4-32. This switch would be a nonshorting switch. In the position shown, the amplifier receives its input from the record changer.

(*a*) Normally closed (*b*) Normally open

Fig. 4-28 Normally open—normally closed symbols.

Ratings

Whenever an electric circuit is broken, an arc (ionization of the air) occurs. The greater either the current or the voltage, the larger the arc will be. Arcing causes switch contacts to erode away. Therefore, switches have both

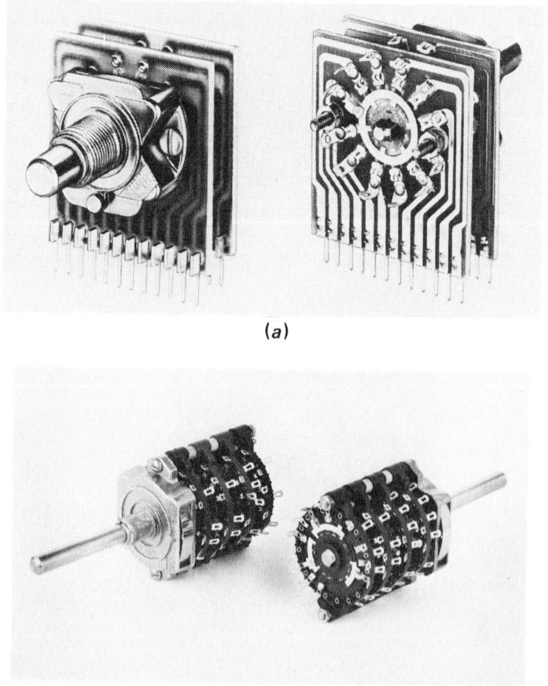

(*a*)

(*b*)

Fig. 4-29 Rotary switches: (*a*) Switch contacts are on a printed circuit. (*b*) Conventional contact arrangement. (Courtesy of Oak Industries, Inc.)

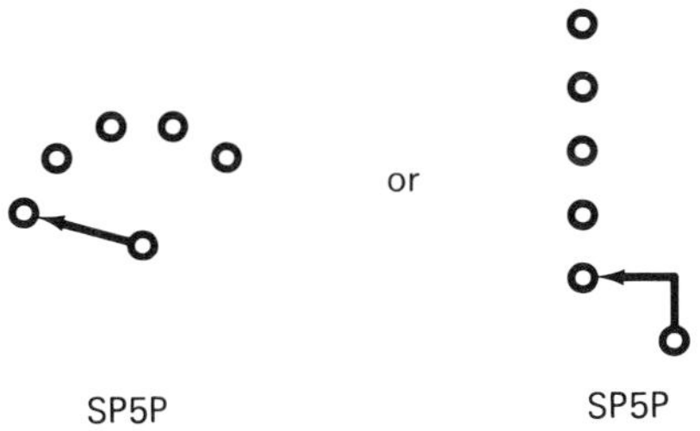

(a) Single-pole, five-position

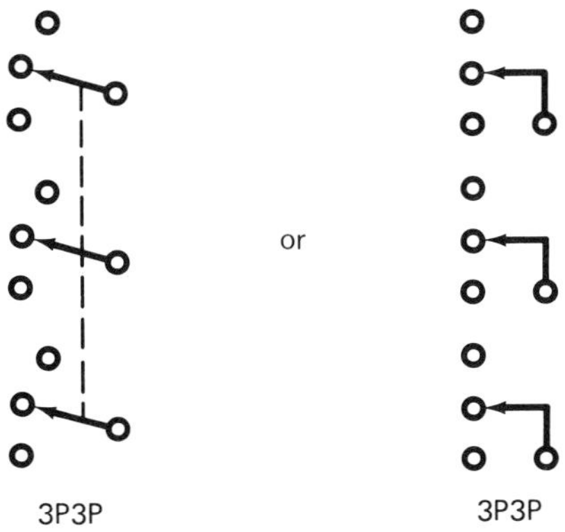

(b) Three-pole, three-position

Fig. 4-30 Rotary switch symbols and names.

a voltage and a current rating. Exceeding the manufacturer's ratings shortens the life of the switch. It can also be dangerous to the operator if the ratings are greatly exceeded. An arc can char and break down the insulation in the switch and thus put the operator in contact with the circuit.

Switches are usually given multiple current and voltage ratings. For example, a switch may be rated for 3 A at 250 V ac, 6 A at 125 V ac, and 1 A at 120 V dc. In general, a switch can handle higher alternating currents than direct currents.

Review Questions

42. What is indicated by dashed (broken) lines in switch symbols?
43. List four common types of switches.
44. What is meant by NC when used in the specifications for a momentary contact push-button switch?
45. What is a shorting rotary switch?
46. Why do switches have a voltage rating?

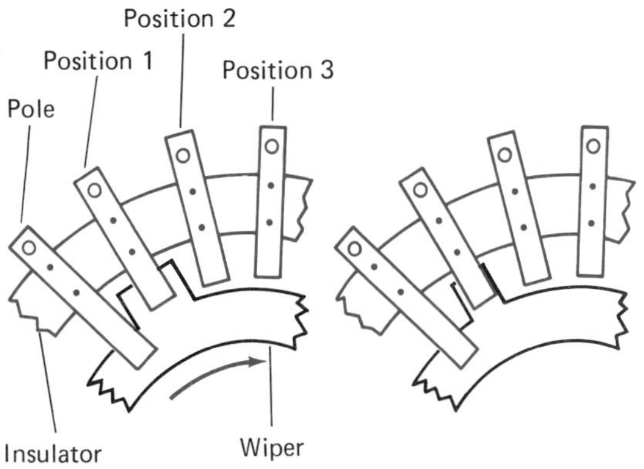

(a) Shorting (b) Nonshorting

Fig. 4-31 Shorting and nonshorting rotary switches.

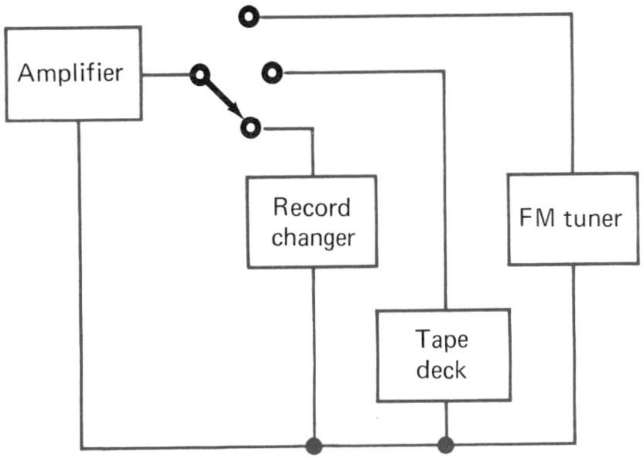

Fig. 4-32 Using a rotary switch to select various input devices.

4-11 WIRES AND CABLES

Wire

Solid conductor

Stranded conductor

Cable

A *cable* is a multiple conductor. The conductors of the cable are insulated from each other and are held together as a single unit. A *wire* is a single conductor. It may or may not be insulated. Typical cables are shown in Fig. 4-33.

A wire may be solid or stranded. A *solid conductor* is made from a single piece of low-resistance material such as copper, aluminum, or silver. A *stranded conductor* is made from a number of wires twisted or braided together to form a single conductor. Stranded wire is more flexible than solid wire.

Smaller-sized conductors used in electric wire and cable are often coated with a thin layer of solder. Bare copper conductors oxidize quite easily. The oxide must be removed before conductors can be joined by soldering. Thus, tinned wires make soldering easier.

Shielding

Coaxial cable

Flat ribbon

Gage number

American Wire Gage

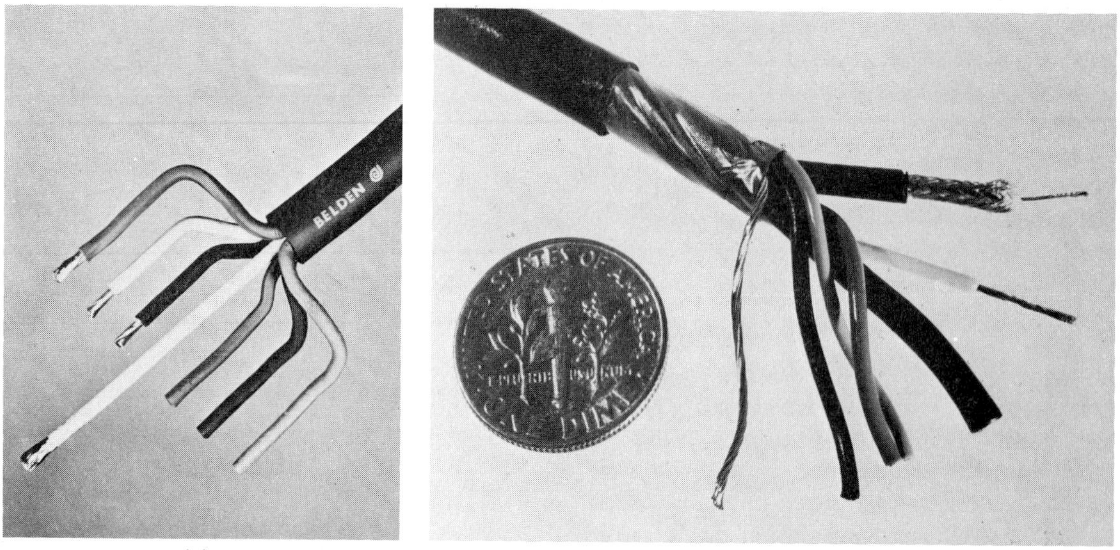

(a) (b)

Fig. 4-33 Electric cables: (*a*) unshielded. (*b*) Shielded. (Courtesy of Belden Corp.)

Electric Cables

Cables may be either shielded or unshielded (Fig. 4-33). The shielding helps to isolate the conductors from electromagnetic fields in the vicinity of the cable. Individual conductors of the cable may be shielded or the whole cable may be shielded.

A wide variety of cables are manufactured for general-purpose and audio use. Shielded cables are used for such things as connecting a record changer or tape deck to an amplifier. Unshielded cables are used for such things as extension cords, telephone cords, and doorbell circuits.

Coaxial cable (Fig. 4-34) is a special-purpose, shielded cable used to connect antennas to receivers or transmitters. These cables are used for amateur radio, citizen-band radio, and some television antenna systems.

Much of the cable used with digital electronic systems is produced in a flat ribbon form. Figure 4-35 shows a typical example. The figure also illustrates how this cable can be connected to printed-circuit boards.

Power cables carrying very high currents often include a hollow tube through which oil is pumped to cool the cables.

Conductor (Wire) Specifications

The material used in making conductors for electric cables and wires is most often copper. However, aluminum is also used. The size of a round conductor (wire) is specified by a *gage number*. Each gage number represents a different diameter of wire. The diameter of wire is specified in mils. A mil is equal to 0.001 in. Thus, each gage number represents a certain diameter in mils. Electric wire is measured with the *American Wire Gage*, which is the same as the Brown and Sharpe gage used by machinists.

The smaller the gage number, the larger the diameter of the wire. A typical wire gage [Fig. 4-36(*a*)] gives both the gage number and the diameter. The diameter is often printed on the reverse side of the gage. Figure 4-36(*b*) illustrates how to use the gage. The width of the slot leading to the hole in the gage indicates the gage size. The hole is larger than the wire

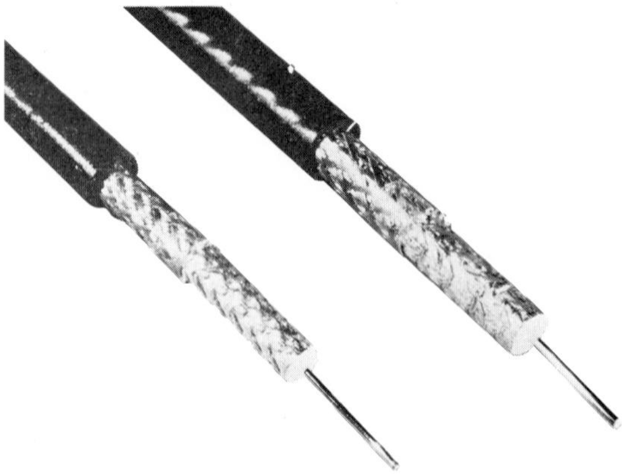

Fig. 4-34 Coaxial cable. (Courtesy of Beldon Corp.)

Fig. 4-35 Flat cable. (Courtesy of Brand-Rex.)

size and merely allows the wire to move freely through the slot. The slot through which the solid, bare, copper wire passes smoothly but snugly represents the wire size. Stranded wire cannot be measured with the wire gage.

Circular mil

The cross-sectional area of a conductor is specified in circular mils (cmil or CM). One circular mil is the area of a circle with a one mil diameter. The cross-sectional area of a conductor is equal to the diameter (in mils) squared. A conductor that is 30 mils in diameter would have a cross-sectional area of 900 cmil ($30 \times 30 = 900$).

Notice from Fig. 4-37 that a circular mil is not equal to a square mil. This could make the calculation of the resistance of a wire a difficult job. However, wire tables which list the resistance per 1000 ft of wire for various gage numbers are readily available. Such a table is provided in Appendix 3. This table also gives the diameter and area of the different gage wires. Notice from the wire table that an increase of three gage numbers results in the area being reduced by about half. Of course, the resistance doubles when the cross-sectional area is halved. This leads to a simple method for remembering wire areas and resistances. Number 10 wire is about 10,000 cmil and 1 Ω per 1000 ft. Thus, no. 13 wire (an uncommon size) is about 5000 cmil and 2 Ω per 1000 ft. Three sizes smaller, no. 16 wire (a common size for extension cords) is about 2500 cmil and 4 Ω per 1000 ft. Check these values against those in the table in Appendix 3.

The amount of current a copper wire can safely carry is dependent upon the cross-sectional area of the wire and how much heat the wire can dissipate before the insulation is damaged and the metal itself is affected. This, in turn, depends on where and how the wire is run. When air is freely circulating around a conductor, it can carry more current than when air circulation is restricted. For example, a no. 12 copper conductor used to wire a house can safely carry 20 A under certain conditions. The same size wire used in a transformer may carry only 4 A. In the house, the wire has about 325 cmil/A, while in the transformer it has over 1500 cmil/A. In the house,

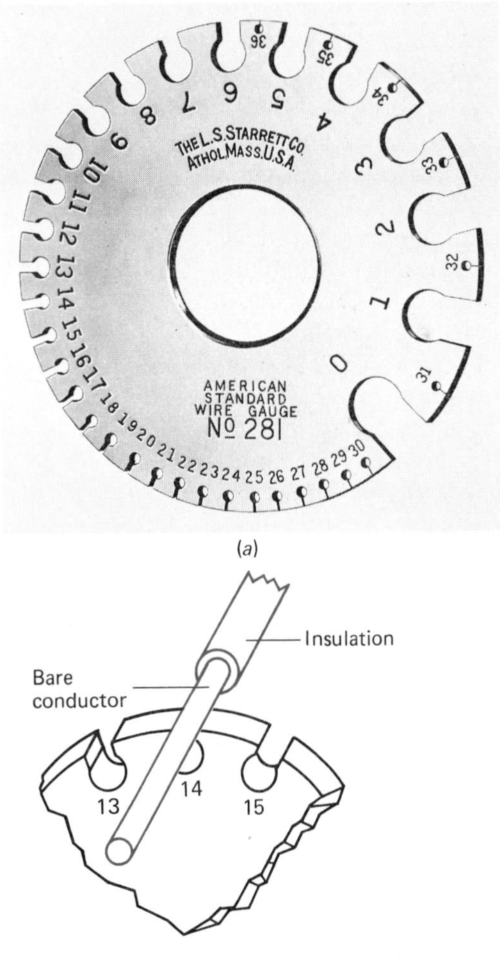

(*a*)

(*b*) Using a gage

Fig. 4-36 American wire gage: (*a*) gage. (Courtesy of L. S. Starret Co.) (*b*) Using a gage.

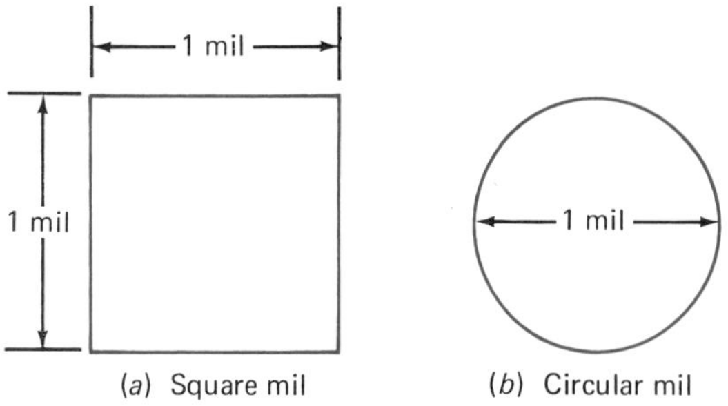

Fig. 4-37 Circular mil and square mil. Notice that a circular mil is smaller than a square mil.

100 ft of the wire is strung throughout the walls and ceilings. There is a lot of air for heat (created by the current in the wire) to transfer into. In the transformer, 100 ft of the wire is coiled into a very small volume. Therefore, less heat can be transferred away from the wire in the transformer, and so the wire must carry less current.

Insulation Specifications

The insulation used on electric wires and cables has a number of important ratings. Each insulating material has a maximum-temperature rating. This is the maximum continuous temperature to which the insulation should be exposed. Typical maximum temperatures range from about 75°C for certain types of rubber to over 200°C for asbestos and mineral insulation.

Maximum-temperature

Voltage ratings for cables and wires depend on both the type of insulation and the thickness of the insulation. The voltage rating of an insulated wire is the maximum voltage which the insulation can continuously withstand. Voltage ratings vary from several hundred volts for appliance cords to several hundred thousand volts for high-voltage cables.

Insulating materials are also rated as to their relative resistance to damage by various materials and environments. They are rated on such things as resistance to water, acid, abrasion, flame, and weather (sun and extreme cold, for example).

Insulating materials used on cables and wires include the following: neoprene, rubber, nylon, vinyl, polyethylene, asbestos, polypropylene, polyurethane, varnished cambric, paper, silicone rubber, and many more.

Electric Wires

Besides building wires, which come in sizes from 18 AWG to as large as 2,000,000 cmil, electric wire is manufactured in an almost endless variety of sizes, styles, and ratings. Some of the common names used in identifying broad groups of electric wires are listed and defined below.

Busbar wire

Test-prod wire

Hookup or lead wire

Magnet wire

Litz wire

1. *Busbar wire*—An uninsulated, usually tinned, solid copper wire. Available in a wide range of sizes.
2. *Test-prod wire*—Primarily used to make test leads. High-voltage insulation (5000 to 10,000 V). Very fine strands for good flexibility.
3. *Hookup or lead wire*—Insulated wire used to connect electric components and devices together. May be either solid or stranded. Usually tinned. Available in a wide variety of wire sizes and types of insulation.
4. *Magnet wire*—Used in magnetic devices such as motors, transformers, speakers, and electromagnets. Always insulated, but the insulation is very thin and often has a copper color. Available in a wide variety of wire sizes and insulating materials.
5. *Litz wire*—A small-size, finely stranded, low-resistance, insulated wire. Used for making certain coils and transformers for electronic circuits.

Review Questions

47. What gage is used to measure wire size?
48. What is a circular mil?
49. List three common insulating materials used on wires and cables.
50. Where is magnet wire used?
51. Will a no. 16 or a no. 19 wire carry more current?

4-12 FUSES AND CIRCUIT BREAKERS

Protection of a circuit from excessive current is provided by a fuse or circuit breaker. Protection is needed because an excessively high

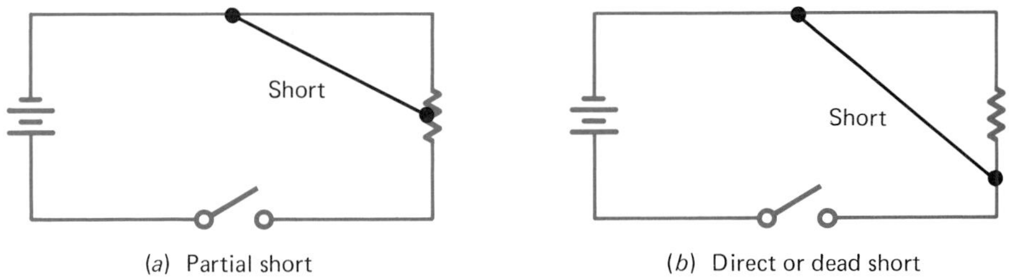

Fig. 4-38 Shorted circuits.

Shorted circuit

Dead short

Open

Fuse

current can damage either the load or the source (or both). Higher than normal currents result from a defective circuit which has lower than normal resistance.

Shorts and Opens

A circuit with *lower than normal resistance* is called a *shorted circuit*. A partially shorted circuit [Fig. 4-38(*a*)] is one in which only part of the load is shorted out (bypassed). Figure 4-38(*b*) shows a circuit with all the load shorted out. This circuit is said to have a *dead short* or a *direct short*. The resistors in Fig. 4-38 could just as well be heating elements in a toaster or coffee pot. The battery could be a 120-V ac outlet.

The circuits in Fig. 4-38 could be protected by a fuse or circuit breaker, as shown in Fig. 4-39. Note the symbols for fuses and circuit breakers. If a short occurs in one of the circuits in Fig. 4-39, the fuse or circuit breaker would open. In electricity, *open* means that the path for current has been interrupted or broken. When a fuse opens, we say, "The fuse has blown." When a circuit breaker opens, we say, "The circuit breaker has tripped."

Fuses

A fuse will open when current through it exceeds its current rating for a period of time. The fuse opens when its conducting element becomes hot enough to melt (Fig. 4-40). The element heats up because it has resistance. A 1-A fuse has approximately 0.13 Ω of resistance. Since $P = I^2R$, more current means more power. More power means more joules (of heat energy) per second converted by the fuse element.

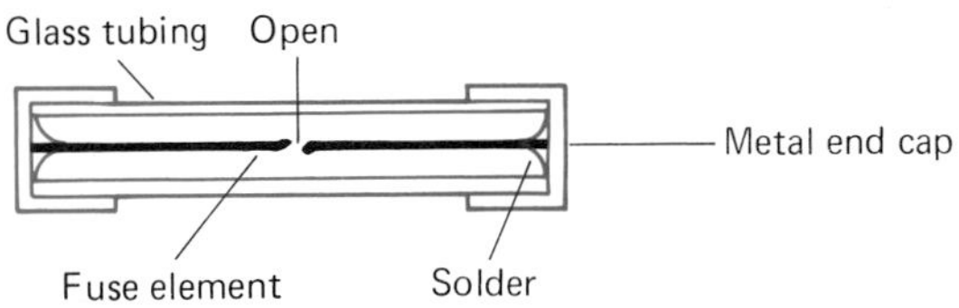

Fig. 4-40 Open fuse.

A variety of fuses and fuse holders are shown in Fig. 4-41. The fuse in Fig. 4-41(*a*) is typical of those used in automobiles, airplanes, and electronic equipment. Figure 4-41(*b*) shows a subminiature fuse and holder of the type used on printed-circuit boards. The two prongs of the holder are to be inserted and soldered in the holes directly below them. Other styles of subminiature fuses and holders are illustrated in Fig. 4-41(*c*). The holder on the right is an indicating holder. When the fuse blows (opens), the lamp glows to indicate the fuse has blown. The fuse pictured in Fig. 4-41(*d*) is a pigtail type. Pigtail

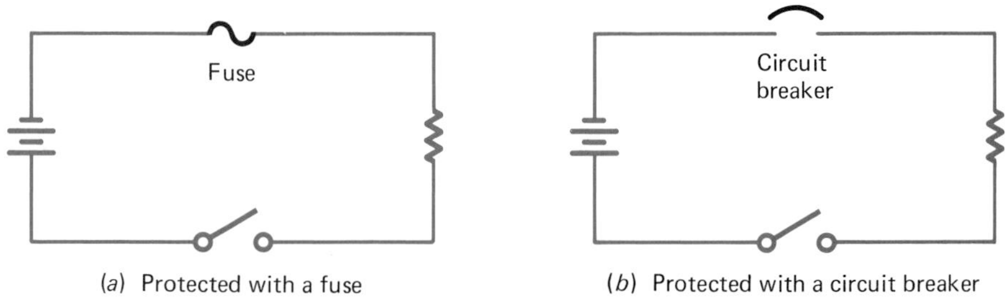

Fig. 4-39 Circuits protected against excessive current.

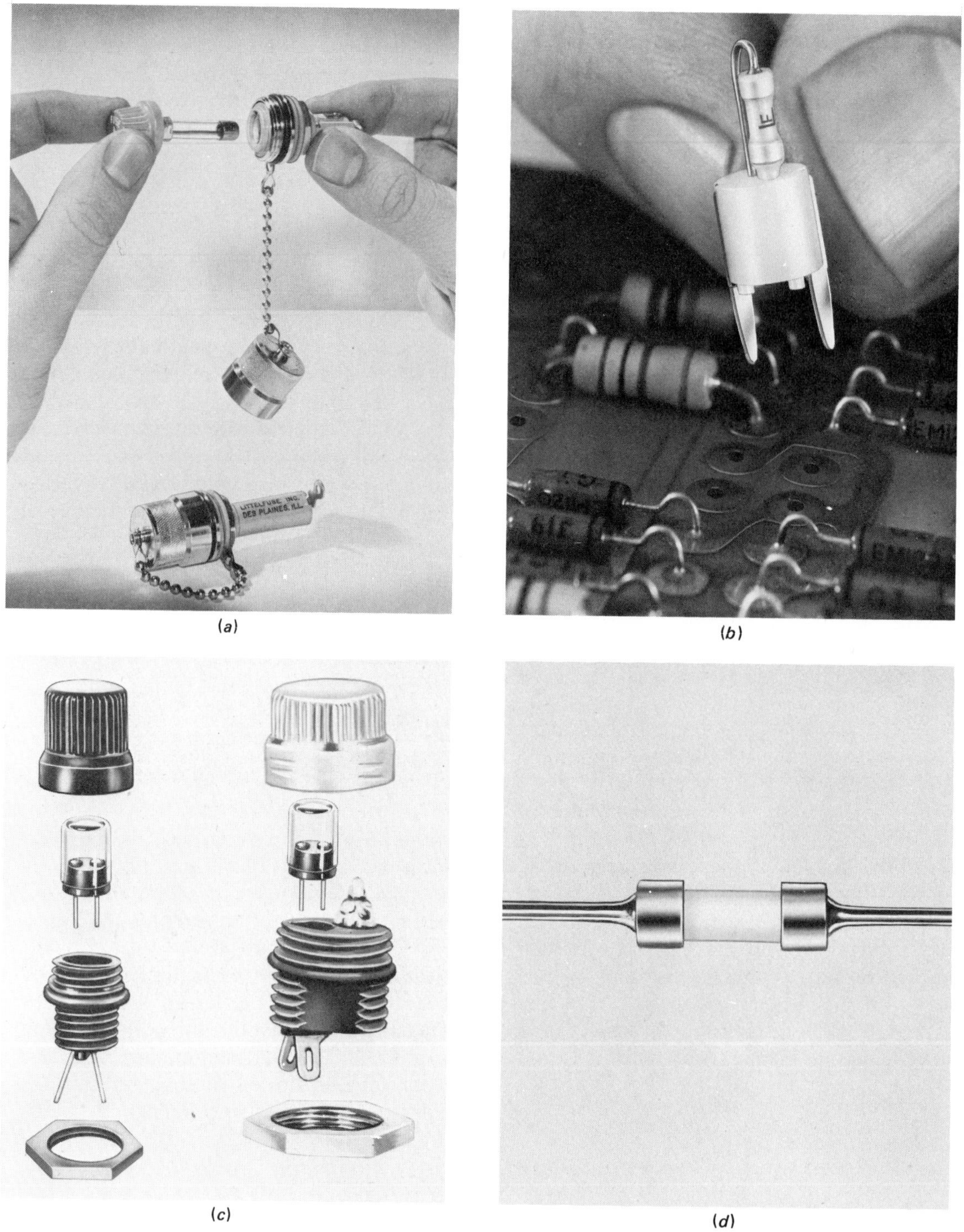

Fig. 4-41 Fuses and fuse holders: (*a*) typical small, glass fuse and holder. (*b*) Circuit-board holder. (*c*) Subminiature fuse and holder. (*d*) Pigtail fuse. (Courtesy of Littlefuse, Inc.)

fuses are directly soldered into a circuit. They do not need a fuse holder.

Fuses have three important ratings or characteristics that need to be considered when buying or specifying fuses. These ratings apply to all fuses regardless of their size, shape, or style. These three characteristics are (1) a current rating, (2) a voltage rating, and (3) a blowing (fusing) characteristic.

Current rating

Current ratings of fuses have already been

mentioned. The current rating of a fuse is given for specific conditions of air circulation and temperature. These conditions are rarely obtained in typical applications of fuses. Therefore, fuses are usually operated well below their rated current. This practice avoids unnecessary blowing of fuses while still providing short-circuit protection.

Voltage rating

Voltage ratings for fuses are necessary for the same reasons that switches have voltage ratings. When a fuse opens, arcing and burning of the element occur. If the current and voltage are high enough, the arcing/burning can continue until fuse caps and holder are melted. Thus, fuses need voltage ratings. Voltage ratings of fuses are based on a power source capable of providing 10,000 A of current when dead shorted. A 250-V fuse will protect any 250-V power source that can provide no more than 10,000 A. Of course, the current rating of the fuse must be less than the current capabilities of the power source. Many power sources (small batteries and transformers, for example) can provide only a few amperes of current when short-circuited. With limited-current power sources, the voltage ratings of fuses are often exceeded. It is fairly common for electronic circuits with more than 400 V to be fused with 250-V fuses.

Blowing characteristic

The blowing characteristic of a fuse indicates how rapidly the fuse blows (opens) when subjected to specified overloads. There are three general categories of blowing characteristics of fuses. These are *fast-blow, medium-blow*, and *slow-blow*. These categories are sometimes referred to as *short-time lag*, *medium-time lag*, and *long-time lag*, respectively. All three categories respond rapidly (about 1 ms) to extreme overloads (more than 10 times the rated current). All three categories respond about the same to very small overloads. At 1.35 times (135 percent) the rated current, they all take a minute or two to open. In between these two extremes, the three categories differ dramatically. For example, at 5 times rated current, a slow-blow may take over 1 s to blow, while a fast-blow will usually open in less than 1 ms. Under the same conditions, a medium-blow would take about 10 ms.

Fast-blow fuse

Instrument fuse

Fast-blow fuses, sometimes called instrument fuses, are used to protect very sensitive devices such as electric meters. The physical appearance of a fast fuse is the same as that of the medium fuse.

Medium-blow fuse

Medium-blow fuses [Fig. 4-42(*a*)] are general-purpose fuses. They are used when the initial (turn on) current of a device is about the same as the normal operating current.

Slow-blow fuse

Slow-blow fuses are used whenever short-duration surge currents are expected. An electric motor is a good example of a device that requires much higher current to start than to continue operating. Its starting current may be 5 or 6 times the normal operating current. Therefore, a motor is often fused with a slow-blow fuse. A typical slow-blow fuse is shown in Fig. 4-42(*b*). When subjected to an

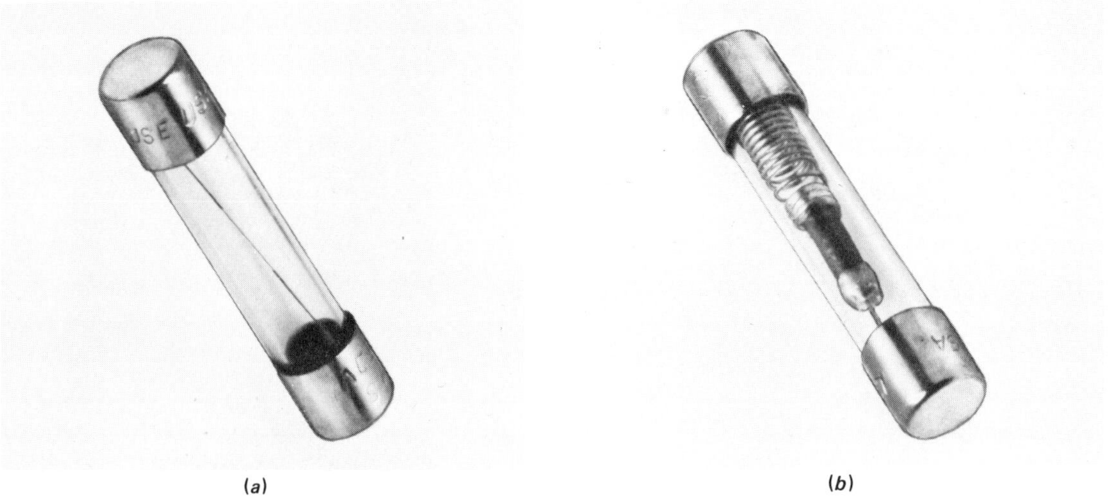

(*a*) (*b*)

Fig. 4-42 Types of fuses: (*a*) medium-blow fuse. (*b*) Slow-blow fuse. (Courtesy of Littlefuse, Inc.)

extreme overload (short circuit), the thin part of the element burns open. When the overload is less severe and of long duration, the solder (holding the spring and element together) melts.

Always replace fuses with the style and ratings specified by the manufacturer of the equipment. Do not ignore the voltage ratings of fuses. If a medium or fast fuse is replaced by a slow-blow fuse, the equipment may be damaged before the fuse can respond. When ordering fuses, either specify current, voltage, blowing characteristic, physical dimensions, and style, or the current and the manufacturer's type number.

Fuses can be tested with an ohmmeter. If the fuse is blown, the ohmmeter will indicate an infinite resistance. The meter will show a very low resistance if the fuse is good. Fuses rated at less than ¼ A should not be checked on the $R \times 1$ range. The ohmmeter may provide enough current to blow the fuse. A higher range of the ohmmeter will provide less current and still indicate if the fuse has continuity, that is, the fuse element still provides a continuous path. When checking a fuse with an ohmmeter, always remove the fuse from the circuit.

Circuit Breakers

Circuit breakers have one big advantage over fuses. When they open (trip), they can be reset. Nothing needs to be replaced. Once the overload that caused the breaker to trip has been corrected, the breaker can be reset and used again. There are two types of trip mechanisms: thermal and magnetic.

Thermal circuit breaker

Thermal circuit breakers are the most common type for such things as small motors, household circuits, and battery chargers. Thermal circuit breakers can have either manual or automatic reset. The automatic reset will reset after the breaker has cooled. Cooling takes a few minutes. Automatic reset is used in applications where overloads may be self-correcting. Thermal breakers use bimetallic strips or disks as their current-sensing element. Therefore, they have long time-lag characteristics. The operating principle of a bimetallic strip is illustrated in Fig. 4-14 and described in Sec. 4-8. For manual reset breakers, a mechanical mechanism locks the bimetallic strip in the open position once the breaker is tripped.

Two styles of manual-reset, thermal breakers are shown in Fig. 4-43. The one on the right [Fig. 4-43(*b*)] is a combination switch/breaker. The breaker is reset by moving the handle all the way to the off position and then to the on position. The breaker in Fig. 4-43(*a*) is a typical push-button breaker. When the breaker is tripped, the button extends out so that the white band shows. Pushing the button in resets the breaker.

Magnetic circuit breaker

The external appearance of a magnetic circuit breaker is like that of the thermal breaker. However, its internal structure, principle of operation, and characteristics are quite different. Magnetic breakers can be made with time lags ranging from a few milliseconds to many seconds. One technique used to control the

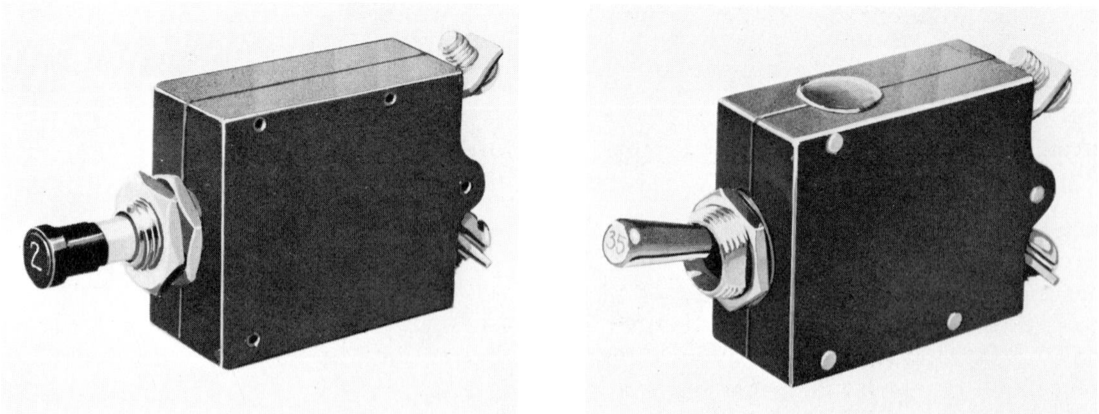

(*a*) (*b*)

Fig. 4-43 Thermal circuit breakers: (*a*) push-button reset. (*b*) Switch/breaker. (Courtesy of AMF, Inc.)

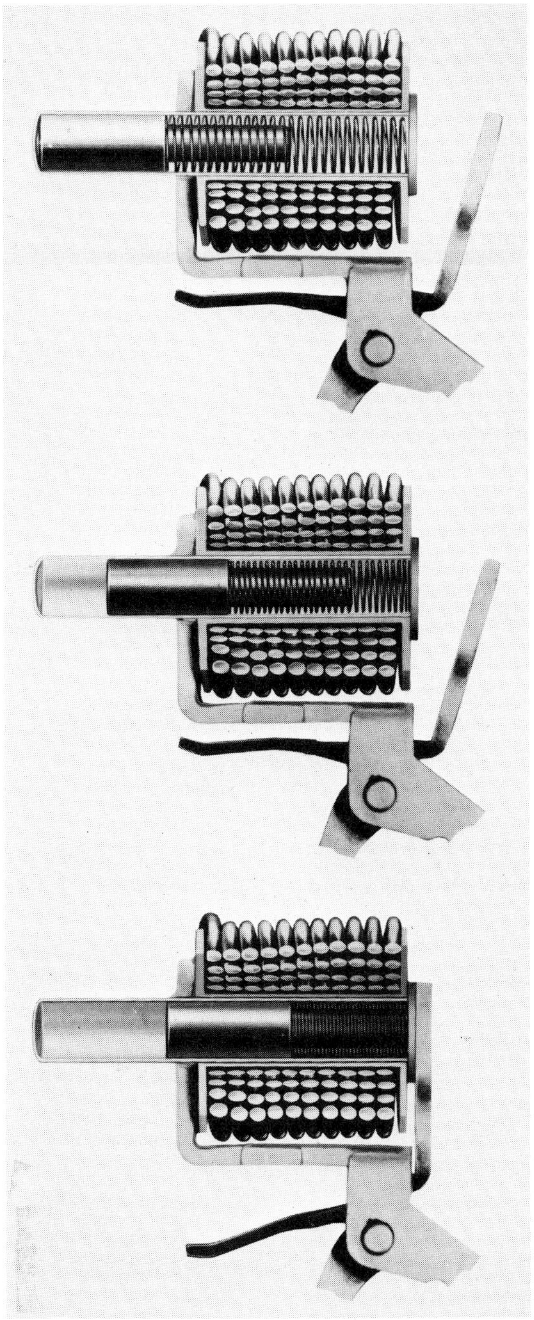

Fig. 4-44 Time-lag mechanism in a magnetic circuit breaker. (Courtesy of Heinemann Electric Co.)

Fig. 4-45 Magnetic circuit breaker. (Courtesy of Heinemann Electric Co.)

time lag is illustrated in Fig. 4-44. The armature in Fig. 4-44 operates a mechanical mechanism which opens the contacts of the breaker when it is tripped. This mechanism is shown in Fig. 4-45, which is a cutaway view of a magnetic circuit breaker. The breaker is shown in the tripped position. Notice the four shields between the breaker contacts (top center of Fig. 4-45). These shields help to reduce the arc or spark that may occur when the contacts are opened. The three small terminals located between the two breaker terminals connect to a single-pole, double-throw switch. The switch is operated by the breaker mechanism, but it is electrically isolated from the circuit protected by the breaker. This switch can be used to control a separate circuit which remotely indicates whether or not the breaker is tripped. The large circuit breakers that protect high-current loads or conductors often have both thermal and magnetic mechanisms. The thermal trip opens when the overload is small and a quick break in the circuit is not needed. A large inrush of current, however, would cause the breaker to open quickly due to the magnetic trip mechanism.

Review Questions

52. Define the following terms:
    a. Open circuit
    b. Short circuit
    c. Slow-blow
53. What is an indicating holder?
54. List three ratings to be considered when replacing fuses.
55. What is the main advantage of a circuit breaker over a fuse?
56. List two types of circuit breakers.

Summary

1. Cells are energy or power sources.
2. Cells provide dc voltage.
3. A cell contains an electrolyte and two electrodes.
4. A battery is made from two or more cells.
5. Primary cells are not rechargeable.
6. Secondary cells are rechargeable.
7. Dry cells may use either a paste or a liquid electrolyte.
8. Gases are produced when cells are charged or discharged.
9. The energy storage capacity of a cell or battery is expressed in ampere-hours.
10. The energy available from a battery is dependent on temperature, rate of discharge, and final voltage.
11. Internal resistance causes a cell's voltage to decrease as the current increases.
12. Polarization refers to the buildup of gas ions around an electrode.
13. Common secondary cells are lead-acid, nickel-cadmium, and rechargeable alkaline.
14. Common primary cells are carbon-zinc, alkaline, mercury, and silver oxide.
15. When a lead-acid cell is discharging, sulfuric acid is being converted to water.
16. Specific gravity is measured with a hydrometer.
17. Safety glasses should be worn when working with lead-acid cells and batteries.
18. Electrolyte on the skin or clothing should be immediately removed by washing with soap and water.
19. Carbon-zinc cells are relatively inexpensive.
20. Alkaline cells store more energy than carbon-zinc cells. They are more efficient at high current drains.
21. Both mercury and silver oxide cells have nearly constant output voltage.
22. Incandescent miniature lamps have tungsten filaments.
23. Miniature lamps have both a current and a voltage rating.
24. Bimetallic strips bend because the two metals have different coefficients of expansion.
25. The resistance of a lamp when hot is greater than its resistance when cold.
26. Only one electrode glows when a neon lamp is operated on direct current.
27. A neon-lamp circuit must have a resistor to limit the current through the lamp.
28. A potentiometer is a variable resistor.
29. A potentiometer can also be used as a rheostat.
30. The power rating of a resistor is independent of its resistance.
31. Common types of resistors are carbon composition, cermet, wirewound, deposited film, and conductive plastic.
32. Resistance values and tolerance can be indicated with a color code on the body of a resistor.
33. Thermistors have high temperature coefficients.
34. Common types of small switches are rotary, toggle, slide, rocker, and push-button.
35. Spring-loaded momentary-contact switches can be either normally open or normally closed.
36. Rotary switches can handle complex switching involving many circuits.
37. Switches have both a current and a voltage rating.
38. Cables are multiple conductors.
39. Wires are single conductors.
40. Conductors may be stranded or solid.
41. Shielded wires and cables are used to help isolate a conductor's electromagnetic fields.
42. Coaxial cables are used to connect antennas to receivers or transmitters.
43. The American Wire Gage is the standard used to specify the size of a conductor.
44. The cross-sectional area of a conductor is specified in circular mils.
45. When the gage of a conductor is decreased by three gage numbers, its cross-sectional area doubles. Its resistance is half as much.
46. Insulation used on conductors has both a temperature and a voltage rating.
47. Circuit breaker mechanisms work on either a magnetic or a thermal principle.
48. Fuses are thermally operated devices.
49. A blown fuse results in an open circuit.
50. Instrument (fast-blow) fuses are used to protect electric meters.
51. Fuses and breakers have both current and voltage ratings.
52. Dry cell characteristics are given in Table 4-2.

Table 4-2 Dry-Cell Characteristics

Type of Cell	Rated Voltage	Voltage during Discharge	Cycle Life	Initial Cost	Internal Resistance	Energy-to-Weight Ratio
Carbon-zinc	1.5	Decreases	—	Low	Med	Med
Alkaline	1.5	Decreases	—	Med	Low	Med
Mercury	1.35	Almost constant	—	High	Low	High
Silver oxide	1.5	Almost constant	—	High	Low	High
Rechargeable alkaline	1.5	Decreases	Low	Med	Very low	Low
Nickel-cadmium	1.25	Decreases slightly	High	High	Very low	Low

Chapter Review Questions

Here is a chance to see what you have learned. For questions 4-1 to 4-21, determine whether the statement is true or false.

4-1. The ampere-hour rating of a cell indicates the amount of energy it stores.

4-2. A battery contains at least two cells.

4-3. All dry cells are primary cells.

4-4. Polarization of a cell is the process by which the negative and positive terminals are created.

4-5. Specific gravity is a number which represents the ratio of the weight of a substance compared with the weight of air.

4-6. When a lead-acid cell is being charged, water is being converted to sulfuric acid.

4-7. Charging a lead-acid cell decreases its specific gravity.

4-8. The resistance of a lamp is greatest when the lamp is cold.

4-9. The filaments of miniature lamps are usually made of carbon.

4-10. Both electrodes of a neon lamp glow when alternating current is used to power the lamp.

4-11. A neon-lamp circuit uses a resistor only if the voltage in the circuit is more than 20 percent above the lamp's rated voltage.

4-12. A bimetallic strip is the control element that causes a miniature lamp to flash.

4-13. Light-emitting diodes operate equally well on alternating current as on direct current.

4-14. When used to describe a switch, NC means no connection.

4-15. Power resistors are usually made by the deposited-film method.

4-16. The greater the resistance of a resistor, the larger its physical size will be.

4-17. The dashed lines between poles of a switch symbol indicate that the poles are mechanically connected.

4-18. With a shorting-type switch, all positions of the switch are momentarily shorted together when the switch is rotated.

4-19. Switches have both a current and a voltage rating.

4-20. A circular mil is equal to a square mil.

4-21. The larger the gage number, the larger the diameter of the wire.

For questions 4-22 to 4-26, choose the letter that best completes the sentence.

4-22. Cells that are designed to be recharged are classified as
(A) Wet cells (B) Dry cells (C) Primary cells (D) Secondary cells

4-23. An example of a rechargeable cell is the
(A) Mercury cell (B) Silver oxide cell (C) Nickel-cadmium cell (D) Carbon-zinc cell

4-24. The cell that provides the least constant output voltage is the
(A) Carbon-zinc cell (B) Nickel-cadmium cell (C) Silver oxide cell (D) Mercury cell

4-25. A three-terminal, variable resistor is called a
(A) Varistor (B) Thermostat (C) Rotary resistor (D) Potentiometer

4-26. A five-pole, four-position switch would most likely be a
(A) Toggle switch (B) Slide switch (C) Rotary switch (D) On-off switch

Answer the following questions.

4-27. A resistor is color coded brown, black, red, and gold. What is its nominal resistance? What is its minimum possible resistance?

4-28. What are the two types of circuit breakers?

4-29. What are the three electrical ratings that are used in specifying fuses?

4-30. What is a short circuit?

4-31. What type of fuse would you use to protect an electric motor?

Answers to Review Questions

1. A cell is a device that converts chemical energy to electric energy. A cell is also known as an energy or power source.
2. A battery is an electric device composed of two or more cells.
3. A cell is composed of two electrodes and an electrolyte.
4. *a*. Primary cell—a cell that is not intended to be rechargeable
   b. Secondary cell—a cell that is rechargeable
   c. Dry cell—a cell that can be operated in any position. It is sealed.
   d. Wet cell—a vented cell that must be operated in an upright position
   e. Ampere-hour—a rating of a cell which indicates the amount of energy it stores

5. Rechargeable alkaline, lead-acid, and nickel-cadmium
6. Alkaline, carbon-zinc, mercury, and silver oxide
7. Its internal resistance
8. Polarization is the accumulation of gas ions around an electrode of a cell.
9. Lead and lead peroxide are changed to lead sulfate while sulfuric acid is changed to water. This process leaves one plate deficient in electrons and the other plate with an excess of electrons.
10. The battery is recharged by using the battery as a load rather than a source. Electrons are forced into the negative terminal and out of the positive terminal. The chemical reaction is also reversed.
11. Charge at, or below, the rate recommended by the manufacturer, and do not overcharge the battery.
12. Specific gravity is a number which represents the ratio of the weight of a substance compared with the weight of water.
13. Specific gravity is measured with a hydrometer.
14. No. It may freeze in cold weather. Also, it is destructive to the plates.
15. It is heavier than water.
16. Wear safety glasses, avoid spilling the electrolyte, and charge in a spark-free ventilated area.
17. No
18. From several hundred to several thousand cycles (depending on operating conditions)
19. *a*. The nickel-cadmium has a more constant output voltage.
    b. The nickel-cadmium provides 3 to 30 times as many cycles.
    c. The alkaline is much cheaper.
20. Initial cost is considerably less.
21. *a*. The alkaline cell is much better for providing high current loads.
    b. Alkaline cells maintain higher voltage under very heavy current loads.
    c. Alkaline cells store more energy for a given weight.
22. More uniform output voltage during discharge, higher energy-to-weight and energy-to-volume ratios, and ampere-hour capacity is less dependent on discharge rates.
23. Tungsten
24. Incandescent and neon glow
25. A bimetallic strip which acts like a switch as it heats and cools
26. The lamp draws 0.25 A.
27. Hot
28. Positive temperature coefficient
29. Both electrodes glow with ac voltage
30. The resistor limits current through the lamp.
31. Do not bend or solder leads within 3 mm of the bulb.
32. The lamp may explode.
33. Variable, adjustable, tapped, and fixed
34. A potentiometer is a three-terminal, variable resistor often used for volume and tone controls.
35. A rheostat is a variable resistor used to adjust current in a circuit to a desired level. It often has only two terminals.
36. The type of material from which it is made and the physical size of the resistor
37. *a*. 10,000 Ω, ±10%
    b. 0.22 Ω, ±5%
    c. 6,800,000 Ω, ±20%
    d. 47 Ω, ±5%
38. Minimum resistance would be 198 Ω, and maximum resistance would be 242 Ω.
39. Carbon composition, conductive plastic, cermet, deposited film, and wirewound
40. The power is 0.18 W.
41. Wirewound resistors
42. The dashed line indicates that two poles are mechanically connected but electrically isolated.
43. Rotary, slide, rocker, toggle, and push button
44. NC means normally closed. The switch is closed except when the button is depressed.
45. A shorting switch is one in which the pole makes contact with a new position before breaking contact with the previous posiiton.
46. Voltage ratings are necessary to prevent excess arcing when the switch is opened.
47. American wire gage
48. A circular mil is equal to the area of a circle with a diameter of 1 mil (0.001 in).
49. Vinyl, rubber, neoprene, nylon, asbestos, and other plastic materials
50. Electromagnets, motors, generators, speakers, transformers, etc.
51. No. 16 wire
52. *a*. Open circuit means the conductive path for current has been broken.
    b. Short circuit means the circuit has an abnormally low resistance.

c. Slow blow means a fuse can stand a short-term overload without blowing.

53. An indicating holder is a fuse holder which lights up when the fuse has blown.
54. Current, voltage, and blowing characteristics
55. Circuit breakers do not have to be replaced when tripped.
56. Thermal and magnetic

Multiple-Load Circuits

- A great majority of electric circuits operate more than one load. Circuits which contain two or more loads are called multiple-load circuits. A multiple-load circuit can be a series circuit, a parallel circuit, or a series-parallel circuit. When you have finished this chapter, you will know the characteristics of each of these circuits. You will be able to measure and calculate current, voltage, and resistance in such complex circuits. Also, you will be able to explain the relationship between power transfer and efficiency.

5-1 SUBSCRIPTS

Notice the symbols R_1, R_2, and R_3 in Fig. 5-1. The subscripts 1, 2, and 3 are used to identify the different load resistors in the circuit. Two-level subscripts are also used in these types of circuits. For example, I_{R_2} is used to indicate the current flowing through resistor R_2. The symbol V_{R_1} indicates the voltage across resistor R_1. Similarly, P_{R_3} is used to specify the power used by R_3.

Electrical quantities for a total circuit are identified by the subscript T. Thus, the voltage of the battery of Fig. 5-1 is indicated by the symbol V_T. The symbol I_T indicates the source (battery) current, and R_T represents the total resistance of all the loads combined together. The power from the battery or source is identified as P_T.

5-2 POWER IN MULTIPLE-LOAD CIRCUITS

The total power taken from a source such as a battery is equal to the sum of the powers used by the individual loads. As a formula, this statement is written

$$P_T = P_{R_1} + P_{R_2} + P_{R_3} + \text{etc.}$$

The "+ etc." means the formula can be expanded to include any number of loads. Regardless of how complex the circuit, the above formula is appropriate.

5-3 SERIES CIRCUITS

A *series circuit* contains two or more loads but only one path for current to flow from the source voltage through the load and back to the source. Figure 5-1 is an example of a series circuit.

Current in Series Circuits

In Fig. 5-1, the battery current I_T flows through the first load R_1, the second load R_2, and the third load R_3. If 1 A flows through R_1, then 1 A also flows through R_2 and R_3. And, of course, the battery provides 1 A of current. In

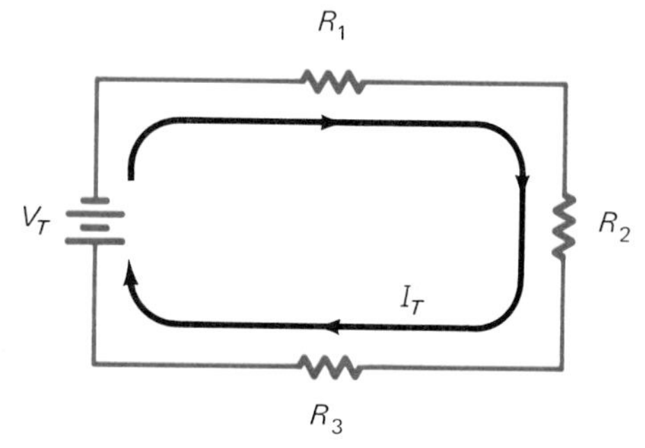

Fig. 5-1 Series circuit. There is only one path for current.

From page 87:
Subscripts

Power in multiple-load circuits

Series circuits

One path for current

Voltage is shared

On this page:
Total resistance

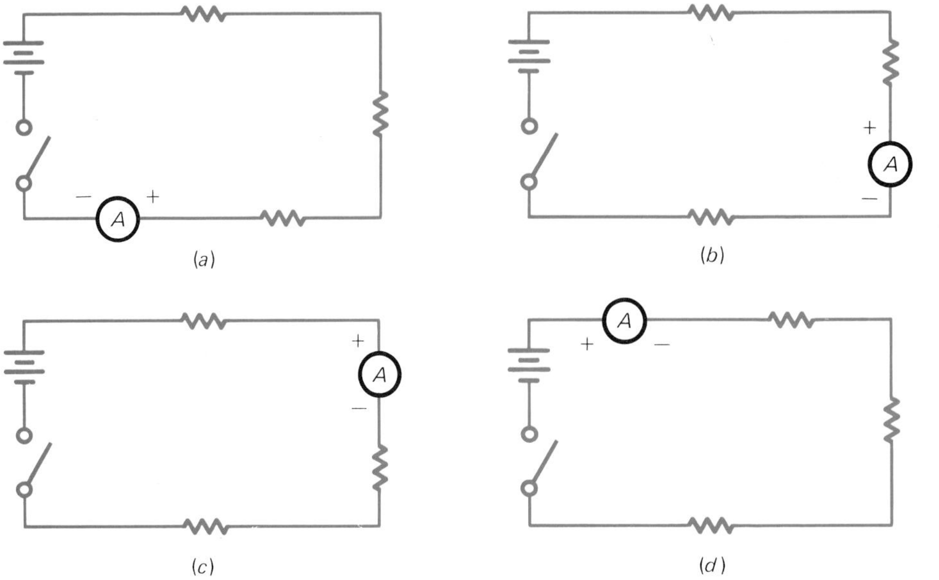

Fig. 5-2 Measuring current in a series circuit. The ammeter will indicate the same value of current in any of the positions shown in diagrams (a) to (d).

symbolic form, the current relationship in a series circuit is

$$I_T = I_{R_1} = I_{R_2} = I_{R_3} = \text{etc.}$$

Current in a series circuit can be measured by inserting the meter in series. Since there is only one path for current, any part of the circuit can be interrupted to insert the meter. All the meters in Fig. 5-2 will provide the same current reading provided the voltage and the total resistance are the same in each case.

Resistance in Series Circuits

The total resistance in a series circuit is equal to the sum of the individual resistances around the series circuit. This statement can be written as

$$R_T = R_1 + R_2 + R_3 + \text{etc.}$$

This relationship is very logical if you remember two things: (1) resistance is opposition to current and (2) current has to be forced through all the resistances in a series circuit.

The total resistance R_T can also be determined by Ohm's law if the total voltage V_T and total current I_T are known. Both methods of determining R_T can be seen by referring to the circuit of Fig. 5-3. In this figure, the 2 A near the ammeter symbol means that the meter is indicating two amperes. The resistance of each resistor is given next to the resistor symbol. Using Ohm's law, the resistance is

$$R_T = \frac{V_T}{I_T} = \frac{90\ \text{V}}{2\ \text{A}} = 45\ \Omega$$

Using the relationship for series resistance yields

$$\begin{aligned} R_T &= R_1 + R_2 + R_3 \\ &= 5\ \Omega + 10\ \Omega + 30\ \Omega \\ &= 45\ \Omega \end{aligned}$$

The total resistance of a series circuit can be measured by connecting an ohmmeter across the loads as in Fig. 5-4(*a*). The power source must be disconnected from the loads. Indi-

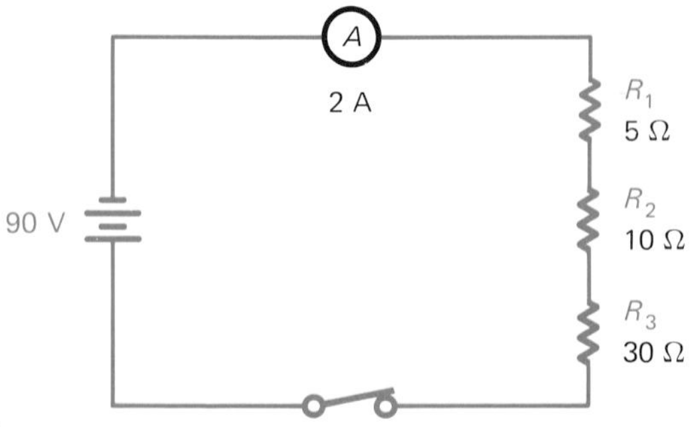

Fig. 5-3 Total resistance can be found by using Ohm's law or by adding individual resistances.

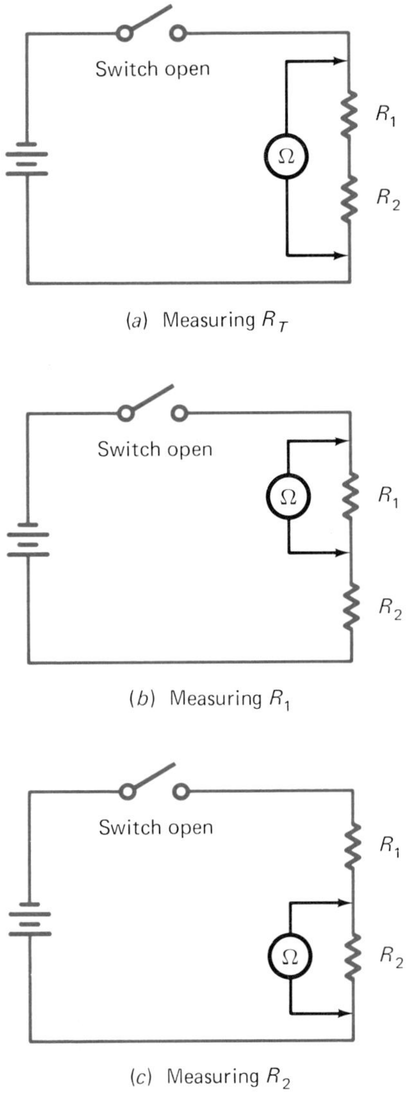

Fig. 5-4 Measuring resistance in series.

vidual resistances can be measured as shown in Fig. 5-4(*b*) and (*c*).

Voltage in Series Circuits

The battery voltage in Fig. 5-3 will divide up across the three load resistors. It will always divide up so that the sum of the individual load voltages will equal the source voltage. That is

$$V_T = V_{R_1} + V_{R_2} + V_{R_3} + \text{etc.}$$

This relationship is often referred to as *Kirchhoff's voltage law*. Kirchhoff's law states that "the sum of the voltages around a circuit equals the applied voltage."

For Fig. 5-3, this relationship can be verified by using Ohm's law to determine the individual voltages.

$$V_{R_1} = I_{R_1} \times R_1 = 2\ \text{A} \times 5\ \Omega = 10\ \text{V}$$

$$V_{R_2} = I_{R_2} \times R_2 = 2\ \text{A} \times 10\ \Omega = 20\ \text{V}$$

$$V_{R_3} = I_{R_3} \times R_3 = 2\ \text{A} \times 30\ \Omega = 60\ \text{V}$$

$$\begin{aligned} V_T &= V_{R_1} + V_{R_2} + V_{R_3} \\ &= 10\ \text{V} + 20\ \text{V} + 60\ \text{V} \\ &= 90\ \text{V} \end{aligned}$$

Kirchhoff's voltage law

Voltage drop

Polarity

Notice from the above calculations that individual voltages are directly proportional to the individual resistances. The voltage across R_2 is twice the voltage across R_1 because the value of R_2 is twice the value of R_1.

Voltage Drop and Polarity

The voltage or potential energy difference across a resistor is referred to as a *voltage drop*. We can say a voltage develops across the resistor. That is, part of the potential energy difference of the source develops or appears across each resistor as a smaller potential energy difference. There is a distinction between source voltage and the voltages across the loads. The source voltage provides the electric energy and the load voltage converts the electric energy into another form.

The voltage drop across a resistor has polarity. However, in this case the polarity does not necessarily indicate a deficiency or excess of electrons. Instead, it indicates the direction of current flow and the conversion of electric energy to another form of energy. Current moves through a load resistor from the negative polarity to the positive polarity. This means that electric energy is being converted to another form. The current *through* a battery moves from the positive to the negative polarity. Thus, the battery is providing the electric energy.

Polarity signs are shown on the resistors in Fig. 5-5. The current through R_1 and R_2 flows from the negative end of the resistor to the positive end. Current inside the batteries flows from the positive to the negative. However, the external current still flows from the negative terminal to the positive terminal of the battery.

Notice in Fig. 5-5 that point *B* is labeled both negative (−) and positive (+). This may seem contradictory, but it is not. Point *B* is

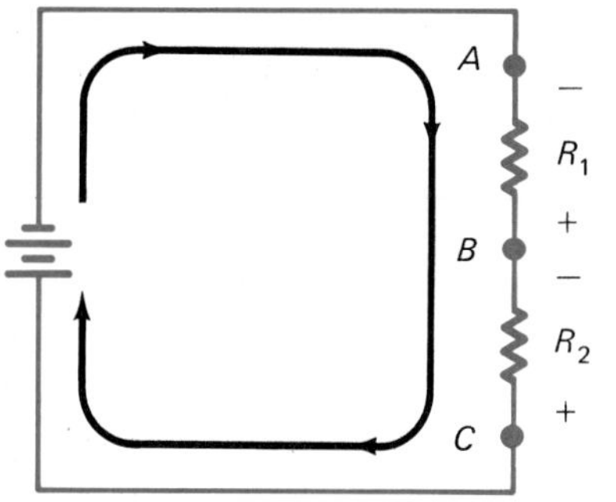

Fig. 5-5 Voltage polarity in a series circuit.

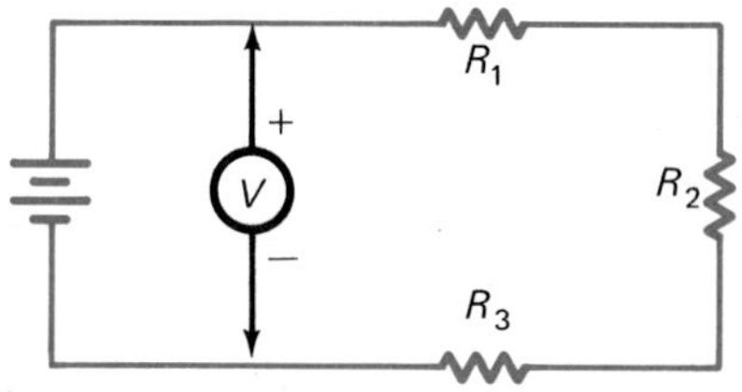

Fig. 5-6 Measuring the total voltage of a series circuit.

positive with respect to point A, but *negative with respect to point* C. It is very important to understand the phrase "with respect to." Remember that voltage was defined as "a potential energy difference between two points." Therefore, it is meaningless to speak of a voltage or a polarity at point *B*. Point *B*, by itself, has neither voltage nor polarity. But, with respect to either point A or point C, point B has both polarity and voltage. Point *B* is positive with respect to point A; this means that point B is at a lower potential energy than point A. It also means that electric energy is being converted to heat energy as electrons move from point A to point *B*.

Measuring Series Voltages

The total voltage of a series circuit must be measured across the voltage source, as shown in Fig. 5-6. Voltages across the series resistors are also easily measured. The correct connections for measuring series voltages are illustrated in Fig. 5-7. The procedure is to first determine the correct function, range, and polarity. Then touch the meter leads to the two points where voltage is to be measured. Voltmeters have a very high internal resistance, so high that connecting them to a circuit has no noticeable effect on many circuits. For now, assume that meters (both ammeters and voltmeters) do not change the circuits in which they are used.

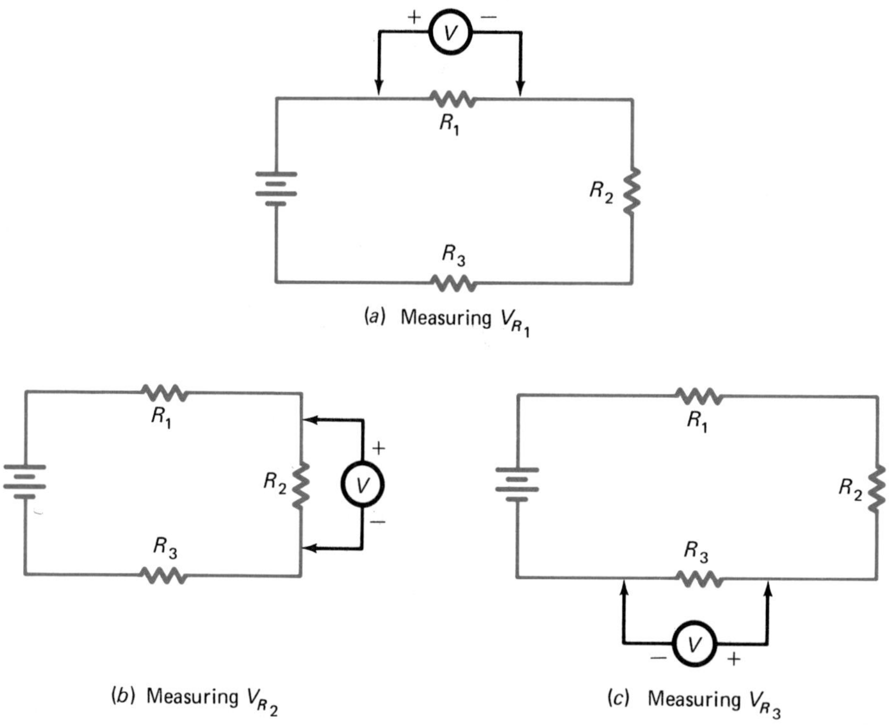

Fig. 5-7 Measuring voltage drops across each resistor in a series circuit.

Open load

Shorts

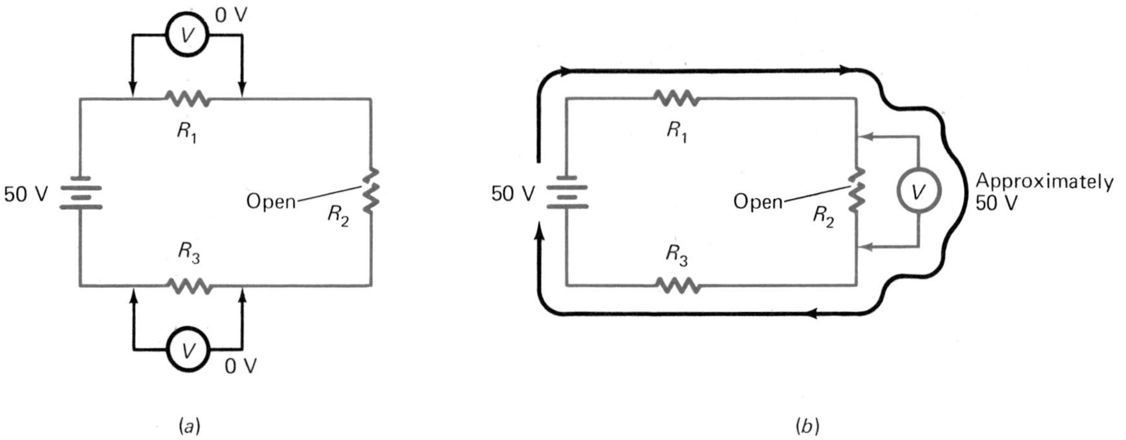

Fig. 5-8 **Voltage-drops across (a) normal loads, (b) open loads.**

Opens in Series Circuits

In a series circuit, if any part of the circuit is open, current stops flowing and voltage and power are removed from all the other loads. This is one of the weaknesses of a series circuit. For example, when one lamp in a series (such as Christmas tree lights) burns out (opens), all lamps go out.

An easy way to determine which load in a series is open is to measure the individual voltage. The load which is open will have a voltage drop equal to the entire source voltage. In Fig. 5-8(*a*) a meter connected across either of the good resistors will read 0 V. Since R_2 is open, no current flows in the series circuit. But, a meter across the open resistor R_2, as in Fig. 5-8(*b*), will read approximately 50 V. When a voltmeter is connected across R_2 as in Fig. 5-8(*b*), a small current will flow through R_1, the voltmeter, and R_3. In most circuits, the internal resistance of the voltmeter will be very, very high compared with the other series loads. Therefore, nearly all the battery voltage will be developed (and read) across the voltmeter. It should be noted that 50 V is also across R_2 in Fig. 5-8(*a*). It has to be to satisfy Kirchhoff's voltage law. That is, $V_{R1} + V_{R2} + V_{R3}$ has to equal the 50 V of the source.

The diagrams of Fig. 5-8 show the open resistor. However, in a real physical circuit, R_2 may look just like R_1 and R_3. There may be no physical evidence to indicate it is open. A technician repairing the circuit would have to interpret the voltmeter readings to conclude that R_2 is open.

Shorts in Series Circuits

When one load in a series circuit is shorted out, the other loads *may continue to operate*. Or, one of the other loads *may open* because of increased voltage, current, and power. In Fig. 5-9(*a*), each lamp is rated at 10 V and 1 A. Each lamp is 10-W ($P = IV = 1\text{ A} \times 10\text{ V}$).

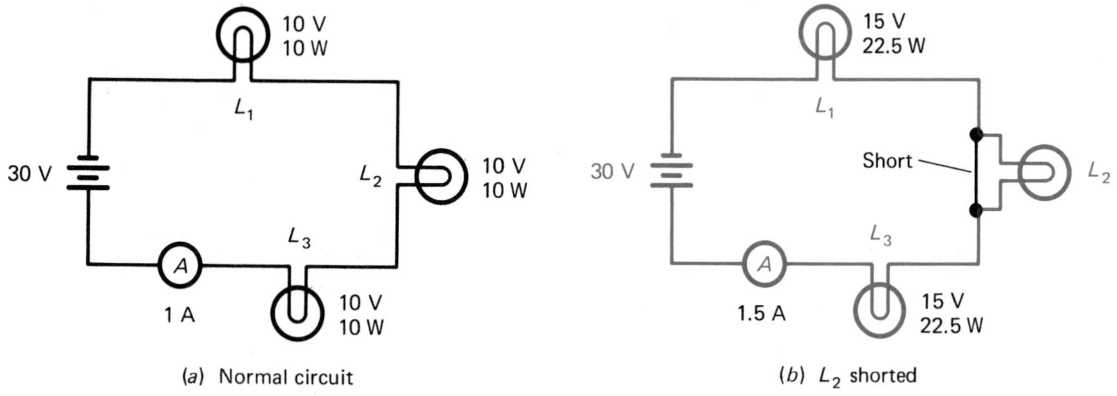

Fig. 5-9 **Effects of a shorted series load.**

When the lamp L_2 shorts out, as shown in Fig. 5-9(*b*), the 30-V source must divide evenly between the two remaining lamps. This means each remaining lamp must drop 15 V. Since the lamp voltages increase 33 percent, the lamp currents will also increase by 33 percent to 1.5 A (this assumes the individual lamp resistance does not change). With 15 V and 1.5 A, each lamp will have to dissipate 22.5 W. One or both of the lamps will soon burn out (open).

The effects of one shorted load in a series can be summarized as follows:

1. The voltage across the remaining loads increases.
2. The total resistance decreases.
3. The total current increases.
4. The power dissipation of the remaining loads increases.
5. The total power increases.
6. The resistance, voltage, and power of the shorted load decreases. If the short is a dead short, these quantities decrease to zero.

Solving Series Circuit Problems

Using subscripts

Proper use of Ohm's law and the series relationships of current, voltage, resistance, and power will solve most series problems. In using Ohm's law, you should develop the habit of using subscripts for voltage, current, and resistance. Without subscripts, it is easy to forget which voltage, current, or resistance to substitute into the formula. If you are using Ohm's law to find the voltage across R_1, you should write

$$V_{R_1} = I_{R_1} R_1$$

If you are calculating the total voltage, you should write

$$V_T = I_T R_T$$

Example 5-1

As shown in Fig. 5-10, an 8-V, 0.5-A lamp is to be operated from a 12.6-V battery. What resistance and wattage rating are needed for R_1?

Given: $V_T = 12.6$ V
$V_{L_1} = 8$ V
$I_{L_1} = 0.5$ A

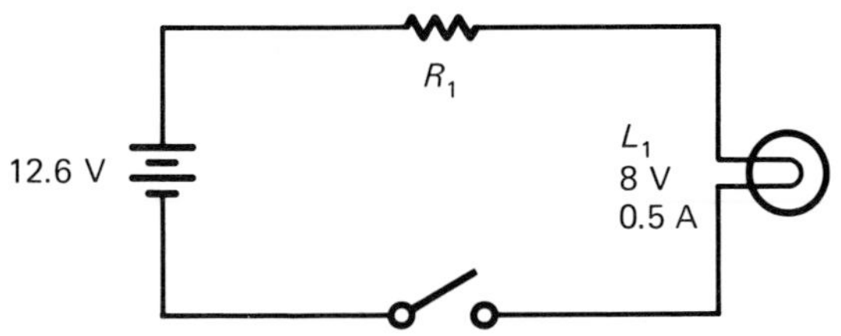

Fig. 5-10 Circuit diagram for Example 5-1.

Find: R_1, P_{R_1}

Known: $R_1 = \dfrac{V_{R_1}}{I_{R_1}}$

$V_T = V_{L_1} + V_{R_1}$

$I_T = I_{R_1} = I_{L_1}$

$P_{R_1} = V_{R_1} I_{R_1}$

Solution: $I_{R_1} = I_{L_1} = 0.5$ A

$V_T = V_{L_1} + V_{R_1}$

Therefore,

$V_{R_1} = V_T - V_{L_1}$

$V_{R_1} = 12.6\text{ V} - 8\text{ V} = 4.6\text{ V}$

$R_1 = \dfrac{4.6\text{ V}}{0.5\text{ A}} = 9.2\ \Omega$

$P_{R_1} = V_{R_1} I_{R_1}$
$= 4.6\text{ V} \times 0.5\text{ A}$
$= 2.3\text{ W}$

Answer: The calculated values for R_1 are 9.2 Ω and 2.3 W. Use a 5-W resistor to provide a safety factor.

Example 5-2

Find the total current and total resistance of the circuit in Fig. 5-11. Also determine the voltage across each resistor.

Given: $V_T = 90$ V
$R_1 = 35\ \Omega$
$R_2 = 70\ \Omega$
$R_3 = 45\ \Omega$

Find: I_T, R_T, V_{R_1}, V_{R_2}, V_{R_3}

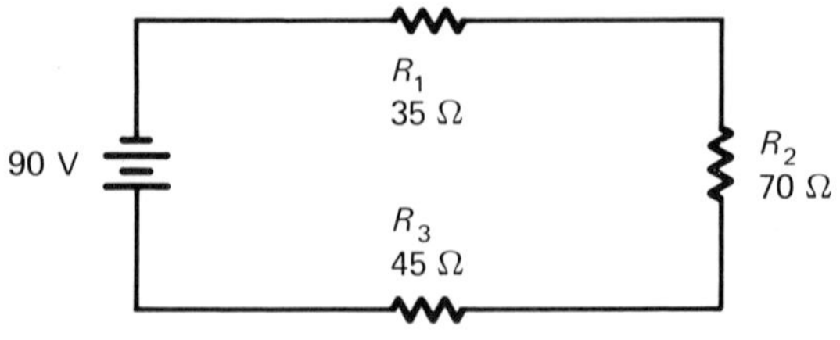

Fig. 5-11 Circuit diagram for Example 5-2.

Known: $I_T = \dfrac{V_T}{R_T}$

$R_T = R_1 + R_2 + R_3$

$V_{R1} = I_{R1} R_1$

$I_T = I_{R1} = I_{R2} = I_{R3}$

Solution: $R_T = 35\ \Omega + 70\ \Omega + 45\ \Omega = 150\ \Omega$

$I_T = \dfrac{90\ \text{V}}{150\ \Omega} = 0.6\ \text{A}$

$V_{R1} = 0.6\ \text{A} \times 35\ \Omega = 21\ \text{V}$

$V_{R2} = 0.6\ \text{A} \times 70\ \Omega = 42\ \text{V}$

$V_{R3} = 0.6\ \text{A} \times 45\ \Omega = 27\ \text{V}$

Answer: $I_T = 0.6\ \text{A}$, $R_T = 150\ \Omega$,

$V_{R1} = 21\ \text{V}$,

$V_{R2} = 42\ \text{V}$,

$V_{R3} = 27\ \text{V}$

After you solve a complex problem, it is a good idea to cross-check the problem for mathematical errors. This can usually be done by checking some relationship *not used* in originally solving the problem. The problem of Example 5-2 can be cross-checked by using Kirchhoff's voltage law:

Cross-check the problem

$$\begin{aligned} V_T &= V_{R1} + V_{R2} + V_{R3} \\ &= 21\ \text{V} + 42\ \text{V} + 27\ \text{V} \\ &= 90\ \text{V} \end{aligned}$$

Since 90 V was specified for V_T, the cross-check verifies that at least the sum of the individual voltages is as it should be.

Estimations, Approximations, and Tolerances

In a series circuit, the resistor with the most resistance dominates the circuit. That is, the highest resistance drops the most voltage, uses the most power, and has the most effect on the total current. Sometimes one resistor is so much larger than the other resistors that its value almost determines the total current. For example, resistor R_1 in Fig. 5-12 dominates the circuit. Shorting out R_2 in the circuit would increase the current from about 18 mA to only 20 mA. Assume R_1 and R_2 are ± 10 percent resistors. Then R_1 could be as low as 90 kΩ and R_2 as low as 9 kΩ. If this were the case, the current in Fig. 5-12 would still be about 20 mA. The presence or absence of R_2 has no more effect on the current than does the tolerance of R_1. Therefore, a good *estimate* of the current in Fig. 5-12 could be obtained by ignoring R_2. This estimate would be close enough for such things as:

Good estimate

1. Determining the power rating needed for the resistors by using $P = I^2R$
2. Determining which range of an ammeter should be used to measure the current
3. Estimating the power required from the battery

When estimating current in a series circuit, ignore the lowest resistance if it is less than the tolerance of the highest resistance.

Applications of Series Circuits

One application of series circuits has previously been mentioned—Christmas tree lights. Other applications include (1) motor-speed control, (2) lamp-intensity control, and (3) numerous electronic circuits.

Motor-speed-control

A simple motor-speed-control circuit is shown in Fig. 5-13. This type of control is used on very small motors such as the motor on a sewing machine. With a sewing machine, the variable resistance is contained in the foot control. The circuit provides continuous, smooth control of motor speed. The lower the resistance, the faster the motor rotates. A big disadvantage of this type of motor control is that it is inefficient. Sometimes the resistance converts more electric energy to heat than the motor converts to mechanical energy.

Inefficient

Fig. 5-12 Dominant resistance. Circuit current and power are largely determined by R_1.

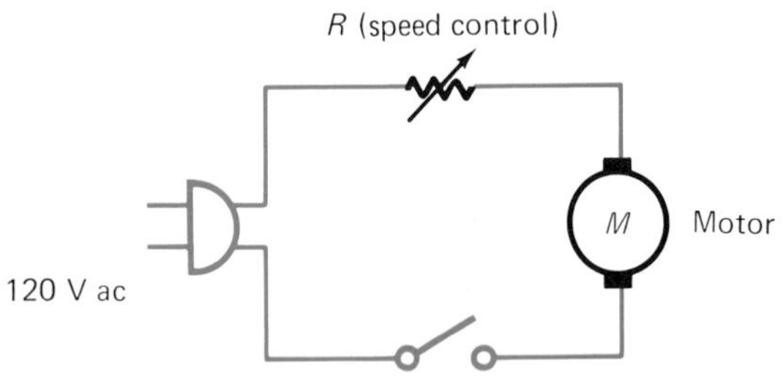

Fig. 5-13 Motor-speed control.

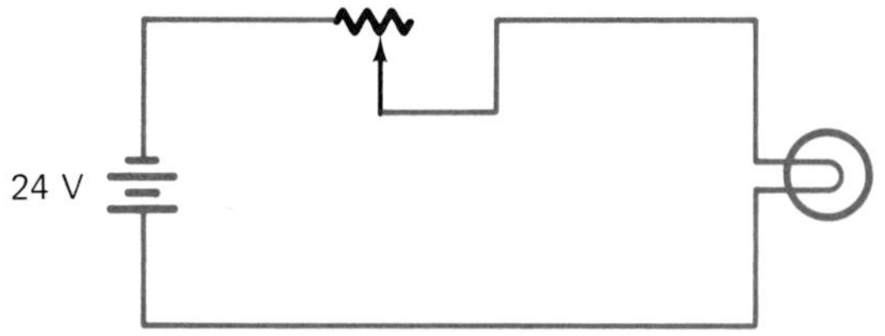

Fig. 5-14 Lamp-intensity control.

Lamp intensity

The intensity of an indicator lamp is often controlled by a variable series resistor. Such circuits are used to illuminate dials and meters on radio and navigation equipment in airplanes. A typical circuit is illustrated in Fig. 5-14. In this circuit, increasing the resistance decreases the total current and the lamp intensity. Notice that the symbol for the variable resistor in Fig. 5-14 is different from the one used in Fig. 5-13. Either symbol is correct. They both are symbols for a rheostat.

A portion of a transistor circuit is shown in Fig. 5-15. Resistor R_1 is in series with the collector and emitter of the transistor. The transistor acts like a variable resistor. Its resistance is controlled by the current supplied to the base. Thus, controlling the current to the base controls the current through the transistor. This makes it possible for a transistor to amplify, or increase, the small current presented to the base.

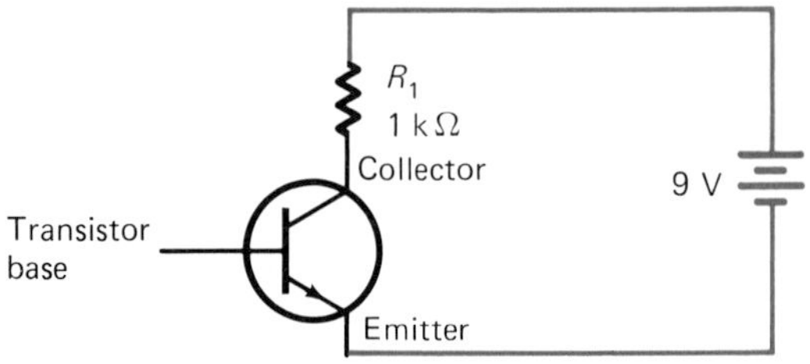

Fig. 5-15 Series resistor in a transistor circuit.

Review Questions

1. Write the symbol for:
   a. Voltage across resistor R_4
   b. Source current
   c. Current through resistor R_2
2. What is the relationship between total power and individual powers in a multiple-load circuit?
3. What is a series circuit?
4. Can the current from a battery be different from the current through any resistor in a series circuit served by the battery? Why?
5. List two formulas for determining the total resistance of a series circuit.
6. Can the resistance of a resistor in a series circuit be measured without completely removing it from the circuit?
7. Write the formula that shows the relationship between the total and the individual voltages in a series circuit.
8. In a series circuit, will a 100- or a 500-Ω resistor have the greater voltage drop?
9. Does the negative (−) polarity marking of a voltage-drop indicate an excess of electrons at that point?
10. A series circuit contains two resistors. One resistor is good, and the other is open. Across which resistor will a voltmeter indicate the most voltage?
11. A series circuit contains three resistors: R_1, R_2, and R_3. If R_1 shorts out, what happens to each of the following?
    a. Voltage across R_2?
    b. Power dissipated by R_3?
    c. Total current?
12. A 100-Ω resistor (R_1) and a 400-Ω resistor (R_2) are series connected to a 120-V source. Determine the following:
    a. Total resistance
    b. Total current
    c. Voltage across R_1
    d. Power dissipated by R_2
13. Refer to Fig. 5-10. Change L_1 to a 6.3-V, 150-mA lamp. What value of resistance would be needed for R_1? How much power would the battery have to furnish?
14. If the resistance in Fig. 5-13 is increased, what will happen to the speed of the motor?
15. If the transistor in Fig. 5-15 opens, how much voltage would be measured across R_1?
16. If the transistor in Fig. 5-15 shorted, how much current would flow through R_1?

5-4 MAXIMUM POWER TRANSFER

Maximum power transfer refers to getting the maximum possible amount of power from a source to its load. The source may be a battery and the load a lamp. Or, the source could be a guitar amplifier and the load a speaker.

Maximum power transfer occurs when the internal opposition to current of the source equals the opposition to current of the load. Resistance is one form of opposition to current; *impedance* is another form of opposition. You will learn more about impedance when you study alternating current. In direct current circuits, maximum power transfer occurs when resistances are matched.

Referring to Fig. 5-16 will aid you in further understanding resistance matching and power transfer. In this figure, the internal resistance

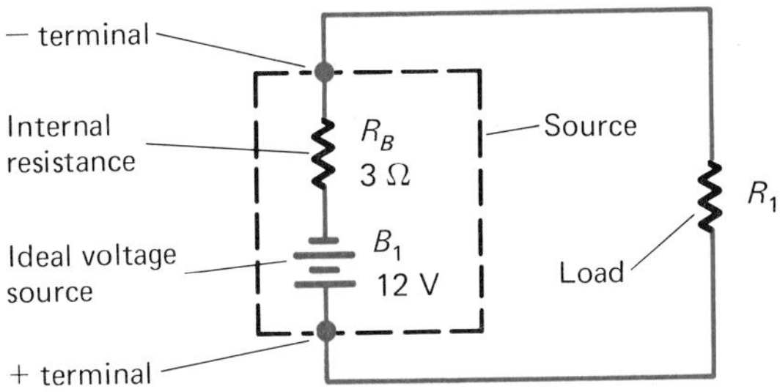

Fig. 5-16 Maximum power transfer occurs when R_B and R_1 are equal.

Table 5-1 Calculated Values for Fig. 5-16

R_1 (Ω)	R_T (Ω)	I_T (A)	P_{R_1} (W)	P_{RB} (W)	P_T (W)
1	4	3.00	9.00	27.00	36.00
2	5	2.40	11.52	17.28	28.80
3	6	2.00	12.00	12.00	24.00
4	7	1.71	11.76	8.82	20.57
5	8	1.50	11.25	6.75	18.00
6	9	1.33	10.67	5.33	16.00

of the battery is represented by R_B. The battery B_1 represents an ideal voltage source. That is, a voltage source with no internal resistance. The dotted lines around R_B and B_1 mean that R_B and B_1 together behave like a real battery. Resistance R_B and B_1 do not actually exist as separate components. They cannot be separated. Thus, they are enclosed in dotted lines. Together they form the source. The load is R_1. When R_1 equals R_B, the maximum amount of power will be transferred from the source to the load. This statement can be proved by an example: First assign fixed values to R_B and B_1. Then calculate the power dissipated by R_1 for various values of R_1. For example, when R_1 is 9 Ω, the series calculations will be

$$R_T = R_1 + R_B = 9\ \Omega + 3\ \Omega = 12\ \Omega$$

$$I_T = \frac{V_T}{R_T} = \frac{12\text{ V}}{12\ \Omega} = 1\text{ A}$$

$$P_{R_1} = I_{R_1}^2 R_1 = (1\text{ A})^2 \times 9\ \Omega = 9\text{ W}$$

$$P_{R_B} = I_R^2 R_B = (1\text{ A})^2 \times 3\ \Omega = 3\text{ W}$$

$$P_T = P_{R_1} + P_{R_B} = 9\text{ W} + 3\text{ W} = 12\text{ W}$$

Calculations for other values of R_1 have been made and recorded in Table 5-1. Notice that the maximum power dissipation occurs when R_1 is 3 Ω.

Also shown in Table 5-1 is the power dissipated within the source, and the total power taken from the source. Notice that when maximum power transfer occurs, the efficiency is only 50%. Of the 24 W furnished by the source, only 12 W is used by the load. As the load gets larger, the efficiency improves and the power transferred decreases.

Maximum power transfer

Impedance

Conductance

Siemens

Review Questions

17. When is maximum power transferred from a source to a load?
18. For high efficiency, should the load resistance be equal to, less than, or greater than the internal resistance of the source?

5-5 CONDUCTANCE

So far in this text, we have considered only a resistor's opposition to current. However, no resistor completely stops current. Therefore, we could just as well consider a resistor's ability to *conduct* current. Instead of considering a resistor's resistance, we could consider a resistor's conductance. *Conductance* refers to the ability to conduct current. It is symbolized by the letter G. The base unit for conductance is the *siemens*, abbreviated S, in honor of the inventor Ernst Von Siemens.

Conductance is the exact opposite of resistance. In fact, the two are mathematically defined as reciprocals of each other. That is:

$$G = \frac{1}{R} \quad \text{and} \quad R = \frac{1}{G}$$

Thus, a 100-Ω resistor has a conductance of $\frac{1}{100}$ or 0.01 siemens (S).

$G = \frac{1}{R}$

$R = \frac{1}{G}$

Conductance is a concept used extensively in electronics. However, we will use it only to show how to solve for total resistance in parallel circuits.

Parallel circuits

More than one path

Branch

Total voltage

5-6 PARALLEL CIRCUITS

Parallel circuits are multiple-load circuits which have more than one path for current. Each different current path is called a *branch*. The circuit in Fig. 5-17 has three branches. Current from the battery splits up among the three branches. Each branch has its own load, and each branch is independent of all other branches. The current and power in one branch is not dependent on the current, resistance, or power of any other branch.

In Fig. 5-18, switch S_2 in the second branch will control the lamp in that branch. However, turning switch S_2 on and off will not affect the lamps in the other branches. Figure 5-18 illustrates the way in which the lamps in your home are connected. In a house, the electric circuits are parallel circuits. The fuse in Fig. 5-18 protects all three branches. It carries the current for all branches. If any one branch draws too much current, the fuse will open. Also, if the three branches together draw too much current, the fuse will open. Of course, when the fuse opens, all branches become inoperative.

Voltage in Parallel Circuits

All voltages in a parallel circuit are the same. In other words, the source voltage appears across each branch of a parallel circuit. In Fig. 5-19(*a*), each voltmeter will read the same voltage. Rearranging the circuit in Fig. 5-19(*a*) yields the circuit shown in Fig. 5-19(*b*). In Fig. 5-19(*b*) it is easier to see that each branch of the circuit receives the total battery voltage. For a parallel circuit, the relationship of source voltage to load voltage is expressed as

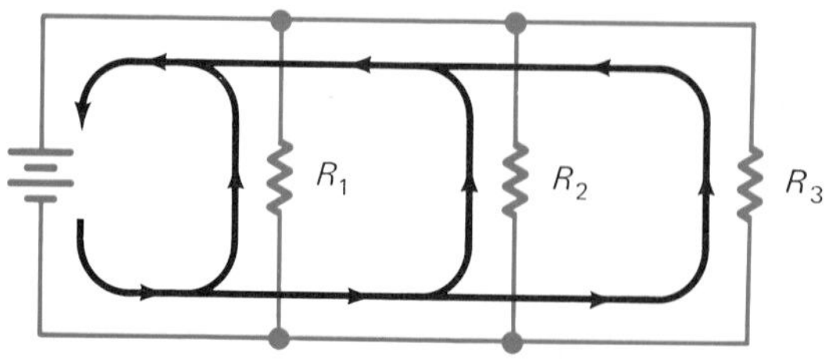

Fig. 5-17 Parallel circuit. There is more than one path for current to take.

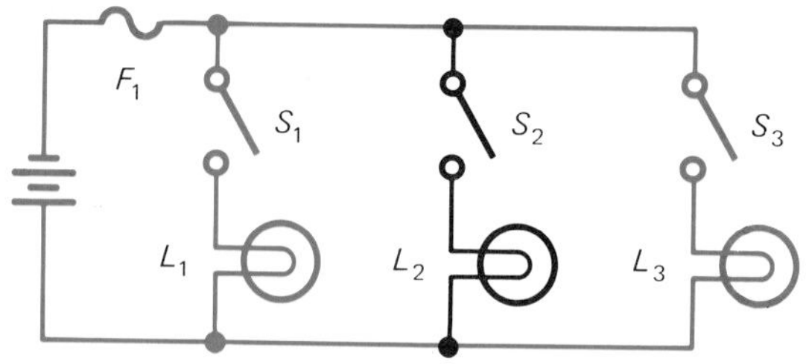

Fig. 5-18 Independence of parallel branches. Opening or closing S_1 has no effect on L_2 or L_3.

$$V_T = V_{R_1} = V_{R_2} = V_{R_3} = \text{etc.}$$

In a parallel circuit, the voltage measured across the load will not change if the load opens. If R_2 in Fig. 5-19(*a*) were open, the voltmeter across R_2 would still measure the voltage of the battery.

Current in Parallel Circuits

The relationship of the currents in a parallel circuit is as follows:

$$I_T = I_{R_1} + I_{R_2} + I_{R_3} + \text{etc.}$$

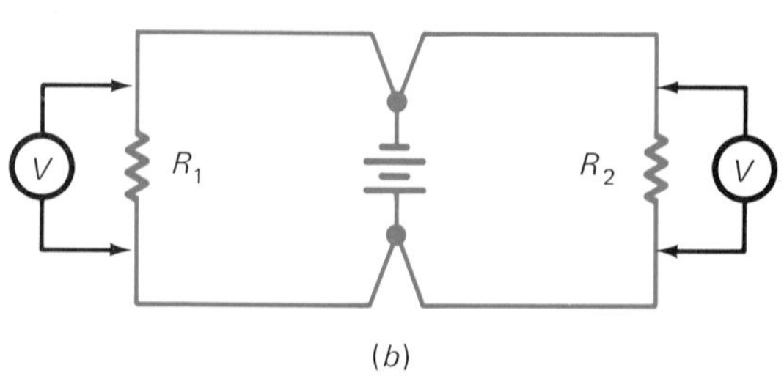

Fig. 5-19 Voltage measurement in parallel circuits. All voltmeters will indicate the same value of voltage.

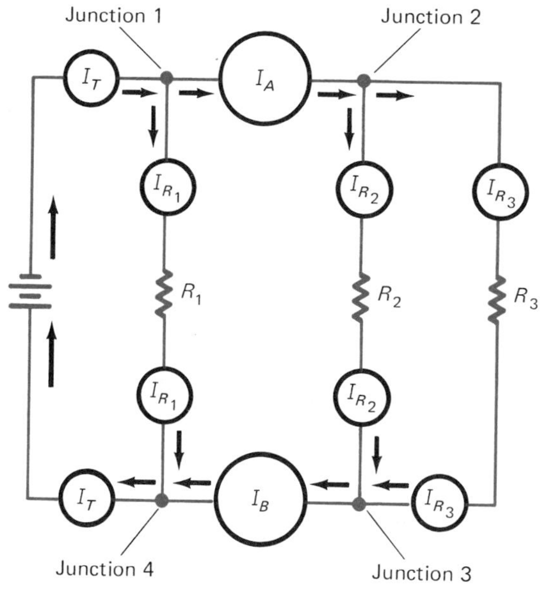

Fig. 5-20 Current measurement in parallel circuits.

In other words, the total current is equal to the sum of the individual branch currents. Figure 5-20 shows all the places where an ammeter could be inserted to measure current in a parallel circuit. The current being measured at each location is indicated inside the meter symbol.

Total current

The total current I_T in Fig. 5-20 splits into two smaller currents at junction 1 (I_{R_1} and $I_{R_2} + I_{R_3}$). A similar division of current occurs in the other three junctions in Fig. 5-20. The various currents entering and leaving a junction are related by *Kirchhoff's current law*. This law states that "the sum of the currents entering a junction equals the sum of the currents leaving a junction." No matter how many wires are connected together at a junction, Kirchhoff's current law still applies. Thus, in Fig. 5-20, the following relationships exist:

Kirchhoff's current law

$$I_T = I_{R_1} + I_A$$

$$I_A = I_{R_2} + I_{R_3}$$

$$I_{R_2} + I_{R_3} = I_B$$

$$I_{R_1} + I_B = I_T$$

In Fig. 5-21, five wires are joined at a junction. If three of the wires carry a total of 8 A into the junction, the other two must carry a total of 8 A out of the junction. Therefore, the unmarked wire in Fig. 5-21 must be carrying 3 A out of the junction.

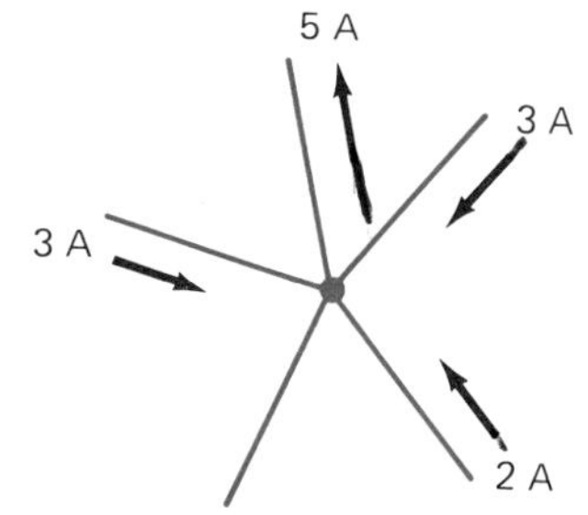

Fig. 5-21 Current at a junction. The unmarked line must carry 3 A of current away from the junction.

Smallest resistance dominates

As shown in Fig. 5-22, the branch of a parallel circuit with the lowest resistance dominates the circuit. That is, the lowest resistance takes the most current and power from the source. Ohm's law shows why this is so:

$$I = \frac{V}{R}$$

With V equal for all branches, a low value of R results in a high value of I. As R increases, the value of I decreases.

Since V is the same for all branches, the branch with the most current uses the most power ($P = IV$).

In Fig. 5-22, removing the 10-kΩ resistor would reduce the total power and current about 10 percent. Following the logic used for series circuits, engineers have developed a rule of thumb for parallel circuits. When *estimating* total current and power, ignore a parallel resistor that is 10 times higher than the other resistor. This rule assumes the resistors have a 10 percent tolerance. If the tolerance is 5 percent the rule is modified to read "20 times higher." In Fig. 5-22, ignoring the 10-kΩ resistor results in an estimated total current of 10 mA. If both resistors were on the high side of their 10 percent tolerance, the actual total current would also be 10 mA. Of course, if both resistors were on the low side of their tolerance, the actual current would then be 12.2 mA.

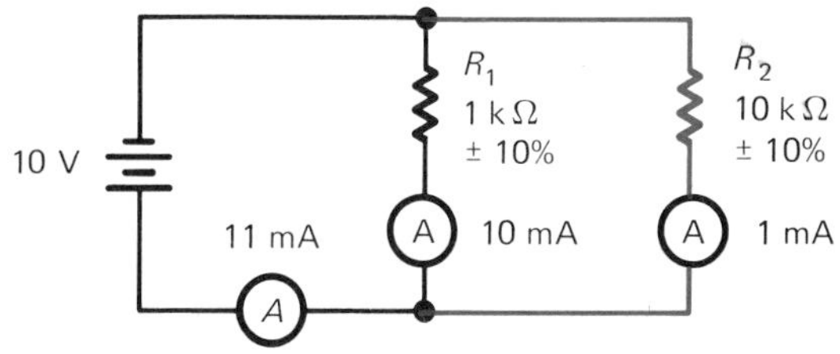

Fig. 5-22 Resistor tolerance and estimated current. Notice that the small resistance carries nearly all the circuit current.

Resistance in Parallel Circuits

Total resistance

The total resistance of a parallel circuit is always less than the lowest branch resistance. It may seem illogical at first that adding more resistors to a parallel circuit decreases the total resistance. The logic of the above statements can be shown through the use of Ohm's law and reference to Fig. 5-23. In Fig. 5-23(*a*) the current in the circuit will be

$$I = \frac{V}{R} = \frac{10\text{ V}}{10{,}000\ \Omega}$$
$$= 0.001\text{ A} = 1\text{ mA}$$

Adding R_2 in parallel, as in Fig. 5-23(*b*), does not change either the resistance of R_1 or the voltage across R_1. Therefore, R_1 still draws 1 mA. The current drawn by R_2 can also be calculated

$$I_{R_2} = \frac{V_2}{R_2} = \frac{10\text{ V}}{100{,}000\ \Omega}$$
$$= 0.0001\text{ A} = 0.1\text{ mA}$$

By Kirchhoff's current law, the total current in Fig. 5-23(*b*) is 1.1 mA ($I_T = I_{R_1} + I_{R_2}$). Now, if the total voltage is still 10 V, and the total current has increased, the total resistance had to decrease. By Ohm's law, the total resistance of Fig. 5-23(*b*) is

$$R_T = \frac{V_T}{I_T} = \frac{10\text{ V}}{0.0011\text{ A}}$$
$$= 9091\ \Omega = 9.09\text{ k}\Omega$$

Notice that the addition of a 100-kΩ resistor in parallel with a 10-kΩ resistor *reduced* the total resistance. Also, notice that the total resistance is less than the lowest (10-kΩ) resistance. In Fig. 5-23(*c*), a 1-kΩ resistor has been added to the circuit of Fig. 5-23(*b*). The total resistance of Fig. 5-23(*c*) is

$$R_T = \frac{V_T}{I_T} = \frac{10\text{ V}}{0.0111\text{ A}}$$
$$900.9\ \Omega = 0.901\text{ k}\Omega$$

Again, notice that R_T has decreased and is less than the lowest resistance (1 kΩ).

In the above examples, the total resistance was calculated by using Ohm's law and the circuit voltage and current. The total resistance can also be calculated with the following formula:

$$R_T = \frac{1}{\frac{1}{R_1} + \frac{1}{R_2} + \frac{1}{R_3} + \text{etc.}}$$

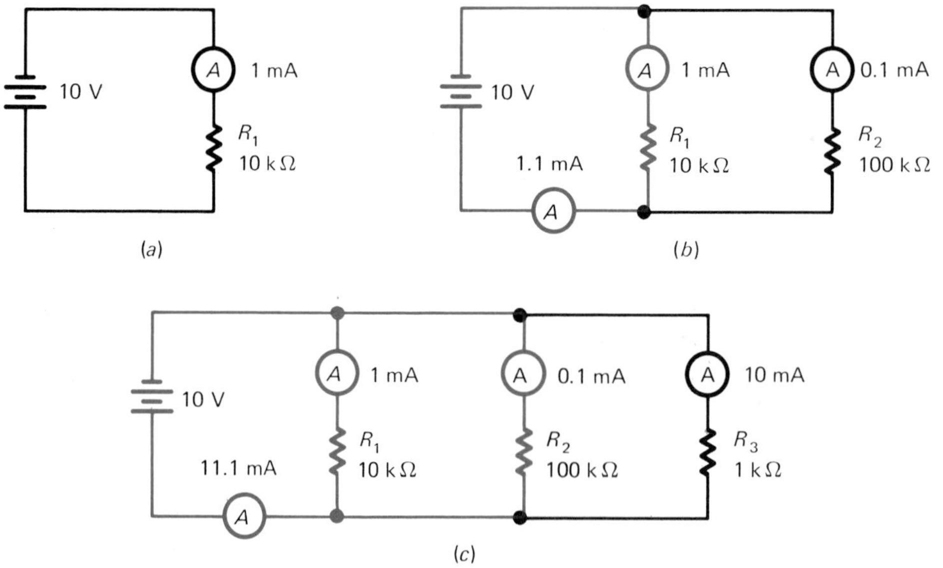

Fig. 5-23 Total resistance in parallel circuits. Adding resistors, as in circuits (b) and (c), increases total current and decreases total resistance.

This formula is often referred to as the *reciprocal formula*. The name reciprocal formula comes from the fact that the reciprocals of the branch resistances are added. Adding a parallel resistor to a circuit provides another path for current. That is, it increases the conductance of the circuit. Therefore, to find the total conductance of a parallel circuit, simply add the branch conductances:

Reciprocal formula

Branch conductances

$$G_T = G_1 + G_2 + G_3 + \text{etc.}$$

This conductance formula can be converted to a resistance formula by substituting 1/R for G (see Sec. 5-5). Therefore, the conductance formula can be rewritten as

$$\frac{1}{R_T} = \frac{1}{R_1} + \frac{1}{R_2} + \frac{1}{R_3} + \text{etc.}$$

Solving for R_T the formula becomes

$$R_T = \frac{1}{\frac{1}{R_1} + \frac{1}{R_2} + \frac{1}{R_3} + \text{etc.}}$$

Example 5-3

What is the total resistance of three resistors, 20 Ω, 30 Ω, and 60 Ω, connected in parallel?

Given: $R_1 = 20\ \Omega$
$R_2 = 30\ \Omega$
$R_3 = 60\ \Omega$

Find: R_T

Known: $$R_T = \frac{1}{\frac{1}{R_1} + \frac{1}{R_2} + \frac{1}{R_3}}$$

Solution: $$R_T = \frac{1}{\frac{1}{20} + \frac{1}{30} + \frac{1}{60}}$$

$$= \frac{1}{\left(\frac{6}{60}\right)} = 10\ \Omega$$

Answer: Total resistance is 10 Ω.

Instead of using fractions to calculate the total resistance, the reciprocals can be converted to their decimal equivalents. Let us solve the problem in Example 5-3 using the decimal equivalents:

$$R_T = \frac{1}{\frac{1}{20} + \frac{1}{30} + \frac{1}{60}}$$

$$= \frac{1}{0.05 + 0.033 + 0.017}$$

$$= \frac{1}{0.100} = 10\ \Omega$$

Two resistors in parallel

When only two resistors are in parallel, a simplified formula can be used to solve parallel resistance problems. This simplified formula is derived from the reciprocal formula. It is

$$R_T = \frac{R_1 \times R_2}{R_1 + R_2}$$

Example 5-4

What is the total resistance of a 27-Ω resistor in parallel with a 47-Ω resistor?

Given: $R_1 = 27\ \Omega$
$R_2 = 47\ \Omega$

Find: R_T

Known: $$R_T = \frac{R_1 \times R_2}{R_1 + R_2}$$

Solution: $$R_T = \frac{27 \times 47}{27 + 47} = \frac{1269}{74}$$
$$= 17.1\ \Omega$$

Answer: Total resistance is 17.1 Ω.

The simpler formula can also be used for circuits containing more than two resistors. The process is to find the equivalent resistance of R_1 and R_2 in parallel. The equivalent resistance will be labeled $R_{1,2}$. Then use this equivalent resistance and R_3 in a second application of the formula to find R_T. As an illustration, let us solve Example 5-3 using this two-step method.

$$R_{1,2} = \frac{R_1 \times R_2}{R_1 + R_2} = \frac{20 \times 30}{20 + 30} = \frac{600}{50} = 12\ \Omega$$

$$R_T = \frac{R_{1,2} \times R_3}{R_{1,2} + R_3} = \frac{12 \times 60}{12 + 60} = \frac{720}{72} = 10\ \Omega$$

When all the resistors in a parallel circuit have the same value, the total resistance can be found very easily. Just divide the value of a resistor by the number of resistors. That is,

$R_T = R/n$

Measuring resistance

$R_T = R/n$, where n is the number of resistors. For example, three 1000-Ω resistors in parallel have a total resistance of

$$R_T = \frac{R}{n} = \frac{1000}{3}$$
$$= 333.3\ \Omega$$

Two 100-Ω parallel resistors would have an equivalent resistance of

$$R_T = \frac{R}{n} = \frac{100}{2}$$
$$= 50\ \Omega$$

Measuring Resistance in Parallel

The total resistance of a parallel circuit is measured in the same way that it is measured in other types of circuits: that is, disconnect the power source and measure across the points where the power had been applied.

To measure an individual resistance in a parallel circuit, one end of the load must be disconnected from the circuit. The correct technique is shown in Fig. 5-24(*a*). When the load is not disconnected, as illustrated in Fig. 5-24(*b*), the meter again reads the total resistance.

Solving Parallel Circuit Problems

Now that we know the relationships between individual and total resistance, current, voltage, and power, we can solve parallel circuit problems. These relationships, plus Ohm's law and the power formula, allow us to solve most parallel circuit problems. The formulas listed below will be the "Known" for the examples that follow.

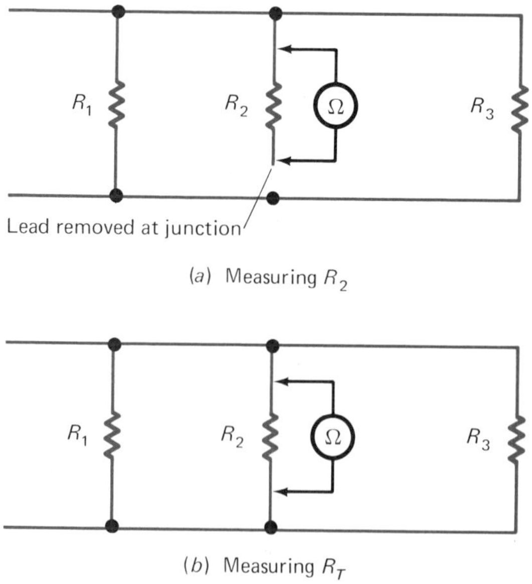

Fig. 5-24 Measurement of resistance in parallel.

Parallel Circuit Formulas

$$I_T = I_{R_1} + I_{R_2} + I_{R_3} + \text{etc.}$$
$$V_T = V_{R_1} = V_{R_2} = V_{R_3} = \text{etc.}$$
$$R_T = \frac{1}{\frac{1}{R_1} + \frac{1}{R_2} + \frac{1}{R_3} + \text{etc.}}$$
$$P_T = P_{R_1} + P_{R_2} + P_{R_3} + \text{etc.}$$
$$I = \frac{V}{R}$$
$$P = IV = I^2R = \frac{V^2}{R}$$

Example 5-5

Find the following for the circuit in Fig. 5-25: I_T, R_T, and P_T.

Given: $V_T = 10$ V
$P_{L_1} = 2$ W
$R_1 = 100\ \Omega$
$I_{R_2} = 0.5$ A

Find: I_T, R_T, P_T

Known: Parallel circuit formulas.

Solution: For branch I:

$$I_{L_1} = \frac{P_{L_1}}{V_{L_1}} = \frac{2\text{ W}}{10\text{ V}} = 0.2\text{ A}$$

For branch II:

$$I_{R_1} = \frac{V_{R_1}}{R_1} = \frac{10\text{ V}}{100\ \Omega} = 0.1\text{ A}$$

For total circuit:

$$I_T = I_{L_1} + I_{R_1} + I_{R_2}$$
$$= 0.2\text{ A} + 0.1\text{ A} + 0.5\text{ A}$$
$$= 0.8\text{ A}$$

$$R_T = \frac{V_T}{I_T} = \frac{10\text{ V}}{0.8\text{ A}} = 12.5\ \Omega$$
$$P_T = I_T V_T = 0.8\text{ A} \times 10\text{ V} = 8\text{ W}$$

Answer: Total current is 0.8 A. Total resistance is 12.5 Ω. Total power is 8 W.

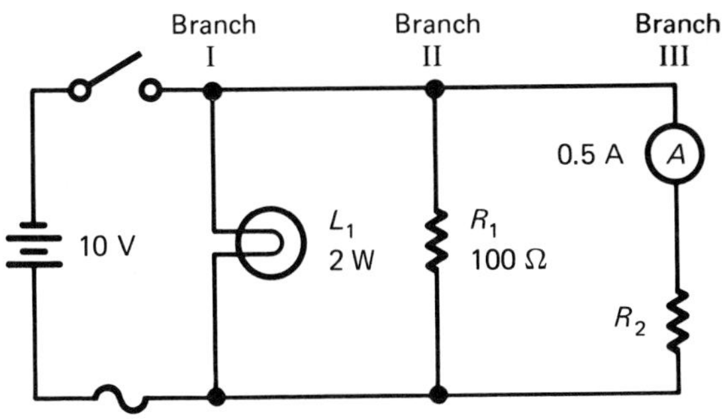

Fig. 5-25 Circuit for Example 5-5.

Example 5-6

With the data found in Example 5-5, find the total resistance using the reciprocal formula

Solution: $R_{L_1} = \frac{V_{L_1}}{I_{L_1}} = \frac{10\text{ V}}{0.2\text{ A}} = 50\ \Omega$

$$R_2 = \frac{V_{R2}}{I_{R2}} = \frac{10\text{ V}}{0.5\text{ A}} = 20\ \Omega$$

$$R_T = \frac{1}{\frac{1}{50} + \frac{1}{100} + \frac{1}{20}}$$

$$= \frac{1}{0.02 + 0.01 + 0.05}$$

$$= \frac{1}{0.08} = 12.5\ \Omega$$

Answer: Total resistance is 12.5 Ω.

Applications of Parallel Circuits

An electric system in which one section can fail and other sections continue to operate has parallel circuits. As previously mentioned, the electrical system used in homes consists of many parallel circuits.

An automobile electric system uses parallel circuits for lights, heater motor, radio, etc. Each of these devices operates independently of the others.

Individual television circuits are quite complex. However, the complete circuits are connected in parallel to the main power source. That is why the audio section of television receivers can still work when the video (picture) is inoperative.

Applications

Review Questions

19. How do you define parallel circuit?
20. How is the voltage distributed in a parallel circuit?
21. How is the current distributed in a parallel circuit?
22. Does the highest or lowest resistance dominate a parallel circuit?
23. List two formulas that could be used to find the total resistance of two parallel resistors.
24. What is conductance?
25. Does adding another resistor in parallel increase or decrease the total resistance?
26. Three resistors, 10 Ω, 20 Ω, and 30 Ω are connected in parallel. Will the total resistance be less than 10 Ω, between 10 Ω and 30 Ω, or greater than 30 Ω?
27. What is the resistance of four 56-Ω resistors connected in parallel?
28. Can the resistance of a parallel resistor be measured without disconnecting one end from the circuit?
29. In a parallel circuit, will a 100-Ω or a 1000-Ω resistor dissipate more power?
30. Can voltage measurements in a parallel circuit determine whether or not a load is open?
31. If one load in a parallel circuit opens, what will happen to each of the following?
    a. Total resistance
    b. Total current
    c. Total power
    d. Total voltage
32. Refer to Fig. 5-26. What is the value of I_{R_1} and I_{R_2}?
33. Refer to Fig. 5-26. What is the voltage of B_1?

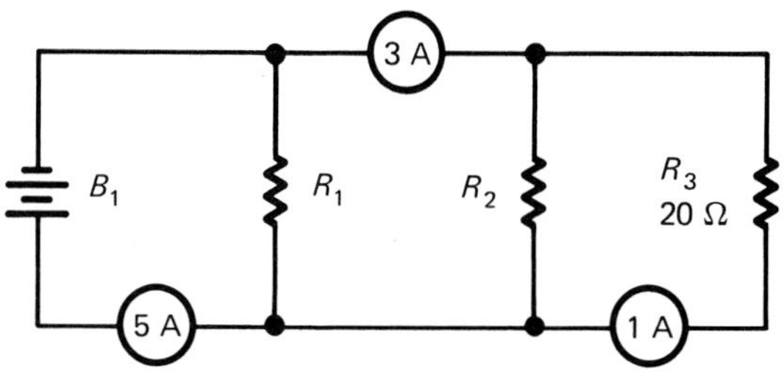

Fig. 5-26 Circuit for Review Questions 32 to 35.

34. Refer to Fig. 5-26. What is the resistance of R_2?

35. Refer to Fig. 5-26. What is the power dissipated by R_1?

5-7 SERIES-PARALLEL CIRCUITS

Series-parallel circuits

Some of the features of both the series circuit and the parallel circuit are incorporated into series-parallel circuits. For example, R_2 and R_3 in Fig. 5-27(*a*) are in parallel. Everything that has been said about parallel circuits applies to these two resistors. In Fig. 5-27(*d*), R_7 and R_8 are in series. All the series circuit relationships apply to these two resistors.

Resistors R_1 and R_2 in Fig. 5-27(*a*) are not directly in series because the same current does not flow through them. However, the equivalent of R_2 and R_3 in parallel [$R_{2,3}$ in Fig. 5-27(*b*)] is in series with R_1. Combining R_2 and R_3 would be the first step in determining the total resistance of this circuit. The result is $R_{2,3}$. Of course, $R_{2,3}$ is not an actual resistor, but represent R_2 and R_3 in parallel. Resistor R_1 is in series with $R_{2,3}$. The final step in finding the total resistance would be to combine R_1 and $R_{2,3}$, as shown in Fig. 5-27(*c*). Referring to Fig. 5-27(*a*), assume $R_1 = 15\ \Omega$, $R_2 = 20\ \Omega$, and $R_3 = 30\ \Omega$. Combining R_2 and R_3 would give:

$$R_{2,3} = \frac{R_2 \times R_3}{R_2 + R_3} = \frac{20 \times 30}{20 + 30} = \frac{600}{50} = 12\ \Omega$$

The total resistance would be:

$$R_T = R_1 + R_{2,3} = 15 + 12 = 27\ \Omega$$

Neither R_7 nor R_8 in Fig. 5-27(*d*) is in parallel with R_9 because they are not across the same voltage source. Therefore, the first step in reducing this circuit is to combine the series re-

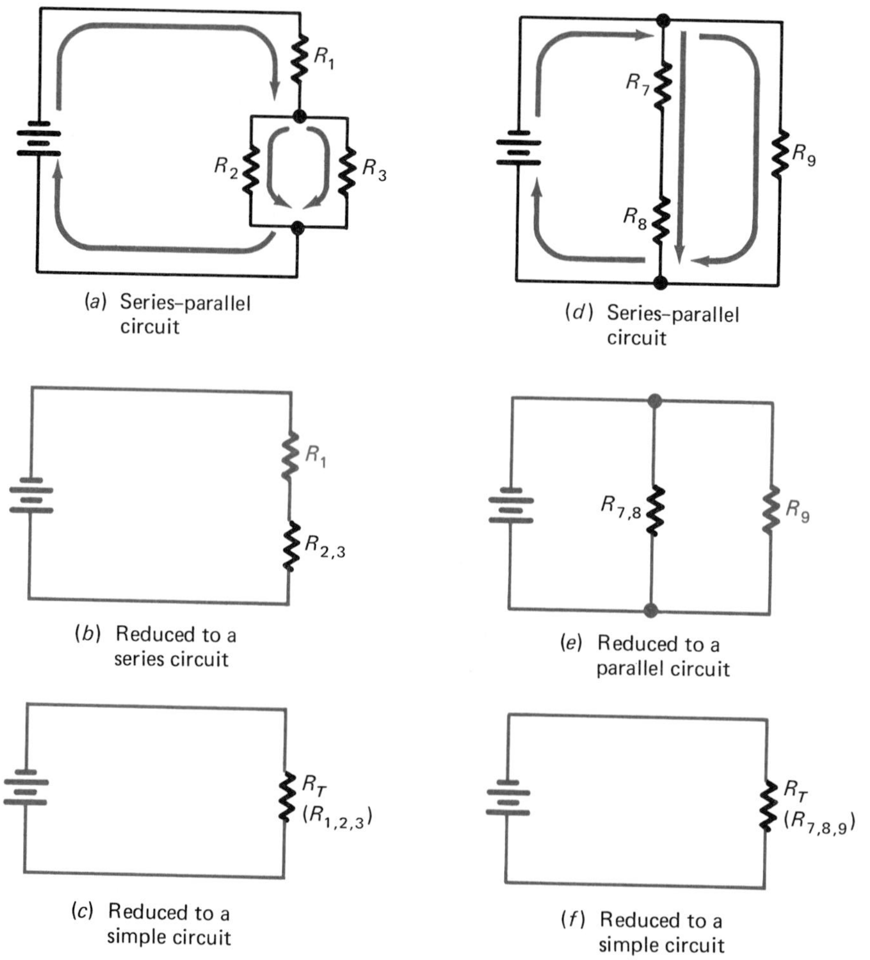

Fig. 5-27 Reducing series-parallel circuits to equivalent, simple circuits.

Kirchhoff's laws

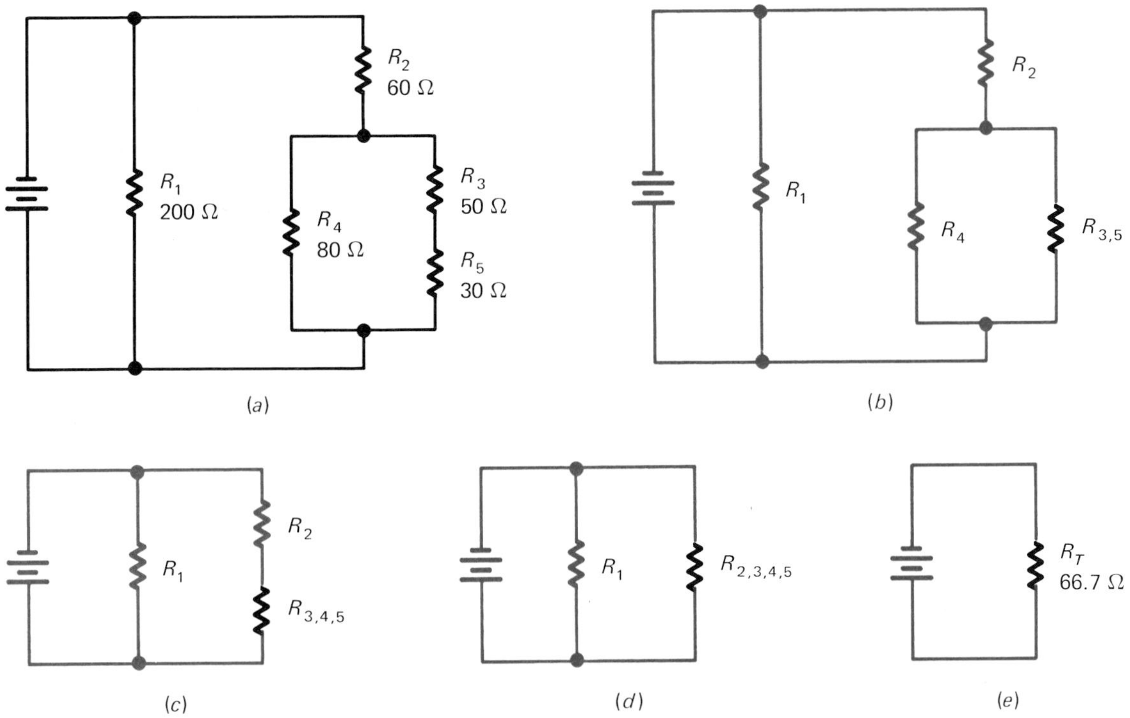

Fig. 5-28 Complex series-parallel circuit (a) can be simplified, as shown in (b) to (d).

sistors of R_7 and R_8. This results in the circuit shown in Fig. 5-27(*e*). Now the parallel circuit of Fig. 5-27(*e*) can be reduced, by a parallel resistance formula, to the simple circuit of Fig. 5-27(*f*). Refer to Fig. 5-27(*d*) and let $R_7 = 40\ \Omega$, $R_8 = 60\ \Omega$, and $R_9 = 20\ \Omega$.

$$R_{7,8} = R_7 + R_8 = 40 + 60 = 100\ \Omega$$

$$R_T = \frac{R_{7,8} \times R_9}{R_{7,8} + R_9} = \frac{100 \times 20}{100 + 20} = \frac{2000}{120} = 16.7\ \Omega$$

A more complex series-parallel circuit is shown in Fig. 5-28(*a*). Determining the total resistance of the circuit is illustrated in Fig. 5-28(*b*) through (*e*). The calculations required to arrive at the total resistance are

$$R_{3,5} = R_3 + R_5 = 50 + 30 = 80\ \Omega$$

$$R_{3,4,5} = \frac{R_4 \text{ or } R_{3,5}}{2} = \frac{80}{2} = 40\ \Omega$$

$$R_{2,3,4,5} = R_2 + R_{3,4,5} = 60 + 40 = 100\ \Omega$$

$$R_T = \frac{R_1 \times R_{2,3,4,5}}{R_1 + R_{2,3,4,5}} = \frac{200 \times 100}{200 + 100} = \frac{20{,}000}{300}$$
$$= 66.7\ \Omega$$

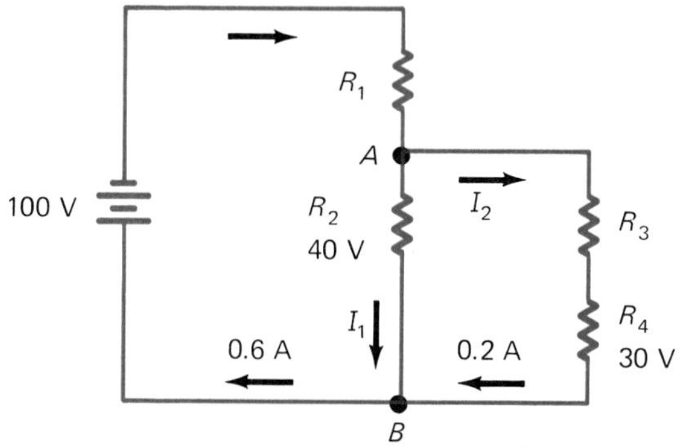

Fig. 5-29 Currents and voltages in series-parallel circuits. Kirchhoff's laws can be used to determine the unknown values.

Using Kirchhoff's Laws in Series-Parallel Circuits

Individual currents and voltages in series-parallel circuits can often be determined by Kirchhoff's laws. In Fig. 5-29, current in some of the conductors is given. The rest of the currents can be found using Kirchhoff's current law. Since 0.6 A enters the battery the same amount must leave. This is the total current I_T. Thus, $I_T = 0.6$ A. Resistors R_3 and R_4 are in series: therefore, $I_2 = 0.2$ A. The current entering point A is 0.6 A. Leaving that point are I_1 and I_2. From Kirchhoff's current law we know $I_1 + I_2 = 0.6$ A. But, $I_2 = 0.2$ A.

Thus, $I_1 + 0.2$ A = 0.6 A and $I_1 = 0.4$ A. We can check this result by examining point B. Here $I_1 + 0.2$ A = 0.6 A. Since I_1 was found to be 0.4 A, we have 0.4 A + 0.2 A = 0.6 A, which agrees with Kirchhoff's current law.

Some of the voltage drops in Fig. 5-29 are indicated beside the resistor symbol. The unspecified voltage drops can be found by using Kirchhoff's voltage law. In a series-parallel circuit, Kirchhoff's law applies to all loops or current paths in the circuit. Thus, for the circuit of Fig. 5-29, we can write two voltage relationships:

$$V_T = V_{R_1} + V_{R_2}$$
$$V_T = V_{R_1} + V_{R_3} + V_{R_4}$$

Studying the above two equations shows that

$$V_{R_2} = V_{R_3} + V_{R_4}$$

In other words, the voltage between points A and B is 40 V, no matter which path is taken. Rearranging the first equation we can solve for the voltage drop across R_1:

$$V_{R_1} = V_T - V_{R_2} = 100 \text{ V} - 40 \text{ V} = 60 \text{ V}$$

The voltage across R_3 can be found by rearranging the second equation:

$$V_{R_3} = V_T - V_{R_1} - V_{R_4} = 100 \text{ V} - 60 \text{ V} - 30 \text{ V} = 10 \text{ V}$$

Also, the voltage across R_3 could be found without knowing the voltage across R_1:

$$V_{R_3} = V_{R_2} - V_{R_4} = 40 \text{ V} - 30 \text{ V} = 10 \text{ V}$$

Solving Series-Parallel Problems

Several problems are solved in the examples that follow. These problems illustrate how to use Ohm's law and Kirchoff's two laws to solve series-parallel problems.

Example 5-7

For the circuit of Fig. 5-30(a), determine all unknown currents and voltages.

Given: $V_T = 60$ V
$V_{R_2} = 40$ V
$I_T = 4$ A
$R_3 = 20\ \Omega$

Find: $V_{R_1}, V_{R_3}, I_{R_1}, I_{R_2}, I_{R_3}$

Known: Ohm's law and Kirchhoff's laws

Solution: $V_{R_3} = V_T - V_{R_2} = 60 \text{ V} - 40 \text{ V} = 20 \text{ V}$
$V_{R_1} = V_T = 60$ V
$$I_{R_3} = \frac{V_{R_3}}{R_3} = \frac{20 \text{ V}}{20\ \Omega} = 1 \text{ A}$$
$I_{R_2} = I_{R_3} = 1$ A
$I_{R_1} = I_T - I_{R_2} = 4 \text{ A} - 1 \text{ A} = 3 \text{ A}$

Answers: $V_{R_1} = 60$ V, $V_{R_3} = 20$ V, $I_{R_1} = 1$ A, $I_{R_2} = 1$ A, $I_{R_3} = 1$ A

All the currents and voltages for Fig. 5-30(a) have now been determined. Using Ohm's law and the power formula, we could easily find the resistance and power of each resistor.

Example 5-8

Refer to Fig. 5-30(b). For this circuit compute the resistance of R_3, the power dissipation of R_4, and the voltage across R_1.

Given: $V_T = 100$ V
$I_{R_1} = 0.8$ A
$I_{R_3} = 0.3$ A
$R_2 = 100\ \Omega$
$V_{R_4} = 30$ V

Find: R_3, P_{R_4}, V_{R_1}

Known: Ohm's law and Kirchhoff's laws

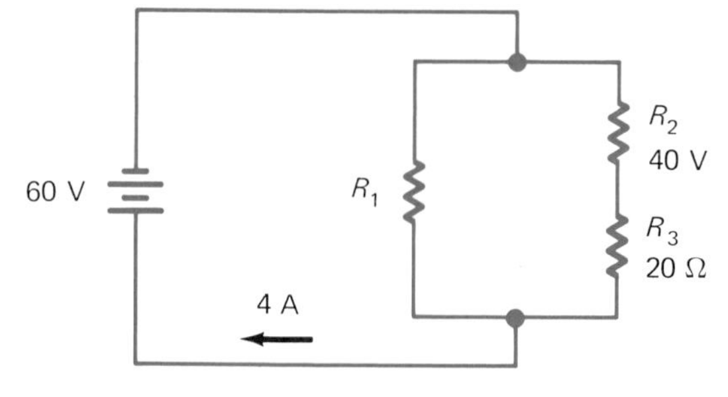

(a)

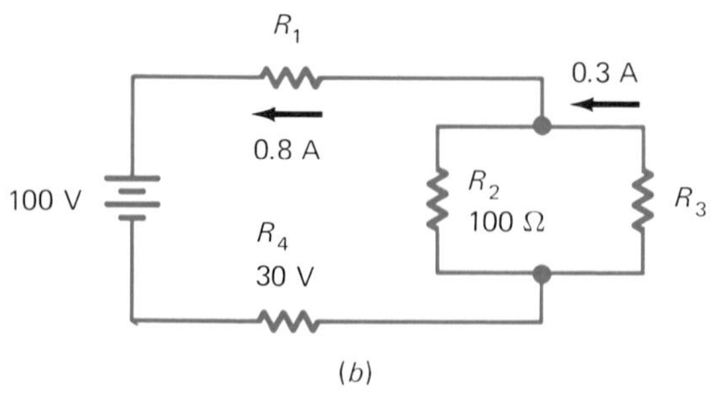

(b)

Fig. 5-30 Circuits for Examples 5-7 and 5-8.

Solution: $I_{R_2} = I_{R_1} - I_{R_3} = 0.8\text{ A} - 0.3\text{ A}$
$= 0.5\text{ A}$

$V_{R_2} = I_{R_2} R_2 = 0.5\text{ A} \times 100\ \Omega$
$= 50\text{ V}$

$V_{R_1} = V_T - V_{R_2} - V_{R_4}$
$= 100\text{ V} - 50\text{ V} - 30\text{ V}$
$= 20\text{ V}$

$V_{R_3} = V_{R_2} = 50\text{ V}$

$$R_3 = \frac{V_{R3}}{I_{R3}} = \frac{50\text{ V}}{0.3\text{ A}} = 166.7\ \Omega$$

$P_{R_4} = I_{R_4} V_{R_4} = 0.8\text{ A} \times 30\text{ V} = 24\text{ W}$

Answers: $R_3 = 166.7\ \Omega$, $P_{R_4} = 24\text{ W}$,
$V_{R_1} = 50\text{ V}$

Relationships in Series-Parallel Circuits

Series-parallel relationships

Like the series circuits, the current, voltage, and power in series-parallel circuits are dependent on one another. That is, changing any one resistance usually changes all currents, voltages, and powers except the source voltage. For example, increasing the resistance of R_3 in Fig. 5-31(*a*) from 40 Ω to 90 Ω will cause the changes listed below:

1. R_T increases (because $R_{3,4}$ increases).
2. I_T decreases (because R_T increases).
3. V_{R_1} decreases (because $I_T = I_{R_1}$).
4. V_{R_2} increases (because $V_{R_2} = V_T - V_{R_1}$).
5. I_{R_2} increases (because $I_{R_2} = V_{R_2}/R_2$).
6. I_{R_4} decreases (because $I_{R_4} = I_{R_1} - I_{R_2}$).
7. V_{R_4} decreases (because $V_{R_4} = I_{R_4} R_4$).
8. V_{R_3} increases (because $V_{R_3} = V_{R_2} - V_{R_4}$).
9. P_{R_1} decreases, P_{R_2} increases, P_{R_4} decreases, and P_T decreases (because of I and V-changes specified above).
10. P_{R_3} increases (because V_{R_3} increases more than I_{R_3} decreases).

The magnitudes of these changes are shown in Fig. 5-31(*b*).

The changes detailed above will occur when any resistor is changed, except in circuits like the one in Fig. 5-28(*a*). In this circuit, changing R_1 will affect only the current and power of R_1 and the battery. This is because R_1 is in parallel with the combination of R_2, R_3, R_4, and R_5 [Fig. 5-28(*d*)].

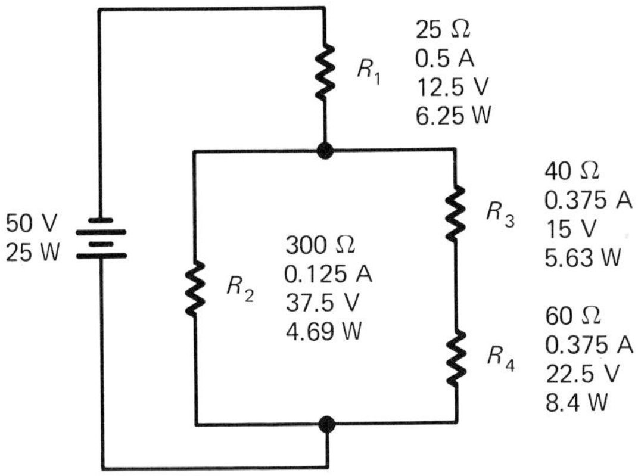

(*a*) Original circuit values

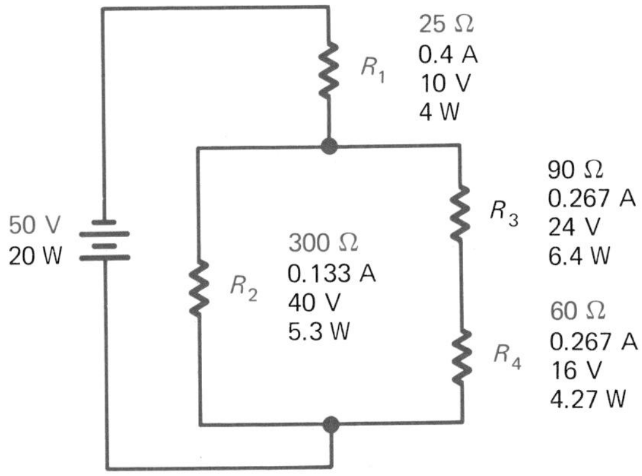

(*b*) Values after R_3 increases

Fig. 5-31 Effects of changing R_3. (a) Original circuit values. (b) Values after R_3 increases.

Review Questions

36. Refer to Fig. 5-28(*a*):
    a. Are any of these resistors directly in series? If so, which ones?
    b. Are any of these resistors directly in parallel? If so, which ones?
37. Refer to Fig. 5-28(*a*). Is each of the following statements true or false?
    a. $I_{R_1} = I_{R_2} + I_{R_3}$
    b. $I_{R_2} = I_{R_4} + I_{R_5}$
    c. $I_{R_1} = I_T - I_{R_2}$
    d. $V_{R_1} = V_{R_4}$
    e. $V_{R_4} = V_T = V_{R_2} + V_{R_3} + V_{R_5}$
38. Refer to Example 5-7. Compute the following:
    a. R_1
    b. P_{R_3}
    c. P_T
39. Refer to Fig. 5-27(*a*). If R_2 is decreased in

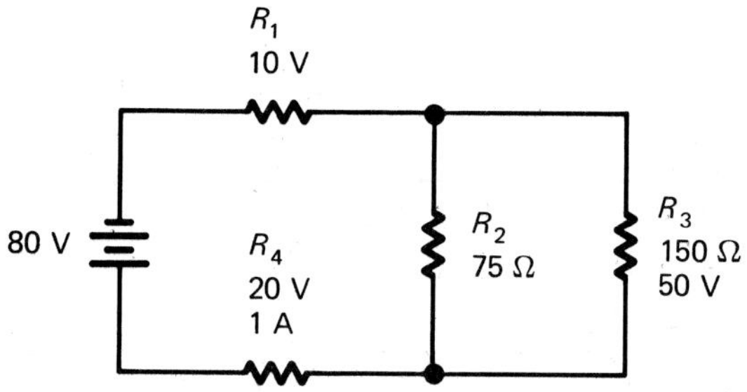

Fig. 5-32 Circuit diagram for Review Question 41.

value, indicate whether each of the following will increase or decrease.

a. I_{R_1}
b. I_{R_3}
c. P_{R_3}
d. V_{R_1}

40. Refer to Fig. 5-31(*a*). Change the value of R_2 to 100 Ω. Then determine the value of each of the following:
    a. R_T
    b. V_{R_1}
    c. I_{R_4}
    d. P_{R_3}

41. For Fig. 5-32, determine the following:
    a. R_1
    b. I_{R_2}
    c. P_{R_3}
    d. I_T
    e. R_T

Summary

1. Multiple-load circuits include series, parallel, and series-parallel circuits.
2. Series circuits are single-path circuits.
3. The same current flows throughout a series circuit.
4. The total resistance equals the sum of the individual resistances in a series circuit.
5. The sum of the voltage drops around a circuit equals the total source voltage (Kirchhoff's voltage law).
6. A voltage drop (voltage across a load) indicates that electric energy is being converted.
7. Polarity of a voltage drop indicates the direction of current flow.
8. The voltage measured across an open, series load will be equal to the source voltage.
9. A shorted load in a series circuit increases the current, voltage, and power of the other loads.
10. When one resistance in a series circuit is smaller than the tolerance of another, the small resistance has little effect on circuit current and power.
11. The highest resistance in a series circuit drops the most voltage.
12. Maximum power transfer occurs when the source resistance equals the load resistance.
13. Conductance is the ability to allow current flow.
14. Parallel circuits are multiple-path circuits.
15. Each branch of a parallel circuit is independent of the other branches.
16. The same voltage appears across each branch of a parallel circuit.
17. The currents entering a junction must equal the currents leaving a junction (Kirchhoff's current law).
18. The total current in a parallel circuit equals the sum of the branch currents.
9. Adding more resistance in parallel decreases the total resistance.
20. The total parallel resistance is always less than the lowest branch resistance.
21. The relationships of both series and parallel circuits are applicable to parts of series-parallel circuits.
22. In all circuits:

$$P_T = P_{R_1} + P_{R_2} + P_{R_3} + \text{etc.}$$
$$R = \frac{1}{G}$$

23. In series circuits:

$$I_T = I_{R_1} = I_{R_2} = I_{R_3} = \text{etc.}$$
$$V_T = V_{R_1} + V_{R_2} + V_{R_3} + \text{etc.}$$
$$R_T = R_1 + R_2 + R_3 + \text{etc.}$$

24. In parallel circuits:

$$I_T = I_{R_1} + I_{R_2} + I_{R_3} + \text{etc.}$$
$$V_T = V_{R_1} = V_{R_2} = V_{R_3} = \text{etc.}$$
$$R_T = \frac{1}{\frac{1}{R_1} + \frac{1}{R_2} + \frac{1}{R_3} + \text{etc.}}$$

For two parallel branches:

$$R_T = \frac{R_1 \times R_2}{R_1 + R_2}$$

For n parallel resistors, R ohms each:

$$R_T = \frac{R}{n}$$

Chapter Review Questions

Here is a chance to see what you have learned. For questions 5-1 to 5-15, determine whether the statement is true or false.

5-1. In a series circuit, the total power is equal to the sum of the individual powers.

5-2. In a series circuit, the total current is equal to the sum of the individual currents.

5-3. In a parallel circuit, the highest resistance will dissipate the most power.

5-4. In a series circuit, the lowest resistance will drop the most voltage.

5-5. Adding more resistances in parallel decreases the total resistance.

5-6. To measure the resistance of an individual resistor in a parallel circuit, one end of the resistor must be disconnected from the circuit.

5-7. Maximum power transfer occurs when the source resistance is very low compared with the load resistance.

5-8. The negative polarity marking on a resistor in a schematic diagram indicates an excess of electrons at that point.

5-9. Voltmeters have a very high internal resistance.

5-10. In a series circuit, an open load drops the total source voltage.

5-11. If one load in a series circuit shorts out, all the other loads will use less power.

5-12. If one load in a parallel circuit opens, all the other loads will use less power.

5-13. Adding more resistors to a parallel circuit will increase the total power used by the circuit.

5-14. The total resistance of parallel resistances is always less than the value of the lowest resistance.

5-15. Changing the value of one resistor in a parallel circuit changes the current through all other resistors in that circuit.

For questions 5-16 and 5-17, choose the letter that best completes the sentence.

5-16. The total resistance of a 30-Ω resistor and a 60-Ω resistor connected in series would be
(A) 20 Ω (B) 30 Ω (C) 45 Ω (D) 90 Ω

5-17. The total resistance of a 30-Ω resistor and a 60-Ω resistor connected in parallel would be
(A) 20 Ω (B) 30 Ω (C) 45 Ω (D) 90 Ω

For questions 5-18 to 5-20, make the indicated computations.

5-18. For the circuit in Fig. 5-33(*a*), compute the following:
a. V_{R_1}
b. R_2
c. P_T

5-19. For the circuit in Fig. 5-33(*b*), compute the following:
a. I_{R_1}
b. R_1
c. I_{R_2}

5-20. For the circuit in Fig. 5-33(*c*), compute the following:
a. V_{R_3}
b. I_{R_1}
c. V_{R_1}

Answer the following question.

5-21. If R_3 in Fig. 5-33(*c*) is decreased to 250 Ω, will V_{R_2} increase, decrease, or remain the same?

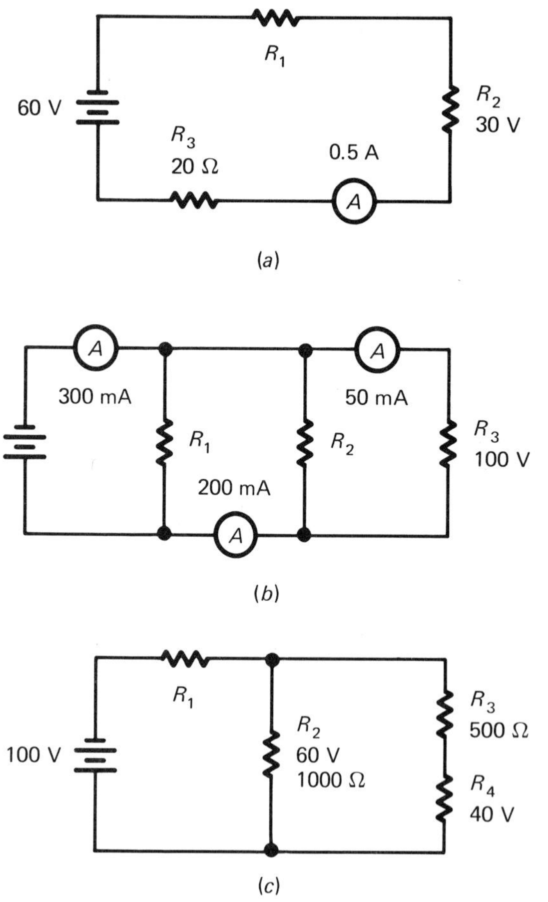

Fig. 5-33 Circuits for Chapter Review Questions 5-18 to 5-20.

Answers to Review Questions

1. a. V_{R4}
   b. I_T
   c. I_{R2}
2. $P_T = P_{R_1} + P_{R2} + P_{R3} +$ etc. Or, the total power is equal to the sum of the individual powers.
3. A series circuit is a multiple-load circuit with only one path for current.
4. No, there is only one path for the current to flow through.
5. $R_T = R_1 + R_2 + R_3 +$ etc., and $R_T = V_T/I_T$
6. Yes
7. $V_T = V_{R_1} + V_{R2} + V_{R3} +$ etc.
8. 500-Ω resistor
9. No
10. The open resistor
11. a. V_{R2} increases.
    b P_{R3} increases.
    c. I_T increases.
12. a. 500 Ω
    b. 0.24 A
    c. 24 V
    d. 23.04 W
13. R_1 would be 42 Ω. The battery would have to furnish 1.89 W.
14. The speed will decrease.
15. 0 V
16. 9 mA
17. When the load resistance is equal to the source resistance
18. Greater than
19. A parallel circuit has two or more loads and two or more independent current paths.
20. The total (source) voltage is applied to each load.
    $V_T = V_{R_1} = V_{R2} = V_{R3}$
21. The total current divides among the branches of the circuit.
    $I_T = I_{R_1} + I_{R2} + I_{R3} +$ etc.
22. The lowest resistor dominates.
23. $R_T = \dfrac{1}{\dfrac{1}{R_1} + \dfrac{1}{R_2}}$ and

    $R_T = \dfrac{R_1 \times R_2}{R_1 + R_2}$
24. Conductance is the ability of a material to support electric current.
25. Decrease
26. Less than 10 Ω
27. 14 Ω
28. No
29. 100 Ω
30. No
31. a. R_T increases.
    b. I_T decreases.
    c. P_T decreases.
    d. V_T remains the same.
32. $I_{R_1} = I_T - (I_{R2} + I_{R3})$
    $= 5\text{ A} - 3\text{ A} = 2\text{ A}$
33. $V_{B_1} = V_{R3} = I_{R3}R_3 = 1\text{ A} \times 20\ \Omega = 20\text{ V}$
34. $I_{R2} = 3\text{ A} - 1\text{ A} = 2\text{ A}$
    $V_{R2} = V_{R3} = 20\text{ V}$
    $R_2 = \dfrac{V_{R2}}{I_{R2}} = \dfrac{20\text{ V}}{2\text{ A}} = 10\ \Omega$
35. $V_{R_1} = V_{R2} = 20\text{ V}$
    $P_{R_1} = V_{R_1} I_{R_1} = 20\text{ V} \times 2\text{ A}$
    $= 40\text{ W}$
36. a. Yes, R_3 and R_5
    b. No
37. a. False
    b. True
    c. True
    d. False
    e. True
38. $R_1 = 20\ \Omega, P_{R3} = 20\text{ W}$, and $P_T = 240\text{ W}$
39. a. Increases
    b. Decreases
    c. Decreases
    d. Increases
40. $R_T = 75\ \Omega$, $V_{R_1} = 16.7\text{ V}$, $I_{R4} = 0.33\text{ A}$, and $P_{R3} = 4.44\text{ W}$
41. $R_1 = 10\ \Omega, I_{R2} = 0.67\text{ A}$, $P_{R3} = 16.7\text{ W}, I_T = 1\text{ A}$, and $R_T = 80\ \Omega$

Magnetism and Electromagnetism

Electricity and magnetism cannot be separated. Wherever an electric current exists, a magnetic field also exists. Magnetism, created by an electric current, operates many devices such as transformers, motors, and loudspeakers.

After completing this chapter, you will know how magnetism is produced and how temporary and permanent magnets are made. You will understand the operation of many common magnetic devices.

6-1 MAGNETISM AND MAGNETS

Magnetism is an invisible force field that acts on some materials but not on other materials. A *magnet* possesses this invisible force. Magnets are devices which will attract pieces of iron and steel because of this invisible force field. Lodestone (an iron compound) is a natural magnet. It was discovered centuries ago.

The magnets we use today are all manufactured. They are made from various alloys containing elements like copper, nickel, aluminum, iron, and cobalt. These magnets are much much stronger than the natural lodestone magnet.

6-2 MAGNETIC FIELDS, FLUX, AND POLES

The invisible force of magnetism is referred to as a *magnetic field*. This field extends out from the magnet in all directions, as illustrated in Fig. 6-1. In this figure the lines extending from the magnet represent the magnetic field.

The invisible lines of force that make up the magnetic field are known as magnetic *flux*. The lines of force in Fig. 6-1 represent the flux. Where the lines of flux are dense, the magnetic field is strong. Where the lines of flux are sparse, the field is weak. The lines of flux are most dense at the ends of the magnet; therefore, the magnetic field is strongest at the ends of the magnet.

Looking again at Fig. 6-1, you can see that the lines of force are farther apart as they get farther from the magnet. This means that the magnetic field gets weaker as the distance from the magnet increases. Assume that flat planes were pushed through the magnetic field, as indicated in Fig. 6-1. The plane farthest from the magnet contacts only four lines of flux. The plane closest to the magnet crosses six lines of flux. Obviously, the plane closest to the end of the magnet is in the stronger field.

The arrows on the lines in Fig. 6-1 indicate

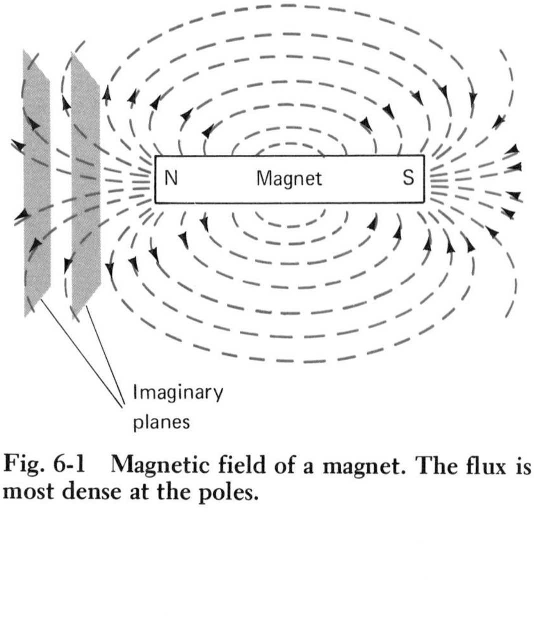

Fig. 6-1 Magnetic field of a magnet. The flux is most dense at the poles.

From page 110:
Magnetism

Lodestone

Magnetic field

Flux

On this page:
North pole

South pole

Like poles repel

Unlike poles attract

the direction of the flux. Lines of force are always assumed to leave the *north pole* (N) and enter the *south pole* (S) of a magnet. North pole and south pole refer to the polarity of the ends of a magnet. When a magnet is suspended on a string and allowed to rotate, its ends will point north and south. The end of the magnet which seeks (or points to) the earth's north magnetic pole is the magnet's north pole. The south pole of the magnet points to the earth's south pole. Assuming that flux (lines of force) leaves the north pole and enters the south pole was an arbitrary decision. However, assigning direction to the lines aids in understanding the behavior of magnetism.

Like magnetic poles repel each other. The two north poles of Fig. 6-2(*a*) and the two south poles of Fig. 6-2(*b*) create a repelling force. The closer the poles are, the more they repel each other. The force of repulsion between magnetic poles varies inversely as the square of the distance between them. That is, if the distance is doubled, the force becomes one-fourth as great. Or, if the distance is halved, the force becomes 4 times as great.

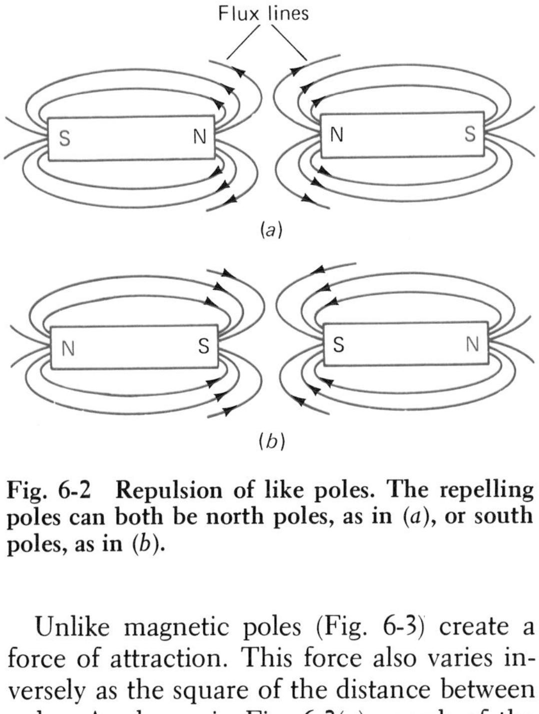

Fig. 6-2 Repulsion of like poles. The repelling poles can both be north poles, as in (*a*), or south poles, as in (*b*).

Unlike magnetic poles (Fig. 6-3) create a force of attraction. This force also varies inversely as the square of the distance between poles. As shown in Fig. 6-3(*a*), much of the flux of the two magnets joins together to form the force of attraction. When the two poles touch, essentially all the flux joins together [Fig. 6-3(*b*)]. The two magnets, when joined

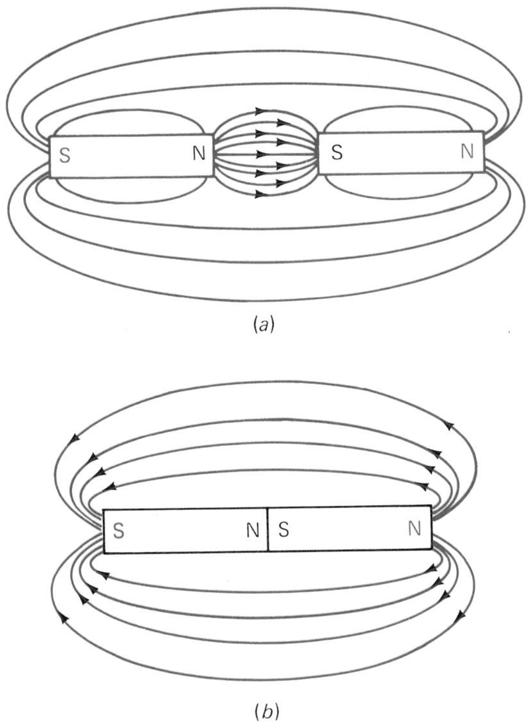

Fig. 6-3 Attraction of unlike poles. The force of attraction is greatest when the poles are touching, as in (*b*).

together, behave as a single magnet. They create a single magnetic field.

Magnetic lines of force (flux) are assumed to be continuous loops. Although not shown in Fig. 6-3, the flux lines continue on through the magnet. They do not stop at the poles. In fact, the poles of the magnet are merely the areas where most of the flux leaves the magnet and enters the air. If a magnet is broken in half (Fig. 6-4) two new poles are created. The one magnet becomes two magnets. A magnet can be broken into many pieces, and each piece becomes a new magnet with its own north pole and south pole.

A magnet does not have to have poles. All the flux lines can be within the magnet itself. This idea is illustrated in Fig. 6-5. In Fig. 6-5(*a*), a typical horseshoe magnet is shown.

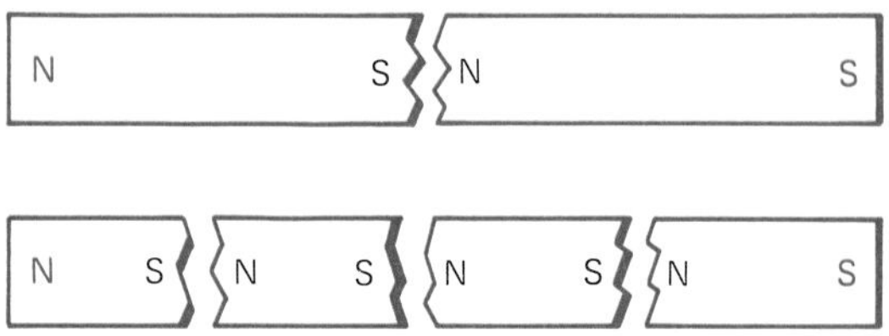

Fig. 6-4 Creation of poles. Each time the magnet is broken a new pair of poles is created.

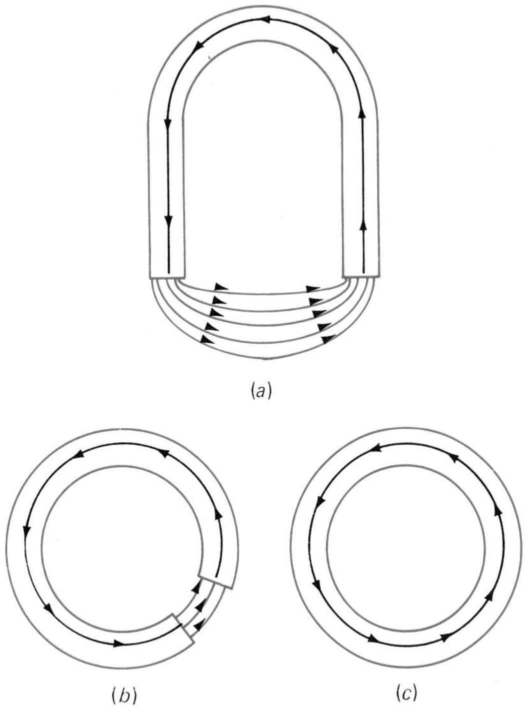

Fig. 6-5 Magnet without poles. Forming a horseshoe magnet (*a*) into a circle (*c*) eliminates the poles.

The dark lines represent the flux within the magnet. In Fig. 6-5(*b*) the horseshoe magnet has been bent around to form a circle with a small gap. And finally, in Fig. 6-5(*c*) the gap has been closed to form a circle. Now all the flux (and the magnetic field) is confined within the magnet itself. The magnet has no poles.

Review Questions

1. What is magnetism?
2. What is a magnetic field?
3. What is magnetic flux?
4. Does magnetic flux flow from the north to the south pole or from the south to the north pole?
5. Will a north magnetic pole be attracted by a south pole or a north pole?
6. Two magnetic poles are 1 cm apart. If they are moved closer together so that they are only 0.5 cm apart, will the force between them double or quadruple?
7. When the flux lines of two poles join together, are the poles of the same or opposite polarities?
8. Are lines of flux continuous loops?
9. Does breaking a magnet in half destroy its magnetic field?
10. Do all magnets have a north pole?

6-3 ELECTROMAGNETISM

Left-hand rule

So far our discussion has centered around the magnetic field and flux possessed by a magnet. However, magnetic fields are also created by electric current. The current-carrying conductor in Fig. 6-6 has a magnetic field around it. The field is always at right angles (perpendicular) to the direction of current. Although the field is shown in only five places along the conductor, it actually exists as a continuous field for the entire length. Notice in Fig. 6-6 that the magnetic field has no poles. The flux exists only in the air. However, the flux still has an assumed direction, just as it does in the circular magnet of Fig. 6-5(*c*). The direction of the flux around a conductor can be determined by using a left-hand rule. Grasp the conductor with your left hand so that your thumb points in the direction of current. Your fingers will then indicate the direction of the flux.

Often the direction of current and flux is indicated as shown in Fig. 6-7. In this figure, you are looking at the ends of current-carrying conductors. Of course, the end of a round conductor looks like a circle. When the circle has an x in it [Fig. 6-7(*a*)], the current is going into the paper. A way to remember this convention is to visualize an arrow or a dart as

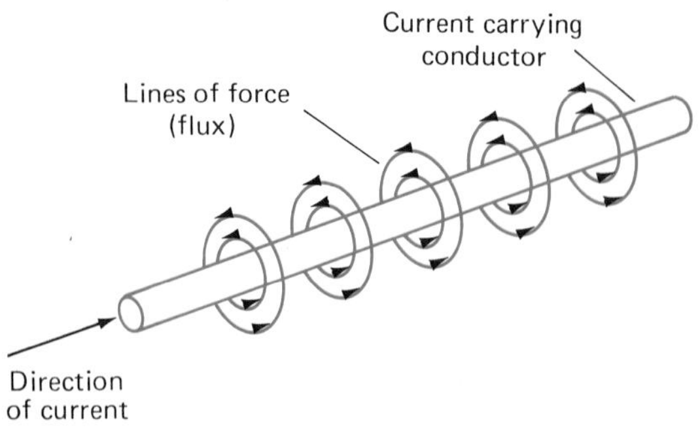

Fig. 6-6 Magnetic field around a conductor. The flux is perpendicular to the direction of current.

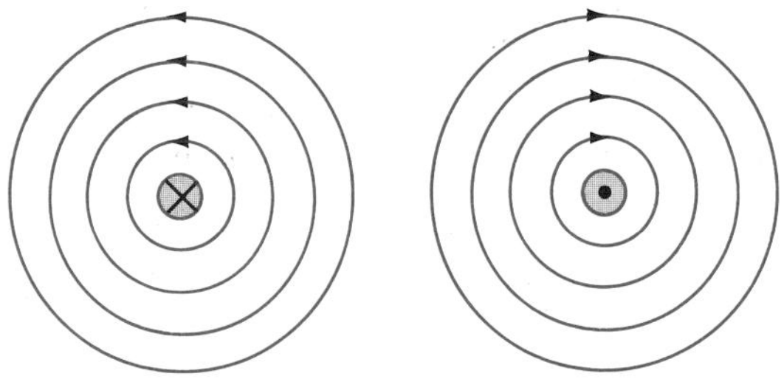

(a) Current entering the paper (b) Current coming out of the paper

Fig. 6-7 Indicating direction of current. (*a*) Current moving toward the page. (*b*) Current moving out from the page.

indicating the direction of current. If the dart were going into the paper, you would see the tail end with feathers arranged in an x shape. If the dart were coming toward you (out of the paper), you would see the front, pointed end. Thus, the dot in a circle represents current flowing toward you.

The strength of the magnetic field around a conductor is determined by the amount of current flowing through the conductor. The strength, at some fixed distance from the conductor, is directly proportional to the current. Doubling the current will double the strength of the magnetic field.

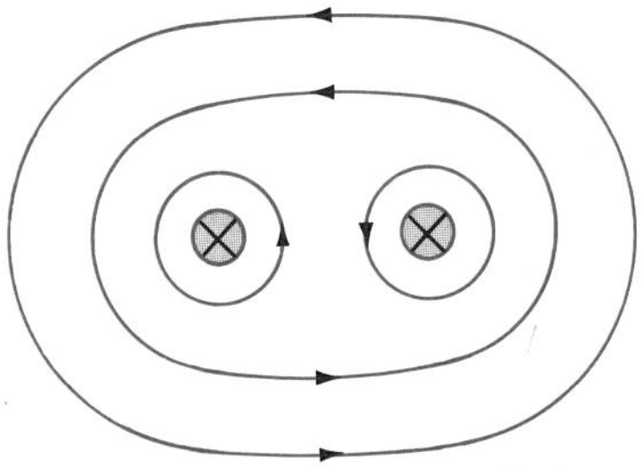

(a) Attraction between conductors

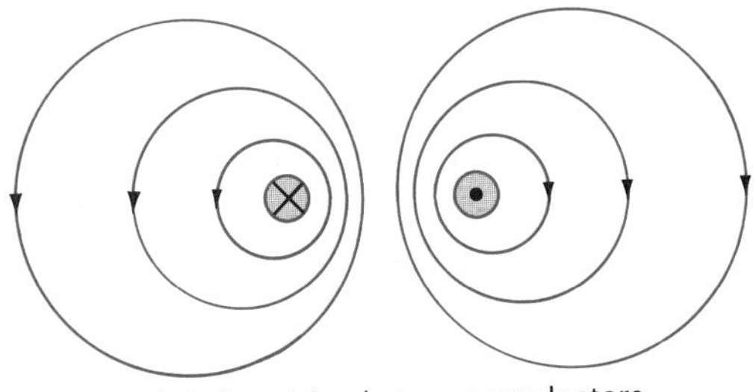

(b) Repulsion between conductors

Fig. 6-8 Forces between parallel conductors. (*a*) Attraction of conductors. (*b*) Repulsion of conductors.

Force between Conductors

Two parallel, current-carrying conductors will attract each other if the currents in the conductors are in the same direction. This is because the fields of the two conductors join together, as shown in Fig. 6-8(*a*). When fields join together, there is a force of attraction. This is the same type of attraction that occurs between unlike poles (Fig. 6-3).

When the currents in parallel conductors are not in the same direction, a force of repulsion exists. The magnetic fields in Fig. 6-8(*b*) are not able to join together because they are opposing each other. Since lines of flux cannot cross each other and cannot share the same space, they repel each other.

Coils

The magnetic field of a single conductor is usually too weak for most practical applications. However, a stronger field can be created by combining the fields around two or more conductors. This is done by coiling a conductor, as shown in Fig. 6-9. A conductor formed into this shape is called a *coil*. Forming a coil out of a conductor creates an electromagnet. Notice that the coils in Fig. 6-9 have poles. The poles are located where the flux is the most concentrated. The flux is the most concentrated in the center of the coil. Thus, the poles are at the ends of the coil where the flux enters and leaves the center of the coil.

The polarity of a coil can be determined by again applying a left-hand rule. This time, wrap your fingers around the coil in the direction that current flows through the turns of the coil. Your thumb will then point to the north pole (it also indicates direction of the flux). Reversing the direction of the current will reverse the polarity of an electromagnet. Of course, the strength of the magnetic field would remain the same.

In Fig. 6-9(*a*) the three turns of the coil have considerable space between them. This spacing of the turns allows some of the flux to loop only a single conductor. The result is a weakening of the magnetic field at the poles of the magnet. This loss of flux at the poles can be minimized by close-winding the coil, as shown in Fig. 6-9(*b*). In this figure all the flux of the individual turns combines to produce one stronger magnetic field.

Coil

Electromagnet

Close-winding

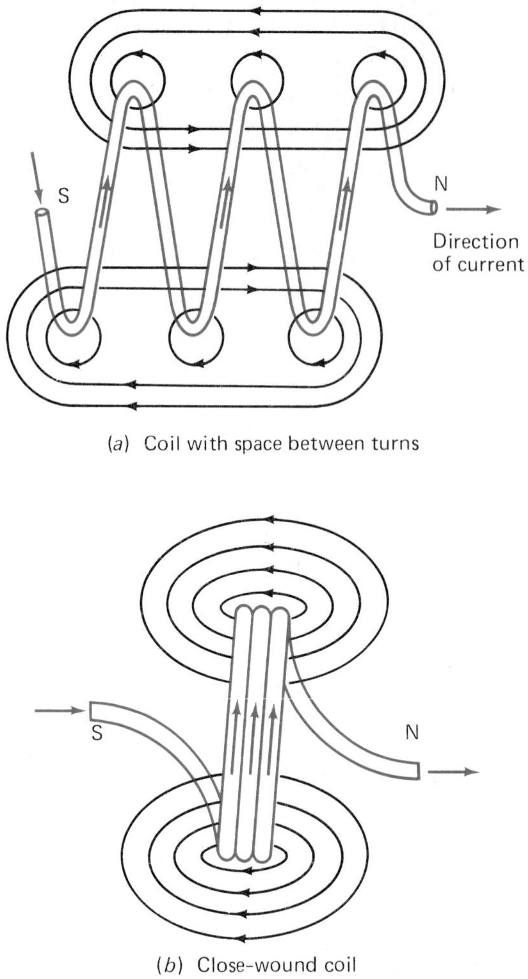

Fig. 6-9 Magnetic fields of coils. (*a*) Coils with space between turns. (*b*) Close-wound coil.

Review Questions

11. Are the magnetic fields around a conductor parallel or perpendicular to the conductor?
12. Does the magnetic field around a straight conductor have a north pole and a south pole?
13. You are looking at the end of a conductor. The current is flowing away from you. Will the direction of the flux be clockwise or counterclockwise?
14. What is a coil?
15. Will the north pole created by the two parallel wires in Fig. 6-8(*b*) be above the two wires or below them?
16. Will increasing the space between the turns of a coil increase or decrease the pole flux?

6-4 MAGNETIC MATERIALS

Magnetic materials

Ferromagnetic materials

Nonmagnetic materials

Materials which are attracted by magnetic fields (and materials from which magnets can be made) are called *magnetic materials*. The most common magnetic materials are iron, iron compounds, and alloys containing iron or steel. Magnetic materials are also called ferromagnetic materials. Ferro is a prefix which means iron, and most magnetic materials contain iron. A few materials, such as nickel and cobalt, are slightly magnetic. They are attracted by strong magnets. But compared to iron, they are only weakly magnetic.

Materials which are not attracted by magnets are called *nonmagnetic materials*. Most materials are in this category. Most metals (except iron) are nonmagnetic. A magnet does not attract copper, brass, aluminum, silver, zinc, or tin. Neither does a magnet attract most nonmetals. Wood, paper, leather, plastic, and rubber are all nonmagnetic. A nonmagnetic material does not stop magnetic flux. Flux goes through nonmagnetic materials about as readily as it goes through air.

Theory of Magnetism

Several complete, rather complex, theories are available to explain magnetism. We will examine some ideas from these theories. Then we can understand why some materials are magnetic and others are not.

It is well known, and easily demonstrated, that an electric current produces a magnetic field. Since current is nothing but the movement of electric charges, any moving charge should create a magnetic field. The electrons of an atom, or molecule, possess electric charges, and they are in motion. Therefore, they should have a magnetic field associated with them. In the molecules of some materials, an equal number of electrons spin or rotate in each direction. This causes the magnetic field of one electron to be canceled by the opposite field of another electron. The result is that the molecule ends up with no overall magnetic field. Nonmagnetic materials behave in this manner. In other materials, more electrons spin in one direction than in the other. Thus, not all the magnetic fields are canceled, and a molecule of the material possesses a weak magnetic field. These molecules then group together to form small *domains* that have a magnetic field with definite poles. Only magnetic materials form these domains.

Figure 6-10 represents the theoretical, internal structure of a piece of magnetic material. In this figure, magnetic domains are represented by the small rectangles. The domains are randomly arranged. They are also randomly moving about within the material. The magnetic fields of the randomly arranged domains cancel each other. Overall, the piece of material in Fig. 6-10 has no magnetic field or poles. It is an *unmagnetized* piece of magnetic material.

Fig. 6-10 Random arrangement of magnetic domains in an unmagnetized material.

Temporary and Permanent Magnets

When a magnetic material is put in the magnetic field of a magnet it becomes *magnetized*. As shown in Fig. 6-11, all the domains are aligned in an orderly fashion. The magnetic field of one domain supports the field of the next domain. The magnetic material in Fig. 6-11 became a magnet when it was magnetized. Whether it is a *temporary* or a *permanent* magnet depends on how it reacts when removed from the original magnetic field. If most of the domains remain aligned, as shown in Fig. 6-11, the magnetic material becomes a *permanent magnet*. Many alloys of iron, especially those that contain over 0.8 percent carbon, become permanent magnets. Most tools such as screwdrivers, pliers, and hacksaw blades contain over 0.8 percent carbon. They can become permanent magnets that are capable of attracting other magnetic materials. Most permanent magnets are made of alloys (such as alnico) which can be highly magnetized. Alnico magnets have very strong fields. They are composed of iron, cobalt, nickel, aluminum, and copper. Ceramic materials also make strong permanent magnets. Permanent magnets are used to make door catches, and the magnetic poles for loudspeakers, electric meters, and motors.

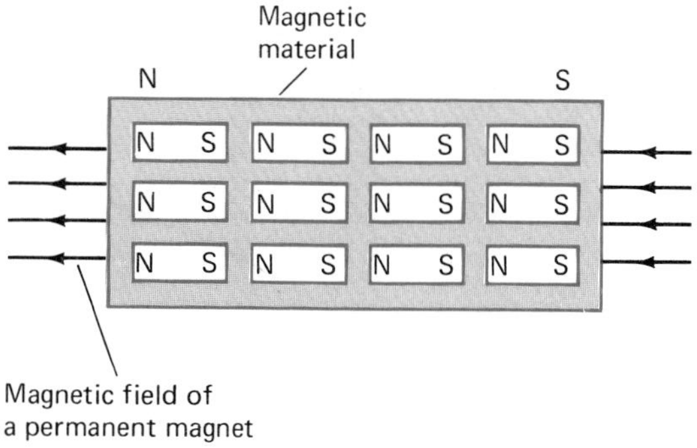

Fig. 6-11 Orderly arrangement of domains in a magnetized material.

Materials such as pure iron, ferrite, and silicon steel make *temporary* magnets. When these materials are removed from a magnetic field, their domains immediately revert to the random arrangement of Fig. 6-10. They lose almost all their magnetism. They no longer attract other magnetic materials. Temporary magnetic materials are used in great quantities in motors, generators, transformers, and electromagnets.

6-5 MAGNETIZING MAGNETIC MATERIALS

Magnetic materials can be magnetized by the magnetic fields of either a permanent magnet or an electromagnet. Magnetizing with an electromagnet is shown in Fig. 6-12(*a*). When the switch is closed, the field from the coil will magnetize the tool steel in the screwdriver. Since tool steel makes a permanent magnet, the screwdriver remains magnetized after the switch is opened. Now the screwdriver can be used to attract other magnetic materials such as a wood screw [Fig. 6-12(*b*)].

When the material inserted in the center of a coil is mild steel or ferrite, a temporary magnet is made. In Fig. 6-13(*a*), a temporary magnet is shown attracting a magnetic material. It will continue to attract the material as long as the switch is closed and current flows. However, as shown in Fig. 6-13(*b*), the magnetic field disappears the instant the switch is opened. The screw is no longer attracted to the ferrite rod.

Domains

Unmagnetized

Magnetized

Permanent magnet

Temporary magnet

Alnico

Magnetomotive force (mmf)

Saturation

Demagnetize

Alternating current

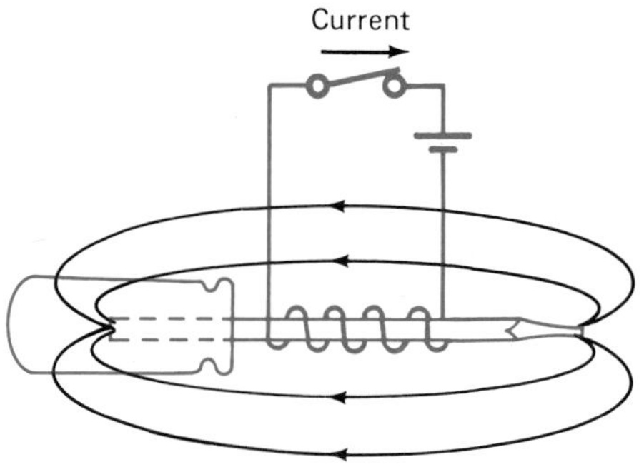

(*a*) Magnetizing a screwdriver

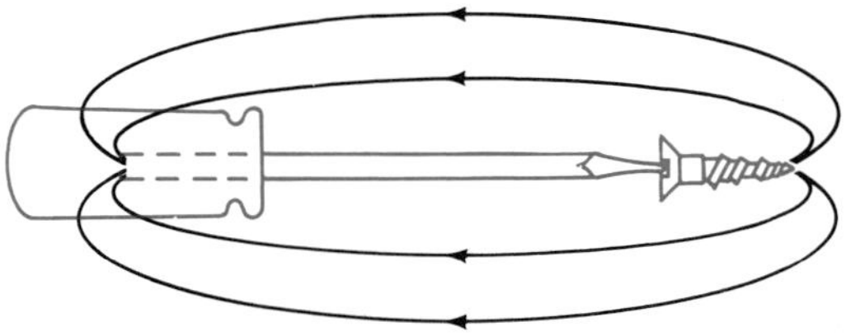

(*b*) The screwdriver becomes a permanent magnet

Fig. 6-12 Making a permanent magnet.

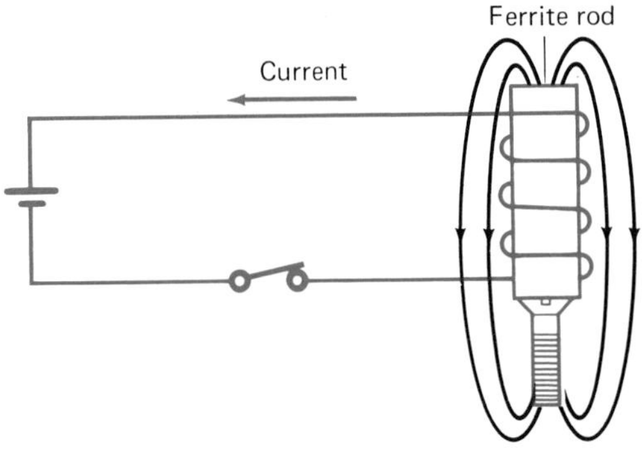

(*a*) Magnetizing a ferrite rod (temporary magnet)

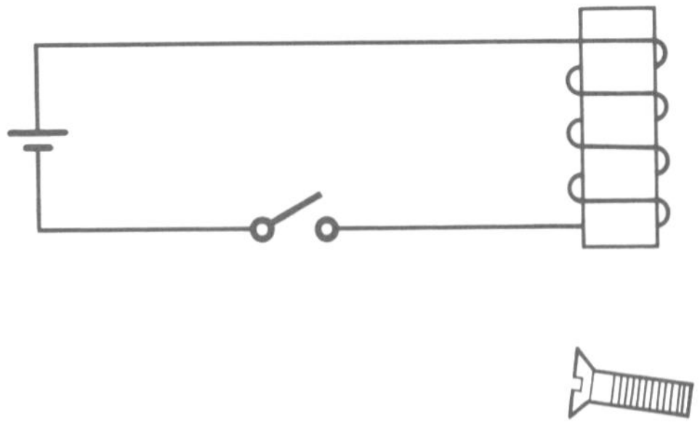

(*b*) When current stops, ferrite rod loses its magnetism

Fig. 6-13 Making a temporary magnet.

Review Questions

17. What element is contained in most magnetic materials?
18. List four metals that are nonmagnetic.
19. What is a magnetic domain?
20. What type of steel is used in transformers?
21. Are permanent magnets made from a single element or from alloys?
22. What happens to the flux in a ferrite rod in a coil when the electric circuit is opened?
23. What happens to the flux in a carbon-steel rod in a coil when the circuit is opened?

6-6 MAGNETOMOTIVE FORCE

The effort exerted in creating a magnetic field (and flux) is called *magnetomotive force*. Magnetomotive force is abbreviated mmf. Increasing either the number of turns or the current in the coil of Fig. 6-13 will increase the mmf. It will also increase the flux *if* the ferrite rod can carry more flux.

6-7 SATURATION

A magnetic material is saturated when an increase in magnetomotive force no longer increases the flux in the material. Increasing the coil current (or turns) in Fig. 6-13(*a*) will increase the flux until the ferrite rod saturates. Once the rod has saturated, increasing either the current or the number of turns will no longer increase the flux.

6-8 DEMAGNETIZING

A permanent magnet can be partially demagnetized by hammering on it. It can also be demagnetized by heating it to a very high temperature. However, neither of these methods is very practical for demagnetizing many materials and devices. Most demagnetizing is done with a coil and an alternating-current (ac) source, as shown in Fig. 6-14(*a*). Alternating current periodically reverses its direction of flow. It is also constantly varying in value [Fig. 6-14(*b*)]. In other words, ac starts at zero current or voltage and slowly

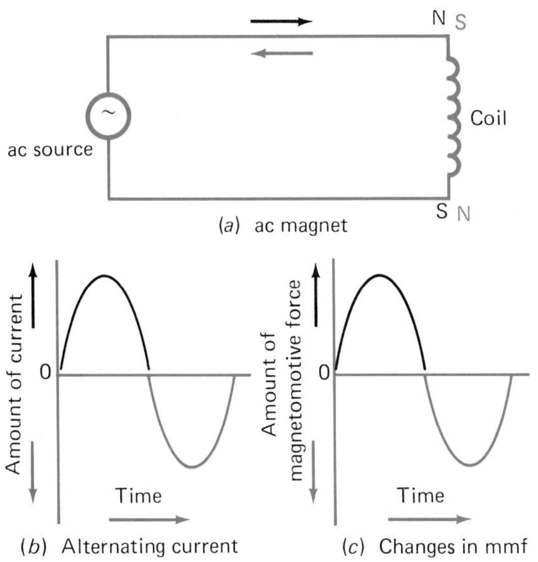

Fig. 6-14 Electromagnetism with alternating current. The polarity of the magnetic field periodically reverses. (*a*) Ac magnet. (*b*) Alternating current. (*c*) Changes in mmf.

increases to maximum current or voltage and then decreases back to zero. At this time it reverses direction, increases to a maximum, and decays back to zero. This sequence is repeated rapidly (every $\frac{1}{60}$ s in the ac system in your home) as long as the circuit is connected. This change in value and direction of the current creates a magnetic field that changes in magnetomotive force and polarity. The changes in magnetomotive force [Fig. 6-14(*c*)] follow the changes in current. Both changes (current and mmf) are indicated in Fig. 6-14(*a*) by the arrows and the pole markings.

When a permanent-magnetic material is put into the coil of Fig. 6-14(*a*), it is magnetized. First it is magnetized in one polarity and then in the reverse polarity. Just as the polarity is reversing there is an instant when no magnetic field exists. If the switch could be opened at that exact instant, the magnet would be demagnetized. However, the chances of opening the switch at that instant are very small. It is very likely that the switch will be opened at some other instant when a magnetic field is present. Then the magnet would still be magnetized. The strength of the magnet would be dependent upon the strength of the field at the instant the switch opened.

To demagnetize with the circuit of Fig. 6-14, the switch must remain closed as the magnet is removed. As the magnet is removed it is being magnetized first in one polarity and then in the other. However, each reversal of polarity yields a weaker magnet. This is because the field of the coil gets weaker as the magnet is moved farther away. When the permanent-magnetic material has been moved several feet away from the coil, it is completely demagnetized. The switch to the coil can then be opened.

A soldering gun like the one shown in Fig. 6-15 can be used to either magnetize or demagnetize. The soldering tip and connecting leads are a single-turn coil. The coil has a large alternating current flowing in it when the gun is on. To magnetize the screwdriver shown in Fig. 6-15, just turn the gun on and off with the screwdriver located as shown. To demagnetize, turn the gun on and pull the screwdriver away from the gun. Then turn off the gun.

6-9 RESIDUAL MAGNETISM

Residual magnetism

The flux that remains in a temporary magnet after it is removed from a magnetic field is called *residual magnetism*. All magnetic materials retain some flux after they have been exposed to a magnetic field. Permanent-magnet materials retain a high magnetism.

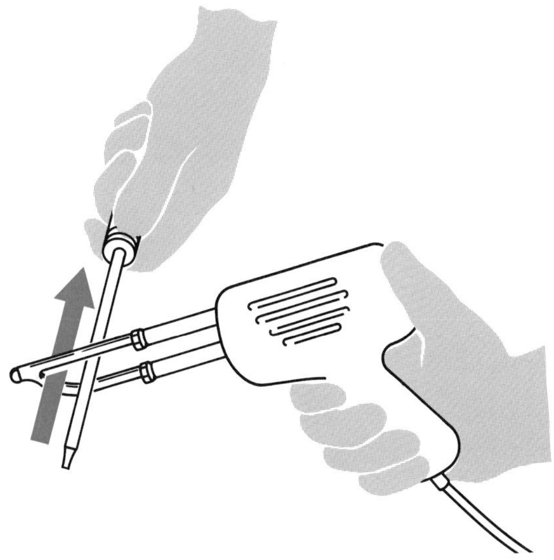

Fig. 6-15 Magnetizing and demagnetizing with a soldering gun. The soldering gun tip is a one-turn coil.

The ideal permanent magnet would retain all the flux.

Temporary-magnet materials have very low residual magnetism. In most temporary magnets, the magnetic field is created by current-carrying conductors. When the current stops, the ideal temporary-magnet material should become completely unmagnetized, but it does not. A small residual magnetism always remains. However, the magnetic field of the residual magnetism is very weak.

When a temporary-magnet device is powered by direct current, the residual magnetism is not very important. However, when the temporary-magnet device is powered by alternating current, the residual magnetism is most important. When alternating current is involved, residual magnetism causes the magnetic material to heat up. Because of residual magnetism, some of the magnetic field's energy is converted into heat energy within the magnetic material. This is part of the reason why electromagnetic devices such as motors and transformers are not 100 percent efficient.

Reluctance

6-10 RELUCTANCE

The opposition to magnetic flux is called *reluctance*. The reluctance of an object depends upon both the material and the dimensions of the material. If the dimensions are assumed to remain constant, then we can compare the reluctance of various materials. Nonmagnetic materials have about the same (within 1 percent) reluctance as does air. On the other hand, air has 50 to 5000 times as much reluctance as do common magnetic materials. For example, silicon steel used in motors and transformers has about $\frac{1}{3000}$ as much reluctance as air. When offered a choice, flux takes the lowest reluctance path. In Fig. 6-16(*a*), the flux extends away from the poles into the surrounding air. However, in Fig. 6-16(*b*), the flux concentrates in the small piece of iron. This is because the iron has much, much less reluctance. The insertion of the iron between the poles has concentrated the flux into a smaller volume. If the iron in Fig. 6-16(*b*) were replaced by a piece of lead, the flux pattern would be the same as it is in Fig. 6-16(*a*).

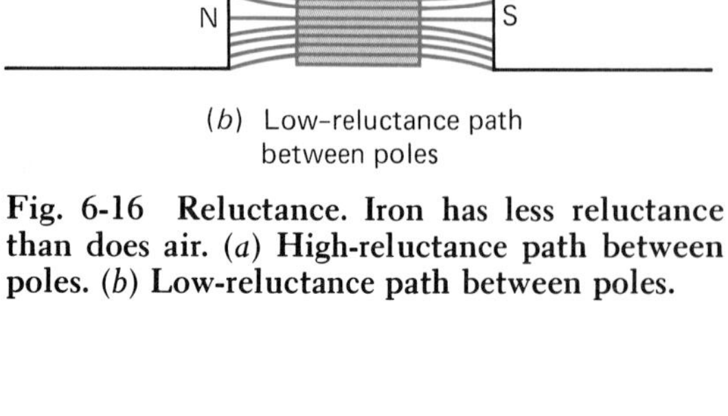

Fig. 6-16 Reluctance. Iron has less reluctance than does air. (*a*) High-reluctance path between poles. (*b*) Low-reluctance path between poles.

Magnetic materials are attracted to a magnet because of their low reluctance. Magnetic flux lines try to take the shortest path between poles of a magnet. They also try to take the path of lowest reluctance. The flux lines can do both if they can pull the surrounding magnetic material to the magnet. This idea is illustrated in Fig. 6-17. The normal lines of force around the poles of a horseshoe magnet are shown in Fig. 6-17(*a*). In Fig. 6-17(*b*), the flux lines have bent to take the path of low reluctance through the iron. There is a force trying to shorten the lines of force. This resembles the force exerted by a stretched rubber band. If either the bar or the magnet is free to move, the two will come together, as shown in Fig. 6-17(*c*). Now the flux has both the shortest path and the lowest-reluctance path.

Magnetic shields

6-11 MAGNETIC SHIELDS

Magnetic flux can be bent, distorted, and guided by inserting low-reluctance materials in the magnetic field. For example, most of the flux in Fig. 6-18 will be bent and guided through the crooked, iron bar. Magnetic shields make use of the tendency for flux to distort and follow the path of lowest reluctance. A magnetic shield is just a very low reluctance material. The material is put around the object which is to be protected from any magnetic field in its vicinity. Figure 6-19 illustrates how a watch movement can be protected by a shield.

Review Questions

24. What is a magnetomotive force?
25. What is magnetic saturation?

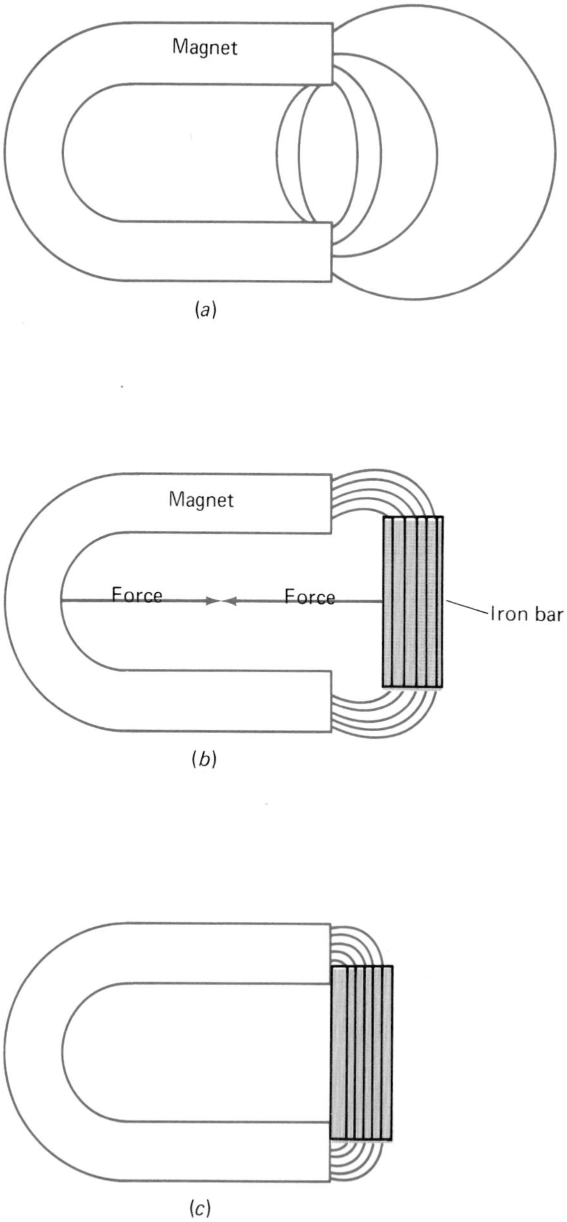

Fig. 6-17 Attraction of magnetic materials. The flux in (*a*) is concentrated and shortened by the magnetic material in (*b*) and (*c*).

26. How is a material demagnetized with a soldering gun?
27. Can a permanent-magnet material be magnetized with an ac coil?
28. What is residual magnetism?
29. Which magnetic material, permanent or temporary, will have the most residual magnetism?
30. Why are magnetic materials attracted to a magnet?

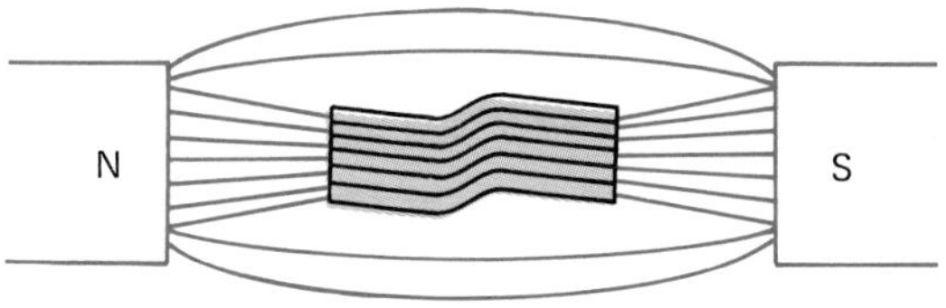

Fig. 6-18 Guiding flux through a low-reluctance path.

6-12 INDUCED VOLTAGE

Induced voltage

Electric generator

We have seen that a current-carrying conductor produces a magnetic field. It is equally true that a magnetic field can induce a voltage (and thus a current) in a conductor. When a conductor moves perpendicularly across lines of flux (Fig. 6-20), a voltage is induced in it. The amount of induced voltage depends on the amount of flux involved and the speed at which it is cut. The greater the flux and the greater the speed, the greater will be the induced voltage. The polarity of the induced voltage depends on both the direction of motion of the conductor and the direction of the flux. Changing the direction of either will change the polarity of the induced voltage.

Generator Action

Induced voltage is the principle on which an electric generator operates. When the shaft of a generator is turned, a conductor loop is forced through a magnetic field. A very simplified generator is shown in Fig. 6-21. As illustrated, the ends of the loop are connected to commutator bars and brushes. The commutator and brushes serve two purposes. The ends of the loop are permanently attached to the commutator bars. As the loop turns, the

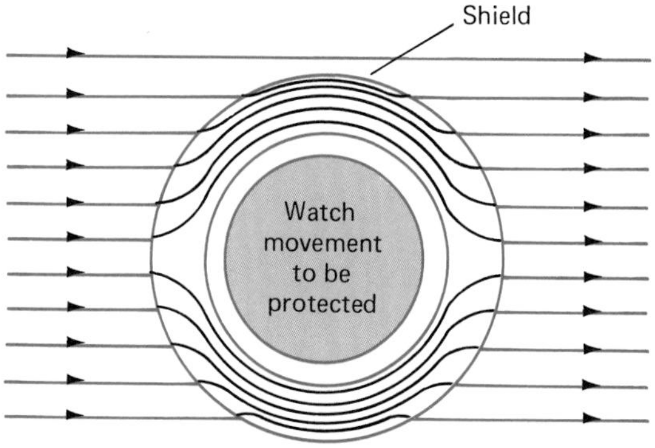

Fig. 6-19 Magnetic shield. The shield provides a low-reluctance path around the protected area.

Dc output

Transformer action

Changing flux

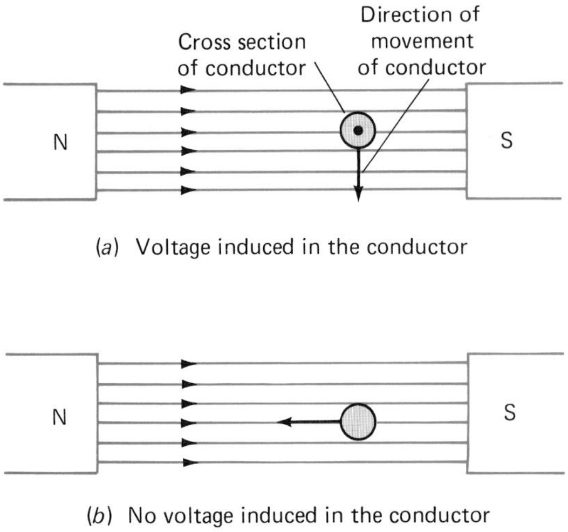

Fig. 6-20 (*a*) **Voltage induced in a moving conductor. The conductor must cut the lines of flux.** (*b*) **No voltage induced in the conductor.**

commutator turns with it. The commutator bars touch the brushes and slide against them as the loop rotates. Thus, the brush makes an electric connection to the rotating loop of wire through the commutator. Second, they reverse connections to the rotating loop every time the polarity of the induced voltage changes. The polarity of the induced voltage changes because the relative direction of motion of the conductor reverses. The top part of the rotating loop in Fig. 6-21 is shown moving from left to right. Ninety degrees later it will start moving from right to left, and its polarity will reverse. At the same time, the brushes will change commutator bars. Thus, one brush

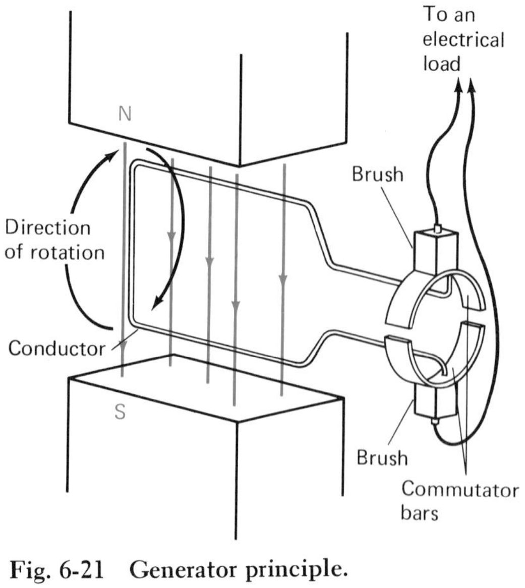

Fig. 6-21 Generator principle.

will always be negative and the other positive. This provides a dc output from the generator.

Transformer Action

Induced voltage is also created when a conductor is in the immediate vicinity of a changing magnetic flux. Again, the conductor must be perpendicular to the lines of flux (Fig. 6-22). When the switch in Fig. 6-22 is opened and closed, the current in the circuit will start to flow and then stop. The changing current in the top coil will produce a changing flux in the iron core. This changing flux in the core induces a voltage into the bottom coil.

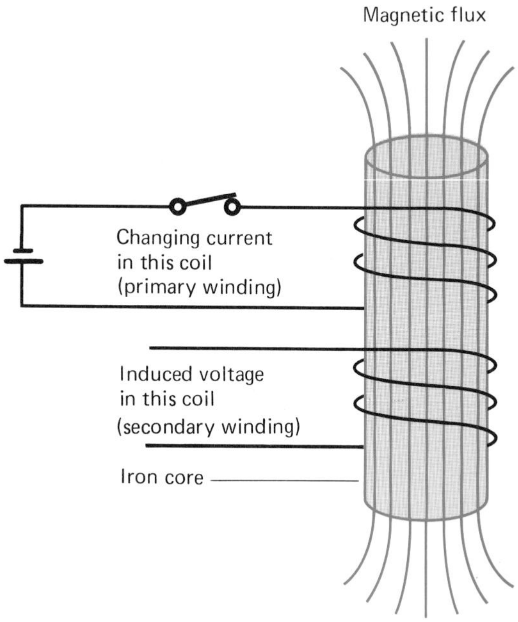

Fig. 6-22 Voltage induced by a changing flux.

Voltage induced by changing flux is the principle on which transformers and other devices operate. The ignition coil (which is a transformer) of an automobile operates similarly to the circuit shown in Fig. 6-22. In the ignition coil (Fig. 6-23) the coils have many turns and one coil is wound on top of the other.

Review Questions

31. What factors determine the amount of induced voltage?
32. What are the functions of the commutator and brushes in a generator?
33. How does an ignition coil operate?

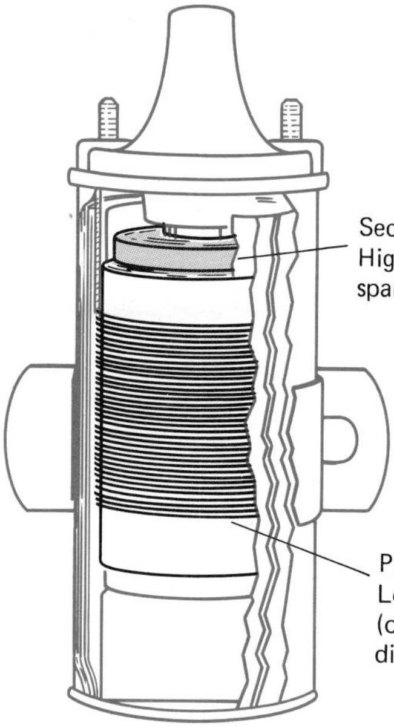

Fig. 6-23 Ignition coil. (Courtesy of Delco-Remy.)

6-13 MAGNETIC QUANTITIES AND UNITS

A number of magnetic quantities, such as flux, have already been discussed. However, we need to know the units in which these quantities are specified and measured.

Magnetomotive Force—The Ampere-Turn

We will use *ampere-turn* for the base unit of magnetomotive force. One ampere-turn is the magnetomotive force created by one ampere flowing through one turn of a coil. Three ampere-turns of magnetomotive force would be created by one ampere through three turns. Three ampere-turns is also equal to three amperes through one turn. Three ampere-turns can be created by any combination in which the product of amperes times turns equals 3.

Example 6-1

What is the mmf of the coil in Figure 6-22 if 1.3 A of current is flowing in the circuit?

Given: Number of turns (t) = 3
current (I) = 1.3 A

Find: mmf

Known: mmf = turns × current

Solution: mmf = 3 t × 1.3 A = 3.9 A·t

Answer: The magnetomotive force of the coil is 3.9 ampere-turns.

Strictly speaking, the base unit of mmf in the metric system is just the ampere. At first, this may seem confusing because the ampere is also the base unit of current in the metric system. The reasonableness of using "ampere" as the unit of mmf is illustrated in Fig. 6-24. In Fig. 6-24(*a*), 4 A of current is moving around the iron core and creating flux. In Fig. 6-24(*b*), 4 A is also moving around the iron core. The fact that the 4 A is split into two paths does not change the total force exerted in creating flux. The same reasoning extends to Fig. 6-24(*c*). The four 1-A currents across the iron produce the same magnetomotive force as one 4-A current. Thus, it is the total current flowing around the iron that determines the magnetomotive force. The number of turns carrying the current really does not matter. However, the easiest way to determine the total current flowing by the iron is to multiply the coil current by the number of turns. Thus, the ampere-turn is a more descriptive unit than ampere. Therefore, we will use ampere-turn as the unit of magnetomotive force.

Ampere-turn

Flux—The Weber

Weber

The base unit of magnetic flux is the weber. A *weber* (Wb) can be defined only in terms of a change in the flux in a single-turn conductor. A weber is the amount of flux change required in one second to induce one volt in the conductor. Refer to Fig. 6-22 for an example of the use of the weber. The bottom coil will have a voltage induced into it when the flux changes. Note that this coil is a three-turn

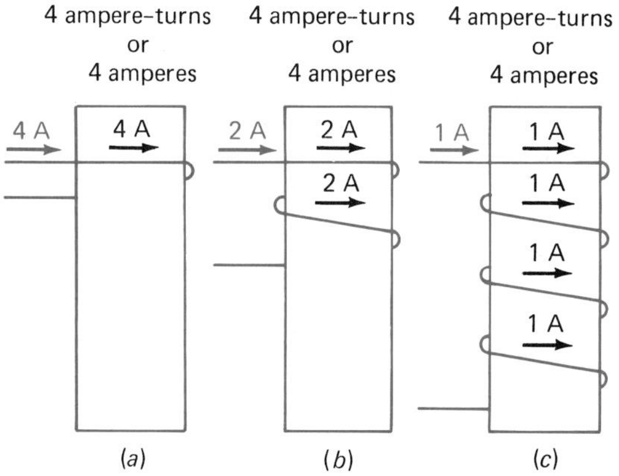

Fig. 6-24 Ampere and ampere-turn.

coil. When the flux in the core changes 1 weber in 1 second, 3 volts will be induced in the coil.

Magnetic Field Strength—Ampere-Turn per Meter

Magnetizing force

Field intensity

Magnetic field strength

Flux density

Tesla

Magnetizing force, *field intensity*, and *magnetic field strength* all mean the same thing. They refer to the amount of magnetomotive force available to create a magnetic field for each unit length of the magnetic circuit. The symbol for magnetic field strength is H. The base unit of magnetic field strength (H) is the ampere-turn per meter. In the magnetic circuit of Fig. 6-25(*a*), there is a four-turn coil carrying 3 A of current. If the average length of the magnetic circuit (iron core) is 0.25 meters (m), the magnetomotive force is

$$3\ \text{A} \times 4\ \text{t} = 12\ \text{A·t}$$

Therefore, the magnetic field strength is

$$\text{Magnetic field strength } (H) = \frac{\text{magnetomotive force (mmf)}}{\text{length of circuit } (l)} = \frac{12\ \text{A·t}}{0.25\ \text{m}} = 48\ \text{A·t/m}$$

In Fig. 6-25(*b*) the mmf is

$$\text{mmf} = 4.8\ \text{A} \times 6\ \text{t} = 28.8\ \text{A·t}$$

The magnetic field strength is

$$H = \frac{\text{mmf}}{l} = \frac{28.8\ \text{A·t}}{0.6\ \text{m}} = 48\ \text{A·t/m}$$

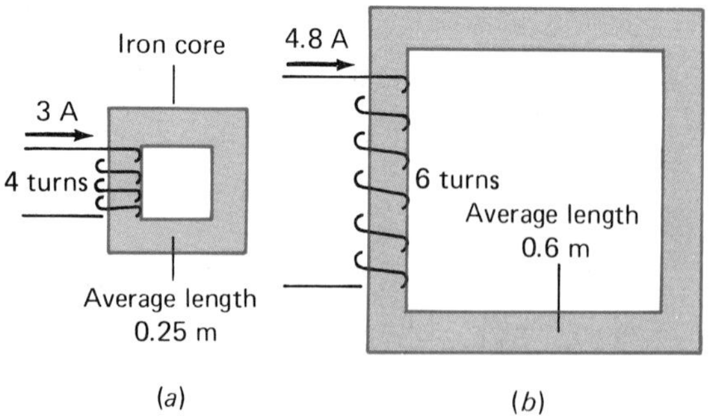

Fig. 6-25 Comparing magnetic circuits. The circuits in (*a*) and (*b*) have the same magnetic field strength (48 A · t/m).

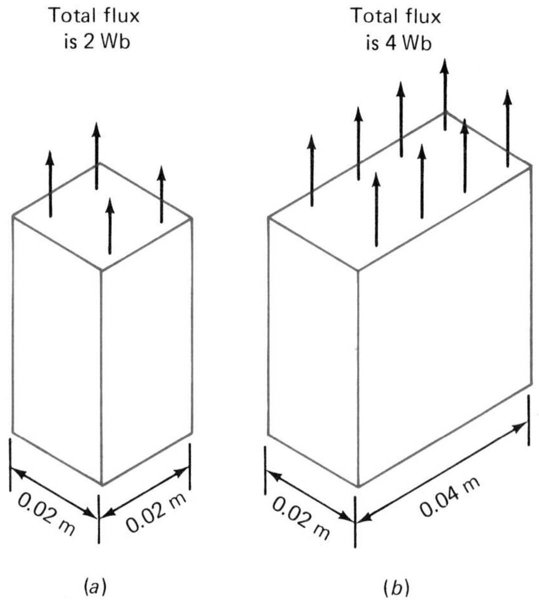

Fig. 6-26 Comparing flux density. Both (*a*) and (*b*) have a flux density of 5000 T.

Notice that the magnetic field strength is the same for both circuits in Fig. 6-25. Yet the two circuits have different currents, different numbers of turns, different magnetomotive forces, and different lengths. Specifying magnetic field strength rather than magnetomotive force makes it possible to compare magnetic circuits without specifying physical dimensions.

Flux Density—Tesla

The amount of flux per unit cross-sectional area is called *flux density* (B). The base unit of flux density is the *tesla* (T). A tesla is equal to a weber per square meter. Specifying flux density rather than flux also makes it easier to compare magnetic circuits of unequal physical sizes. For example, the iron cores in Fig. 6-26 have equal flux density. Yet they have different cross-sectional areas. Flux density (B) can be calculated by dividing the total flux by the cross-sectional area. In Fig. 6-26(*a*), the cross-sectional area is

$$\text{Area} = 0.02\ \text{m} \times 0.02\ \text{m} = 0.0004\ \text{m}^2$$

The flux density, therefore, is

$$\text{Flux density} = \frac{\text{flux}}{\text{area}} = \frac{2\ \text{Wb}}{0.0004\text{m}^2} = 5000\ \text{T}$$

If you calculate the flux density for Fig. 6-26(*b*), you will find it is also 5000 T.

Permeability and Relative Permeability

Some materials conduct flux more readily than others. *Permeability* refers to the ability of a material to conduct or carry flux. Its symbol is μ. Permeability is technically defined as the ratio of flux density (B) over magnetic field strength (H). That is

$$\mu = \frac{B}{H}$$

This makes the weber per meter ampere-turn the base unit of permeability. Since both flux density (B) and field strength (H) are independent of physical size, permeability is, too. Thus, permeability is a convenient way of describing the magnetic properties of a material.

Often the permeability of a material is specified as relative permeability. *Relative permeability* compares the permeability of the material with that of air. Suppose a piece of iron has a relative permeability of 600. This means the iron will carry 600 times as much flux as an equal amount of air. Since relative permeability is a ratio of two permeabilities, the units cancel out. Thus, relative permeability is a unitless (pure) number.

The relative permeability of all nonmagnetic materials is very close to 1. Permeabilities of magnetic materials range from about 30 to over 6000.

The permeability of a magnetic material decreases as the flux density increases. As a material approaches saturation, its permeability decreases rapidly. When the material is fully saturated, the permeability is only a small fraction of its value at lower flux densities.

Review Questions

34. What is the magnetomotive force of a 200-t coil with 0.3 A of current flowing through it?
35. List the base unit of the following quantities:
    a. Magnetic field strength
    b. Flux
    c. Flux density
36. List two other names for magnetic field strength.
37. Why is it easier to compare magnetic circuits when flux density rather than flux is specified?
38. What is relative permeability?

6-14 ELECTROMAGNETS

We have already discussed many of the ideas of an electromagnet in Secs. 6-3 and 6-5. A coil of wire with a current through it is an electromagnet. However, most useful electromagnets are wound on, and partially encased in, a material that can become a temporary magnet.

The strength of an electromagnet depends upon four factors:

1. The type of core material (the magnetic material)
2. The size and shape of the core material
3. The number of turns on the coil
4. The amount of current in the coil

In general, the magnet will be strongest when:

1. The core material has the highest permeability
2. The core material has the largest cross-sectional area and the shortest length for flux lines. This results in a low reluctance.
3. The number of turns and amount of current are greatest. This provides a large magnetomotive force.

Electromagnets are used extensively in industry for a variety of jobs. They are used to hold steel in position while it is being machined. They sort magnetic from nonmagnetic materials. They are used for lifting and moving heavy iron and steel products. Figure 6-27 shows a large electromagnet lifting a number of heavy iron pipes.

6-15 DC MOTORS

Physically and electrically, a dc motor resembles a dc generator (Fig. 6-21). In fact, in some cases a single machine may be used as either a generator or a motor. Whereas the shaft of a generator turns because of some outside mechanical force, the shaft of a motor rotates

Permeability

$\mu = \frac{B}{H}$

Relative permeability

Strength of an electromagnet

Dc motors

Brushes

Commutator

Armature

Field poles

Solenoid

Fig. 6-27 Large industrial electromagnet. (Courtesy of O. S. Walker Co.)

because of the interaction of two magnetic fields within the motor itself. The principle of a simple dc motor is illustrated in Fig. 6-28. Current from an external source flows through the brushes, commutator, and armature coil. This produces a magnetic field in the armature iron. The armature poles are attracted by the field poles of the permanent magnet. The result is a rotational force called *torque*, as shown in Fig. 6-28(*a*).

The commutator and brushes change the direction of current in the armature coil every 180°. This change reverses the magnetic polarity of the armature so that it is always attracted or repelled by the field poles. Thus, the direction of the torque (rotational force) is always the same. When rotated 90°, the armature's south pole will be in line with the field's north pole [Fig. 6-28(*a*)]. However, the force of the moving armature carries the armature just past the vertical position. At this time, the current in the armature coil changes, and the armature field reverses. Now the north pole of the armature is repelled by the north pole of the field [Fig. 6-28(*b*)].

Practical motors have more than two commutator segments and many more armature coils. With more commutator segments, it is not necessary to rely on inertia as the armature polarity is reversed. As seen in Fig. 6-29, the segments change an instant before the poles of the fields are aligned. During the time that each brush contacts two segments of the commutator, four poles are created. However, as shown in Fig. 6-29(*b*), all four poles provide torque in the same direction.

The field poles in a dc motor can be either permanent magnets or electromagnets. When electromagnets are used, the current in the field never changes direction. Thus dc motors

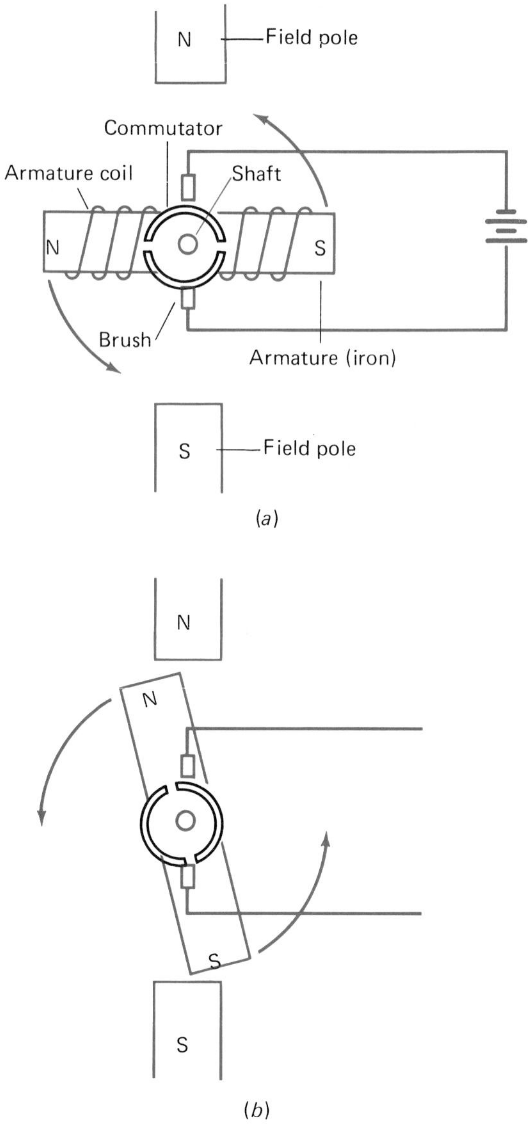

Fig. 6-28 Motor principle. The armature is rotated by alternate attraction (*a*) and repulsion (*b*) of magnetic poles.

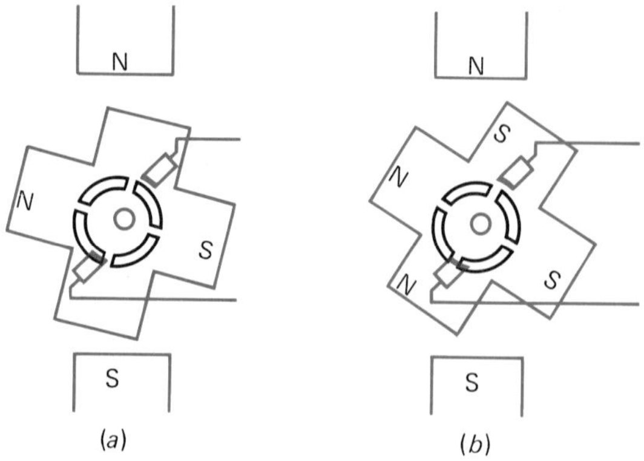

Fig. 6-29 Four-segment commutator.

with electromagnetic fields operate in the same way as those with permanent-magnet fields.

6-16 SOLENOIDS

A *solenoid* is an electromagnetic device that enables an electric current to control a mechanical mechanism. The movable plunger in Fig. 6-30 can be used to operate a mechanical brake or clutch. Or, it could operate a valve to control water flow.

When the solenoid is de-energized [Fig. 6-30(*b*)], there is no magnetic force to hold the plunger in the center of the coil. When current is supplied to the coil, the solenoid is energized and the plunger is pulled into the coil because of its high permeability. The plunger has much less reluctance than the air it replaces. As shown in Fig. 6-30(*c*), the flux path is almost entirely through iron when the solenoid is energized.

The pull of the plunger of a solenoid is dependent on its magnetic properties. In general, increasing its magnetomotive force will increase the pull. This means increasing the number of turns in the coil or the amount of current (or both). Decreasing the reluctance will also increase the pull of the solenoid. Decreasing the reluctance will increase both the flux and the flux density. The reluctance can be decreased by using a higher permeability iron. Shortening the length of the air gap (when the solenoid is de-energized) would also increase the flux density.

Most solenoids are made to operate from an ac supply. However, dc solenoids are also available. The two types (ac and dc) are not interchangeable.

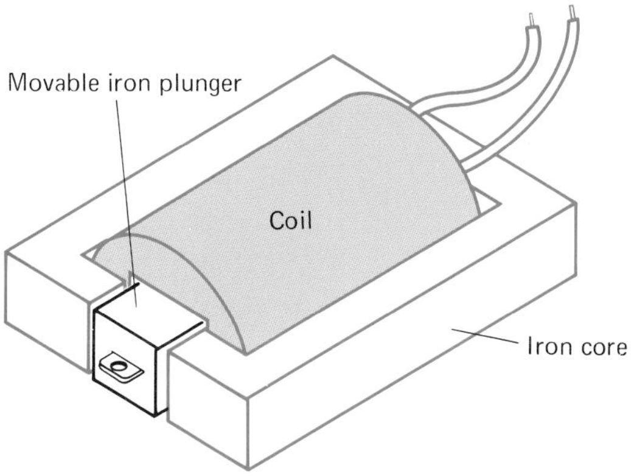

(*a*) Parts of a solenoid

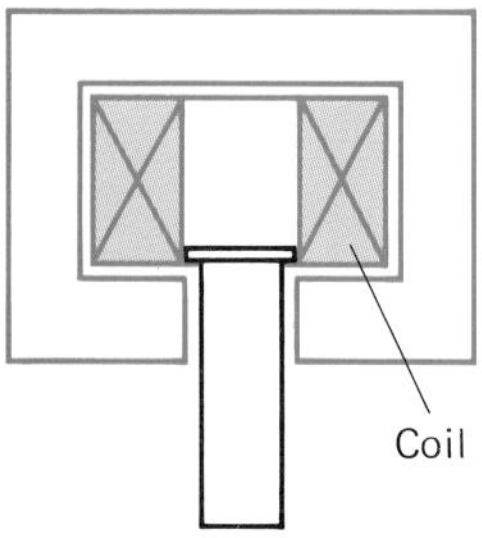

(*b*) No current in coil (de-energized)

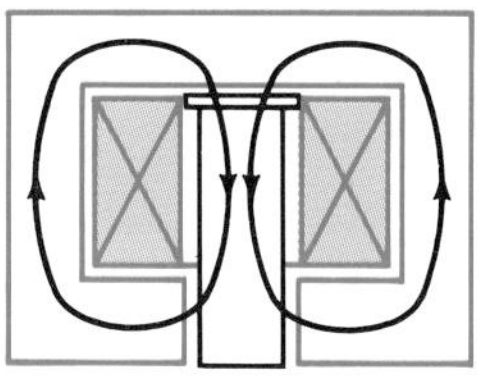

(*c*) Current in coil (energized)

Fig. 6-30 (*a*) **Parts of a solenoid.** (*b*) **No current in coil (de-energized).** (*c*) **Current in coil energizes solenoid.**

6-17 RELAYS

Magnetic relay

A magnetic relay uses the attraction between an iron armature and an energized coil to operate a pair of electric contacts. The cross section of a relay in Fig. 6-31 is shown in the de-energized position. When current flows through the coil, the armature is pulled down against the iron core. This closes the set of contacts which can complete another electric circuit. When current stops flowing in the coil in Fig. 6-31, the spring pulls the armature up and opens the contacts.

Relays are manufactured in a wide variety of ratings for both the coil and the contacts. Coils are rated for both the current required to energize the relay and the voltage required to produce that current. The contacts are also rated for both current and voltage, just like any switch is rated. Relays often have more than one set of contacts. The contacts may be either normally open or normally closed or a mixture of both. The relay may be energized by either ac or dc. However, ac and dc relays are not interchangeable.

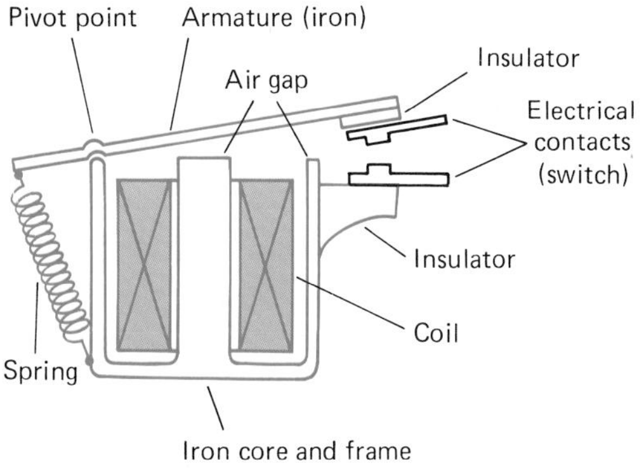

Fig. 6-31 Relay. The armature is attracted to the iron core when current flows through the coil.

Remote operation

One of the chief advantages of a relay over a simple switch is that it allows for remote operation. A low-voltage, low-current supply can control the relay coil. Then the relay contacts can control a high-voltage and/or high-current circuit. The switch that operates the coil can be located in a remote place. Only a small two-conductor cable has to run between the switch and the relay.

Review Questions

39. Would a high-permeability or a low-permeability material make the strongest electromagnet?
40. Should the core of an electromagnet have a high or low reluctance?
41. How can the magnetomotive force of an electromagnet be increased?
42. In a dc motor, is it the armature or the field that periodically changes magnetic polarity?
43. What is the purpose of the commutator and brushes in a dc motor?
44. How can the pull of a solenoid be increased?
45. Why is the plunger of a solenoid pulled into the center of the coil?
46. Can a relay control two or more electric circuits simultaneously?
47. What electrical ratings are specified for a relay?

Summary

1. Magnetism is an invisible force field.
2. A magnetic field exists around a magnet.
3. The lines of force of a magnetic field are called magnetic flux.
4. The denser the flux the stronger the magnetic field.
5. The magnetic field is strongest at the poles of a magnet.
6. The direction of flux is from the north pole to the south pole.
7. Like magnetic poles repel, while unlike poles attract.
8. When magnetic poles attract each other, their flux lines join together.
9. Lines of flux are continuous loops.
10. Magnetic fields exist around all current-carrying conductors.
11. The magnetic field around a straight conductor has no poles.
12. The left-hand rule is used to determine the direction of flux around a conductor.
13. The strength of the magnetic field around a conductor is directly proportional to the current in the conductor.
14. Two parallel conductors will attract each other if their currents are in the same direction. They repel if the currents are in opposite directions.
15. A coil of wire carrying current forms magnetic poles.
16. The north pole of a coil can be determined by use of the left-hand rule.
17. Decreasing the space between turns of a coil increases the flux at the poles of a coil.
18. Iron, nickel, and cobalt are all magnetic materials.
19. Iron is the most magnetic of the elements.
20. Nonmagnetic materials do not prohibit flux from passing through them.
21. The domains of a permanent magnet remain aligned after removal from the magnetizing field.
22. Magnetic materials are magnetized by putting them in the fields of other magnets.
23. Magnetomotive force can be increased by increasing the current or the turns in a coil.
24. When a material is saturated, increasing the magnetomotive force will not increase the flux.
25. Residual magnetism refers to the flux that remains in a temporary magnet after it is removed from a magnetic field.
26. Reluctance is the opposition to magnetic flux.
27. Magnetic materials have less reluctance than nonmagnetic materials.
28. A changing magnetic flux can induce a voltage within a conductor.
29. The ampere-turn is the base unit of magnetomotive force (mmf).
30. The base unit of magnetic flux is the weber (Wb).
31. The base unit of magnetic field strength (H) is the ampere-turn per meter.

32. Magnetic field strength is also known as field intensity and magnetizing force.
33. Flux density (B) is the flux per unit cross-sectional area of a magnetic material.
34. The tesla (T) is the base unit of flux density. A tesla is a weber per square meter.
35. Permeability refers to the ability of a material to pass or carry magnetic flux. It is equal to flux density divided by magnetic field strength.
36. Relative permeability is the ratio of the permeability of a material to the permeability of air.
37. The strength of an electromagnet is determined by the reluctance and permeability of the core and the magnetomotive force.
38. Direct-current motors rotate because of attraction and repulsion between the magnetic fields of the field poles and the armature.
39. The plunger of a solenoid is made from high-permeability material.
40. The pull of a solenoid depends on the magnetomotive force and the reluctance of its magnetic circuit.
41. A relay is an electromagnetic switch.

Chapter Review Questions

Here is a chance to see what you have learned. For questions 6-1 to 6-15, determine whether the statement is true or false.

6-1. Magnetic flux is assumed to leave a north pole and enter a south pole.

6-2. Breaking a bar magnet into two pieces completely demagnetizes the magnet.

6-3. Either like poles or unlike poles will attract each other.

6-4. All magnets have both a north and a south pole.

6-5. Lines of flux are continuous loops.

6-6. The pole flux of a coil can be increased by decreasing the space between the turns of the coil.

6-7. Motors and transformers always use permanent-magnet materials.

6-8. Nonmagnetic materials have orderly arranged domains.

6-9. An ac coil can be used either to magnetize or to demagnetize a permanent-magnet material.

6-10. Permanent-magnet materials have very little residual magnetism.

6-11. Whenever a conductor cuts a magnetic line of force, a voltage is induced in the conductor.

6-12. Magnetic shields are generally made from nonmagnetic materials.

6-13. The rotating part of a dc motor is called the armature.

6-14. The weber is the base unit of flux density.

6-15. The ampere-turn per meter is the base unit of field intensity.

For questions 6-16 to 6-23, choose the letter that best completes the sentence.

6-16. Most magnetic materials contain considerable amounts of
(*A*) Iron (*B*) Lead (*C*) Copper (*D*) Tin

6-17. When all other factors are equal, the material with the highest permeability will
(*A*) Have the least reluctance (*B*) Saturate most easily (*C*) Have the most residual magnetism (*D*) Have the lowest flux density

6-18. Ampere-turn is the base unit of
(*A*) Flux (*B*) Magnetizing force (*C*) Flux density (*D*) Magnetomotive force

6-19. The pull of a solenoid would be increased by
(*A*) Forcing less current through the coil (*B*) Using fewer turns of wire in the coil (*C*) Using higher permeability core material (*D*) Increasing the air gap around the plunger

6-20. A magnet is a device which
(*A*) Attracts materials like paper and glass (*B*) Always has two poles (*C*) Produces an invisible force field (*D*) Creates reluctance in a conductor

6-21. The invisible lines of force associated with a magnet are known as
(*A*) Flux (*B*) Reluctance (*C*) Permeability (*D*) Domains

6-22. Doubling the distance between two magnetic poles will
(*A*) Double the force between the poles (*B*) Quadruple the force between the poles (*C*) Make the force between the poles one-half as great (*D*) Make the force between the poles one-fourth as great

6-23. If you are looking at the cross section of a conductor and the current is flowing toward you, the direction of the flux will be
(*A*) Toward you (*B*) Away from you (*C*) Clockwise (*D*) Counterclockwise

Answers to Review Questions

1. Magnetism is an invisible force field that exists around certain materials and devices.
2. A magnetic field is the area of invisible force extending around a magnet.
3. Magnetic flux consists of all the invisible lines of force which make up a magnetic field.
4. North to south pole
5. South pole
6. Quadruple
7. Opposite polarities
8. Yes
9. No
10. No
11. Perpendicular
12. No
13. Counterclockwise
14. A coil is a conductor which is shaped into one or more circular turns.
15. Above them
16. Decrease
17. Iron
18. Aluminum, tin, zinc, silver, copper, brass, or other nonferrous metals
19. A magnetic domain is a group of molecules that cluster together to create a small magnetic field.
20. Silicon steel
21. Alloys
22. The flux disappears.
23. The steel becomes a permanent magnet, and the flux remains as its field.
24. It is the force that creates a magnetic field and magnetic flux.
25. Saturation means that no additional flux will be created by an increase in magnetomotive force.
26. By removing the material from the soldering gun tip while the gun is on
27. Yes
28. Residual magnetism is the flux that remains in a material after it has been removed from the magnetizing force.
29. Permanent
30. Because the reluctance of the magnetic material is much less than that of the air which it replaces

31. The amount of flux and the speed at which the conductor moves through it
32. The functions are to make connections to the rotating conductor and to provide dc output.
33. A changing flux caused by one coil induces a voltage within another coil.
34. 60 A·t
35. *a*. Ampere-turn per meter
    b. Weber
    c. Tesla (weber per square meter)
36. Magnetizing force and field intensity
37. Because the physical dimensions of the circuit are unimportant when flux density is specified
38. It is the ability of a material to pass flux compared with the ability of air to pass flux.
39. High-permeability material
40. Low reluctance
41. By increasing the coil current or the number of turns (or both)
42. The armature
43. They periodically reverse the direction of current in the armature.
44. By increasing the magnetomotive force or decreasing the reluctance of the magnetic circuit
45. Because the plunger has much higher permeability than does air. When the plunger is pulled into the coil, the reluctance of the circuit is greatly reduced.
46. Yes
47. Current and voltage ratings for both the coil and the contacts. Also, the switching arrangement of the contacts is specified.

Basic Measuring Instruments

- In Chap. 3 you learned how to use meters to measure electrical quantities. In this chapter you will learn how a meter measures such things as current, voltage, and resistance.

  When you have completed this chapter, you will understand how a basic meter movement operates. You will be able to extend the range of a voltmeter and construct a wheatstone bridge. You will be able to interpret major specifications of meters and select the appropriate meter for a given measurement. Finally, you will be able to determine the amount of circuit loading caused by meters.

7-1 D'ARSONVAL METER MOVEMENT

Meters with moving pointers are called *analog meters*. They measure quantities by moving through an infinite number of points on a scale.The major part of any analog multimeter is the basic meter movement. The most common type of meter movement is the d'Arsonval movement (named after its inventor). This meter movement is based on the interaction of the fields of a permanent magnet and an electromagnet.

The armature assembly in Fig. 7-1 contains a coil of very thin wire wound on a lightweight supporting form. The coil, coil form, and attached pointer rotate on pivot points. When current flows through the coil, a magnetic field is created as seen in Fig. 7-1. The coil's magnetic field reacts with the permanent magnet's field to create a clockwise rotational force. This rotational force reacts against either a spiral spring or a suspension band.

A meter movement with a suspension band is shown in Fig. 7-2. When current flows in the coil it rotates the coil and puts a twist in the suspension band. When the coil current decreases, the reverse torque in the twisted suspension band rotates the coil counterclockwise. When the coil current is zero, the suspension band is untwisted and the pointer indicates zero. Meter movements which use a suspension band (often called a taut band) are quite rugged. They can withstand severe mechanical shock.

The more traditional d'Arsonval movement uses spiral springs to counteract the rotational force of the armature. As shown in Fig. 7-3, these springs attach to the pivot shaft on which the armature rotates. The points of the shaft rest in jewel bearings to keep friction as low as possible. This arrangement is known as a jewel-and-pivot suspension system.

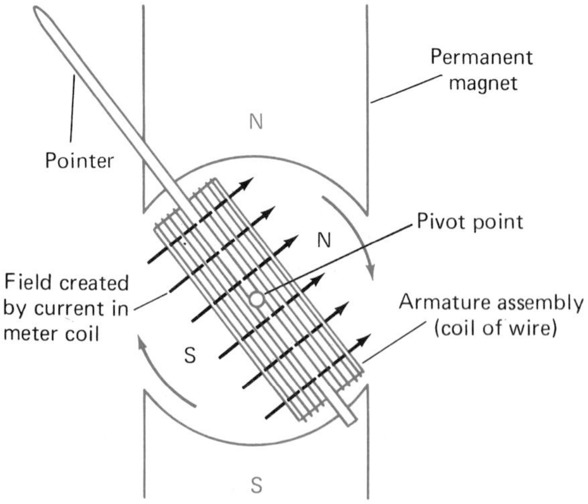

Fig. 7-1 D'Arsonval movement principle. The electromagnetic field of the armature tries to line up with the field of the permanent magnet.

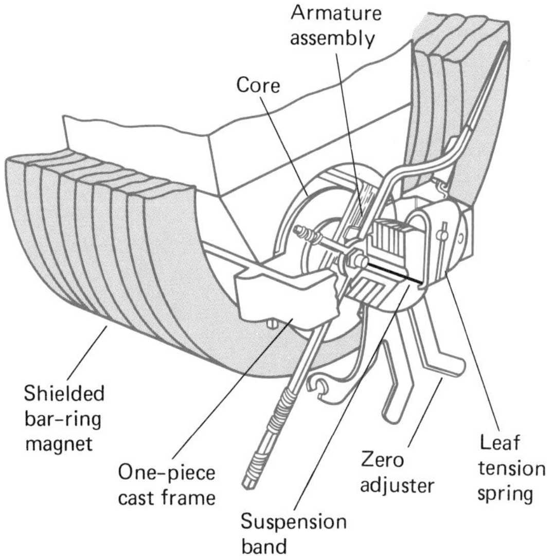

Fig. 7-2 Taut-band meter. (Courtesy of Triplett Corporation.)

D'Arsonval meter movements are polarized. If reverse current flows in the coil, a reverse torque is applied to the armature. This reverse torque can damage the meter movement and the pointer if it is excessive.

Ratings of Meter Movement

All basic meter movements have a *full-scale current rating*. This is the coil current needed to cause the meter to deflect to the maximum value of the scale. Meter movements with full-scale deflection currents as low as 5 μA (5×10^{-6} A) are commonly available.

Another important rating of a meter movement is its *internal resistance*. Basic meter movements have appreciable resistance because of the small-diameter wire used in the coil. In general, the lower the full-scale current rating, the higher the internal resistance. A typical 1-mA meter will have less than 100 Ω of resistance. A typical 50-μA meter will have more than 900 Ω of resistance.

Since a meter movement has both a current and a resistance rating, it must also have a voltage rating. Usually, the manufacturer specifies only two of the three possible ratings. However, the third rating can easily be determined by using Ohm's law. The voltage across a meter movement (V_m) must be equal to the product of the full-scale current (I_m) and the internal resistance (R_m). Thus, the voltage rating of a 1-mA, 100-Ω meter movement is

$$\begin{aligned} V_m &= I_m \times R_m \\ &= 0.001 \text{ A} \times 100\ \Omega \\ &= 0.1 \text{ V or } 100 \text{ mV} \end{aligned}$$

This means that a 1-mA, 100-Ω meter movement requires 100 mV across its terminals to cause full-scale deflection.

Accuracy of Meter Movements

Like all other electric components, meters have tolerances. A typical panel meter has an accuracy of $\pm$ 2 percent of full-scale value. Higher-quality meters will have $\pm$ 1 percent or $\pm$ 0.5 percent tolerances.

Suppose a 100-μA meter has a tolerance of $\pm$ 2 percent. Then the current indicated by the meter could be in error by as much as 2 μA ($0.02 \times 100\ \mu\text{A} = 2\ \mu\text{A}$). If the meter indicated 5 μA, the true current could be anywhere between 3 and 7 μA. If the meter indicated 87 μA, the actual value of current might be as low as 85 or as high as 89 μA. Obviously, the *percent error* is less when a meter is measuring a current close to its full-scale rating.

Review Questions

1. What is the most common type of analog meter movement?
2. What force counteracts the rotational force of the armature in a meter movement?
3. List three meter ratings.
4. A 200-mA meter has an accuracy of $\pm$ 3 percent. If the meter indicates 130 mA, what are the possible minimum and maximum values of the actual current?
5. A meter movement has a full-scale current of 100 μA and a voltage drop of 200 mV. What is its internal resistance?
6. You need to monitor the current in an electric circuit. The current varies between

From page 130:
D'Arsonval movement

Suspension band

Taut band

Jewel-and-pivot

On this page:
Full-scale current rating

Internal resistance

Tolerance

Shunt

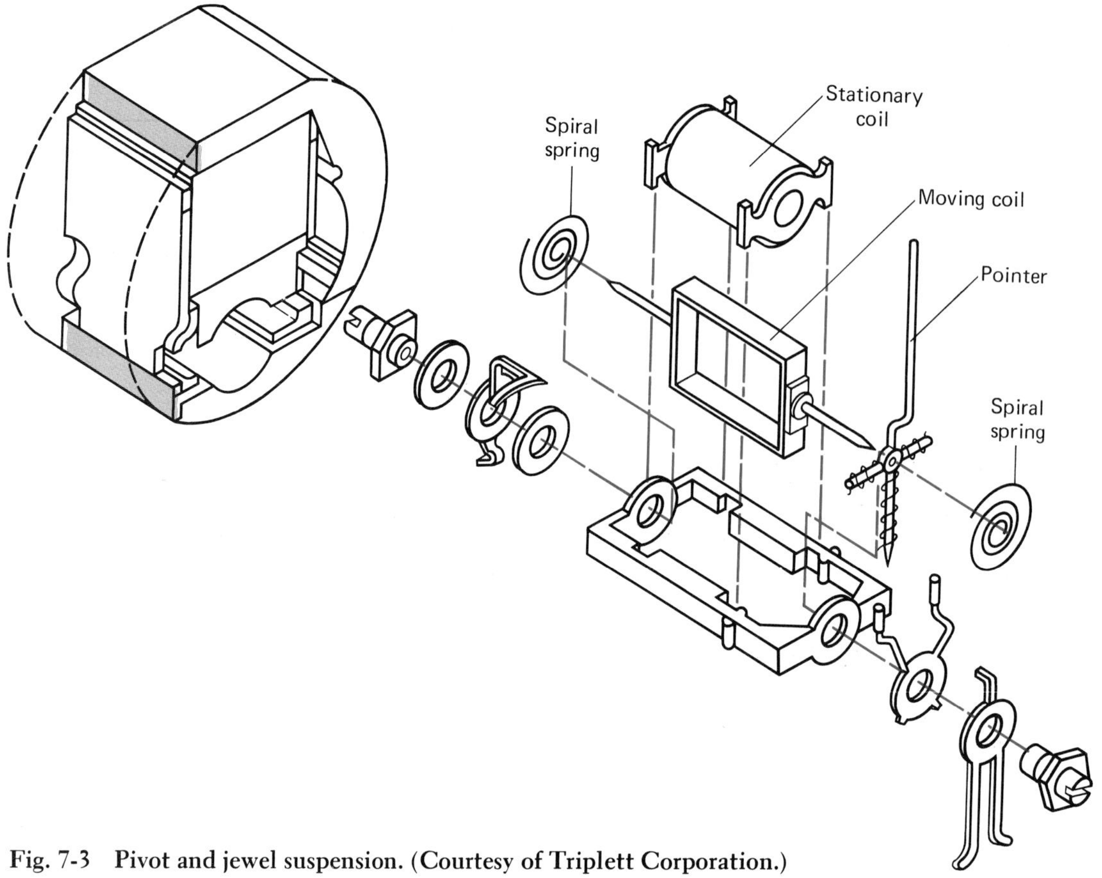

Fig. 7-3 Pivot and jewel suspension. (Courtesy of Triplett Corporation.)

6 and 9 mA. Would you use a ± 2 percent 10-mA meter or a ± 1 percent 15-mA meter? Why?

7-2 AMMETERS

A basic meter movement is an ammeter with a range equal to the full-scale deflection current. The range of an ammeter can be made larger by adding a shunt to the meter. A *shunt* is nothing but a very low resistance resistor connected in parallel with the basic meter movement (Fig. 7-4). Shunts are usually made from materials with very low temperature coefficients. They are generally precision, low-tolerance, (± 2 percent or less) resistors.

A shunt extends the range of an ammeter by diverting most of the current around the meter movement. For example, a 100-μA movement is converted to a 1-mA ammeter by shunting 900 μA around the movement. As shown in Fig. 7-4, the 1-mA ($I_{circuit}$) splits at the junction of the meter movement and the shunt. One hundred microamperes (I_m) goes through the movement and causes full-scale deflection of the pointer. The other 900 μA (I_s) goes through the shunt. If the current entering the ammeter is reduced to 0.5 mA, then 50 μA goes through the movement. The remaining 450 μA goes through the shunt. The pointer would move up halfscale. Thus, it would indicate 0.5 mA of current.

Calculating Shunt Resistance

The value of a shunt resistor is easily determined by using Ohm's law and parallel circuit

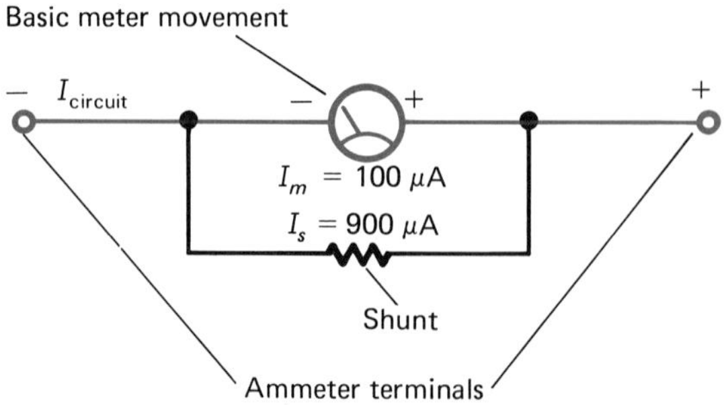

Fig. 7-4 Ammeter with shunt. The shunt extends the range of the basic meter movement.

relationships. Assume the meter movement in Fig. 7-4 has 1000 Ω of internal resistance. Therefore, the full-scale voltage would be

$$V_m = I_m R_m = 100 \times 10^{-6} \text{A} \times 1000\Omega = 0.1 \text{ V}$$

Since the shunt and movement are in parallel, the shunt must also have 0.1 V across it. Thus, the resistance of the shunt would have to be

Shunt in parallel

$$R_s = \frac{V_s}{I_s} = \frac{0.1 \text{ V}}{0.0009 \text{ A}} = 111\ \Omega$$

The power dissipated by the shunt would be

$$P_s = I_s V_s = 0.0009 \text{ A} \times 0.1 \text{ V} = 0.00009 \text{ W or } 90\ \mu\text{W}$$

The power dissipated by the shunt would be only 90 μW, so even a ⅛-W resistor would suffice.

Example 7-1

It is desired to make a 1-A ammeter from a 1-mA, 80-Ω movement. Determine the resistance and power dissipation of the required shunt.

Given: $I_m = 1$ mA
$R_m = 80\ \Omega$

Find: R_s, P_s

Known: $R_s = \frac{V_s}{I_s}$, $V_s = V_m$, $V_m = I_m R_m$,
$I_s = I_{circuit} - I_m$, $P_s = I_s V_s$

Solution: $I_s = 1 \text{ A} - 0.001 \text{ A} = 0.999 \text{ A}$
$V_m = 0.001 \text{ A} \times 80\ \Omega = 0.08 \text{ V}$

$$R_s = \frac{0.08 \text{ V}}{0.999 \text{ A}} = 0.08\ \Omega$$

$$P_s = 0.999 \text{ A} \times 0.08 \text{ V} = 0.0799 \text{ W} = 79.9 \text{ mW}$$

The shunt's resistance would be 0.08 Ω, and its power dissipation would be 79.9 mW.

The total internal resistance of an ammeter is less than the shunt resistance because the shunt is in parallel with the movement. In general, the internal resistance of an ammeter should be as low as possible. When an ammeter is inserted into a circuit, it should not add a significant amount of resistance to the circuit.

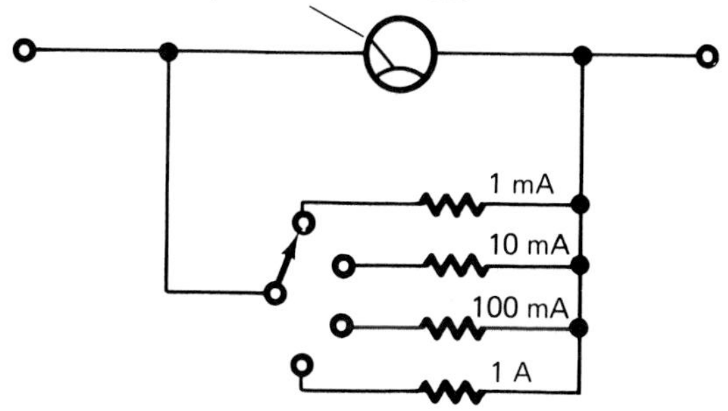

Fig. 7-5 Multirange ammeter. Switching to a smaller shunt increases the range.

Multirange Ammeters

Shorting-type switch

A multirange ammeter (or the ammeter section of a multimeter) is illustrated in Fig. 7-5. It consists of a meter movement, a number of shunts, and a rotary switch. The switch is a shorting (make-before-break) type. A shorting-type switch is necessary so that the meter movement is shunted even while switching ranges. With a nonshorting switch, the meter movement would carry all the current for the instant it takes to change ranges. This instant would be sufficient time to damage or burn out the meter movement.

Review Questions

7. What is a shunt?
8. List two desirable characteristics of a shunt.
9. Refer to Fig. 7-5. How much current is carried by the shunt for the 10-mA range?
10. The meter movement shown in Fig. 7-5 has 800 Ω of resistance.
   a. What is the resistance of the shunt for the 100-mA range?
   b. What is the power used by the shunt?
11. What type of rotary switch is used in a multirange ammeter?
12. Should the internal resistance of an ammeter be as low as or as high as possible?

7-3 EXTERNAL SHUNTS

The shunts discussed in the previous section are enclosed in the same housing as the basic meter movement. However, for high currents

External shunts

Multipliers

Series circuit relationships

(above 10 A) an external shunt is often used. External shunts, as shown in Fig. 7-6, are built to handle large currents without heating up or changing resistance. Therefore, they are physically large (5 to 15 cm long).

External shunts have both a current rating and a voltage rating. Current ratings ranging from 10 A to several thousand amperes are common. Voltage ratings are usually either 50 or 100 mV. The voltage rating of a shunt specifies how much voltage the shunt drops when it is carrying its rated current. The rated voltage appears between the screw terminals shown on the shunts in Fig. 7-6 (the small screws in the shunt shown on top). When these shunts are used, leads are run from the screw terminals to a basic meter movement. Of course, the basic meter movement must have the same voltage rating as the external shunt. The heavier terminals provided on some shunts are bolted into the circuit in which the current is to be measured.

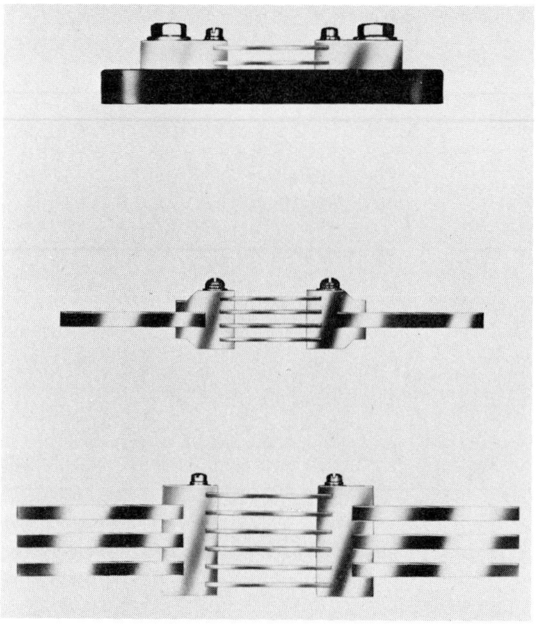

Fig. 7-6 External shunts. (Courtesy of Triplett Corporation.)

Review Questions

13. When are external shunts used?
14. How are external shunts rated?

7-4 VOLTMETERS

The same basic meter movements used to make ammeters are also used to make voltmeters. However, voltmeters have multipliers rather than shunts. As shown in Fig. 7-7, a *multiplier* is a resistor connected in series with a meter movement. The resistor in Fig. 7-7 is called a multiplier because it multiplies (increases) the voltage range of the meter movement. Remember that basic meter movements have a full-scale voltage rating as well as a full-scale current rating. Like shunts, multipliers are precision, low-temperature-coefficient resistors. Unlike shunts, multiplier resistors have very high resistance values.

Calculating Multiplier Resistance

The voltmeter consisting of the basic meter movement and a multiplier is a series circuit. Therefore, the resistance of the multiplier can be determined by using series circuit relationships and Ohm's law. As an example, let us determine the multiplier resistance needed for a 1-V voltmeter using the meter movement in Fig. 7-7. First, we know that the current through the multiplier (I_{mt}) is the same as the current through the meter movement (I_m). Second, we know that the multiplier voltage (V_{mt}) plus the meter movement voltage (V_m) equals the voltmeter voltage (V_t). Therefore, for a 1-V voltmeter

$$V_{mt} = V_t - V_m = 1\text{ V} - 0.1\text{ V} = 0.9\text{ V}$$

The resistance of the multiplier will be

$$R_{mt} = \frac{V_{mt}}{I_{mt}} = \frac{0.9\text{ V}}{0.0001\text{ A}} = 9000\ \Omega$$

The power dissipated by the multiplier would be

$$P_{mt} = I_{mt} V_{mt} = 0.0001\text{ A} \times 0.9\text{ V} = 0.00009\text{ W} = 90\ \mu\text{W}$$

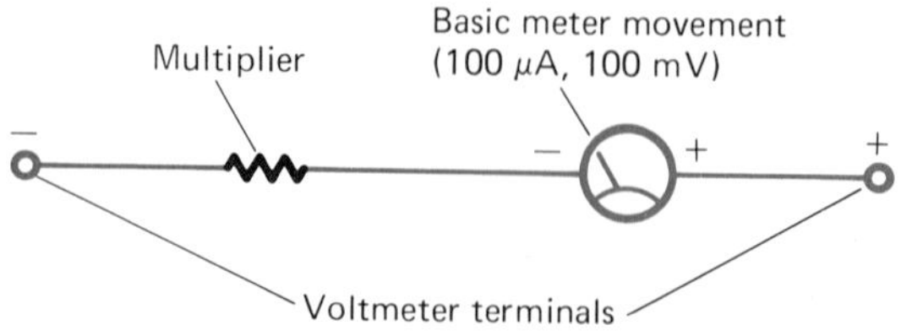

Fig. 7-7 The voltmeter multiplier extends the voltage range of the basic meter movement.

Example 7-2

What value multiplier is needed to make a 5-V voltmeter from a 1-mA, 100-Ω meter movement?

Given: $I_m = 1$ mA
$R_m = 100\ \Omega$

Find: R_{mt}, P_{mt}

Known: $P_{mt} = I_{mt}V_{mt}$, $R_{mt} = \frac{V_{mt}}{I_{mt}}$, $I_{mt} = I_m$,
$V_{mt} = V_t - V_m$, $V_m = I_m R_m$

Solution: $V_m = 0.001\text{ A} \times 100\ \Omega = 0.1\text{ V}$

$V_{mt} = 5\text{ V} - 0.1\text{ V} = 4.9\text{ V}$

$$R_{mt} = \frac{4.9\text{ V}}{0.001\text{ A}} = 4900\ \Omega$$

$P_{mt} = 0.001\text{ A} \times 4.9\text{ V}$
$= 0.0049\text{ W} = 4.9\text{ mW}$

Answer: The multiplier's resistance would be 4900 Ω and its power dissipation would be 4.9 mW.

The procedure used in Example 7-2 is always a correct procedure. However, on higher voltage voltmeters, the voltage across the voltmeter can be ignored. As an illustration, assume the meter movement in Fig. 7-7 was used to make a 100-V voltmeter. The voltage across the meter movement is only 0.1 percent (0.1 V/100 V) of the total voltage. Therefore, assuming V_{mt} is 100 V instead of 99.9 V results in only a 0.1 percent error. This is an insignificant error when the tolerance of the multiplier is ± 1 percent. When you are using 1 percent multipliers, ignore the meter movement voltage if it is less than 1 percent of the total voltage. If you are using 2 percent multipliers, ignore the movement voltage if it is less than 2 percent of the total.

Multirange Voltmeters

A typical multirange voltmeter is illustrated in Fig. 7-8. The same type of circuitry is used in the voltmeter section of a multimeter such as a VOM. Notice in Fig. 7-8 that with the exception of the 1-V range all ranges are multiples of 5 and 10. Thus, only two voltage scales are needed for the six ranges.

The selector switch in Fig. 7-8 is the nonshorting type. When switching ranges, the voltmeter is momentarily opened. A shorting switch would not be appropriate for a voltmeter. It would momentarily connect two multipliers in parallel. The total resistance of the paralleled multipliers would be less than the lowest valued multiplier. This could cause the meter movement to be overloaded.

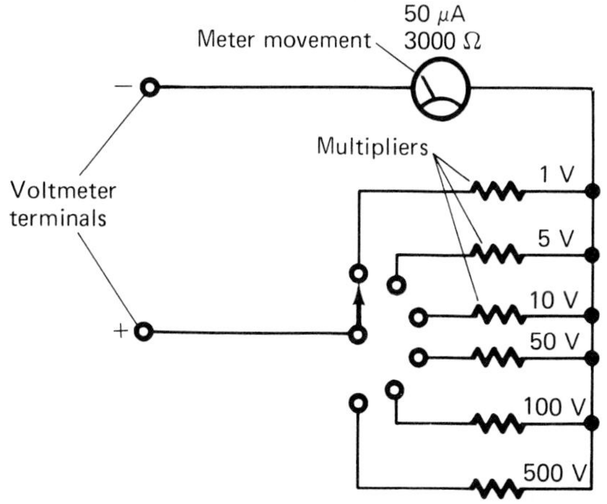

Fig. 7-8 Multirange voltmeter. Switching to a larger multiplier increases the range.

Input resistance

Ohms per volt

Sensitivity

Voltmeter Rating

The total internal resistance of a dc voltmeter is an important rating of the voltmeter. The internal resistance of a voltmeter is called its *input* resistance. For a multirange meter like the one in Fig. 7-8, the input resistance is different for each range. Rather than an input resistance for each range being specified, one general rating for all ranges is used. This rating is specified in *ohms per volt*. The ohms-per-volt rating indicates the input resistance for each volt of that range. For example, a 1000 Ω/V voltmeter would have 1000 Ω of input resistance on the 1-V range. On the 10-V range, it would have 10,000 Ω of input resistance. The input resistance of a voltmeter on a given range is found by multiplying the range by the ohms-per-volt rating. The ohms-per-volt rating of a voltmeter is often referred to as its *sensitivity*.

Example 7-3

What is the input resistance of a 20,000 Ω/V voltmeter on the 50-V range?

Given: 20,000 Ω/V voltmeter
Find: Input resistance on 50-V range
Known: Input resistance
= sensitivity × range

Input resistance

$$= 20{,}000\ \Omega/\text{V} \times 50\ \text{V}$$
$$= 1{,}000{,}000\ \Omega$$
$$= 1\ \text{M}\Omega$$

The input resistance on the 50-V range is 1 MΩ.

The sensitivity (ohms-per-volt rating) of a voltmeter is determined by the full-scale current of the meter movement. Mathematically, the relationship is

$$\text{Sensitivity} = \frac{1}{\text{full-scale current}}$$

Therefore, the sensitivity of the voltmeter of Fig. 7-8 would be

$$\text{Sensitivity} = \frac{1}{0.00005\ \text{A}} = 20{,}000\ \Omega/\text{V}$$

Notice that voltmeters have very high internal resistances, while ammeters have very low internal resistances. Voltmeters should not significantly change the voltage distribution (or the load) of the circuit in which they are used. Therefore their internal resistance must be high compared with the resistance of the load.

Electronic voltmeters have the same input resistance for all voltage ranges. They usually have higher input resistance than nonelectronic voltmeters on the lower ranges. However, on the higher ranges the opposite is often true. For example, a 100 kΩ/V meter on the 500-V range has 50 MΩ of input resistance. This exceeds the input resistance of most electronic voltmeters (500-V range).

Review Questions

15. What is the name of the resistor used to extend the voltage range of a meter movement?
16. Do voltmeters have resistors in series or in parallel with the meter movement?
17. Do ammeters or voltmeters have the higher internal resistance?
18. Is the internal resistance of a nonelectronic multirange voltmeter the same on all voltage ranges?
19. What is the sensitivity of a voltmeter which uses a 10-μA meter movement?
20. A 500-μA, 200-Ω meter movement is used in a 50-V voltmeter. What is the power dissipation and the resistance of the multiplier?
21. What is the internal resistance of a 100-V voltmeter which uses a 50-μA meter movement?
22. Does the internal resistance of an electronic voltmeter change from range to range?

7-5 OHMMETERS

Ohmmeter

A simple ohmmeter (Fig. 7-9) consists of a series-connected cell, meter movement, and two resistors. The rheostat (R_1 in Fig. 7-9) is the ohms-adjust control. Adjustment of this control compensates for changes in the voltage of the cell. When the ohmmeter terminals are shorted together as in Fig. 7-10(*a*), current will flow through the meter movement. The ohmmeter is then measuring zero resistance. If R_1 is adjusted until the meter movement indicates full scale, then full scale indicates 0 Ω. This is why the ohmmeter scale of a VOM is reverse reading.

Reverse reading

Determining Center-Scale Resistance

When the ohmmeter of Fig. 7-10(*a*) is ohms-adjusted, the internal resistance will be

$$R_{\text{int}} = \frac{V}{I_m} = \frac{1.5\ \text{V}}{0.0001\ \text{A}} = 15{,}000\ \Omega = 15\ \text{k}\Omega$$

This internal resistance includes the resistance of the meter movement and both resistors (R_1 and R_2). Suppose a 15-kΩ resistor is

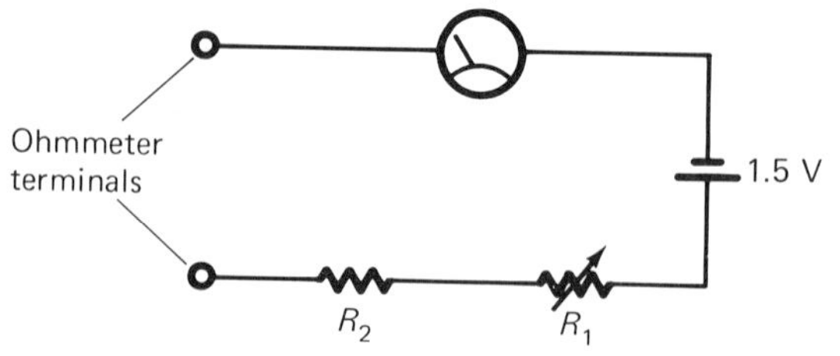

Fig. 7-9 Basic ohmmeter circuit. R_1 is adjusted to provide full-scale current when the terminals are shorted together.

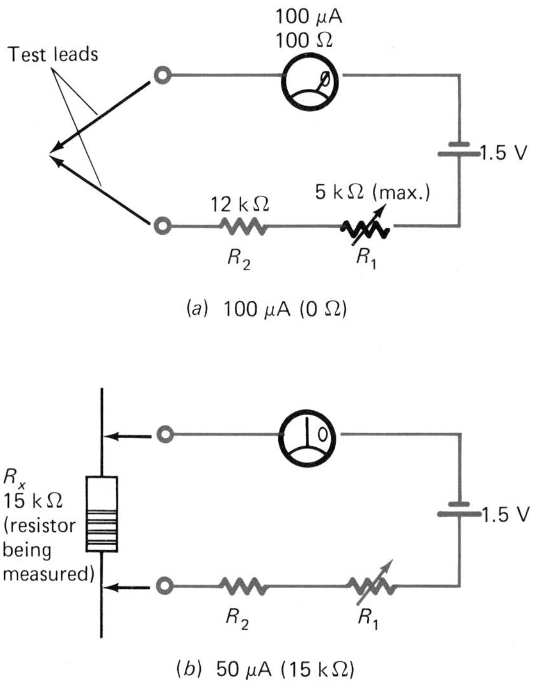

Fig. 7-10 Measuring resistance. (*a*) 100 μA (zero ohms) and (*b*) 50 μA (15 kΩ).

placed between the ohmmeter terminals, as in Fig. 7-10(*b*). Now the total resistance of the circuit will be 30 kΩ ($R_t = R_{\text{int}} + R_x$). The current through the meter movement will be cut in half because the resistance doubled:

$$I_m = \frac{V}{R_t} = \frac{1.5 \text{ V}}{30{,}000 \ \Omega}$$

$$= 0.00005 \text{ A} = 50 \ \mu\text{A}$$

Thus, the half-scale point of the ohmmeter will be 15 kΩ. This half-scale point is known as the center-scale resistance of the ohmmeter. The center-scale resistance is always equal to the internal resistance of the ohmmeter.

Center-scale resistance

When R_x in Fig. 7-10 is replaced by a 45-kΩ resistor, the total resistance will be 60 kΩ (R_t = 15 kΩ + 45 kΩ). Now the meter current will be

$$I_m = \frac{1.5 \text{ V}}{60{,}000 \ \Omega}$$

$$= 0.000025 \text{ A} = 25 \ \mu\text{A}$$

This is one-fourth of the full-scale current of the meter. Therefore, the quarter-scale point on the ohmmeter will be 45 kΩ.

Nonlinear ohmmeter scale

The above calculations illustrate how nonlinear an ohmmeter scale is. The first half of the scale covers 0 to 15 kΩ (Fig. 7-11). The next quarter of the scale covers 15 to 45 kΩ. And, the last quarter has to cover 45 kΩ to infinity.

Multirange Ohmmeters

The range of the ohmmeter circuit in Fig. 7-9 can be changed by either of two methods. First, the voltage of the cell *and* the resistance of R_2 can be increased. Second, the full-scale current of the movement can be increased by shunting it *and* the resistance of R_2 can be decreased. The first method will increase the range of the ohmmeter, and the second method will decrease the range.

Let us see how the first method increases the ohmmeter range. Suppose the cell voltage in Fig. 7-10(*a*) were increased to 15 V and R_2 were increased to 120 kΩ. Now the internal resistance of the ohmmeter (when ohms-adjusted) would be

$$R_{\text{int}} = \frac{15 \text{ V}}{0.0001 \text{ A}}$$

$$= 150{,}000 \ \Omega = 150 \text{ k}\Omega$$

And R_x in Fig. 7-10(*b*) would have to be 150 kΩ for the meter movement to indicate half scale. Thus, the range of the ohmmeter has been increased by a factor of 10. The center of the scale in Fig. 7-11 would be 150 kΩ rather than 15 kΩ.

The second method of changing ranges is shown in Fig. 7-12. The meter movement has been shunted so that the total current for full-scale deflection will be 1 mA (1000 μA). Resistor R_2 has been reduced to 1200 Ω. The internal resistance of the ohmmeter is

$$R_{\text{int}} = \frac{1.5 \text{ V}}{0.001 \text{ A}} = 1500 \ \Omega$$

Notice this is $\frac{1}{10}$ as great as the internal resistance of the ohmmeter in Fig. 7-10(*a*). Therefore, the ohmmeter in Fig. 7-12 will have $\frac{1}{10}$ as great a range as the ohmmeter in Fig. 7-10(*a*).

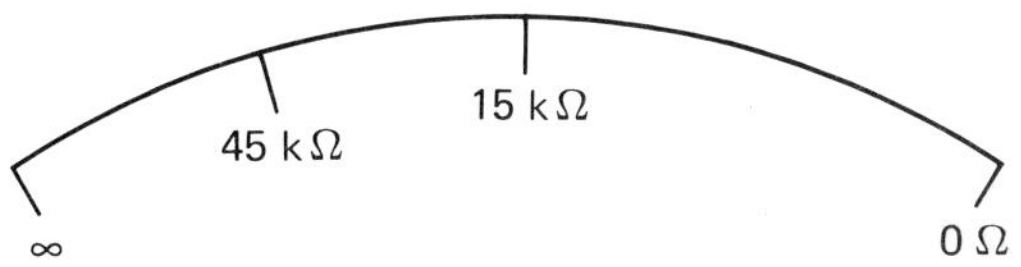

Fig. 7-11 Nonlinear ohmmeter scale.

Four-range ohmmeter

Accuracy

Percentage of the scale length

Insulation tester

Megger

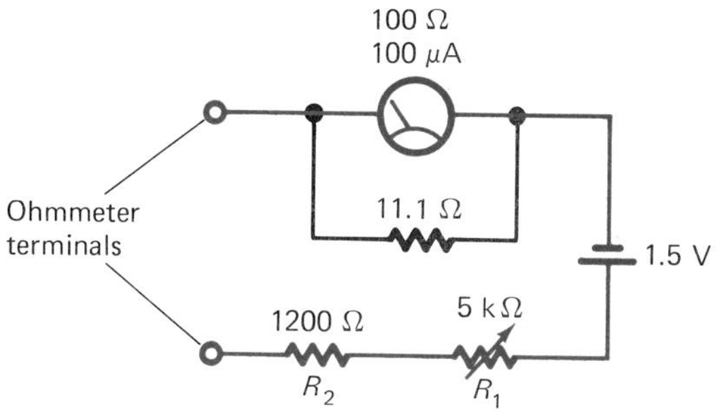

Fig. 7-12 Decreasing the range of the ohmmeter. A shunt is added to the meter movement, and the value of R_2 is decreased.

The center-scale reading for the ohmmeter of Fig. 7-12 will be 1500 Ω.

A four-range ohmmeter is shown in Fig. 7-13. The ganged, rotary switch changes shunts and series resistors simultaneously. The ohmmeter is shown on its highest range (no shunt on the meter movement). When the switch is moved to a lower range, it selects a smaller shunt and a smaller series resistor.

Most multirange ohmmeters use a combination of the two methods discussed above. This allows a wide range of resistances to be measured. The shunt method provides low ranges, and the higher-voltage battery provides very high ranges.

Other types of ohmmeter circuits are used in electronic VOMs. However, all of them use a voltage source to force current through the resistor being measured. Also, they all have nonlinear scales.

Accuracy of Ohmmeters

Ohmmeter accuracy cannot be specified in the same way as voltmeter and ammeter accuracy. This is because the full-scale value of any range of an ohmmeter is infinity. Therefore, ohmmeter accuracy is specified as a percentage of the scale length. A typical ohmmeter has an accuracy of ±2 or 3 percent of scale length. Two percent of scale length represents a lot of resistance near the infinity end of the scale. It represents much less resistance near the zero end of the scale.

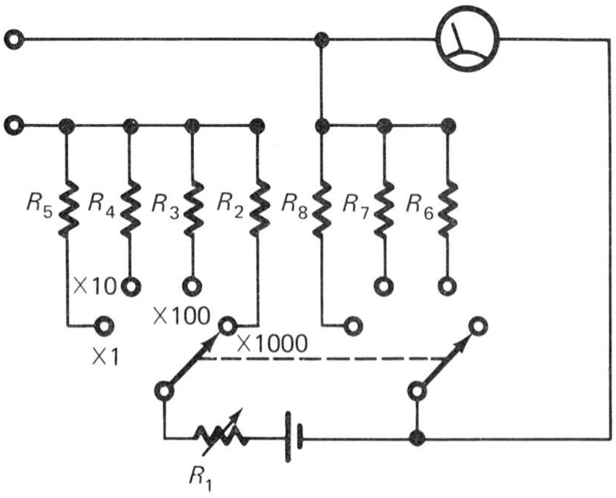

Fig. 7-13 Multirange ohmmeter.

Review Questions

23. What components are used in a simple ohmmeter?
24. What is the function of the rheostat in an ohmmeter?
25. If the internal resistance of an ohmmeter is 5 kΩ, what will its center-scale resistance equal?
26. An ohmmeter has a 10 mA meter movement and a 1.5-V cell. What is its internal resistance when properly ohms-adjusted?
27. Refer to Fig. 7-13.
    a. Does ohms-adjusting on the ×1 or the ×10 range require the most current?
    b. Does R_7 have more or less resistance than R_6?
    c. Does R_3 have more or less resistance than R_5?
28. What are two methods of changing the center-scale resistance (ranges) of an ohmmeter?
29. How is the accuracy of an ohmmeter specified?

7-6 INSULATION TESTERS

Insulating materials which must withstand high voltages have extremely high resistances. They must also be able to stand the electric stress created by the high voltage. An ordinary ohmmeter will not do an adequate job of checking the condition of insulation which is subjected to high voltages. Therefore, the insulation in some equipment, such as transformers and motors, is checked with *insulation testers* or *meggers*. These instruments test the insulation's resistance by applying a high voltage across the insulation and the frame or wiring of the equipment. Very often the test is made between ground and insulation. The insulation is tested at a voltage greater than the highest voltage to which the insulation is normally subjected. Engineering test procedures often specify a method for determining the test voltage. For example, certain equipment must be tested at twice the operating voltage plus 5000 V.

The insulation tester in Fig. 7-14 is capable of measuring up to 20,000 MΩ of resistance. It provides test voltages up to 5000 V. Because of the high voltage, insulation testers must be used with great care. Only the test leads supplied (or recommended) by the manufacturer should be used with an insulation tester. The leads should be checked regularly for signs of deterioration (checking and cracking). The manufacturer's directions and safety rules should be followed when using an insulation tester or megger.

Review Questions

30. What is an insulation tester?
31. Why is more care needed in using an insulation tester than an ohmmeter?

7-7 WHEATSTONE BRIDGES

The wheatstone bridge (Fig. 7-15) is a versatile circuit. It can be used to measure many different electrical quantities. One of its common uses is to measure resistance with great precision.

The meter movement used in the bridge is a zero-center-scale meter. When the current or voltage is zero, the meter pointer is at the center of the scale. The type of meter used is called a *galvanometer*. It can measure very low current flow in either direction, or voltage of either polarity.

The bridge circuit in Fig. 7-15 is balanced. When the bridge is balanced, no current flows through the meter. Therefore, there is no potential difference (voltage) between points *A* and *B*. The voltage across R_1 must equal the voltage across R_2 and, of course, V_R must equal V_{R_3}. Also notice that

$$I_{R_1} = I_{R_u} \quad \text{and} \quad I_{R_2} = I_{R_3}$$

when the bridge is balanced. Because of these voltage and current relationships, a definite relationship exists between the resistances of

Wheatstone bridge

Zero-center-scale meter

Galvanometer

Balanced

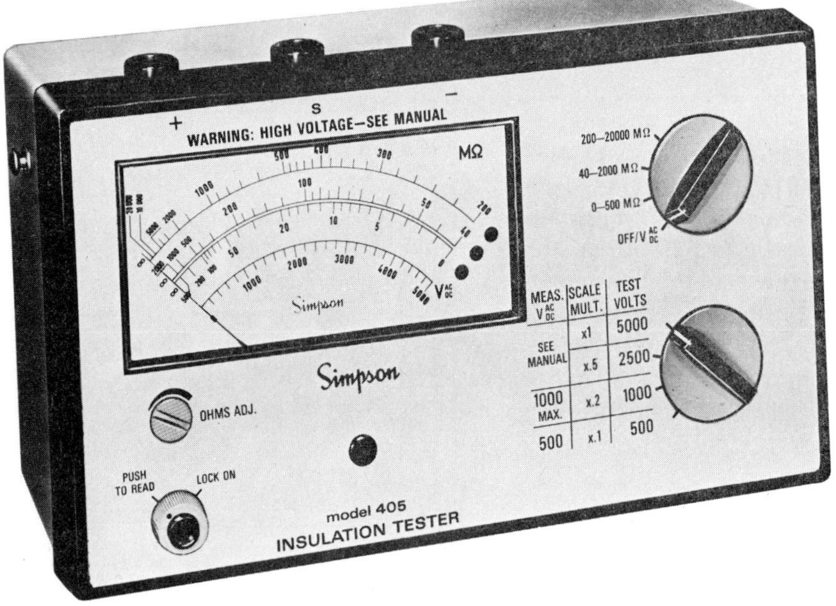

Fig. 7-14 Insulation tester. (Courtesy of Simpson Electric Co.)

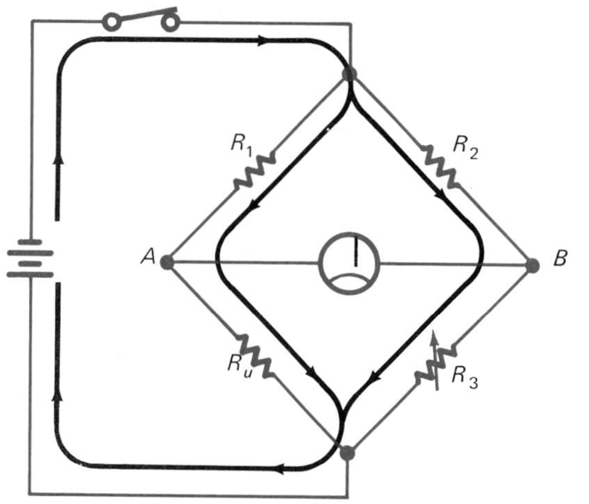

Fig. 7-15 Wheatstone bridge circuit (balanced). There is no voltage between points *A* and *B* when the bridge is balanced.

Balanced bridge formula

the circuit. The ratio of R_1/R_u must be equal to the ratio of R_2/R_3. Therefore, for a balanced bridge, we can write

$$\frac{R_1}{R_u} = \frac{R_2}{R_3}$$

Rearranging the formula to solve for R_u yields

$$R_u = \left(\frac{R_1}{R_2}\right) R_3$$

This formula can be used in measuring an unknown resistance (R_u). All we need to know is the ratio of R_1/R_2 and the value of R_3.

When the bridge is used, the resistance to be measured (R_u) is connected into the circuit, and power is applied. Then, R_3 is adjusted to achieve a balance (zero current through the meter). Resistor R_3 is a precision, calibrated rheostat. Its value can be read from a dial attached to its shaft. Resistors R_1 and R_2 are precision resistors of known value. The value of R_u is then found by multiplying the dial reading of R_3 by the ratio of R_1/R_2. The precision with which R_u can be measured is determined by the precision (tolerances) of the other three resistors.

Example 7-4

Refer to Fig. 7-15. Resistor $R_1 = 10{,}000\ \Omega$, $R_2 = 1000\ \Omega$, $R_3 = 250\ \Omega$ when the bridge is balanced. What is the value of R_u?

Given: $R_1 = 10{,}000\ \Omega$
$R_2 = 1000\ \Omega$
$R_3 = 250\ \Omega$

Find: R_u

Known: $R_u = \left(\frac{R_1}{R_2}\right) R_3$

Solution: $R_u = \frac{10{,}000}{1000} \times 250 = 10 \times 250$
$= 2500\ \Omega$

Answer: The unknown resistance is $2500\ \Omega$.

When a bridge is not in balance, none of the voltage, current, and resistance relationships discussed above are valid. When R_3 is too low, current flows through the meter from point A to point *B* [Fig. 7-16(*a*)]. The voltage across R_2

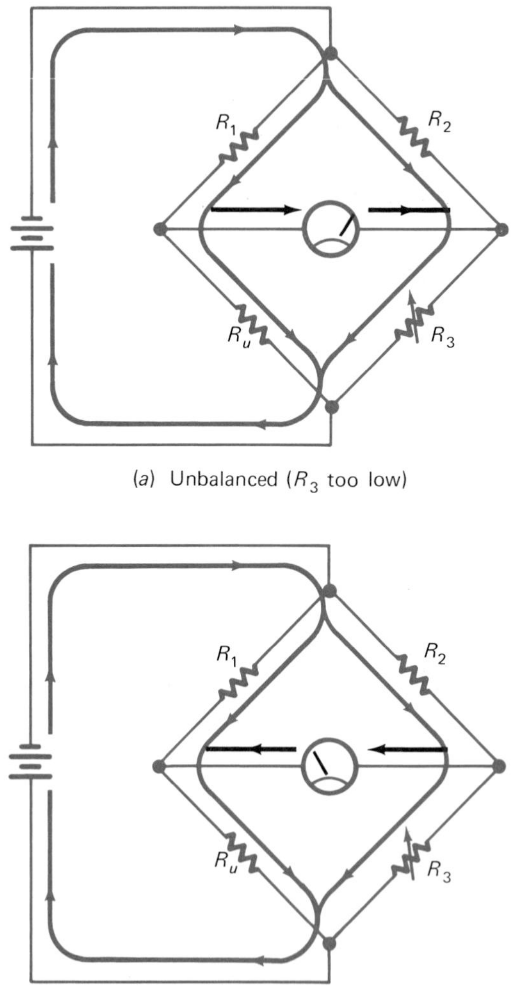

(*a*) Unbalanced (R_3 too low)

(*b*) Unbalanced (R_3 too high)

Fig. 7-16 Unbalanced bridge. (*a*) R_3 too low and (*b*) R_3 too high.

will be greater than the voltage across R_1, and the current through R_3 will be greater than the current through R_2.

If R_3 is too high [Fig. 7-16(*b*)] current flows through the meter from point *B* to point *A*. Now, V_{R_1} is greater than V_{R_2}, and I_{R_2} is greater than I_{R_3}. Obviously, the formula for finding R_u is no longer appropriate.

Review Questions

32. What type of meter is used in a wheatstone bridge?
33. Does current flow through the meter of a wheatstone bridge when it is balanced or unbalanced?
34. Refer to Fig. 7-15. Which resistors must drop equal voltages when the bridge is balanced?
35. What determines the accuracy of a measurement made with a bridge?
36. Refer to Fig. 7-15. Assume $R_1 = 300\ \Omega$, $R_2 = 450\ \Omega$, and $R_3 = 1500\ \Omega$ at balance. What is the value of R_u?

7-8 METER LOADING

Ammeter loading

The ideal voltmeter would have an infinite internal resistance and the ideal ammeter would have no resistance. If ideal meters existed, we would not have to be concerned with meter loading. However, most meters do not even approach the ideal. Therefore, we must deal with meter loading.

Whenever a meter is connected to a circuit, the circuit currents (and often the circuit voltages) change. This is what is meant by *meter loading*. Significant meter loading occurs when the current or voltage changes exceed those that might be caused by component tolerances.

Ammeter Loading

Ammeter loading is illustrated in Fig. 7-17. In Fig. 7-17(*a*), the nominal current in the circuit is

$$I = \frac{1.5\ \text{V}}{1500\ \Omega} = 0.001\ \text{A} = 1\ \text{mA}$$

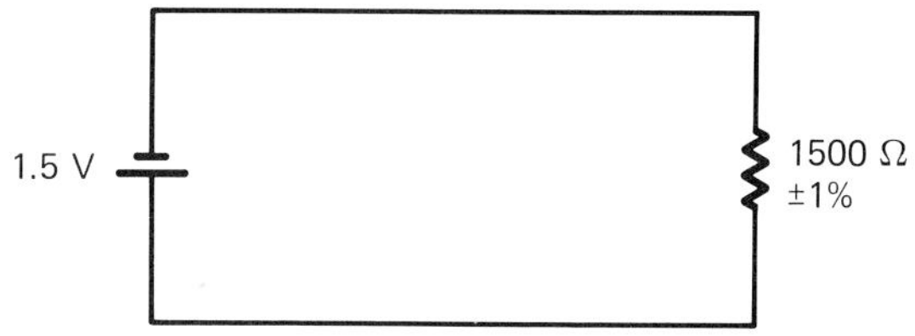

(*a*) Exected current is 0.99 to 1.01 mA

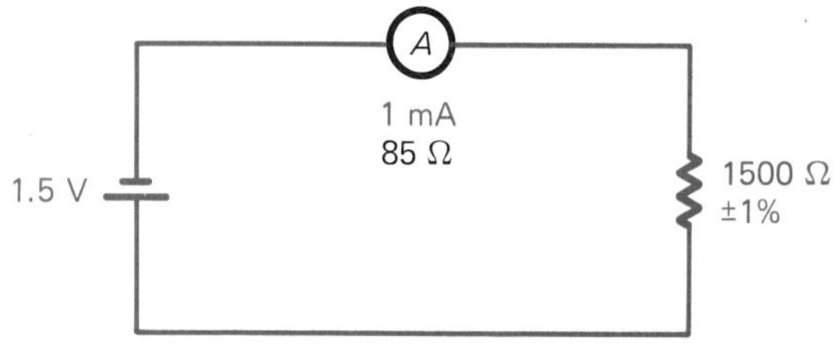

(*b*) Measured current is 0.95 mA

Fig. 7-17 Ammeter loading. (*a*) Expected current is 0.99 to 1.01 mA. (*b*) Measured current is 0.95 mA.

If the resistor is on either end of its tolerance, the current would be only 1 percent more or less (0.99 to 1.01 mA). In Fig. 7-17(*b*), the ammeter has been inserted into the circuit. The measured current, if both meter and resistor were completely accurate, would be 0.95 mA. The meter has loaded the circuit.

Low-voltage, low-resistance circuit

Significant ammeter loading does not occur very often. If the resistor in Fig. 7-17 had 10 percent tolerance, the loading would not be noticed. Also, many 1 mA ammeters have less than 85 Ω of internal resistance. When ammeter loading does occur, it is usually in a low-voltage, low-resistance circuit.

Voltmeter Loading

High-resistance circuit

Voltmeter loading

Voltmeter loading is much more common than ammeter loading. It can occur in series and series-parallel circuits. It does not occur in simple parallel circuits because the voltmeter is connected across the power source. That is, it is in parallel with all other parts of the circuit. In general, voltmeter loading occurs in high-resistance circuits like those found in many electronic devices.

The circuits in Fig. 7-18 illustrate voltmeter loading. In Fig. 7-18(*a*), the source voltage is split into equal parts by the equal resistors. However, the voltmeter in Fig. 7-18(*b*) only indicates 2 V across R_2. This is an obvious case

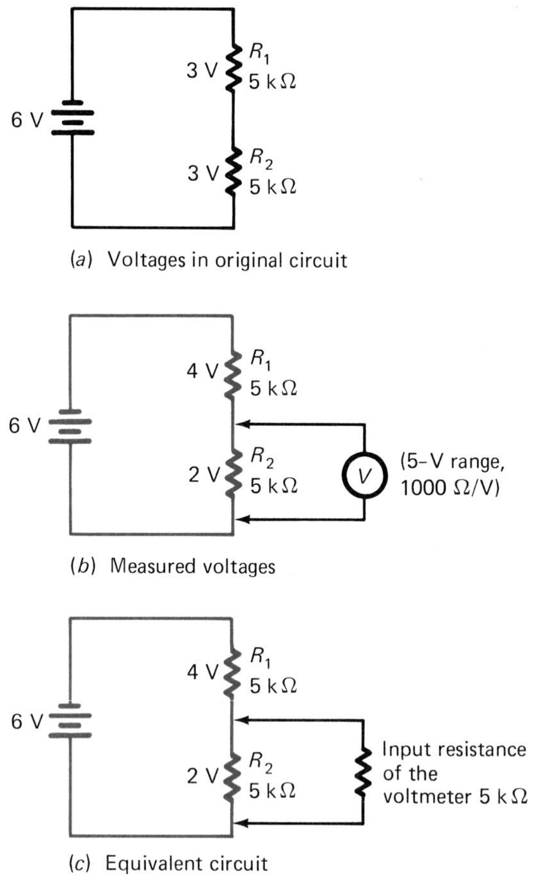

Fig. 7-18 Voltmeter loading. (*a*) Voltages in original circuit. (*b*) Measured voltages. (*c*) Equivalent circuit.

of voltmeter loading. The input resistance of the voltmeter in Fig. 7-18(*b*) (on the 5-V range) can be calculated as follows:

$$\begin{aligned}\text{Input resistance} &= \text{sensitivity} \times \text{range} \\ &= 1000\ \Omega/\text{V} \times 5\ \text{V} \\ &= 5000\ \Omega = 5\ \text{k}\Omega\end{aligned}$$

As shown in the equivalent circuit of Fig. 7-18(*c*), this 5 kΩ is in parallel with the 5 kΩ of R_2. These two 5-kΩ parallel resistances have a combined resistance of 2.5 kΩ. Therefore, R_1 drops twice as much voltage as R_2.

If the voltmeter in Fig. 7-18 had been a 20 kΩ/V voltmeter, the loading would have been minimal. The input resistance would have been 100 kΩ, and the parallel resistance of R_2 and the meter would have been 4.76 kΩ. Even a 1 percent resistor at R_2 could have a lower resistance than this.

Manufacturers of electrical-electronic equipment often specify the voltages at various points in their circuits. Usually, they specify either the sensitivity or the input resistance of the meter used in measuring these voltages. If a technician uses a meter different from the one specified, he must be aware of possible meter loading. He must be sure that any observed low voltages are not due to meter loading rather than component failures. This can be done by analyzing the circuit diagram to determine the value of the resistances in the circuit. The voltmeter's input resistance should be 20 times greater than the resistance across which the voltage is measured. Under these conditions, the loading will change the resistance of the circuit less than 5 percent. If 10 percent loading can be tolerated, a 10:1 resistance ratio is adequate.

Review Questions

37. What is meter loading?
38. Is ammeter loading or voltmeter loading more common?
39. Does voltmeter loading occur in parallel circuits?
40. A 15-kΩ resistor in a series circuit drops 6 V. Will significant voltmeter loading occur if the voltage is measured with:
    a. A 10-kΩ/V voltmeter on the 10-V range?
    b. A 100-kΩ/V voltmeter on the 10-V range?
    c. An electronic voltmeter with 10 MΩ input resistance?
41. A 10-MΩ resistor in a complex electronic circuit has 400 V across it. Which of the following will give the most accurate voltage reading?
    a. A 20 kΩ/V voltmeter on the 1000-V range
    b. An electronic voltmeter with 10 MΩ input resistance

Summary

1. The d'Arsonval meter movement is commonly used in dc voltmeters, ammeters, and ohmmeters.
2. In a meter movement, angular motion results from the interaction of two magnetic fields.

3. Basic meter movements are rated for voltage, current, and resistance.
4. Shunts extend the range of ammeters.
5. Multipliers extend the range of voltmeters.
6. Values for shunts and multipliers can be calculated if the meter movement rating and the ranges are known.
7. Ammeters have very low internal resistance.
8. Voltmeters have very high internal resistance.
9. External shunts are rated in amperes and millivolts.
10. Shunts and multipliers are precision, low-temperature-coefficient resistors.
11. Shunts are in parallel with the meter movement.
12. Multipliers are in series with the meter movement.
13. Nonelectronic voltmeters have an ohms-per-volt rating.
14. Electronic voltmeters have an input-resistance rating.
15. Input resistance = sensitivity × range
16. $\text{Ohms per volt (sensitivity)} = \dfrac{1}{\text{full-scale current}}$
17. Ohmmeters have a power source, rheostat, meter movement, and fixed resistor.
18. Ohmmeter scales are nonlinear and often reverse reading.
19. Ohmmeter ranges are changed by switching voltages or shunts.
20. Insulation testers and meggers use high voltages to measure very high resistances (insulation).
21. A wheatstone bridge uses a galvanometer which indicates zero when the bridge is balanced.
22. A bridge is balanced when the ratio of rheostat to unknown resistance equals the ratio of the two known resistors.
23. The accuracy of a bridge is determined by the tolerances of the resistors used in it.
24. Meter loading causes changes in circuit currents and voltages when measurements are made.

Chapter Review Questions

Here is a chance to see what you have learned. For questions 7-1 to 7-15, determine whether the statement is true or false.

7-1. Voltmeter loading is most likely to occur in a parallel circuit.

7-2. An analog multimeter uses a galvanometer meter movement.

7-3. D'Arsonval meter movements are polarized.

7-4. Rotational force in a d'Arsonval meter movement is created by the interaction of two electromagnetic fields.

7-5. The range of an ammeter can be extended by using a multiplier resistor.

7-6. The range switch of a multirange ammeter should be a nonshorting-type switch.

7-7. The internal resistance of an electronic voltmeter changes from range to range.

7-8. A 20-μA meter movement will make a more sensitive voltmeter than a 50-μA meter movement.

7-9. Ohmmeters have a nonlinear scale.

7-10. Ohmmeter tolerance is expressed as a percentage of full-scale value.

7-11. The range of an ohmmeter can be changed only by changing the battery voltage.

7-12. Insulation testers measure resistance under high-voltage conditions.

7-13. When a wheatstone bridge is in balance, no current flows in any part of the circuit.

7-14. The accuracy of a bridge measurement is determined by the accuracy of the meter used in the bridge.

7-15. Ammeter loading occurs more often than voltmeter loading.

Answer the following questions.

7-16. What is the input resistance of a 20 kΩ/V voltmeter on the 25-V range?

7-17. Would the above meter cause significant loading if used to measure the voltage across a 100-kΩ resistor in a series circuit?

7-18. What is the purpose of the rheostat in an ohmmeter circuit?

7-19. What is the value of the internal resistance of a meter movement which is rated 100 μA, 200 mV?

7-20. A 100-mA meter indicates 36 mA. Its accuracy is ± 2 percent. What are the minimum and maximum values of the actual current?

7-21. You need to convert a 50-μA, 1000-Ω meter movement to a 100-mA meter. What value of resistor will you need?

7-22. You need to convert the meter movement in question 7-21 to a 100-V meter. What value of resistor will you need?

7-23. You need to convert the meter movement in question 7-21 to a series-type ohmmeter with a 30,000-Ω center-scale resistance. How much battery voltage will be needed?

7-24. What is the ohms-per-volt rating of a voltmeter which uses a 100-μA meter movement?

Answers to Review Questions

1. D'Arsonval
2. The force of spiral springs or a taut band
3. Full-scale current, internal resistance, full-scale voltage
4. 124 mA and 136 mA
5. 2000 Ω
6. 15-mA meter, because 1 percent of 15 is less than 2 percent of 10.
7. A shunt is a resistor in parallel with a basic meter movement.
8. Low temperature coefficient and low-resistance tolerance
9. 9.9 mA
10. *a*. 0.8 ohms
    b. 8 mW
11. A shorting type (make before break)
12. As low as possible
13. When measuring currents of 10 A or higher
14. By current capacity and voltage drop
15. A multiplier resistor
16. Series
17. Voltmeters
18. No
19. 100,000 Ω/V
20. P_{mt} = 25 mW, R_{mt} = 100,000 Ω
21. 2 MΩ
22. Yes
23. A meter movement, a cell, a fixed resistor, and a variable resistor
24. The rheostat adjusts the ohmmeter for zero-resistance readings.
25. 5 kΩ
26. 150 Ω
27. *a*. ×1 range
    b. Less
    c. More
28. Change the voltage and the series resistance; or change the series resistance and the shunt.

29. As a percentage of scale length
30. A device that measures resistance at a very high voltage
31. Because the tester uses high voltage
32. Galvanometer
33. Unbalanced
34. R_1 and R_2 or R_3 and R_u
35. The tolerance of the resistors in the bridge
36. 1000 ohms
37. Meter loading means that the meter's internal resistance appreciably changes the circuit's current or voltage.
38. Voltmeter loading
39. No
40. *a*. Yes
    b. No
    c. No
41. A 20 kΩ/V voltmeter on the 1000-V range

Alternating Current and Voltage

- **Most electrical energy is distributed in the form of alternating current and voltage. Our homes and our factories are both powered by alternating current electricity. These, of course, are the major consumers of electrical energy.**

  After completing this chapter, you will be able to specify the characteristics of alternating current using correct terminology. You will understand how single-phase and three-phase alternating current can be produced. You will also understand the characteristics of wye and delta connections.

8-1 ALTERNATING CURRENT TERMINOLOGY

Alternating current (ac) is a current that periodically changes the direction in which it is flowing. It also changes its magnitude (amount) either continuously or periodically. With most types of ac, the magnitude is changing continuously.

Of course, if there is an alternating current, there must also be an alternating voltage and power. Although it is not proper English, we often refer to alternating voltage as ac voltage. When written or spoken in the abbreviated form, the term "ac voltage" looks and sounds all right. However, when spelled out as alternating current voltage, the term can be confusing. Ac voltage refers to a voltage which produces an alternating current when used to power a circuit. Likewise, ac power refers to power that is produced by alternating current and alternating voltage.

8-2 WAVEFORMS

An electrical waveform is represented by a line on a graph as shown in Fig. 8-1(*a*). The line is produced by plotting points on a graph and then connecting the points together. The points represent the value of some electrical quantity at different times. Magnitude and direction of the electrical quantity (voltage or current) are indicated on the vertical axis of

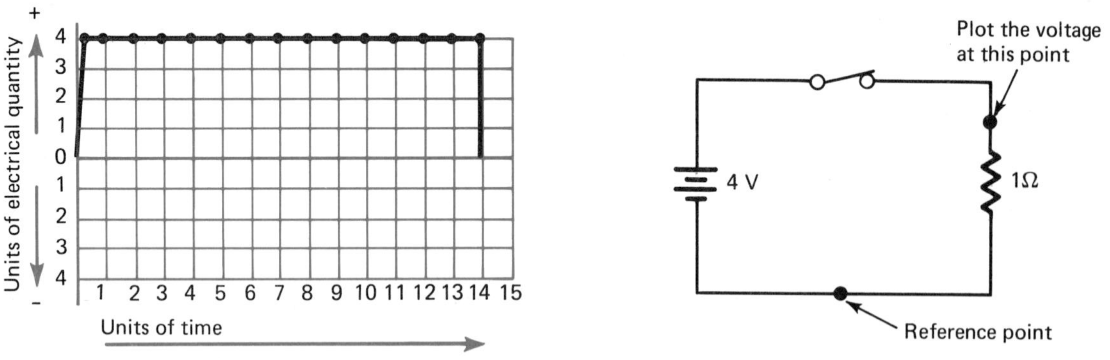

(a) Waveform

(b) Circuit which produces the waveform

Fig. 8-1 Plotting a waveform. An electrical quantity, such as voltage, is plotted against time.

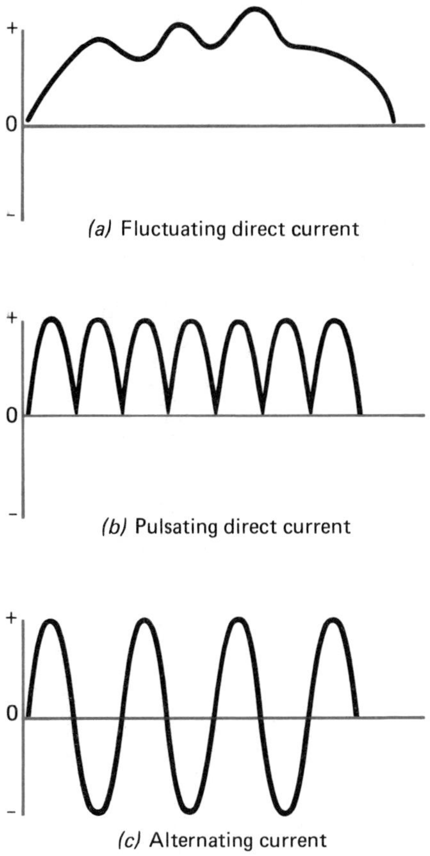

Fig. 8-2 Electrical waveforms.

the graph. Time is marked on the horizontal axis.

The waveform shown in Fig. 8-1(*a*) could represent the voltage across the resistor in Fig. 8-1(*b*). In this case the units on the vertical axis would be volts (V). The waveform shows that the voltage rapidly increases to its maximum value (4 V) when the circuit is energized. The voltage remains at its maximum value until the circuit is opened. When the circuit is opened, the voltage across the resistor suddenly drops to zero.

The waveform of Fig. 8-1(*a*) could just as well represent the current in Fig. 8-1(*b*). All that would change would be the units on the vertical axis. For the current waveform the units would be amperes (A).

Notice in Fig. 8-1(*a*) that the waveform is always above the zero reference line. This means that the polarity of the voltage is always positive with respect to some reference point. Or, in the case of a current waveform, it means that the direction of the current never reverses. In other words, Fig. 8-1(*a*) is the waveform of a pure (steady) direct current or voltage. The magnitude of the voltage or current changes only when the circuit is turned on or off.

Other common waveforms are shown in Fig. 8-2. The fluctuating dc in Fig. 8-2(*a*) is the type of current produced in an amplifying transistor. The pulsating waveform of Fig. 8-2(*b*) represents the type of current (or voltage) produced by a battery charger. Notice that pulsating direct current periodically reduces to zero while fluctuating direct current does not. Figure 8-2(*c*) shows one type of alternating current. Notice the alternating current waveform goes below the zero reference line. This means that the polarity of the voltage reverses, and that the direction of the current changes.

From page 146:
Alternating current
Alternating voltage
Ac power
Electrical waveform
On this page:
Fluctuating dc
Sine wave

8-3 TYPES OF AC WAVEFORMS

The most common type of ac waveform is the *sine wave* shown in Fig. 8-3(*a*). Ac with this type of waveform is referred to as *sinusoidal* alternating current. The alternating current (and voltage) supplied to homes and factories is sinusoidal. A pure sine wave has a very specific shape. Its shape can be precisely defined by mathematics. All of the formulas we will be using to solve ac circuit problems will be based

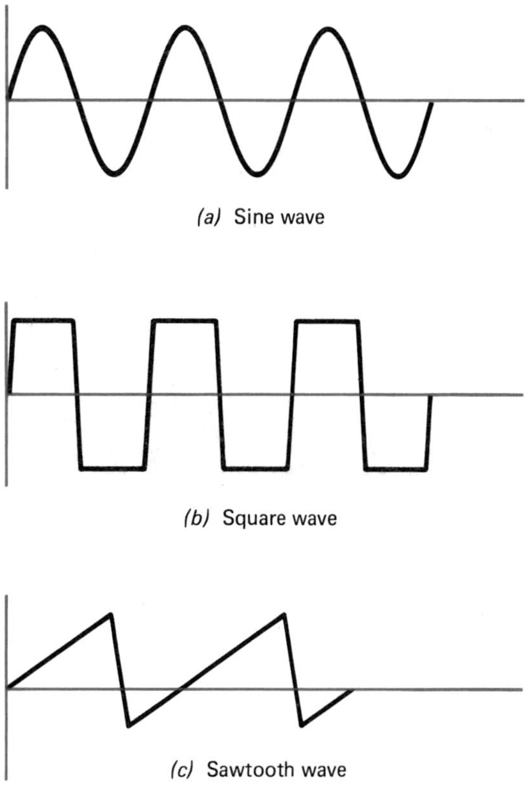

Fig. 8-3 Common types of ac waveforms.

on the sine wave. Therefore, these formulas will be appropriate only when working with sinusoidal ac.

Square wave

Figure 8-3(*b*) shows a square wave. This type of ac waveform is used extensively in computer circuits. With a square wave the magnitude of the current (or voltage) is not continuously varying. However, both amplitude and direction do periodically change.

Sawtooth wave

The sawtooth waveform of Fig. 8-3(*c*) is used in television receivers, radar receivers, and other electronic devices. Sawtooth voltages and currents are used in the circuits which produce the picture on a television screen.

Wave shapes other than those shown in Fig. 8-3 are possible. In fact, ac can be electronically produced in an almost infinite variety of waveforms. Electronic music is created by producing and mixing together a wide variety of waveforms.

Review Questions

1. What quantity is marked on the horizontal axis of a waveform graph?
2. Would the vertical axis of a waveform graph be marked in units of voltage or time?
3. What does a waveform graph show?
4. Does any type of dc waveform ever cross the zero reference line?
5. What is the difference between fluctuating and pulsating dc?
6. What is the most common type of ac waveform?
7. List two other common ac waveforms.

8-4 QUANTIFYING AC

Fully describing alternating current (or voltage) requires the use of a number of terms. Some of these terms have general meanings, but when used in electricity they have a very specific meaning.

Cycle

Cycles

The waveform in Fig. 8-4 shows four *cycles* of ac. A cycle is that part of a waveform which does not repeat or duplicate itself. Each cycle in Fig. 8-4 is a duplicate of every other cycle in the figure. During the time it takes to complete one cycle, the current does the following:

1. Builds from zero to maximum amplitude in one direction
2. Decays back to zero
3. Builds to maximum in the opposite direction
4. Again decays to zero

Alternation

The part of the cycle above the horizontal line in Fig. 8-4 is called the positive half cycle. A half cycle is also called an *alternation*. Therefore, the positive half cycle could be called the positive alternation. The negative half cycle, of course, is that part below the horizontal reference line.

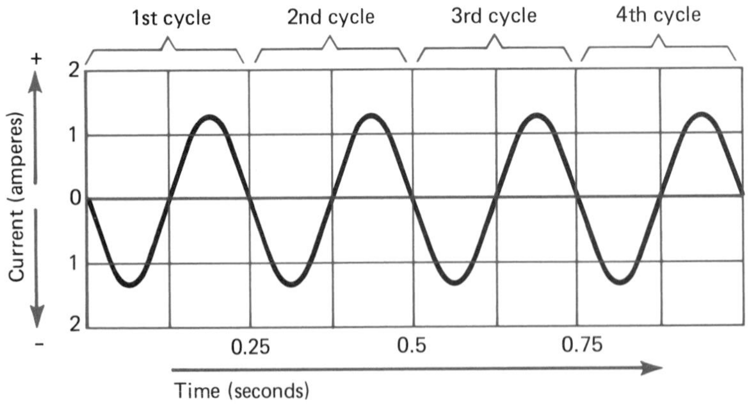

Fig. 8-4 Cycle, period, and frequency. The waveform has a period of 0.25 seconds and completes four cycles per second.

Period
Frequency
Hertz
$T = \frac{1}{f}$
$f = \frac{1}{T}$
Peak value
Peak-to-peak value
$V_{p\text{-}p} = 2V_p$

Period

The time required to complete one cycle is the *period* (T) of a waveform. In Fig. 8-4 it takes 0.25 seconds (s) to complete one cycle. Therefore the period (T) of that waveform is 0.25 s.

Frequency

The rate at which cycles are produced is called the *frequency* (f) of an alternating current or voltage. Frequency, then, refers to how rapidly the current reverses, or how often the voltage changes polarity.

Unit of Frequency—the Hertz

The base unit of frequency is the *hertz* (Hz). One hertz is equal to one cycle per second. The sine wave in Fig. 8-4 goes through four cycles in one second. Therefore, it has a frequency (f) of four hertz (Hz). In abbreviated form we could write $f = 4$ Hz.

Commercial power in North America is distributed at a frequency of 60 Hz. It is sinusoidal ac. In many of the European countries a frequency of 50 Hz is used for electrical power.

Electronic circuits use a wide range of frequencies. For example, an audio amplifier must amplify all frequencies between 20 hertz and 20 kilohertz (20 kHz). Frequencies between 540 kHz and 1600 kHz are used for AM radio, and TV uses frequencies up to 890 megahertz (MHz).

Frequency (f) and period (T) are reciprocally related. That is:

$$T = \frac{1}{f} \quad \text{and} \quad f = \frac{1}{T}$$

The period will be in seconds when the frequency is in hertz. (Remember, one hertz is equal to one cycle per second.)

Example 8-1

How much time is required to complete one cycle if the frequency is 60 Hz?

Given: $f = 60$ Hz
Find: T
Known: $T = \frac{1}{f}$
Solution: $T = \frac{1}{60 \text{ Hz}} = 0.0167$ s
Answer: The time required is 0.0167 seconds or 16.7 milliseconds (ms).

Example 8-2

What is the period of a two-megahertz (2-MHz) sine wave?

Given: $f = 2$ MHz
Find: T
Known: $T = \frac{1}{f}$
Solution: $T = \frac{1}{2{,}000{,}000} = 0.0000005$ s
Answer: The period is 0.0000005 seconds or 0.5 microseconds (μs).

As a review, let us rework the solution to Example 8-2 using powers of ten:

$$T = \frac{1}{2 \times 10^6} = 0.5 \times 10^{-6} \text{ s}$$

Notice how much easier it is to solve the problem when using powers of ten.

Example 8-3

What is the frequency of a waveform that requires 0.01 seconds to complete one cycle?

Given: $T = 0.01$ s
Find: f
Known: $f = \frac{1}{T}$
Solution: $f = \frac{1}{0.01} = 100$ Hz
Answer: The frequency is 100 hertz.

Specifying Amplitude

The amplitude of an ac wave form can be specified several different ways as shown in Fig. 8-5. These ways of specifying amplitude are appropriate for either voltage or current.

For symmetrical waveforms the *peak* value (also called the maximum value) of the negative and positive half cycle are equal. Therefore, the *peak-to-peak* value is twice as great as the peak value. For a sinusoidal ac voltage we can write:

$$V_{p\text{-}p} = 2V_p$$

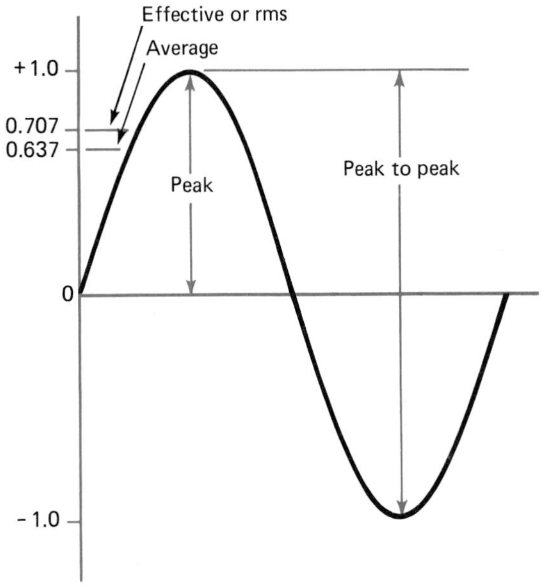

Fig. 8-5 Values of a sine wave. The amplitude of a sine wave can be specified in any one of four ways.

Many of the waveforms in electronic circuits are not symmetrical. Such waveforms can be specified in either peak or peak-to-peak values. When specifying a peak value, one must indicate whether it is the positive or negative peak.

Average value

$V_{av} = 0.637\ V_p$

$V_p = 1.57\ V_{av}$

The *average* value of a waveform is the arithmetical mean. Finding the arithmetical mean of a waveform is illustrated in Fig. 8-6(*a*). The instantaneous value of the waveform is determined at a number of equally spaced intervals along the horizontal axis. Then, all the instantaneous values are added together and divided by the total number of intervals. The closer the intervals, the more accurate will be the estimate of the average value of the waveform. Figure 8-6(*b*) provides a further idea of the meaning of the average value of a waveform. The area of the waveform above the average value is equal to the "missing" areas below the average value. The area labeled A is equal to the area labeled A′ [see Fig. 8-6(*b*)]. Area *B* is equal to area *B*′.

For a sine wave, the relationship between peak value and average value is indicated in Fig. 8-5. For voltage, the relationship can be written as:

$$V_{av} = 0.637V_p$$

Or, solving for V_p, the relationship is:

$$V_p = 1.57V_{av}$$

The relationship between average and peak voltage is used when analyzing circuits which convert ac voltage to pulsating dc voltage.

Example 8-4

What is the average voltage of a sinusoidal waveform if its peak-to-peak voltage is 300 volts?

Given: $V_{p\text{-}p} = 300\text{ V}$

Find: V_{av}

Known: $V_p = \frac{V_{p-p}}{2}$ and

$V_{av} = 0.637\ V_p$

Solution: $V_p = \frac{300\text{ V}}{2} = 150\text{ V}$

$V_{av} = 0.637 \times 150\text{ V} = 95.6\text{ V}$

Answer: The average voltage is 95.6 volts.

Rms value

Effective value

The most common way of specifying the amount of ac is by stating its *effective* or *rms* value. The effective value of alternating cur-

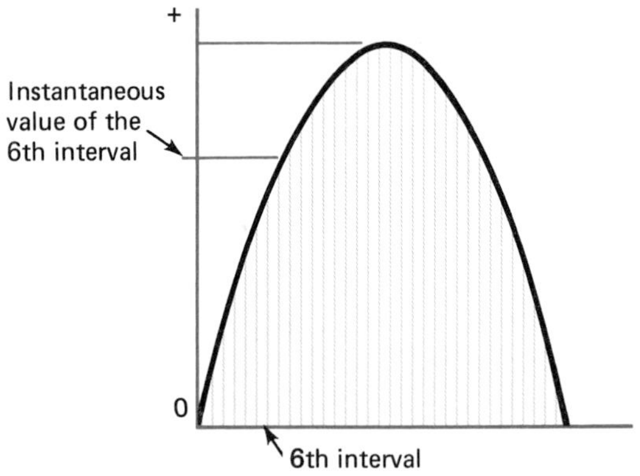

(a) Method of determining the average value

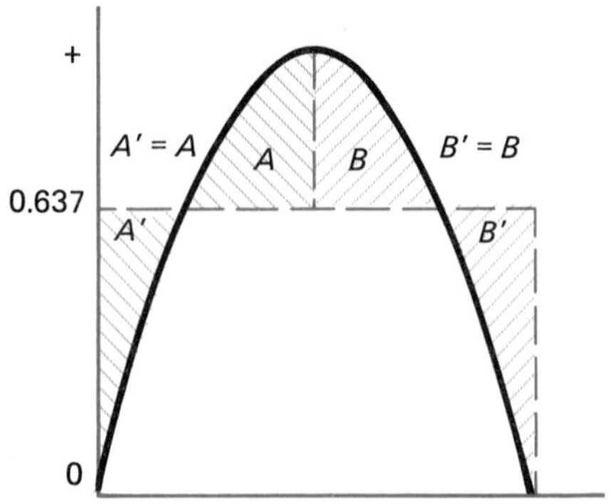

(b) Average area

Fig. 8-6 Average value of a waveform. The combined area of A and B is equal to the area of A′ plus B′.

rent has the same effect as direct current in producing heat in a resistive circuit. One ampere of effective ac and one ampere of dc produce equal power when flowing through equal value resistors. Also, equal amounts of dc voltage and effective ac voltage will produce equal power across equal value resistors.

In Fig. 8-7(*a*), 20 V of effective ac voltage forces 2 A of effective alternating current through a 10-ohm (Ω) resistor. This causes the 10-Ω resistor to dissipate 40 watts (W).

$$P = IV$$
$$= 2\text{ A} \times 20\text{ V} = 40\text{ W}$$

As shown in Fig. 8-7(*b*), 20 V of dc voltage across a 10-Ω resistor also dissipates 40 W.

Now that you know what "effective value" means, let us examine why it is also the "rms value." The effective value of a waveform can be determined by a mathematical process known as root mean square (rms). With this process the waveform is divided into many intervals, as was done in Fig. 8-6(*a*). The instantaneous value of each interval is squared. The mean (average) of the squared values is then determined. Finally, the square root of the mean is calculated. When many intervals are used, this process (rms) yields the effective value. Thus, we can use the terms "rms value" and "effective value" interchangeably.

For sinusoidal ac, the rms (effective) value and peak value are related by the following formulas:

$$V_{rms} = 0.707V_p \quad \text{and} \quad V_p = 1.414V_{rms}$$

Example 8-5

What is the peak voltage of 120 volts rms?

Given: $V_{rms} = 120$ V

Find: V_p

Known: $V_p = 1.414\ V_{rms}$

Solution: $V_p = 1.414 \times 120\text{ V} = 169.7\text{ V}$

Answer: The peak voltage is 169.7 volts.

Alternating currents and alternating voltages are usually specified in rms values. It is common practice to assume that ac is in rms units unless the subscript of p-p or p or av is included. If no subscripts are used, the ac voltage (or current) is the rms value. For example, the voltage at the outlets in your home may be specified as 120 volts (120 V). This is the rms value of the ac voltage.

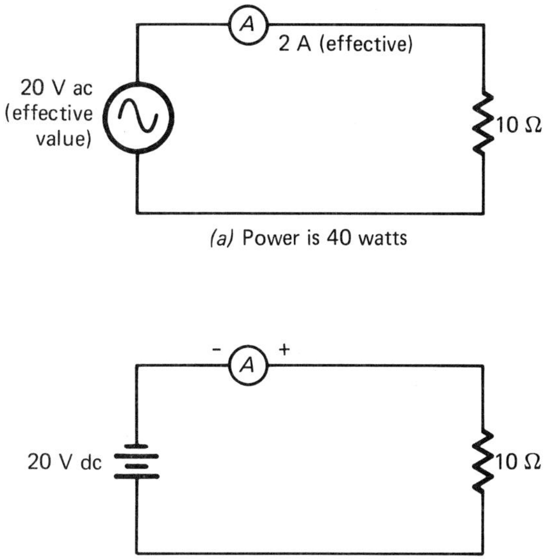

(a) Power is 40 watts

(b) Power is 40 watts

Fig. 8-7 Effective current and voltage. Direct current and effective alternating current produce the same amount of heat.

Review Questions

8. Define the following terms:
   a. Cycle
   b. Positive half cycle
   c. Period
   d. Frequency
9. What is the base unit of frequency?
10. Define the base unit of frequency.
11. What is the frequency and the type of waveform used for power distribution in the United States?
12. What is the period of a 400 Hz square wave?
13. What is the frequency of a waveform that has a period of 4 ms?
14. What is the average voltage (V_{av}) of 10 V peak-to-peak?
15. What is the peak current (I_p) of a 5-A average current?
16. What is the rms value of a 100-V peak voltage?

$V_{rms} = 0.707\ V_p$

$V_p = 1.414\ V_{rms}$

17. What is the peak voltage of 100 V, effective value?
18. What is the peak value of a 10-A alternating current?
19. Which is greater, the effective or the average value of a 20-V sine wave?
20. What is the symbol for frequency?
21. What is the abbreviation for the base unit of frequency?
22. What is an alternation?

8-5 THE SINE WAVE

Mastery of the principles of induced voltage is essential in understanding how a sine wave is produced.

Induced Voltage and Current

Induced voltage

When a conductor cuts across magnetic flux, a voltage is induced in the conductor. This voltage, of course, causes a current to flow if there is a complete circuit. Therefore, one can also say that a current is induced in the conductor.

Flux cut per unit of time

The amount of induced voltage is a function of the amount of flux cut by the conductor per unit of time. The amount of flux cut per unit of time is in turn determined by the following factors:

The speed at which the conductor is moving
The flux density
The angle at which the conductor crosses the magnetic flux

As shown in Fig. 8-8(*a*), no voltage (or current) is induced when the conductor moves parallel to the flux. This is because no lines of flux are crossed or cut. Some current is induced when the motion is at 45 degrees (45°) to the flux [Fig. 8-8(*b*)]. At 45° some flux is being cut. But, for a given speed of conductor motion, more flux is cut as the angle increases toward 90°. When the direction of movement is perpendicular (90°) to the flux, maximum current is induced [Fig. 8-8(*c*)]. The **x** in the center of the conductors in Fig. 8-8 indicates that the current is flowing into the paper.

The direction of an induced current (or the polarity of an induced voltage) is determined by two factors:

1. The direction in which the conductor is moving
2. The polarity of the magnetic field or the direction of the flux

Figure 8-9 shows the four possible combinations of conductor movement and polarity. The resulting direction of current is indicated by a • or an **x** in the center of the conductor. The dot means the current is coming out of the paper.

Left-hand rule

The direction of induced current can be determined by using the "left-hand rule" which is illustrated in Fig. 8-10. In this figure, the thumb indicates the direction of conductor movement. The index (first) finger (pointing straight out) is aligned with the direction of the flux. The second finger (bent at 90° to the palm) indicates the direction the current flows. Apply the left-hand rule to the diagrams in Fig. 8-9. Your second finger should point

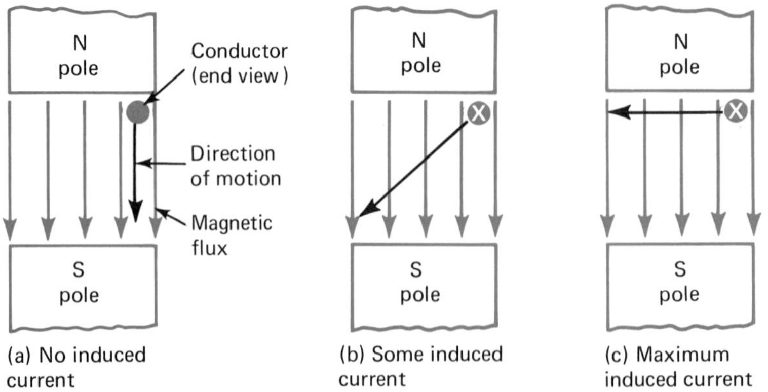

Fig. 8-8 Moving conductor in magnetic field. The amount of induced current depends on the angle at which the conductor cuts the flux.

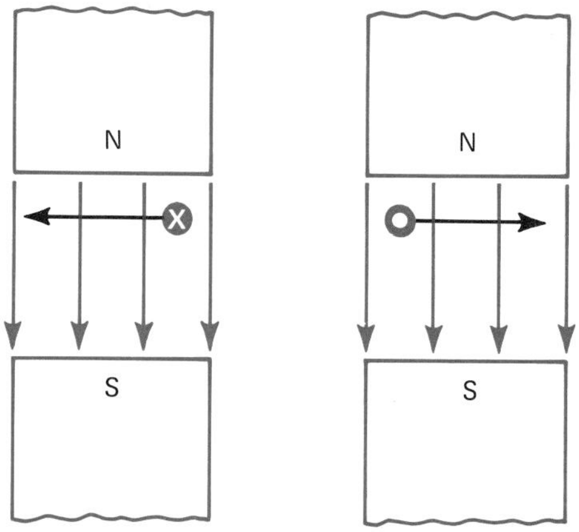

(a) Changing direction of conductor movement

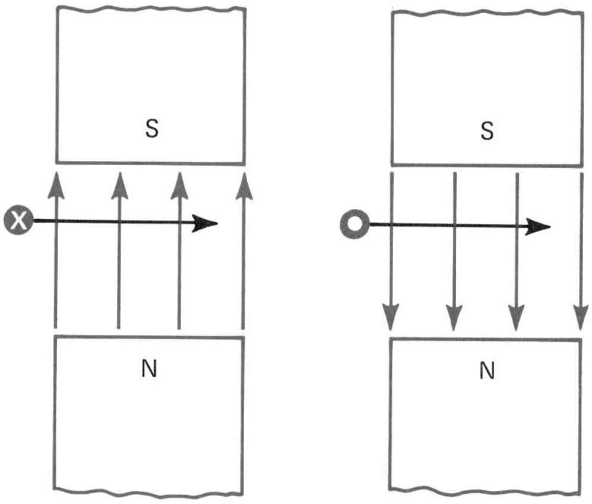

(b) Changing the polarity of the field

Fig. 8-9 Changing the direction of movement or the polarity of the field changes the direction of the induced current.

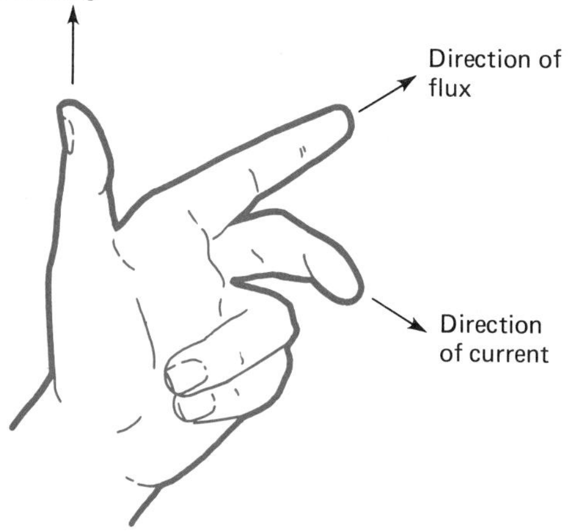

Fig. 8-10 Determining the direction of induced current using the left-hand rule.

into the page for those diagrams with an **x** in the conductor.

Producing the Sine Wave

Now that you are familiar with induced voltages and currents, let us see how we can produce a sine wave. A sine wave is produced by rotating a conductor in a magnetic field. The conductor must rotate in a perfect circle and at a constant speed. The magnetic field has to have a uniform flux density. Production of a sine wave is illustrated in Fig. 8-11. At position 1, the conductor is moving parallel to the flux and no voltage is induced in the conductor. (The instantaneous direction of the conductor is indicated by the arrow.) As the conductor rotates from position 1 to position 2, it starts to cut the flux at a slight angle. Thus, a small voltage is induced in the conductor. A current produced by this voltage would be flowing out of the page as indicated by the dot in the conductor. As the conductor rotates on through positions 3 and 4, the angle at which the conductor cuts the flux increases. At position 4, the conductor is moving perpendicular to the flux. Maximum flux per unit of time is now being cut. Therefore, maximum voltage (V_p) is being produced as shown in Fig. 8-11(*b*). The positive peak of the sine wave is produced. Now as the conductor moves on to position 5 and 6, less flux is cut. This is because the conductor's direction of movement again starts to parallel the flux. (For a given length of movement of the conductor, less flux is being cut.) By the time the conductor reaches position 7, no voltage is being induced. The first alternation (positive half cycle) of the sine wave has been produced. Notice [Fig. 8-11(*a*)] that the direction in which the conductor cuts the flux reverses as the conductor leaves position 7. Thus, the polarity of the induced voltage is reversed and the negative half of the cycle is started. At position 10, the conductor is producing its peak negative voltage. Finally, as the conductor returns to position 1, the induced voltage drops to zero. The first cycle is completed; a sine wave has been produced. Each additional revolution of the conductor produces another cycle of a sine wave.

Electrical and Mechanical Degrees

Notice in Fig. 8-11(*a*) that the conductor ro-

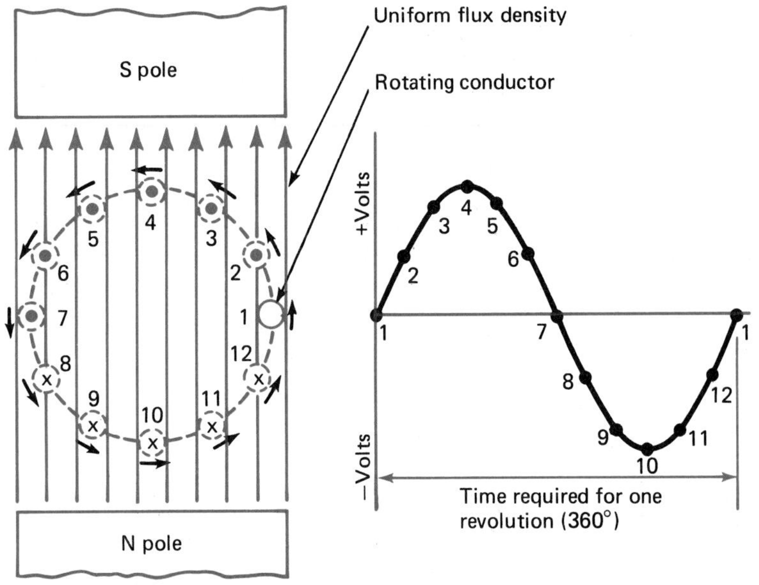

(a) Conductor rotating in magnetic field
(b) Resulting sine wave

Fig. 8-11 Generating a sine wave by rotating a conductor through a magnetic field.

Electrical degrees

Mechanical degrees

tated 360° (one revolution) in producing one cycle of a sine wave. The horizontal axis of the waveform can, therefore, be marked off in degrees rather than units of time. This has been done in Fig. 8-12 for each of the 12 positions of the conductor in Fig. 8-11. It can be seen from Fig. 8-12 that a sine wave reaches its peak values at 90° and 270°. Its value is zero at 0° (or 360°) and 180°. When degrees are used to mark off the horizontal axis of a waveform, they are referred to as *electrical degrees*. One alternation of a sine wave contains 180 electrical degrees; one cycle has 360 electrical degrees.

The advantage of marking the horizontal axis in degrees (rather than seconds) is that electrical degrees are independent of frequency. A cycle has 360° regardless of its frequency. However, the time required for a cycle (that is, the period) is totally dependent on the frequency. When the horizontal axis is marked off in seconds, the numbers are different for each frequency. Thus, you can see that the use of electrical degrees is very convenient and practical.

In Fig. 8-11 the number of electrical degrees is equal to the number of *mechanical degrees*. That is, the conductor rotates 360 mechanical degrees to produce the 360 electrical degrees of the sine wave. This one-to-one relationship of mechanical and electrical degrees is not essential. In fact, most ac generators produce more than one cycle per revolution.

Figure 8-13 shows the magnetic field produced by two pairs of magnetic poles. This field produces two cycles per revolution of the conductor. Every 45° of rotation (position 1 to position 2, for example) produces 90 electrical degrees. The conductor in Fig. 8-13 cuts flux under a south pole and a north pole for each

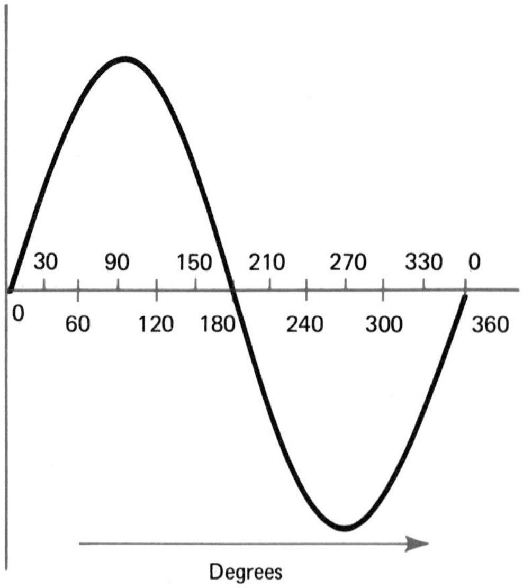

Fig. 8-12 Degrees of a waveform. One cycle has 360 electrical degrees.

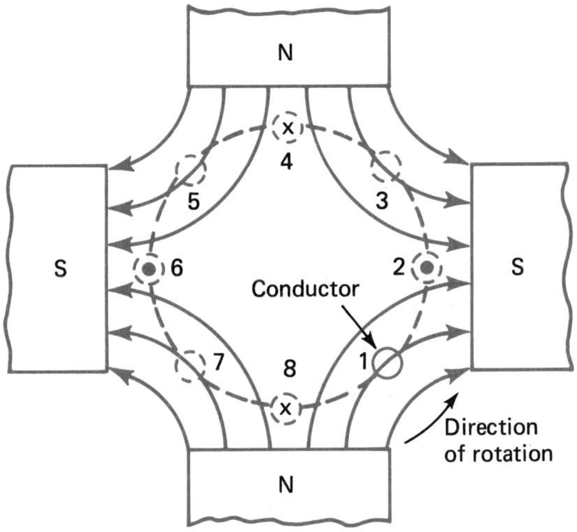

(a) Four-pole magnetic field

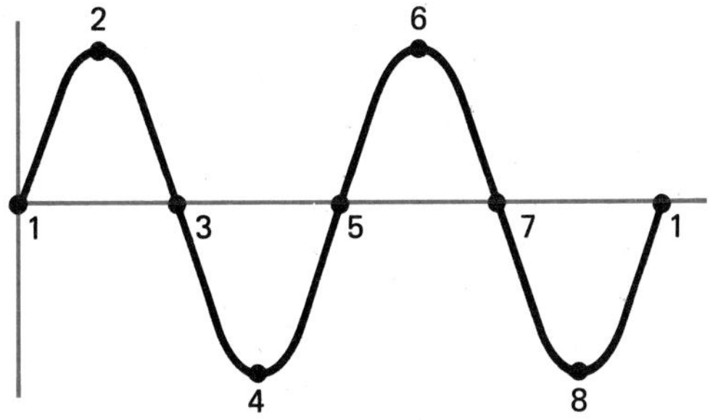

(b) 720 electrical degrees per revolution

Fig. 8-13 Electrical and mechanical degrees. The mechanical and electrical degrees of a generator need not be equal.

180° of rotation. This produces one cycle (360 electrical degrees) for 180 mechanical degrees.

8-6 ALTERNATING CURRENT GENERATOR

In the preceding section we saw how a conductor rotating in a magnetic field produces an ac voltage. Now, let us see how this voltage can be connected to a load.

A drawing showing the essential electrical parts of an ac generator is shown in Fig. 8-14(*a*). In the drawing only one loop of wire (conductor) is shown. In a generator a coil consisting of many loops of wire would be used. This multilooped coil is called a *winding*. The winding is wound on a silicon steel core called an *armature*. The armature winding is composed of many coils (armature coils).

As seen from Fig. 8-14(*a*), both the top and the bottom of the loop move through the magnetic field. The voltage induced into the top of the loop aids the voltage induced in the bottom of the loop. Both voltages force current in the same direction through the load. Figure 8-14(*b*) shows the top and bottom conductors of the loop using the symbols developed in the previous section. This sketch again shows that the voltages (and currents) induced into the conductors of the rotating loop will reinforce each other.

Notice in Fig. 8-14(*a*) that the bottom conductor always makes contact with slip ring 1. This ring always rotates on brush 1. Notice further, that brush 1 is negative with respect to the other brush when the loop is in the position shown. It will remain negative until the loop rotates another 90°. After 90° it will become positive as the bottom loop starts to pass under the south pole. It will remain positive for the next 180° while the conductor is moving under the south pole. Then, it will again become negative while the conductor moves across the face of the north pole. Thus, the polarity of the terminals of an ac generator reverses every 180 electrical degrees.

In Fig. 8-14 the magnetic field for the generator is created by a permanent magnet. In many generators the magnetic field is created by an electromagnet. The coils which create the electromagnet are called *field coils*. The field coils are powered by a direct current so that the polarity of the field never changes. Also, in many large generators the conductors (loops) do not rotate. Instead, the loops are held stationary and the magnetic field is rotated around the conductors. The end result is, of course, the same; voltage is induced into the conductors.

Generator Voltage

As previously stated, the voltage induced in a rotating conductor is determined by the flux density and the rotational speed. In a generator, the rotating winding consists of multiturn coils (many loops). The voltage induced in each turn of a coil is aiding the voltage induced in every other turn. Thus, the output voltage of a generator is dependent on:

1. The number of turns in the rotating (armature) coils

Ac generator
Winding
Armature
Slip ring
Brush
Field coils

Generator frequency

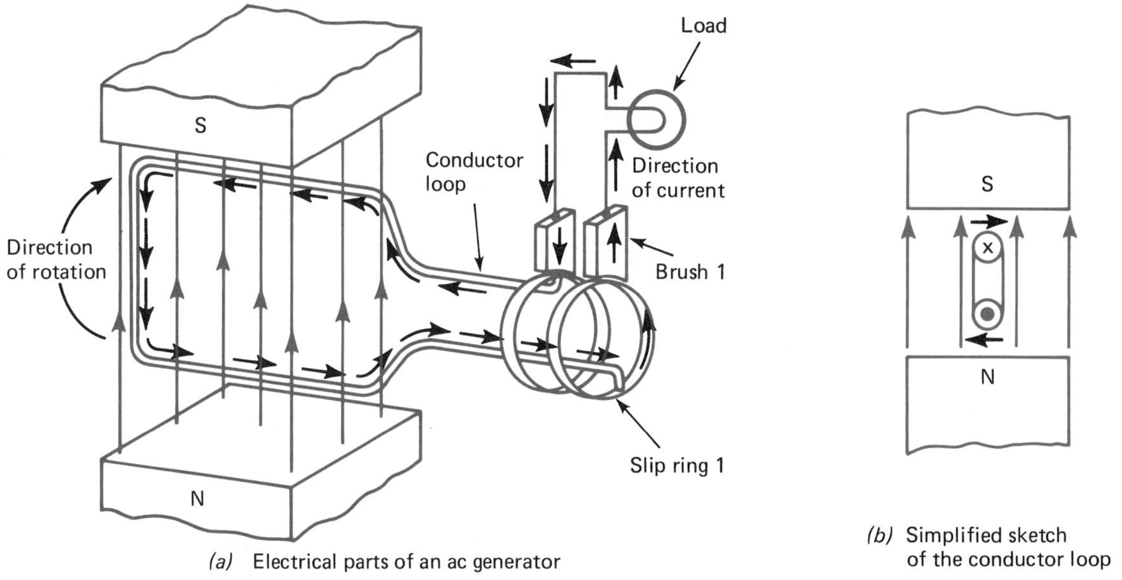

Fig. 8-14 Ac generator. The rotating conductor is connected to the load through the slip rings and brushes.

2. The speed at which the coils rotate
3. The flux density of the magnetic field

Generator Frequency

The speed at which the generator coils rotate influences the frequency as well as the output voltage of a generator. The frequency of the sine wave produced by a generator is determined by (1) the number of pairs of magnetic poles and (2) the rotational speed of the coils (the revolutions per minute of the generator shaft).

With a north pole and a south pole (two poles or one pair), one cycle is produced for each revolution of the generator shaft. If the generator rotates at 60 revolutions per minute (r/min) or 1 r/s, the frequency would be 1 Hz (1 cycle per second). With a four-pole generator, 60 r/min would produce 2 Hz. Stated as a formula, the frequency f of a generator's output will be:

$$\text{Frequency } f = \frac{\text{r/min}}{60} \times \text{pairs of poles}$$

Using this formula, the frequency f will be in base units of hertz (Hz).

Example 8-6

What is the frequency f of a six-pole generator rotating at 1200 r/min?

Given: Pairs of poles = 3 and r/min = 1200

Find: f

Known: $f = \frac{\text{r/min}}{60} \times \text{pairs of poles}$

Solution: $f = \frac{1200}{60} \times 3 = 60 \text{ Hz}$

Answer: The frequency of the generator's output will be 60 hertz.

8-7 ADVANTAGES OF ALTERNATING CURRENT

Compared to direct current, alternating current is easier to transform (change) from one voltage level to another. That is why electrical energy is distributed in the form of alternating current. In the country, power transmission lines can be operated at very high voltages (over 400,000 volts). When the power lines approach a city, the voltage can be reduced (by a transformer) to a few thousand volts. This lower voltage is then distributed within the city.

Another advantage of ac is that ac motors are less complex than dc motors. Alternating current motors can be built without brushes and commutators. This greatly reduces the amount of maintenance needed on an ac motor.

Review Questions

23. What factors determine the amount of voltage induced in a conductor?

24. Is voltage induced when a conductor moves parallel to lines of flux?

25. Is more voltage induced when a conductor moves at 45° or 90° to the lines of flux?

26. What two factors determine the polarity of an induced voltage?

27. What are the necessary conditions for inducing a sine wave in a conductor?

28. How many electrical degrees are produced by one revolution of a conductor rotating in a four-pole magnetic field?

29. Is a generator's output voltage induced in its field coils or armature coils?

30. What is the name of the part of the ac generator with which the brush makes contact?

31. Do the field coils of a generator require ac or dc?

32. What factors determine the output voltage of a generator?

33. What factors determine the frequency of a generator's output?

34. A four-pole generator rotates at 600 r/min. What is its output frequency?

35. Two generators rotate at the same speed. One has eight poles, the other six poles. Which generator produces the higher frequency?

36. List two advantages of ac over dc.

8-8 THREE-PHASE ALTERNATING CURRENT

Three-phase ac

Electricity is produced at an electrical power plant in the form of *three-phase* ac. Phase is often abbreviated with the symbol ϕ. Therefore, three-phase is sometimes written as 3-ϕ. In three-phase ac, see Fig. 8-15, each phase is a sine wave. Each phase is displaced from the other two phases by 120 electrical degrees. Phase 2, Fig. 8-15, starts its positive alternation 120° after phase 1 but 120° before phase 3.

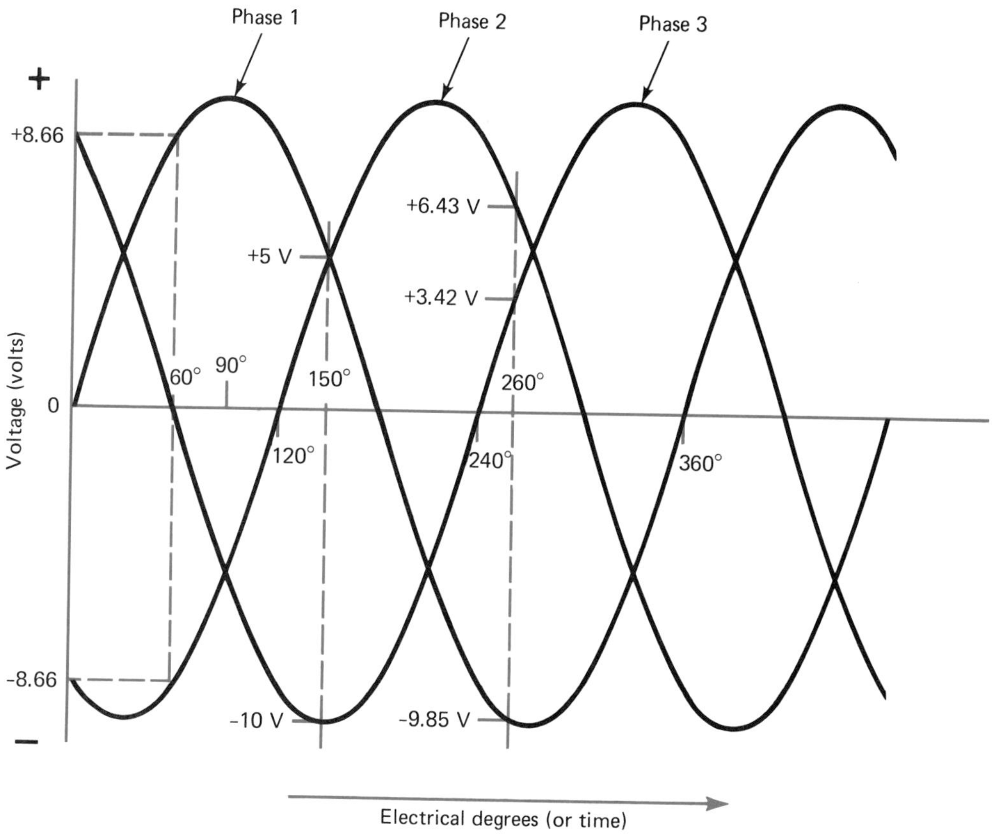

Fig. 8-15 Three-phase ac waveform. Each phase is separated by 120 electrical degrees.

Balanced system

Three-phase generator

Displaced 120 electrical degrees

Delta connection

Reference end

One continuous loop

Phase winding

Inspection of Fig. 8-15 will show that the algebraic sum of the three phases at any instant is zero. For instance, at 60°, phase 1 is +8.66 V, phase 2 is −8.66 V and phase 3 is zero; the sum of the three is zero*. At 150°, phases 1 and 2 are both at +5 V, while phase 3 is at −10 V. Again the algebraic sum is zero. At 260° the positive voltages of phases 2 and 3 are exactly canceled by the negative voltage of phase 1. Pick any instant of time you desire to compare the three phases. You will always find that the sum of two of the phases will be equal and opposite to the remaining phase.

Generation of Three-Phase AC

Suppose three loops of wire (as illustrated in Fig. 8-16) were spaced 120° apart and rotated in a magnetic field. If each loop was electrically isolated from the others and had its own slip rings and brushes, then each would produce a sine wave. Each sine wave would be displaced 120 electrical degrees from the other two sine waves. In other words, we would have a three-phase generator.

Figure 8-16 is drawn to represent the beginning (zero electrical degrees of phase 1) of the waveforms in Fig. 8-15. In Fig. 8-16 the colored end of a loop is considered to be the reference point for that phase. Note that each phase has its own reference point. Thus, phase 2 is now negative and will become more negative as the generator turns a few more degrees. Phase 1 is presently zero and will become positive. And, phase 3 is positive, but will decrease in value. Notice how this agrees with Fig. 8-15. When phase 1 in Fig. 8-15 goes positive, phase 3 becomes less positive and phase 2 becomes more negative. After the generator in Fig. 8-16 has rotated 90°, phase 1 will be at peak positive voltage. Phase 3 will be negative (and increasing in value), while phase 2 will be less negative.

The three phases of a generator can be interconnected so that the load can be carried on only three conductors. Thus, only three conductors are needed to carry three-phase power from a power plant to its place of use. The three phases of a generator can be connected in either *delta* or *wye* connections.

* These instantaneous values can also be calculated with a mathematical formula based on the number of electrical degrees and the peak value of the voltage.

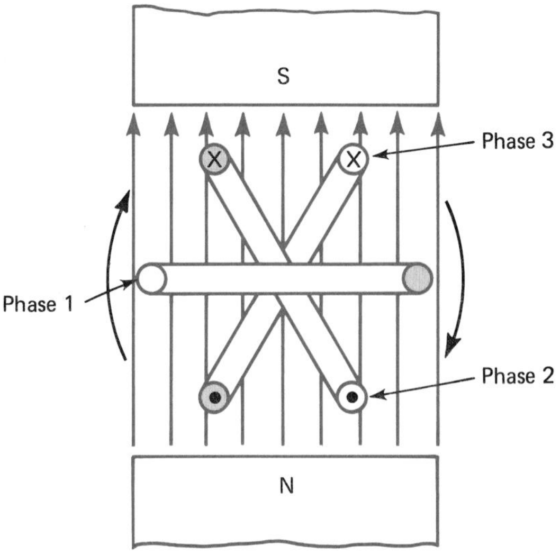

Fig. 8-16 Three-phase generator. The rotating conductors are mechanically locked together but electrically isolated.

Delta Connection

A delta connection is shown in Fig. 8-17. In this figure, each of the generator symbols represents one of the phases of a three-phase generator. The dot on one end of the generator symbol indicates the reference end of the phase winding. Notice that the reference ends of the generators are never connected together for the delta connections. The polarities and voltages indicated for the generators are momentary polarities and voltages. They are the ones which occur at 150° in Fig. 8-15. The following discussion assumes a balanced system. That is, all phase voltages are equal and all loads connected across them are equal.

The three phases are connected in one continuous loop for a delta connection. Yet, no current flows in the loop until a load is applied to the generator. This is because the sum of the voltages of any two of the phases is equal and opposite to the other phase. For instance, the voltages of phase 1 and phase 2 are aiding each other. The sum of these two voltages is 10 V (see Fig. 8-17), which exactly cancels the 10 V of phase 3.

Although the three phase voltages cancel each other within the continuous loop, each individual phase voltage is readily available. The full sinusoidal voltage produced by phase 1 of the generator is available between lines 1 and 3 in Fig. 8-17. Phase 2 is available between lines 1 and 2, and phase 3 exists between lines

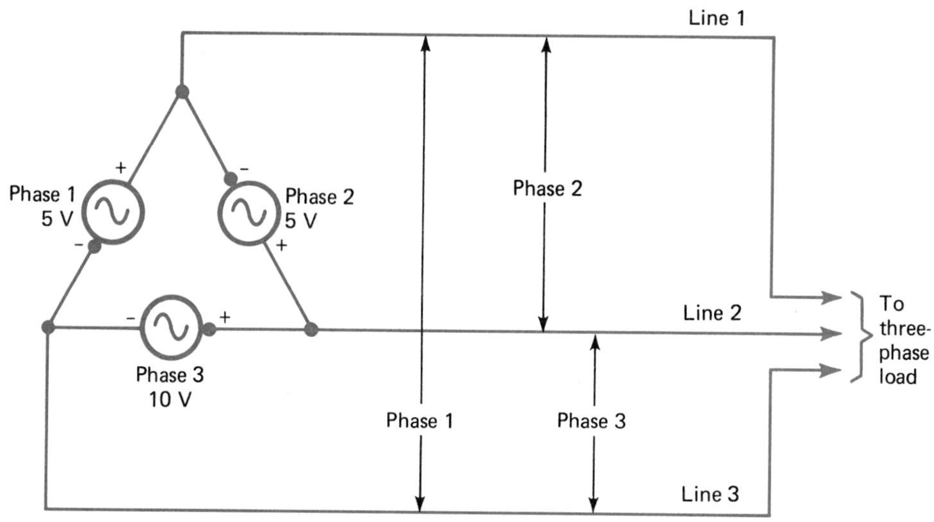

Fig. 8-17 Delta connection of generator. The line-to-line voltage is equal to the phase voltage.

Line voltage

Phase voltage

Line current

Phase current

2 and 3. Thus, the three phases can serve a load with a total of three lines.

Notice in Fig. 8-17 that the voltage between any two lines is exactly equal to one of the phase voltages. In other words, the line voltages and the phase voltages are the same. For a delta connection

$$V_{line} = V_{phase}$$

When a delta-connected generator is loaded, Fig. 8-18, currents flow in the loads, lines, and phase windings. However, the line currents and the phase currents are not equal. With equal loads connected to each phase, the line current will be 1.732 times as great as the phase current. For a delta connection

$$I_{line} = 1.732\ I_{phase}$$

The generator's phase currents and the load currents are equal. Therefore, the line currents will also be 1.732 times as great as the load current. Line currents are not equal to phase currents because each line carries current from two phases. The currents from any two phases are 120° out of phase.

Notice in Fig. 8-18 that neither voltage polarity nor current direction is indicated. This is because it is an ac system in which polarity and direction periodically reverse. The current and voltage are in rms values since they are not specified otherwise.

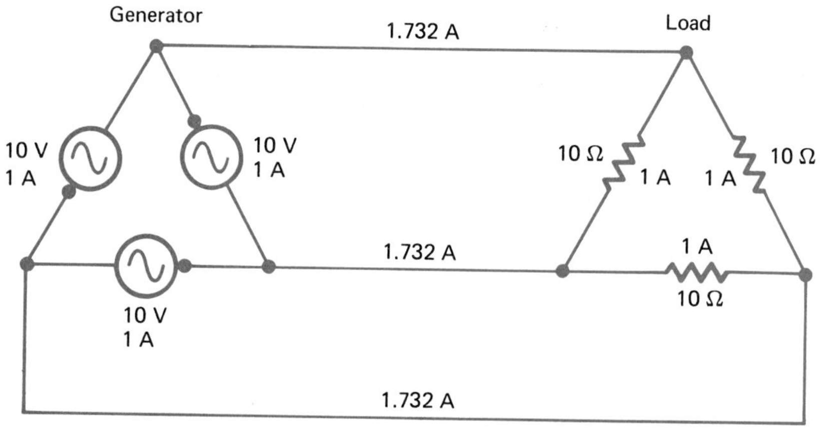

Fig. 8-18 Currents in a three-phase delta system. The line current is 1.732 times as great as the phase current.

Balanced delta
Wye connection
Reference points
Balanced wye
Phase voltage
Line voltage
Phase current
Line current

Remember the following points about a balanced delta connection:

1. No current flows in the phase windings until a load is connected
2. The line voltage and the phase voltage are equal
3. The line current is 1.732 times as great as the phase current or the load current

Wye Connection

Figure 8-19 shows a wye-connected, three-phase generator and load. Again in this diagram, each generator symbol represents one phase of the three-phase generator. The voltages and currents are rms (effective) values. With the wye connection, and balanced (equal) loads, all currents are the same. For a wye connection

$$I_{line} = I_{phase}$$

However, notice in Fig. 8-19 that the line voltages are 1.73 times as great as the phase voltages. For a wye connection

$$V_{line} = 1.732V_{phase}$$

The voltage between any two lines is the result of two phase voltages. Since the two phase voltages are separated by 120 electrical degrees, their values cannot be simply added together.

Adding phase voltages to obtain line voltages is further complicated by the different reference points used for these voltages. Referring to Fig. 8-20 will help you understand reference points and addition of phase voltages. Figure 8-20(*b*) shows that phase 1 is +0.5 V at 30 electrical degrees. What does +0.5 V mean? Nothing, unless we have a reference point. (Remember, voltage is a potential energy difference *between two points*.) The reference point for all three phase voltages is the center of the wye [Fig. 8-20(*a*)]. With this reference point established, the voltages read from the waveform graph in Fig. 8-20(*b*) have specific meaning. At 30° both phase 1 and phase 3 are 0.5 V positive with respect to the center of the wye. Phase 2 is 1 V negative with respect to the center of the wye. These voltages and polarities are marked on the diagram in Fig. 8-20(*a*). Notice that the instantaneous voltages of phase 1 and phase 2 are in series between line 1 and line 2. These two voltages (phase 1 and phase 2) are aiding each other at this instant so they can be added together. Together they produce 1.5 V between line 1 and line 2. Referring back to Fig. 8-20(*b*), notice that phase 1 and phase 2 at 30° are of opposite polarities. But, when connected in a wye configuration, the two voltages aid each other. Also notice in Fig. 8-20(*a*) that phase 1 and phase 3 are series opposing. Their momentary voltages cancel each other to provide a net voltage of zero between lines 1 and 3.

In Fig. 8-20 we have only looked at instantaneous values taken at 30° on the three-phase waveform. Of course, these instantaneous values would be different for each degree (or part of a degree) of the waveform. If all possible instantaneous values between two lines

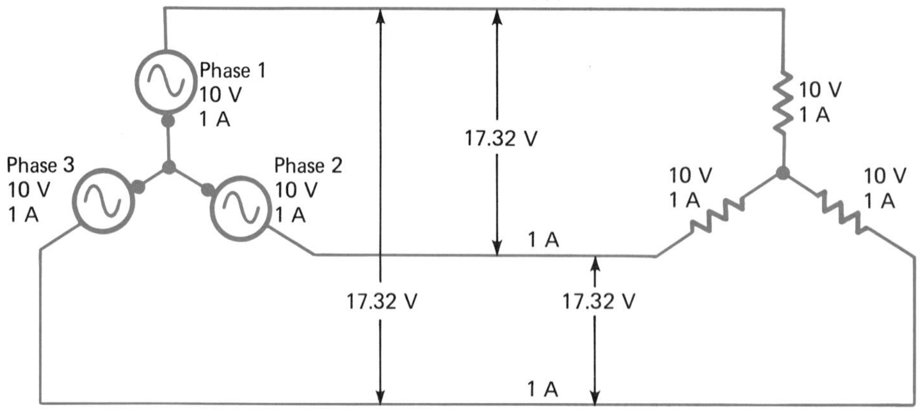

Fig. 8-19 Three-phase wye system. The line voltage is 1.732 times as great as the phase voltage.

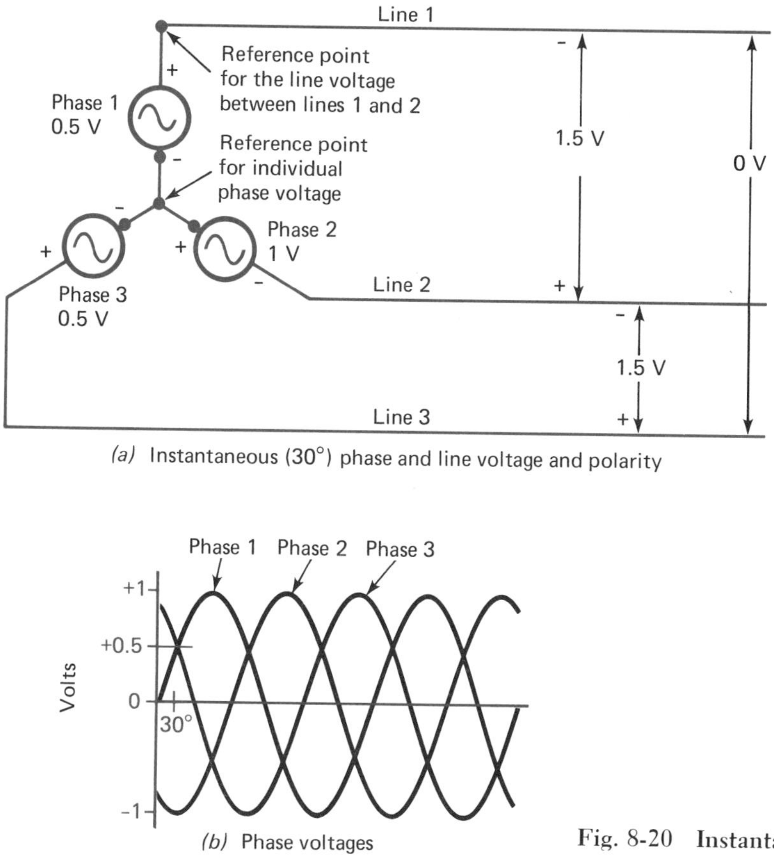

Fig. 8-20 Instantaneous voltages in a wye system.

were plotted on a graph, a sine wave would result. The sine wave would be 1.732 times as large as the sine wave of the phases which produced it. In plotting a waveform of this line voltage, a reference point would be needed. The selection of this point is arbitrary. Let us select line 1 as the reference point. Then, we can plot a waveform of the line voltage between lines 1 and 2. This waveform is shown in Fig. 8-21(*a*) superimposed on the waveforms of phases 1 and 2. Figure 8-21(*b*) lists some of the instantaneous values used to plot the waveform of the line voltage. Notice, from Fig. 8-21, that the line voltage is displaced 30° from the phase-2 voltage. The line voltage lags the phase-2 voltage by 30°. That is, the negative peak of the line voltage occurs 30° after the negative peak of the phase voltage.

Suppose line 2 had been used as the reference point for the line voltage in Fig. 8-20. Then the polarity of the line voltage would have started out positive rather than negative. The line voltage waveform in Fig. 8-21 would be completely inverted. It would start at +0.866 V rather than −0.866 V. The line voltage would then lead the phase-1 voltage by 30°.

In summary, the major characteristics of the balanced, wye-connected, three-phase system are:

1. The phase and line currents are equal
2. The line voltage is 1.732 times greater than the phase voltage

Four-Wire Wye System

Star point

Neutral wire

Figure 8-22 illustrates a four-wire wye system. In this figure, the generator symbol for each phase has been replaced by a coil symbol. The coil symbol represents one of the phases in a three-phase generator. This is the more traditional way of representing a three-phase generator.

The fourth wire of the four-wire system comes from the common, center connection of the phase windings. (This common connection is sometimes referred to as the *star* point or connection.) This fourth wire is often called the neutral wire because it is electrically connected to ground (earth). Thus, the fourth wire is neutral (has no voltage) *with respect to ground*.

Balanced, three-phase load

Rectification

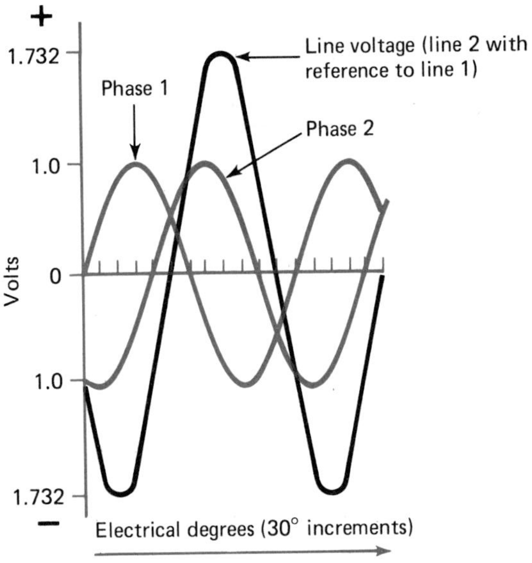

(a)

Electrical degrees	Instantaneous values (volts)		
	Phase 1	Phase 2	Line
0	0	-0.866	-0.866
30	+0.5	-1.	-1.5
60	+0.866	-0.866	-1.732
90	+1	-0.5	-1.5
120	+0.866	0	-0.866
150	+0.5	+0.5	0
180	0	+0.866	+0.866
210	-0.5	+1.	+1.5
240	-0.866	+0.866	+1.732

(b)

Fig. 8-21 Adding wye-connected phase voltages to obtain the line voltages.

With a balanced, three-phase load, the neutral wire carries no current. Therefore, it is needed only when unequal phase loading is likely to occur. Unequal loading usually occurs when both single-phase and three-phase loads are connected to the wye system.

Inspection of Fig. 8-22 will reveal the main feature of the four-wire wye system. This system can supply (with only four wires) all of the voltage requirements of a small manufacturing facility. Single-phase, 120-V circuits are connected between the neutral and any line. Single-phase, 208-V circuits are connected between any two of the three lines. And, 208-V, three-phase devices are connected to the three lines. Common four-wire wye system voltages are 120/208 V and 277/480 V. The first value in each case is the phase, or single-phase, voltage. The second value in each case is the three-phase system voltage. Notice that the three-phase system voltage is 1.732 times the single-phase voltage.

Advantages of Three-Phase Systems

Compared to single-phase, the three-phase system has the following advantages:

1. Makes more efficient use of copper
2. Provides a more constant load on the generator
3. Operates a motor that is less complex and provides more constant torque
4. Produces a smoother dc voltage and current when rectified.

Suppose a three-phase system and a single-phase system are delivering the same amount of power over the same length of lines. The single-phase system requires about 1.15 times as much copper for power lines as does the three-phase system.

The power required by a resistive load is equal to I^2R ($P = I^2R$). With single-phase current, the power required from the generator will follow the current variations. Thus, the load on the generator goes from zero to maximum power and back to zero each alternation. With three-phase current, at least two of the phases are providing current (and thus power) at any instant. The load on the three-phase generator never reduces to zero. The more uniform load provides for smoother mechanical operation of the generator.

Most ac motors can start only if the magnetic field behaves as if it were rotating. With single-phase ac, this rotating magnetic field is created through an auxiliary circuit in the motor. Once the motor has started, this auxiliary circuit is disconnected by a mechanically operated switch. In a three-phase motor, the rotating field is created by having each phase produce part of the magnetic field of each pole. No internal switch (or mechanical mechanism to operate it) is needed with the three-phase motor. Of course, the three-phase motor has more uniform torque since at least one phase is always producing a magnetic field.

Alternating current can be converted to direct current by a process known as *rectification*. The results of rectifying both single-

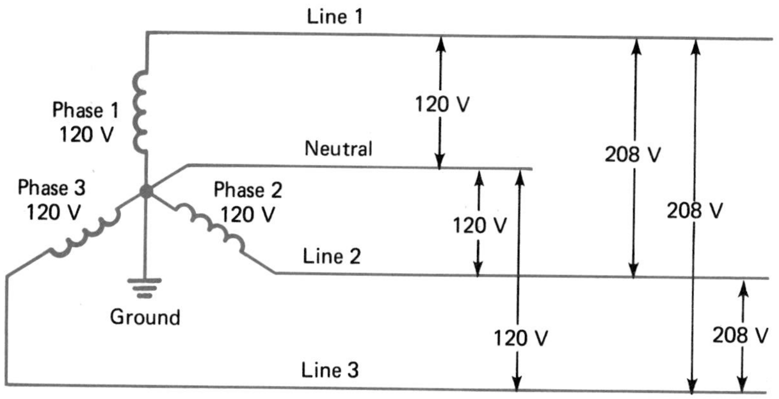

Fig. 8-22 Four-wire wye system. Four wires provide 120 V single-phase, 208 V single-phase, and 208 V three-phase.

phase ac and three-phase ac are illustrated in Fig. 8-23. Rectified single-phase ac produces pulsating dc where the current periodically reduces to zero. However, when three-phase ac is rectified, the direct current never reduces below 0.866 of its peak value. Obviously the three-phase system produces a dc output with much less fluctuation.

Review Questions

37. How many electrical degrees are there between the sine waves of a three-phase system?
38. In a three-phase system the instantaneous voltage of phase 1 is +14 V and phase 2 is +6 V. What is the instantaneous voltage of phase 3?
39. List two ways in which the windings of a three-phase generator can be connected.
40. How many conductors are required to transmit power from a three-phase generator to its load?
41. How are the phase windings of a three-phase generator connected for a delta connection?
42. For a delta connection, what is the relationship between the following:
    a. The phase voltage and the line voltage?
    b. The phase current and the line current?
43. What is the name of the point that connects the three phases together in a wye connection?
44. Will generator windings connected in the wye or delta configuration provide the higher line voltage?
45. Are the line current and the phase current equal in a wye system?
46. Are the line voltage and the phase voltage equal in a wye system?
47. What is the name of the fourth wire in a four-wire wye system?

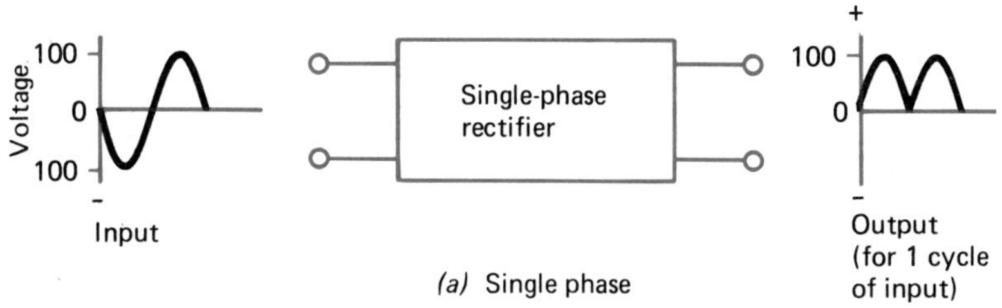

(a) Single phase

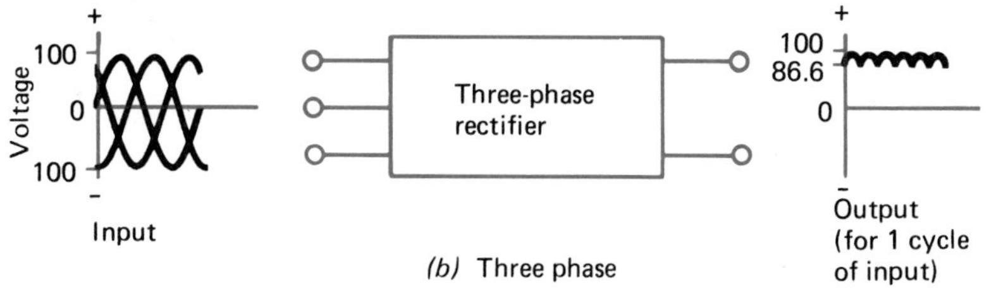

(b) Three phase

Fig. 8-23 Rectification of alternating current. When rectified, three-phase ac produces a smoother dc than does single-phase ac.

48. When does the fourth wire in a wye system carry current?

49. What voltages are available from a four-wire, three-phase, 208 V wye system?

50. A wye-connected, three-phase generator produces a phase voltage of 277 V. What is the line voltage?

Summary

1. The great majority of electrical energy is used in the form of alternating current and voltage. A sine wave is the most common waveform for ac.
2. Alternating current (ac) periodically reverses the direction in which it flows.
3. Alternating current periodically or continuously changes magnitude.
4. A dc waveform never crosses the zero reference line.
5. A cycle is that part of a periodic waveform which occurs without repeating itself.
6. A cycle is composed of two alternations: a negative alternation and a positive alternation.
7. The period is the time required to complete one cycle. The symbol for period is T.
8. Frequency is the rate at which cycles are produced. The symbol for frequency is f.
9. The base unit of frequency is the hertz. The abbreviation for hertz is Hz.
10. One hertz is equal to one cycle per second.
11. Power in North America is distributed at a frequency of 60 Hz.
12. The reciprocal of the frequency is equal to the period.

$$T = \frac{1}{f}$$

13. The reciprocal of the period is equal to the frequency.

$$f = \frac{1}{T}$$

14. Unless otherwise indicated, ac quantities are assumed to be specified in rms (effective) values.
15. Effective or rms values of ac produce the same heating effect as the same value of dc.
16. A conductor rotating in a perfect circle, at constant speed, in a uniform magnetic field, produces a sine wave.
17. There are 180 electrical degrees per alternation; 360 electrical degrees per cycle.
18. Generator voltage is determined by (1) speed of rotation, (2) number of turns in the coils, and (3) flux density.
19. Generator frequency is determined by (1) speed of rotation, and (2) number of pairs of magnetic poles.

$$f = \frac{\text{r/min}}{60} \times \text{pairs of poles}$$

20. The magnitude of ac voltage is easier to change than is the magnitude of dc voltage.
21. Electrical power is usually transmitted from the main power plant as three-phase ac.
22. The algebraic sum of the instantaneous phase values of three-phase voltage or current is always zero.
23. The voltages in a three-phase system are separated by 120 electrical degrees.
24. In a delta connection the line and phase voltage are the same.

$$V_{\text{line}} = V_{\text{phase}}$$

25. In a delta connection the line current is 1.732 times as large as is the phase current.

$$I_{\text{line}} = 1.732\ I_{\text{phase}}$$

26. In a wye connection the line and phase current are equal, but the line voltage is 1.732 times the phase voltage.

$$I_{\text{line}} = I_{\text{phase}}$$
$$V_{\text{line}} = 1.732 V_{\text{phase}}$$

27. The fourth wire in a four-wire wye system is the neutral wire.
28. With a balanced load, the fourth wire in a wye connection carries no current.
29. The advantages of three-phase ac compared to single-phase ac are (1) more efficiency, (2) more constant load on generator, (3) more constant torque from motors, (4) less fluctuation when rectified to dc.

Chapter Review Questions

For questions 8-1 to 8-15, determine whether each statement is true or false.

8-1. A half cycle is also known as an alternation.

8-2. One cycle per second is equal to one hertz.

8-3. Fluctuating dc periodically reduces its value to zero.

8-4. The average value of a sine wave is larger than the effective value.

8-5. The number of mechanical degrees and the number of electrical degrees are always the same for an ac generator.

8-6. The rotational speed of a generator influences both its frequency and its output voltage.

8-7. Ac motors require less maintenance than dc motors.

8-8. The algebraic sum of the instantaneous phase values of the voltages in a three-phase system is always zero.

8-9. A three-phase system requires at least six conductors to transfer power between two locations.

8-10. In a delta-connected system the line voltages and the phase voltages are equal.

8-11. In a wye-connected system the line currents and the phase currents are equal.

8-12. When line currents and load currents are not equal in a balanced three-phase system, they are related by a factor of 1.732.

8-13. The line voltage in a wye-connected system is always displaced 30° from one of the phase voltages.

8-14. In a 208-V, four-wire system, 120 V is available between the neutral and any line.

8-15. The reference ends of the phase windings of a generator are all connected together for a delta connection.

For questions 8-16 to 8-18, choose the letter that best completes each statement.

8-16. The majority of electrical energy is distributed in the form of a
(A) Sawtooth wave (B) Square wave (C) Sine wave (D) Pulsating dc wave

8-17. The base unit of frequency is the
(A) Hertz (B) Period (C) Cycle (D) Alternation

8-18. The abbreviation for the base unit of frequency is
(A) T (B) f (C) Rms (D) Hz

Answer the following questions.

8-19. An electrical waveform goes from zero to maximum positive and back to zero in 0.001 s. Find its period and frequency.

8-20. What is the period of a 200-Hz sine wave?

8-21. What is the rms value of a sine wave which has a peak-to-peak value of 400 V?

8-22. What is the average value of the 120 V available at an electrical outlet?

8-23. What determines the amount of voltage induced in a conductor moving through a magnetic field?

8-24. List three factors that determine the output voltage of a generator.

8-25. An eight-pole generator is turned at 1800 r/min. What is the frequency of its output voltage?

8-26. List four advantages of a three-phase system over a single-phase system.

Answers to Review Questions

1. Time
2. Voltage
3. It shows variation in an electrical quantity with the passing of time.
4. No
5. Pulsating dc periodically reduces to zero. Fluctuating dc reduces to zero only when the circuit is interrupted (turned off).
6. A sine wave
7. Square wave and sawtooth wave
8. *a.* The largest part of a periodic waveform which doesn't repeat itself
   b. The part of the cycle that is above the zero reference line
   c. The time required to complete one cycle
   d. The number of cycles occurring in a given period of time, usually one second
9. Hertz
10. A hertz is one cycle per second.
11. 60 Hz, sinusoidal
12. 0.0025 s or 2.5 ms
13. 250 Hz
14. 3.185 V
15. 7.85 A
16. 70.7 V
17. 141.4 V
18. 14.14 A
19. Effective value
20. *f*
21. Hz
22. An alternation is one-half of a cycle.
23. Amount of flux cut by the conductor per second
24. No
25. 90°
26. Polarity of the magnetic field and direction of conductor movement
27. Uniform magnetic field, constant speed of rotation of the conductor, and rotation in a perfect circle.
28. 720°
29. Armature coils
30. Slip ring
31. dc
32. Number of turns in the armature coils, strength of the poles, and the speed at which the armature rotates.
33. Number of magnetic poles and revolutions per minute of the armature
34. 20 Hz
35. The eight-pole generator
36. Easier to change voltage and ac motors are easier to construct and maintain
37. 120°
38. −20 V
39. Delta and wye
40. Three
41. In a continuous ring or loop. The reference end of one winding is connected to the nonreference end of the next winding.
42. *a.* The phase and line voltages are equal.
    b. The line current is equal to 1.732 times the phase current.
43. Star point
44. Wye
45. Yes
46. No, the line voltage equals 1.732 times the phase voltage.
47. The neutral wire
48. When the phases are unequally loaded
49. 120-V, single-phase; 208-V, single-phase; and 208-V, three-phase.
50. 480 V

Power In Ac Circuits

- **The most common form of electrical power is the power produced by alternating current and alternating voltage.**

  When you have finished this chapter you will be able to use phasor diagrams to represent current and voltage. You will be capable of solving electrical problems using the properties of a right triangle. You will understand the importance of phase relationships and their effect on power. Finally, you will be able to calculate power and power factors for various types of ac circuits.

9-1 POWER IN RESISTIVE AC CIRCUITS

When the load on an ac source contains only resistance, the current and voltage are *in phase*. "In phase" means that two waveforms arrive at their peak positive values, zero values, and peak negative values together. The current in Fig. 9-1 is in phase with the voltage. The resistive load in Fig. 9-1(*a*) might represent an electric heater, a clothes iron, or a stove. As with dc, power is the product of the current and voltage. With ac, the values of current and voltage vary with time. At any instant, the power is equal to the current at that instant multiplied by the voltage at that instant. As shown in Fig. 9-1(*b*), the product of their instantaneous current and voltage creates an instantaneous power. Plotting all the instantaneous powers produces a power waveform. However, notice that the power waveform is always positive for in-phase currents and voltages. A negative current multiplied by a negative voltage yields a positive power. This means that the resistive load is converting electrical energy into heat energy during the complete cycle. For in-phase circuits, that is, resistive circuits, we can state:

$$P = IV = I^2R = \frac{V^2}{R}$$

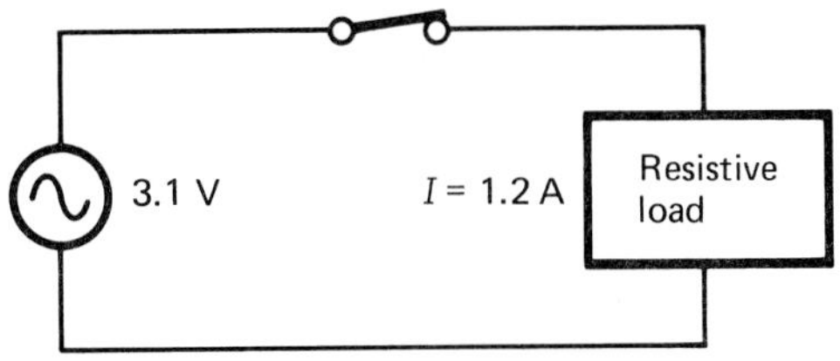

(a) Resistive ac circuit

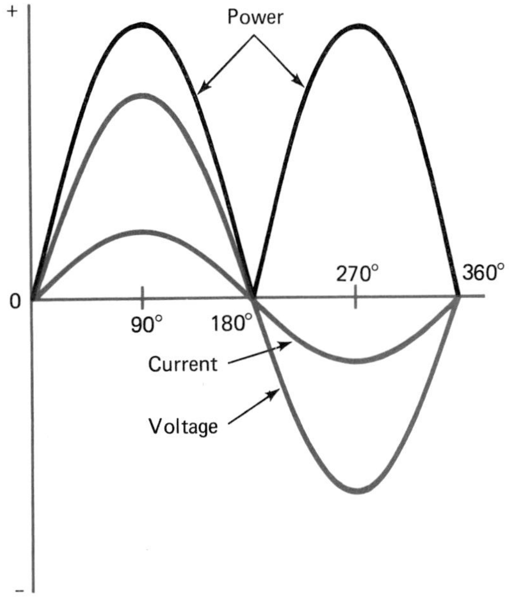

(b) Current, voltage, and power waveforms

Fig. 9-1 Power in a resistive ac circuit.

From page 167:
In phase

Resistive load

On this page:
Out of phase

Reactance

Phase shift

Capacitive reactance

Inductive reactance

where I and V are the rms current and voltage respectively.

Example 9-1

How much power is used by the resistive load in Fig. 9-1(*a*)

Given: V = 3.1 V
I = 1.2 A
Find: P
Known: $P = IV$
Solution: $P = 1.2 \text{ A} \times 3.1 \text{ V} = 3.72 \text{ W}$
Answer: The power used is 3.72 watts.

Example 9-2

A 400-Ω resistor is connected to a 100-V ac source. How much power is used?

Given: V = 100 V
$R = 400\ \Omega$
Find: P
Known: $P = \dfrac{V^2}{R}$
Solution: $P = \dfrac{V^2}{R} = \dfrac{(100)\ (100)}{400} = 25 \text{ W}$
Answer: The resistor dissipates 25 W.

Notice that power calculations in resistive ac circuits are no different than in dc circuits.

9-2 POWER IN OUT-OF-PHASE CIRCUITS

An ac circuit may contain a load that is not entirely composed of resistance. The load may also contain *reactance*. Reactance is the name given to the opposition to current caused by capacitors and inductors. Although these devices were mentioned briefly in the discussion of direct current, their behaviors in ac circuits are quite different. For now, we will study their effect on power: we will discuss the devices themselves later. Although reactance opposes current in an ac circuit, it is different from resistance. It causes a *phase shift* to occur. That is, it causes the current to be out of phase with the voltage. Figure 9-2 shows the current and voltage waveforms for circuits containing capacitors, inductors, and resistors.

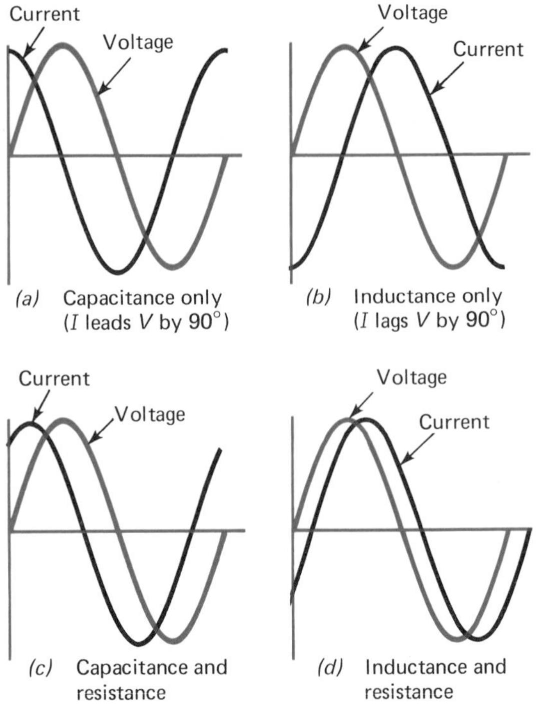

Fig. 9-2 Phase shift in ac circuits.

Capacitors produce capacitive reactance. The effect of capacitive reactance is to have the current "lead" the voltage. In Fig. 9-2(*a*) the voltage is zero at time zero. At this point, however, the current has already reached its maximum value.

Inductors produce inductive reactance. The effect of inductive reactance is to have the current "lag" the voltage. In Fig. 9-2(*b*) the voltage has already reached its peak positive value, while the current is only beginning to go positive. Loads containing only capacitance, or only inductance, cause the current and voltage to be exactly 90° out of phase. In other words, such loads cause a 90° phase shift between current and voltage. Loads containing a combination of resistance and reactance (either capacitive or inductive) cause a phase shift of less than 90°. The exact amount of phase shift is determined by the relative amounts of resistance and reactance.

A good example of a combination load is an electric motor. It contains both inductance and resistance. When a small ac motor has no load on its shaft, the phase shift is about 70°. When the motor is loaded, it acts more like a resistance. When loaded, the phase shift reduces to about 30°. Of course, in a purely resistive circuit, the phase shift is 0°.

For circuits in which phase shift occurs, calculating power is no longer a simple matter of multiplying current and voltage. In Fig. 9-3, three examples of phase shift and the resulting power are shown. Notice that in some cases the current and voltage are of opposite polarities. This yields a negative power for that part of the cycle. What does it mean to have negative power? It means that the load is not taking power from the source during that period of time. In fact, it means that the load is returning power to the source for that part of the cycle. Therefore, the net power used by the load is the difference between the positive power and the negative power. Notice also in Fig. 9-3 that the greater the phase shift the smaller the net power taken from the source. At 90° phase shift, no net power (thus no energy) is taken from the source.

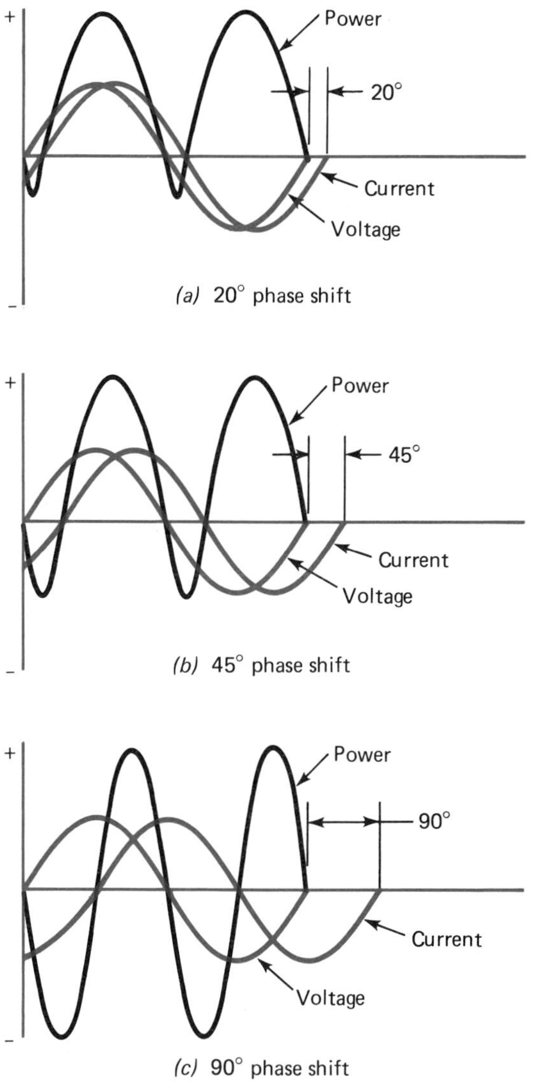

Fig. 9-3 Power in phase-shifted circuits. At 90° of phase shift the power is zero.

Since capacitive and inductive reactances cause a 90° phase shift, we can conclude that reactance uses no power. Only resistance uses power. Therefore, in circuits with combination loads we can conclude several things (see Fig. 9-4). First, the resistance part of the load uses power. In the resistance part the current and voltage are in phase and the power is simply $P = IV$. Second, the reactance part of the load uses no power. The current and voltage are a full 90° out of phase. Third, the phase relationship between the total voltage and current is determined by a combination of the resistance and reactance.

We need a way in which to determine what part of the total current and voltage is associated with the resistance of the load. Then we can determine the net (or real) power used by any combination load with any amount of phase shift. Therefore, let us next develop a way of breaking a current or voltage into a resistive part and a reactive part. We will do this through the use of *phasors* and *right triangle* relationships.

Review Questions

1. What is the phase relationship between current and voltage in a circuit that contains only resistance?
2. What is the phase relationship between current and voltage in a circuit that contains only reactance?
3. How much power is used by a 100-Ω resistor which has 0.3 A of alternating current flowing through it?
4. Would the power in the above question be the same if the current were dc?
5. What is reactance?
6. If the load of a circuit contains only reactance, does it use any power?
7. If the load of a circuit contains both resistance and reactance, does it use any power?
8. What type of reactance does a typical ac motor contain?

Phasor

Voltage phasor

Current phasor

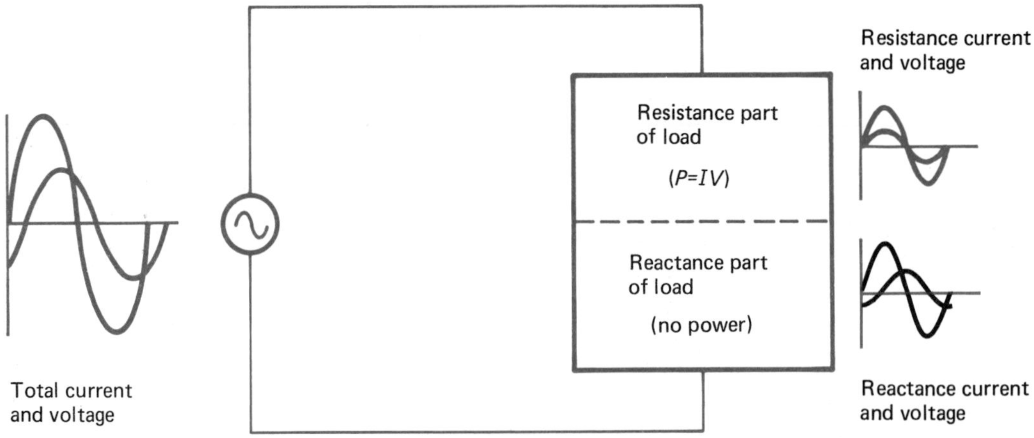

Fig. 9-4 Current and voltage in a combination load.

9. Does the current lead or lag the voltage in the motor of question 8?

Phasors

Drawing sinusoidal waveforms to show amplitude and phase shift is slow and tedious work. A much simpler technique of showing amplitude and phase shift is through the use of phasors.

A phasor is a line whose direction represents the phase angle in electrical degrees and whose length represents the magnitude or value of the electrical quantity. You may have studied *vectors* in mathematics or physics classes. From the definition of a phasor, you can see that phasors and vectors are very similar. In electricity and electronics, the term *phasor* is used to emphasize the fact that the position of the line changes with time as the wave moves through its cycle from 0° to 360°. In Fig. 9-5 the phasor represents 10 A of current at 60°. By convention, we agree that zero electrical degrees is on the right hand side of the horizontal axis. Figure 9-6 shows the phasors of two voltages that are 45° out of phase. In this figure the two phasors have to use the same scale since they both represent voltage. For convenience in drawing the phasors, V_1, is shown on the 0° line. Typically, one phasor of a group is drawn on this line. This phasor, then, becomes the reference to which the other phasors are related.

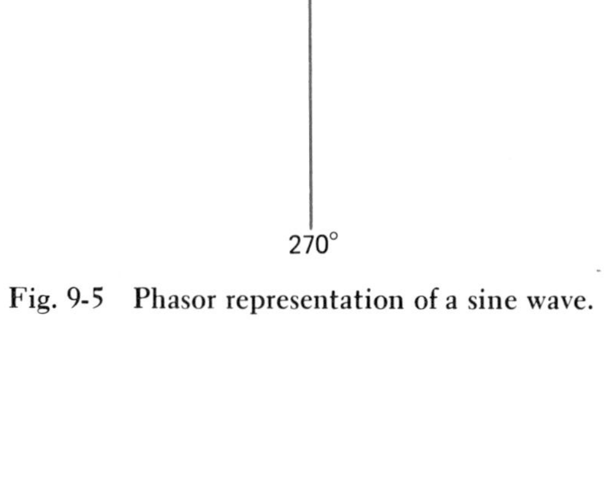

Fig. 9-5 Phasor representation of a sine wave.

Remember that phasors are actually moving with time. By convention, phasors are assumed to rotate in a counterclockwise direction. By showing V_1 on the 0° line, we stopped its motion at that instant. Of course, the motion of V_2 was also stopped at the same instant. Thus, the phase relationship between V_1 and V_2 did not change. Since the motion of phasors is in a counterclockwise direction, V_2 in Fig. 9-6 is lagging behind V_1 which is the reference phasor. Since V_2 is out of phase with V_1 by 45°, we can conclude from Fig. 9-6 that V_2 lags V_1 by 45°.

A voltage phasor and a current phasor are shown in Fig. 9-7. Since current and voltage are different quantities, these two phasors can use different scales for their magnitude. The phasor diagram shows that the current leads the voltage by 45°. The power resulting from this current and voltage can be found by breaking the current phasor into two smaller phasors. One of the smaller phasors will represent the current in the resistive portion of the load. It will be in phase with the voltage phasor. The other smaller phasor will repre-

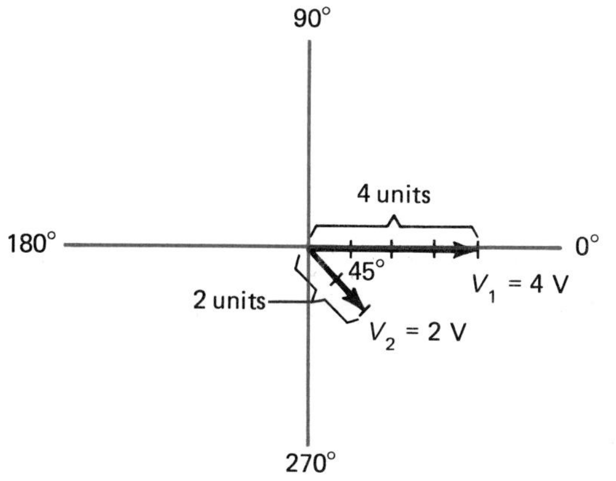

Fig. 9-6 Out-of-phase voltages. Amplitude and phase shift can be shown by phasors.

sent the current in the reactive portion of the load. It, of course, will be 90° out of phase with the voltage. The two smaller phasors will be separated by 90°.

The idea of breaking a phasor into two smaller phasors is illustrated in Fig. 9-8. In this figure, we are using vectors because the positions and magnitudes do not change with time. One vector of 500 newtons (N), approximately 112 lb, force is exerted on the spring in the northeast direction. In the first case [Fig. 9-8(*a*)] the force is created by one person pulling in a northeast direction with a force of 500 N. In the second case [Fig. 9-8(*b*)] one person pulls north with 300 N of force and another person pulls east with 400 N. The net result is 500 N of force in the northeast direction. The effect on the spring is exactly the same in both cases.

The vectors in Fig. 9-8(*b*) represent two forces at a right angle (90°) to each other. However, they could equally well represent

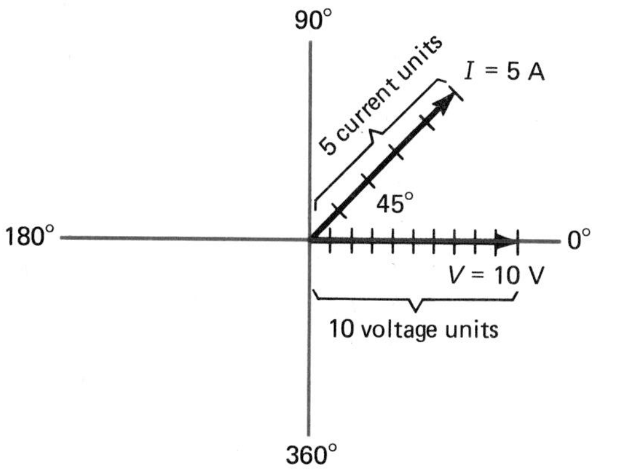

Fig. 9-7 Current and voltage phase relationship represented by phasors. *I* leads V by 45°.

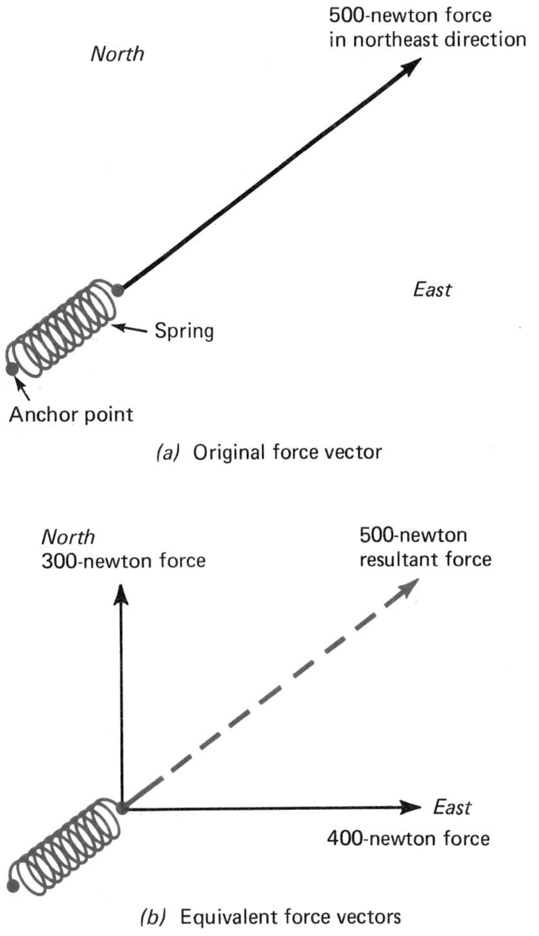

Fig. 9-8 Equivalency of vectors. A single vector can be represented by two smaller vectors separated by 90°.

any other quantity that has magnitude and direction. Thus, they could be phasors representing resistive current and reactive current in an electric circuit.

The magnitude of resistive and reactive current can be determined graphically. The process is shown in Fig. 9-9. In this diagram, V is at the 0° reference line. The total current I_T = 5 A and is leading the voltage by 36.9°. The first step is to draw the original phasor (I_T) to scale at the correct angle. Then draw a vertical line (line 1) from the tip of the I_T phasor through the horizontal axis. Line 1 must be parallel to the vertical axis. The intersection of line 1 and the horizontal axis locates the end of the resistive current phasor (I_R). The end of the reactive current phasor (I_X) is found by drawing line 2 from the tip of I_T to the intersection of the vertical axis. Again, line 2 must be parallel to the horizontal axis. Finally, the

Graphical method

Theta

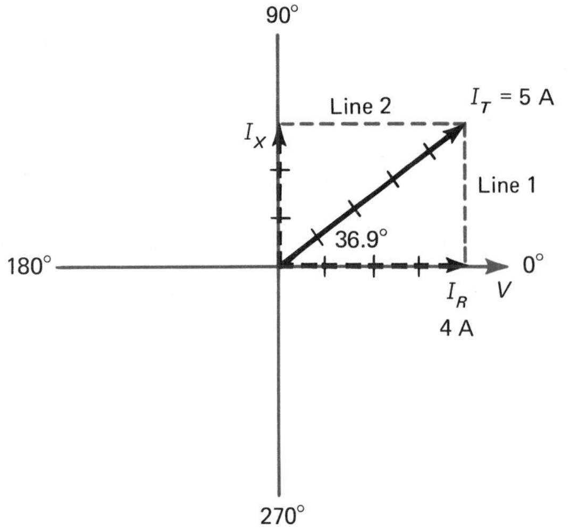

Fig. 9-9 Graphical analysis of a phasor.

magnitude of the resistive and reactive currents are measured with the scale used to measure I_T. The 4-A resistive current in Fig. 9-9 is the current that is in phase with the voltage. It is the current that can be used to calculate the power when 5 A of total current leads the total voltage by 36.9°.

The currents determined by the graphical method illustrated in Fig. 9-9 are as accurate as the drawing itself. For most applications the graphical method is accurate enough. However, it is also a slow and tedious method. Let us develop a faster, easier way of determining these currents. This easier way is through the use of the properties of a right triangle. Notice in Fig. 9-9 that the reactive current phasor is exactly the same length as line 1. Also, notice that it is parallel to line 1. Therefore, line 1 can be replaced by the reactive current phasor. This has been done in Fig. 9-10. It can be clearly seen in Fig. 9-10 that the three current phasors form a right triangle. Now can you see why we need to know how to work with right triangles?

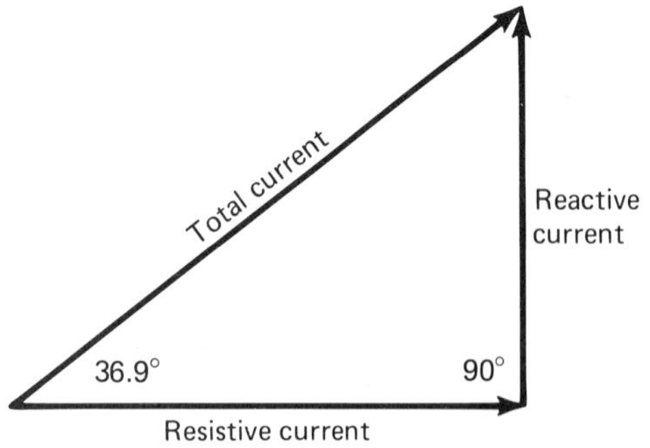

Fig. 9-10 Right triangle formed by current phasors.

Review Questions

10. What is a phasor?
11. Draw a phasor to represent 10 V at 280°.
12. Draw an 8-A current phasor at 45°.
13. Draw a phasor diagram of a current lagging a voltage by 90° if $I = 6$ A and $V = 100$ V.
14. Refer to Fig. 9-5. Draw a sketch that shows the resistive part of the 10-A phasor.
15. Refer to Fig. 9-8. Draw a right triangle which represents the three force vectors shown in this diagram.
16. Must all voltage phasors in a phasor diagram use the same scale?
17. Must all phasors in a phasor diagram use the same scale?

Functions of a Right Triangle

The right triangle of Fig. 9-10 is redrawn and relabeled in Fig. 9-11. The new labels are terms used to describe any right triangle. The symbol θ (which is the Greek letter "theta") is used to indicate the angle that will be used to analyze the triangle. In the phasor diagram of Fig. 9-9, theta (θ) is the angle by which the total current leads the voltage. The adjacent side is always the shortest side that makes up angle θ. The side across the triangle from angle θ is always called the opposite side. The longest side of a right triangle is always the hypotenuse.

The three angles of any triangle always contain a total of 180°. This means that specifying θ for a right triangle also determines the other angle. Thus, the only possible difference between right triangles with the same angle θ is their size. This is illustrated in Fig. 9-12. Triangles ABG, ACF, and ADE all have the same angle θ. Only their sizes differ.

Notice in Fig. 9-12 that the sides and the hypotenuse increase proportionately. That is, when the hypotenuse is doubled in length, the opposite and adjacent sides must also be doubled. For example, all of the sides of triangle ACF are twice as great as the corresponding

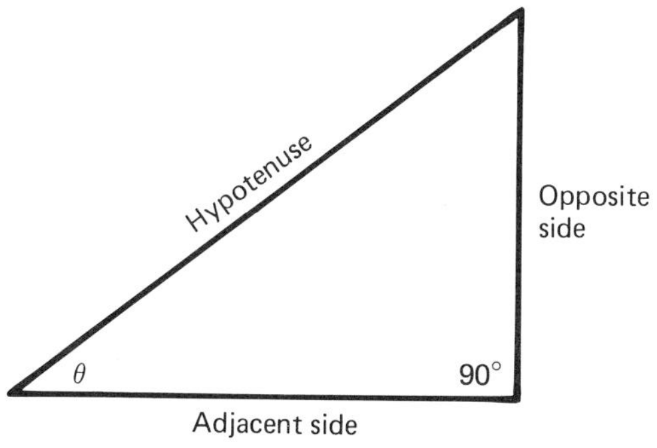

Fig. 9-11 Right triangle terminology.

sides of triangle ABG. The sides of triangle ADE are four times as large as the corresponding sides of triangle ABG.

Since the sides of a right triangle increase proportionately, the ratio of any two sides must remain constant. In Fig. 9-12, side BG divided by side AG is exactly equal to side CF divided by side AF. The opposite side divided by the adjacent is a constant for all right triangles that have the same angle θ. This same statement can be made for any other combination of sides (including the hypotenuse) of a right triangle.

The ratios of the sides of a right triangle are called the *trigonometric functions*. Three of the common ones used in electricity are shown in Fig. 9-12. The ratio of the opposite over the adjacent is called the tangent of angle θ and abbreviated tan θ. Adjacent over hypotenuse is called cos θ which is the abbreviation for the cosine of θ. Opposite over hypotenuse is called sin θ which stands for the sine of θ.

Trigonometric functions

Tangent

Cosine

Sine

For each possible value of angle θ there will be a specific value for the sine, the tangent, and the cosine. A table of trigonometric functions for even angles between 0 and 90° is given in Appendix 6. Using this table (or a calculator) and the formulas in Fig. 9-12, we can easily solve right triangle problems. For example, if we know the angle and the length of the hypotenuse we can solve for the length of the adjacent side.

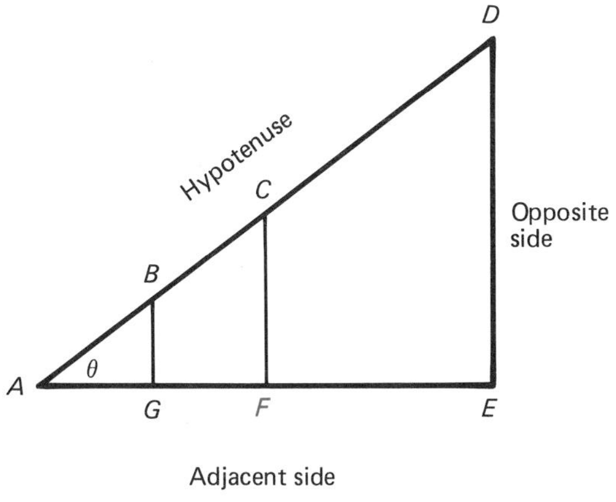

$$\tan\theta = \frac{\text{opposite}}{\text{adjacent}} = \frac{BG}{AG} = \frac{CF}{AF} = \frac{DE}{AE}$$

$$\sin\theta = \frac{\text{opposite}}{\text{hypotenuse}} = \frac{BG}{AB} = \frac{CF}{AC} = \frac{DE}{AD}$$

$$\cos\theta = \frac{\text{adjacent}}{\text{hypotenuse}} = \frac{AG}{AB} = \frac{AF}{AC} = \frac{AE}{AD}$$

Fig. 9-12 Trigonometric functions of a right triangle.

Example 9-3

How long is the adjacent side if the hypotenuse is 7 centimeters (cm) and the angle is 40°?

Given: Hypotenuse = 7 cm
$\theta = 40°$

Find: Adjacent side

Known: $\text{Cos}\,\theta = \dfrac{\text{adjacent}}{\text{hypotenuse}}$

Solution: From the cosine formula we can solve for the adjacent side:

$$\text{Adjacent} = \text{hypotenuse} \times \cos\theta$$

From a table of trigonometric functions (Appendix 6) we can find the value of cos 40°:

$$\text{Cos } 40° = 0.766$$

Then,

$$\text{Adjacent} = 7 \text{ cm} \times 0.766 = 5.362 \text{ cm}$$

Answer: The adjacent side is 5.362 centimeters long.

In an electrical circuit the hypotenuse might represent the total current and the adjacent side represent the resistive current (Fig. 9-10). In that case,

$$\text{Cos}\,\theta = \frac{\text{resistive current}}{\text{total current}}$$

Let us now use the functions of a right triangle to solve an electrical problem.

Example 9-4

The current in a circuit leads the voltage by 25°. The total current is 8.4 A. What is the value of the resistive current?

Given: $\theta = 25°$
$I_T = 8.4$ A

Find: Resistive current

Known: $\text{Cos } \theta = \dfrac{I_R}{I_T}$

Solution: From the cosine formula:

$$I_R = I_T \times \cos \theta$$

From Appendix 6:

$$\text{Cos } 25° = 0.906$$

Then,

$$I_R = 8.4 \text{ A} \times 0.906$$
$$= 7.610 \text{ A}$$

Answer: The resistive current is 7.61 amperes.

Notice in the trigonometric table Appendix 6 that the cosine varies from 1 to 0 as the angle changes from 0 to 90°. In other words, at 0° the adjacent side and the hypotenuse are the same length. The cosine of angle θ will be 1. At 90° the adjacent side will reduce to 0 and cos θ will be 0. Referring back to Fig. 9-10, we can see that there is no resistive current when the angle is 90°. When θ is 0°, there is no reactive current. Notice, then, that the cosine of the θ tells what part, or percentage, of the total current is in phase with the voltage. That is, cos θ is the portion of the total current that is resistive. For example, when θ is 45° the cosine is 0.707. Therefore, 0.707 of the total current is resistive current. That portion (0.707) of the current uses power.

Because cos θ determines the amount of resistive current, it is included in the formula for calculating power. The power formula is:

$P = IV \cos \theta$

$$P = IV \cos \theta$$

This is the general formula for power. It is the correct formula for all types of circuits. However, when a circuit contains only resistance, cos θ will be 1. Therefore, the cos θ can be dropped when working only with resistance.

Example 9-5

A 208-V motor draws 12 A. The current lags the voltage by 35°. What is the power input to the motor?

Given: $V = 208$ V
$I = 12$ A
$\theta = 35°$

Find: P

Known: $P = IV \cos \theta$

Solution: From Appendix 6:

$$\text{Cos } 35° = 0.819$$

Then,

$$P = IV \cos \theta$$
$$= 12 \text{ A} \times 208 \text{ V} \times 0.819$$
$$= 2044 \text{ W}$$

Answer: The power required by the motor is 2044 watts.

So far in developing our power equation we have always shown the current out of phase with the voltage on the 0° reference line. Thus, we broke the total current into two parts: the resistive and the reactive. It is also possible (and common) for the current to be shown on the 0° line and the voltage shown at some other angle. In that case, the total voltage would be separated into a resistive voltage and a reactive voltage. The phasor diagram for such a case is illustrated in Fig. 9-13. In this case, we still use cos θ. Only now we use it to determine the part of the total voltage that is resistive. Then we will know the resistive voltage and current, and from that we can find the power.

Example 9-6

In an electric circuit the voltage leads the current by 73°. The voltage is 80 V, and the current is 4 A. What is the power used by this circuit?

Given: $\theta = 73°$
$V = 80$ V
$I = 4$ A

Find: P

Known: $P = IV \cos \theta$

Solution: $\text{Cos } 73° = 0.292$
$P = 4 \text{ A} \times 80 \text{ V} \times 0.292 = 93.4 \text{ W}$

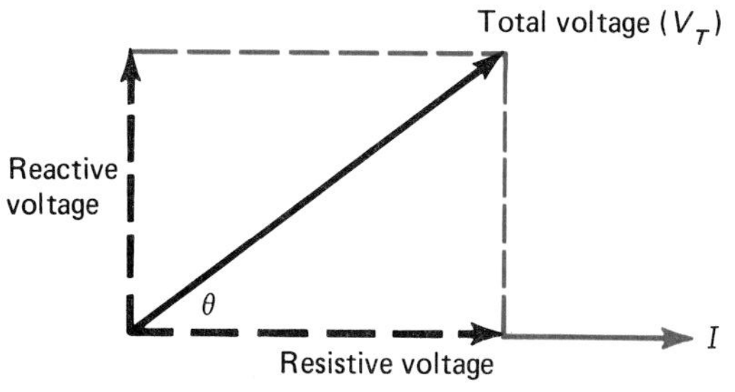

(a) Phasor diagram

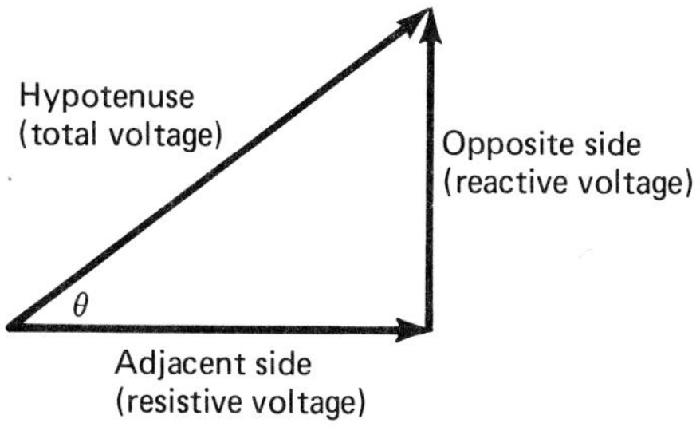

(b) Voltage triangle

Fig. 9-13 Determining the resistive voltage. The total voltage phasor is represented by its two equivalent phasor.

Answer: The power used by the circuit is 93.4 watts

In the above example the voltage was leading the current by 73°. However, the power would have been the same if the voltage had been lagging the current.

Review Questions

18. In electrical circuits, what does the symbol θ represent?
19. Draw a right triangle in which θ equals 30°. Mark the number of degrees of the other two angles. Label all sides of the triangle.
20. Does the opposite side of a right triangle represent the resistive, reactive, or total current?
21. Does the adjacent side of a right triangle represent the resistive, reactive, or total voltage?
22. Which trigonometric function (sin, cos, or tan) is calculated by dividing the adjacent side by the hypotenuse?
23. Which trigonometric function is obtained by dividing the resistive current by the total current?
24. An 8-A current lags the source voltage by 40°. How much of the current is resistive?
25. A voltage of 208 V leads the current in a circuit by 15°. What is the value of the resistive voltage?
26. A 5-A current in a circuit leads the voltage by 78°. The voltage is 120 V. How much power is used?
27. The voltage in a motor leads the current by 40°. The current is 20 A and the voltage is 440 V. What is the power used by the motor?

9-3 TRUE POWER AND APPARENT POWER

True power

Wattmeter

Apparent power

Voltampere

So far we have been calculating the *true power* of ac circuits. That is, we have determined the *actual* power used by the circuit. We have done this by taking into account the phase shift between the total current and the total voltage. We can also determine true power in a circuit by measuring it with a *wattmeter.* A wattmeter is internally constructed so that it takes into account any phase difference between current and voltage.

Sometimes it is as important to know the *apparent power* in a circuit as it is to know the true power. The apparent power is obtained by ignoring the fact that the current and voltage are out of phase. It is the power that *appears* to be present when the voltage and current in a circuit are measured separately. The apparent power, then, is the product of the voltage and the current regardless of the phase angle θ. In a circuit containing both resistance and reactance the *apparent power* is not equal to the *true power.*

Apparent power is calculated by the formula:

$$P_{app} = IV$$

It has a base unit of *voltamperes* (VA) to indicate that it is not true power which has units of watts. One watt of true power means electrical energy is converted into heat energy at a rate of one joule (J) per second. One voltampere of apparent power converts *less*

$\cos\theta = \frac{P}{P_{app}}$

Power factor

than one joule per second. How much less depends on how far the voltage and current are out of phase.

Look closely at the formulas for apparent power and true power given below.

$$P = IV \cos\theta$$
$$P_{app} = IV$$

Notice that the only difference between the two is that the true power includes the cos θ. Combining these two formulas yields the following:

$$\cos\theta = \frac{P}{P_{app}}$$

This relationship makes it relatively easy to determine cos θ and thus the phase relationship between current and voltage. All we need to know is the value of the current, the voltage, and the true power. All three of these are easily measured as shown in Fig. 9-14. With the measured current and voltage we can calculate the apparent power. Then, with the measured power and the calculated apparent power we can calculate the cos θ. After calculating the value of cos θ, we can use the table of trigonometric functions to find θ.

Example 9-7

Refer to Fig. 9-14. Suppose the wattmeter reads 813 W, the voltmeter reads 220 V, and the ammeter reads 6 A. What is the phase angle (θ) between the current and the voltage?

Given: $P = 813$ W
$V = 220$ V
$I = 6$ A

Find: θ

Known: $P_{app} = IV$, $\cos\theta = \frac{P}{P_{app}}$

Solution: $P_{app} = 6\text{ A} \times 220\text{ V} = 1320\text{ VA}$

$$\cos\theta = \frac{813\text{ W}}{1320\text{ VA}} = 0.616$$

Using Appendix 6 we find 0.616 in the cosine column. The angle whose cosine is 0.616 is 52°.

Answer: The current and voltage are 52° out of phase ($\theta = 52°$).

If the reactive part of the load in the above example were inductance, the current would lag the voltage by 52°. If it were capacitance, the current would lead the voltage by 52°. Phasor diagrams for both cases are drawn in Fig. 9-15.

Review Questions

28. What is "apparent power"?
29. Can the true power ever be greater than the apparent power?
30. What is the base unit of apparent power?
31. The measured voltage, current, and power in a circuit are 120 V, 8 A, and 604 W. Determine the apparent power, cos θ, and angle θ.

9-4 POWER FACTOR

The ratio of the true power to the apparent power in a circuit is known as the *power factor* (PF). Since this ratio also yields the cos θ, PF is just another way of specifying the cos θ. Mathematically we can then write:

$$\text{Power factor} = \cos\theta = \frac{P}{P_{app}}$$

Power factor, therefore, is a way of indicating that portion of the total current and voltage that is producing power. When current and voltage are in phase, the power factor will be 1; the true power will be the same as the apparent power. When current and voltage are 90° out of phase, the power factor will be 0. Thus, the power factor of a circuit can be any value between 0 and 1. Sometimes power factor is

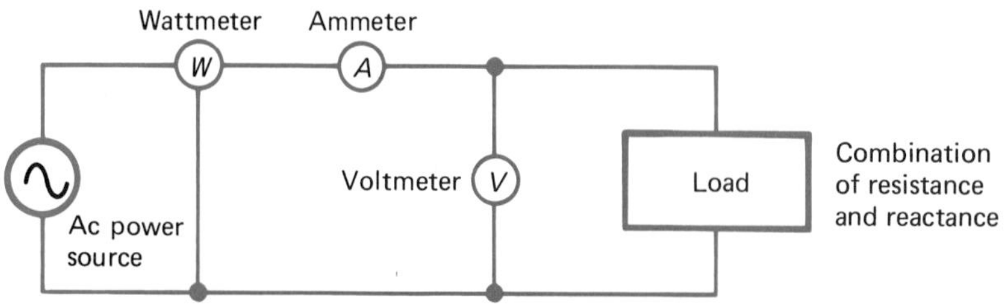

Fig. 9-14 Measuring power and apparent power. The apparent power is found by multiplying the current by the voltage. The true power is indicated by the wattmeter.

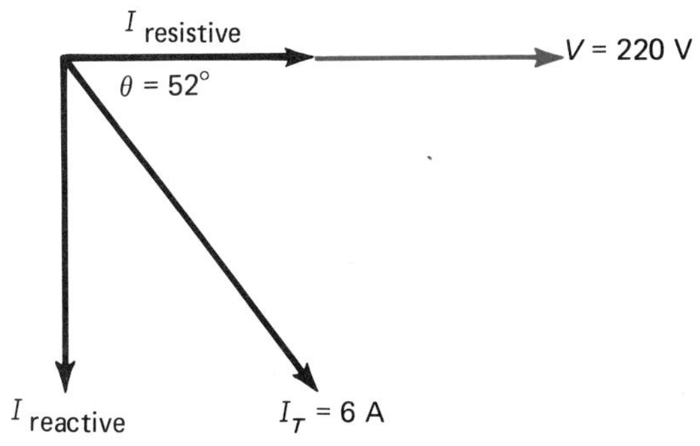

(a) If load has inductance

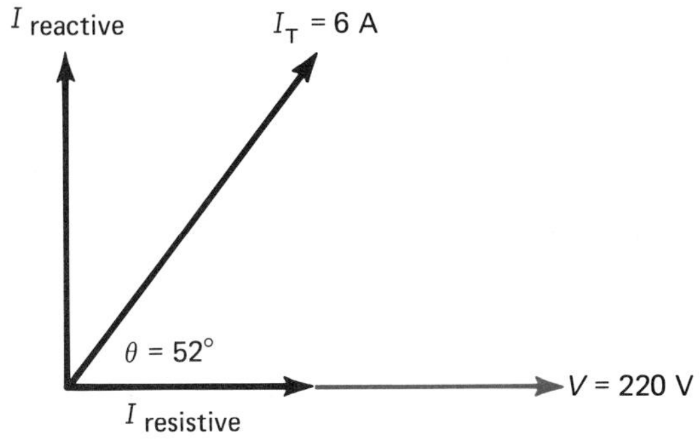

(b) If load has capacitance

Fig. 9-15 Phasor diagrams for Example 9-7.

expressed as a percentage. Then the power factor can vary from 0 to 100 percent.

Example 9-8

An electric motor draws 18 A of current from a 240-V source. A wattmeter connected to the circuit indicates 3024 W. What is the power factor of the circuit?

Given: $I = 18$ A
$V = 240$ V
$P = 3024$ W

Find: Power factor (PF)

Known: $\text{PF} = \dfrac{P}{P_{\text{app}}}, P_{\text{app}} = IV$

Solution: $P_{\text{app}} = 18 \text{ A} \times 240 \text{ V} = 4320 \text{ VA}$

$$\text{PF} = \frac{3024 \text{ W}}{4320 \text{ VA}} = 0.7$$

Answer: The power factor is 0.7 or 70 percent.

Power factor is very important to the utility company that sells electrical energy. Its importance can be best shown by an example. Suppose a utility company is selling energy to two different manufacturers. The two manufacturers are located the same distance from the utility's distribution center (Fig. 9-16). Both manufacturers receive power at the same voltage (4700 V) and both require the same true power (1.5 MW). However, manufacturer A uses a lot of reactive loads (motors) and operates with a power factor of 60 percent. Manufacturer *B* uses mostly resistive loads (heaters) and operates with a power factor of 96 percent. In order to serve manufacturer A, the utility company must provide the following number of voltamperes:

$$P_{\text{app}} = \frac{P}{\text{PF}} = \frac{1{,}500{,}000}{0.6}$$
$$= 2{,}500{,}000 \text{ VA} = 2.5 \text{ MVA}$$

In order to serve this load, the utility company feeders (conductors) must be able to carry the following current:

$$I = \frac{P_{\text{app}}}{\text{V}} = \frac{2{,}500{,}000 \text{ VA}}{4700 \text{ V}}$$
$$= 532 \text{ A}$$

Manufacturer *B* uses the same true power as manufacturer A, and thus pays the same amount to the utility company. Manufacturer *B*, however, requires the following apparent power:

$$P_{\text{app}} = \frac{P}{\text{PF}} = \frac{1{,}500{,}000}{0.96}$$
$$= 1{,}562{,}500 \text{ VA}$$

or about 1.56 MVA. This is almost 1 MVA less than was required by manufacturer A. In

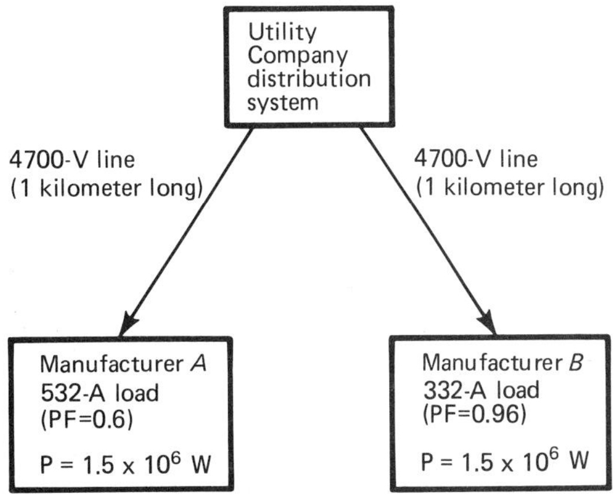

Fig. 9-16 Importance of power factor. It costs more to deliver a unit of power when the power factor is low.

addition, the current drawn by manufacturer B would be:

$$I = \frac{P_{app}}{V} = \frac{1,560,000}{4700}$$
$$= 332 \text{ A}$$

Again, manufacturer B draws less current than manufacturer A (332 A to 532 A) to obtain *exactly the same true power.*

The utility company (and the customer, too) designs its transmission and distribution systems according to the *apparent power* and current it must deliver. Since the customer is billed for the *true power* used, the utility company encourages the use of high-power-factor systems. Power factor can be improved (or corrected) by inserting a reactance opposite to the one causing the low power factor. Thus, a lagging power factor can be improved by inserting a leading-power-factor device, such as a capacitor, in the system.

Review Questions

32. Define power factor.
33. The current and the voltage in a circuit are 30° out of phase. What is the power factor?
34. The apparent power in a circuit is 400 VA. The power is 300 W. What is the power factor expressed in percentage?
35. Two industries use equal amounts of true power for the same length of time. Industry X operates with a PF of 0.95 and industry Y operates with a PF of 0.80. Which industry uses:
    a. More current?
    b. More energy?
36. Which industry in question 35 is utilizing the distribution equipment more efficiently?

Summary

1. In a pure resistance circuit, current and voltage are in phase.
2. When I and V are in phase, $P = IV$.
3. Reactance causes current and voltage to be 90° out of phase.
4. Circuits with inductive reactance cause current to lag voltage.
5. Circuits with capacitive reactance cause current to lead voltage.
6. A phasor represents the magnitude and direction of a quantity.
7. A single phasor can be broken into two smaller phasors which are at right angles to each other.
8. Only the resistive part of a current or a voltage can use power (convert electrical energy).
9. The cos θ is the ratio of the adjacent side over the hypotenuse.
10. The cos θ can be the ratio of the resistive current over the total current.
11. The cos θ can be the ratio of the resistive voltage over the total voltage.
12. In any electrical circuit, power can be calculated with the formula $P = IV \cos \theta$.
13. Apparent power does not take into account the phase relationship between current and voltage.
14. The unit for apparent power is the voltampere (VA).
15. The ratio of power over apparent power is equal to the cos θ. It is also equal to the power factor.
16. Power factor (PF) varies from 0 to 1 or from 0 to 100%.
17. Energy used at a low power factor is more costly than the same energy used at a high power factor.

Chapter Review Questions

For questions 9-1 to 9-14, determine whether each statement is true or false.

9-1. A load that contains both resistance and reactance uses no power.

9-2. A typical motor is an inductive load.

9-3. Reactance causes electrical energy to be converted into heat energy.

9-4. In a circuit containing only resistance, the current and voltage are in phase.

9-5. Capacitance causes current to lead voltage by 90°.

9-6. All voltage phasors in a phasor diagram need not be drawn to the same scale.

9-7. Any phasor can be represented by two shorter phasors which are 90° apart.

9-8. The cosine of an angle is equal to the adjacent side divided by the hypotenuse.

9-9. A current that leads its voltage source uses more power than a current that lags its voltage source.

9-10. The base unit of apparent power is the voltampere.

9-11. The apparent power and the true power are equal in a circuit containing capacitance.

9-12. The true power in a circuit is often greater than the apparent power.

9-13. Power factor and cos θ have the same numerical value.

9-14. Industry pays more for a unit of energy used at a high power factor than one used at a low power factor.

Answer the following questions.

9-15. An ac circuit with a pure capacitance load draws 10 A from a 100-V source. How much power does it use?

9-16. What does a phasor represent in an electrical phasor diagram?

9-17. The current in a circuit is 10 A. It lags the voltage by 30°. How many amperes of this current is resistive current?

9-18. It requires 6 A to operate a 120-V motor. The motor causes the current and the voltage to be 50° out of phase. How much power does the motor require?

9-19. What is the apparent power of the motor in question 9-18?

9-20. The measured current, voltage, and power in a circuit are 15 A, 240 V, and 2875 W. Determine θ.

9-21. Determine the power factor for the circuit of question 9-20.

Answers to Review Questions

1. The current and voltage are in phase.
2. The current and voltage are 90° out of phase.
3. 9 W
4. Yes
5. Reactance is the opposition to ac produced by inductors and capacitors.
6. No
7. Yes

8. Inductive reactance
9. Lags
10. A phasor is a line which shows direction and magnitude of a quantity.
11.

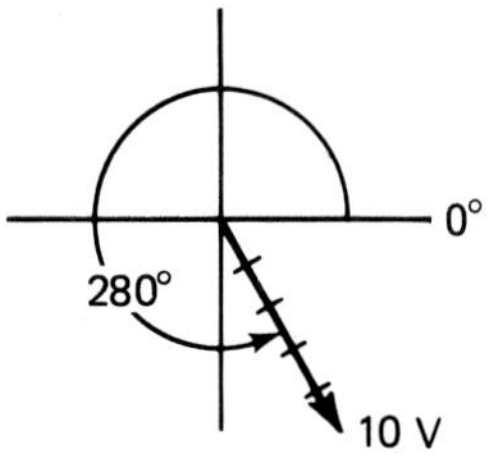

12.

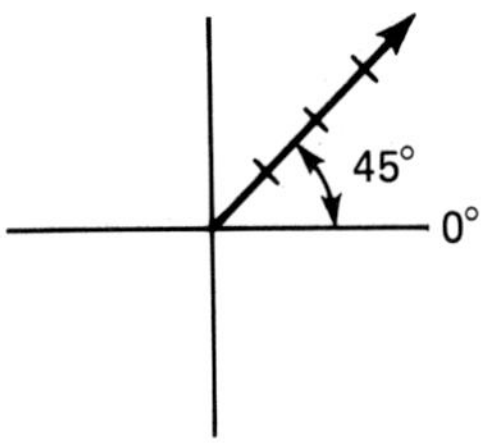

13.

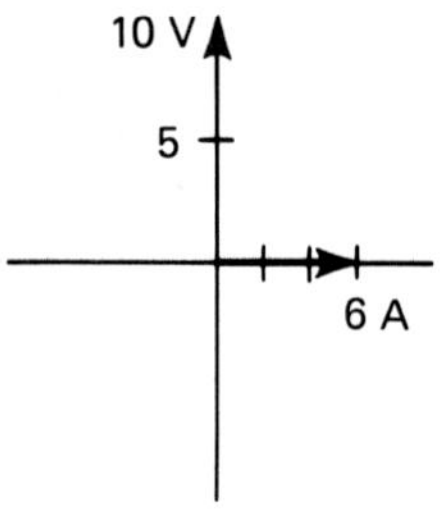

14.

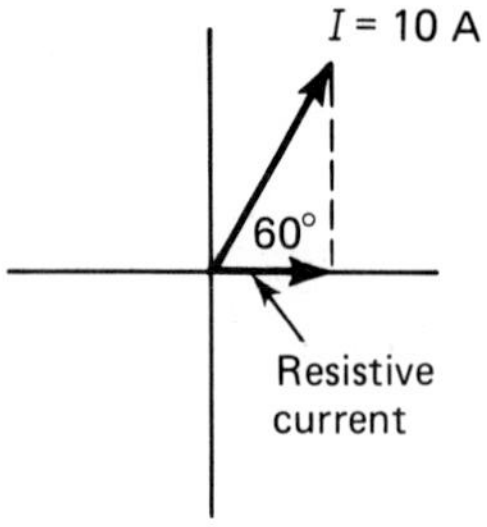

15.

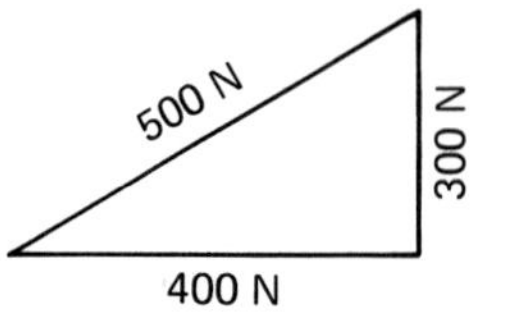

16. Yes
17. No
18. The angle between the total current and the total voltage
19.

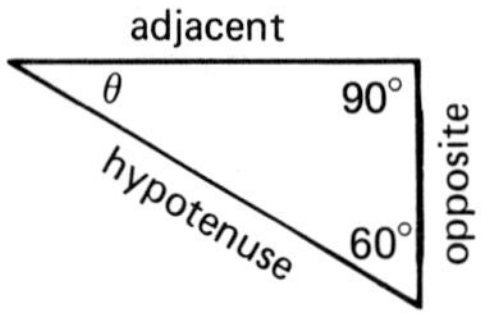

20. Reactive current
21. Resistive voltage
22. Cos θ
23. Cos θ
24. 6.13 A
25. 200.9 V
26. 124.7 W
27. 6741.2 W
28. Apparent power is the product of the current times the voltage in a reactive circuit.
29. No
30. Voltampere
31. $P_{app} = 960$ VA
    $\cos\theta = 0.625$
    $\theta = 51°$
32. Power factor is the ratio of the power over the apparent power.
33. 0.866 or 86.6 percent
34. 75%
35. *a*. Industry *Y*
    b. They use the same.
36. Industry *X*

Inductance

- **Many electrical and electronic devices operate on the principle of inductance. Therefore, it is important that you develop a sound understanding of inductance and inductors.**
  In this chapter you will learn to identify various types of inductors and write complete specifications for them. You will learn to work with series and parallel inductances. Also, you will learn the relationships between inductance, frequency, and reactance. Finally, you will learn about some of the many uses of inductors.

10-1 CHARACTERISTICS OF INDUCTANCE

Inductance is the electrical property which opposes any change in the magnitude of current in a circuit. The letter L is the symbol used to represent inductance.

Physical devices which are used to provide the inductance in a circuit are called *inductors*. Inductors are also known as *chokes*, *coils*, and *reactors*. These last three names are all descriptive of the way in which inductance behaves in a circuit. Inductance, and thus an inductor, "chokes-off" and restricts sudden changes in current. Inductance reacts against (resists) changes, either increases or decreases, in current. Finally, inductors are usually coils of wire.

Inductance is the result of a voltage being induced in a conductor. The magnetic field that induces the voltage in the conductor is produced by the conductor itself. In Chap. 6 we discussed how a magnetic field is formed around a current-carrying conductor. A brief review will be given here. You will recall that when current begins to flow in a conductor, magnetic flux rings start to expand out from the conductor, as in Fig. 10-1(*a*). This expanding flux induces a small voltage into the conductor itself. The induced voltage has a polarity that opposes the increasing source voltage which is creating the increasing current. Thus, the inductance of the conductor opposes the rising current and tries to keep it constant. Of course, it cannot completely stop the increase in current because the induced voltage is caused by the increasing flux. And, the increasing flux depends on the increasing (rising) current. The inductance of the conductor, therefore, only restricts the rate at which the current can increase.

When the current in a conductor starts to decrease, as in Fig. 10-1(*b*), the flux starts to collapse. The collapsing flux reverses the polarity of the induced voltage from what it was when the flux was increasing. Thus, the voltage induced by a decreasing flux (and current) *aids* the source voltage and tends to keep the current from decreasing. Again, the inductance restricts the rate at which the current can change.

The amount of voltage induced in a single conductor like that in Fig. 10-1 is very small. So small, in fact, that it has no practical significance in most low frequency electrical-electronic devices. However, at high frequencies, like those used in television systems, the inductance of a single conductor can be very significant.

The inductance of a conductor can be greatly increased by forming the conductor into a coil as in Fig. 10-2. Now, the flux produced by one turn of the coil induces voltage in not only itself, but also in adjacent turns. The long, closed, flux loops in Fig. 10-2 are the result of the magnetic fields of all three turns of the coil. They are stronger than the

From page 181:
Inductance

Chokes

Coils

Reactors

On this page:
Self-inductance

Mutual inductance

Counter electromotive force

Back electromotive force

Lenz's law

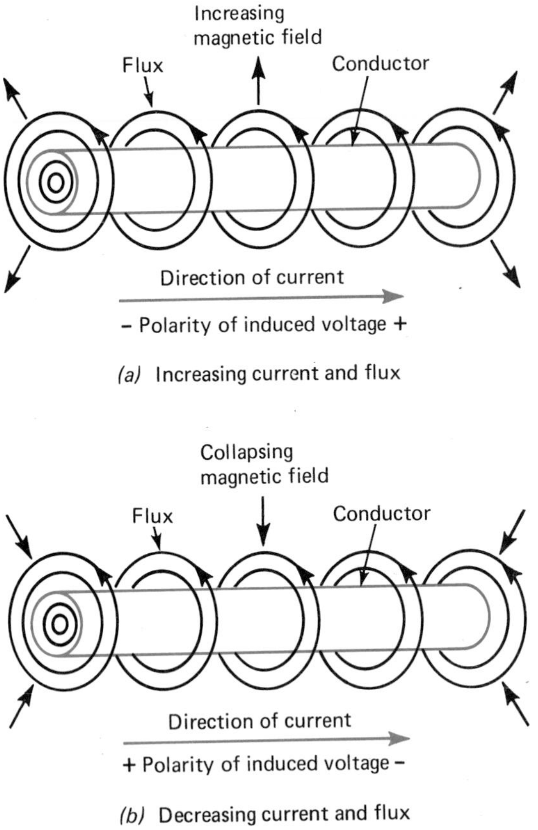

Fig. 10-1 Induced voltage in a conductor. The polarity of the induced voltage is dependent on whether the current is increasing or decreasing.

flux created by any one of the turns. Yet, they induce a voltage into each of the three turns. The inductance of the coil is much greater than the inductance of the straight length of conductor from which it was made.

Self-Inductance

The inductance of an inductor is called *self-inductance*. It is given this name because the inductor induces voltage in itself. That is, its own changing magnetic field induces voltage in its own turns of wire. In the case of a single straight conductor, its own field induces a voltage in it.

Mutual Inductance

When the magnetic flux from one conductor induces a voltage in another electrically isolated conductor, it is called *mutual inductance*. With mutual inductance, circuits that are electrically separated can be magnetically coupled together. A transformer uses the principle of mutual inductance. It will be fully discussed in the next chapter.

Lenz's Law and Cemf

The voltage induced in a conductor or coil by its own magnetic field is called a *counter electromotive force* (cemf). Recall that electromotive force (emf) is another name for voltage. Since the induced emf (voltage) is always opposing, or countering, the action of the source voltage, it is known as a cemf. Counter electromotive force is sometimes referred to as back electromotive force (bemf). The term "back" implies that the induced voltage is backward, or working against, the effort of the source voltage.

The concept contained in *Lenz's law* was used to explain how inductance behaves. *Lenz's law* states that a cemf will always have a polarity which will oppose the force that created it. This idea is illustrated in Fig. 10-3(*a*) which shows, in schematic form, an inductor and an ac voltage source. When the voltage is increasing, as shown in the graph, the cemf opposes the source voltage. When the voltage is decreasing [Fig. 10-3(*b*)] the cemf aids the source voltage and tries to keep the current constant.

Energy Storage and Conversion

Another way to look at inductance is in terms of energy conversion and storage. When current flows through an inductor, the inductor builds up a magnetic field. In the process of building its magnetic field the inductor converts electrical energy into magnetic energy. When the current increases, more electrical

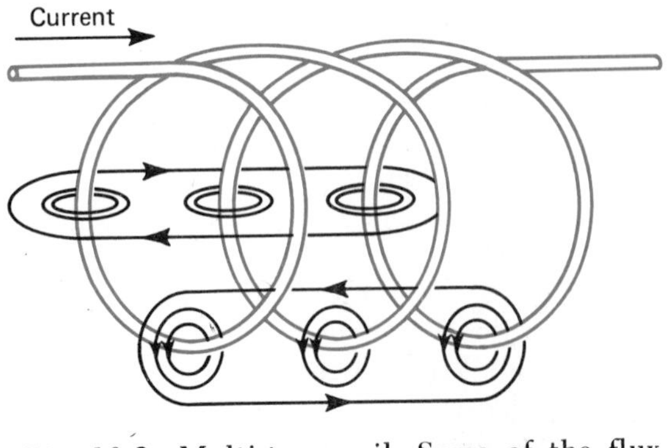

Fig. 10-2 Multi-turn coil. Some of the flux created by one turn links to all the other turns.

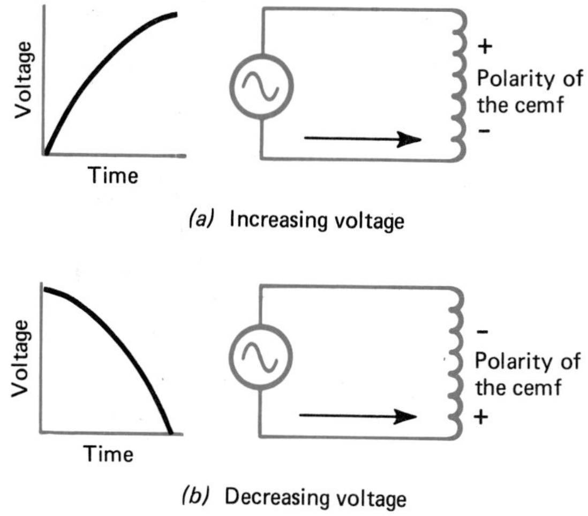

Fig. 10-3 Polarity of counter electromotive force.

energy is converted into magnetic energy. The inductor's magnetic field now processes more energy than it had before the current increased. When current through an inductor decreases its magnetic field decreases. Magnetic energy from the field is converted back into electrical energy within the inductor. Thus, an inductor stores energy when its current increases and returns stored energy when its current decreases. *Inductance converts no electrical energy into heat energy.* Only resistance is capable of converting electrical energy into heat energy. Thus, if an inductor with absolutely no resistance could be constructed, its net use of energy would be zero. For half of the ac cycle, it would take energy from the system. For the other half of the cycle, it would return exactly the same amount of energy to the system.

Review Questions

1. What is inductance?
2. What is an inductor?
3. List three other names for inductors.
4. Does a straight length of wire possess inductance?
5. Would a straight length of wire be called an inductor? Why?
6. When the current in an inductive circuit is decreasing, will the cemf aid or oppose the source voltage?
7. What is bemf the abbreviation for?
8. How does mutual inductance differ from self-inductance?
9. What is the name of the law that is used to determine the polarity of the cemf in a coil?
10. What energy conversion occurs when the current in an inductor is increasing?
11. What is the symbol for inductance?

10-2 UNIT OF INDUCTANCE—THE HENRY

Henry

The base unit of inductance is the *henry*. Henry, named in honor of an American scientist, is abbreviated H. A henry is defined in terms of the amount of cemf produced when the current through an inductor is changing amplitude. One henry (H) of inductance will develop one volt of cemf when the current changes at a rate of one ampere per second. This definition of a henry is shown graphically in Fig. 10-4.

A wide range of inductance is used in electrical-electronic circuits. Inductors used in very-high-frequency circuits are often less than one microhenry (μH). For low-frequency circuits, inductors with five henries or more of inductance are common.

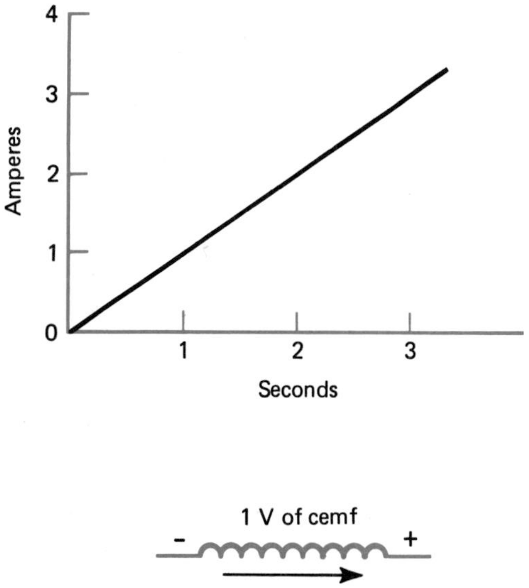

Fig. 10-4 One henry of inductance. A 1-henry inductor produces 1 V of cemf when the current changes at a rate of 1 A/s.

10-3 FACTORS DETERMINING INDUCTANCE

The inductance of an inductor is primarily determined by four factors. These four factors are:

1. The type of core material
2. The number of turns of wire
3. The spacing between turns of wire
4. The diameter of the coil (or core)

The amount of current in an iron-core inductor also influences its inductance. This is because the magnetic properties of the iron core change as the current changes.

Ultimately, the amount of inductance is determined by the amount of cemf produced by a specified current change. Of course, the amount of cemf depends upon the amount of flux which interacts with the conductors of the coil.

If all other factors remain constant, an iron-core conductor will have more inductance than an air-core inductor. This is because the iron has a higher permeability, that is, it is able to carry more flux. With this higher permeability there is more flux change, and thus more cemf, for a given change in current.

Adding more turns to an inductor increases its inductance because each turn adds more magnetic field strength to the inductor. Increasing the magnetic field strength results in more flux to cut the turns of the inductor.

When the distance between the turns of wire in a coil is increased, its inductance decreases. Figure 10-5 illustrates why this is so. With widely spaced turns [Fig. 10-5(*a*)] many of the flux lines from adjacent turns do not link together. Those lines that do not link together produce a voltage only in the turn which produced them. As the turns come closer together [Fig. 10-5(*b*)] fewer lines of flux fail to link up.

When other factors are equal, the inductor with the largest diameter core will have the most inductance. The reason for this is because all of the flux has to go through the core of an inductor. Thus a large-diameter core can handle more flux, at a specified flux density, than can a small-diameter core.

Review Questions

12. What is the base unit of inductance?
13. What is the abbreviation for the base unit of inductance?
14. Define the base unit of inductance in terms of the base units of voltage, current, and time.
15. List four main factors that determine the amount of inductance.
16. Explain what could be done with *each* of these four factors to increase inductance.

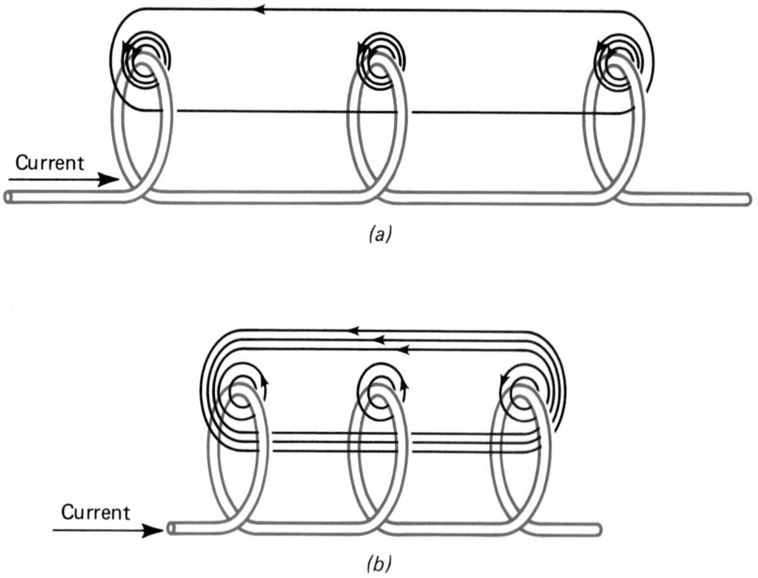

Fig. 10-5 Effects of turn spacing. Widely spaced turns (*a*) provide less inductance than do closely spaced turns (*b*).

(a) Iron core (magnetic)

(b) Air core (nonmagnetic)

Fig. 10-6 Fixed-value inductor symbols.

10-4 TYPES OF INDUCTORS

One way of classifying inductors is by the type of material used for the core of the inductor. The "core of an inductor" refers to the material which occupies the space enclosed by the turns of the inductor.

Symbols

The core of an inductor may be either a magnetic material or a nonmagnetic material. The symbols for inductors with these materials are shown in Fig. 10-6.

Inductors are also classified as either fixed or variable. Figure 10-7 shows two symbols used to indicate a variable inductor. The most common way of varying inductance is by adjusting the position of the core material. In Fig. 10-8 the position of the ferrite core material (called a *slug*) is adjustable within the coil form. Maximum inductance occurs when the slug (ferrite material) is positioned directly in line with the coil of wire. Some variable inductors use a brass slug. Brass has more reluctance (opposition to flux) than does air. Therefore, the brass slug decreases inductance when it is centered in the coil.

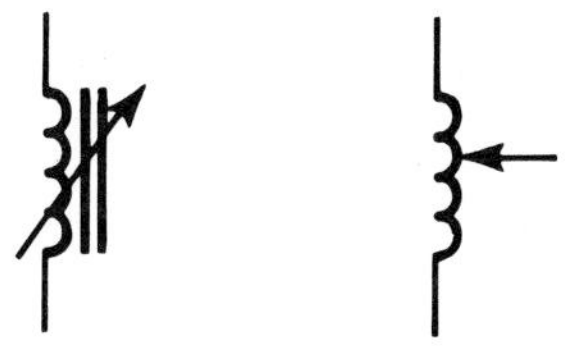

Fig. 10-7 Variable inductor symbols. Either symbol can be used for either magnetic or nonmagnetic cores.

Variable inductor

Slug

Air-core inductor

Iron-core inductor

Air-Core Inductors

An air-core inductor, used as part of a high-frequency circuit, is shown in Fig. 10-9. This inductor is self-supporting and requires no coil form. However, many inductors which are represented by the air-core symbol are wound on a coil form. The form may be either solid or hollow. These forms have about the same reluctance (opposition to magnetic flux) as does air. Therefore, the inductor is much like an air-core inductor; its core is nonmagnetic. These inductors may be wound on such core materials as ceramic or phenolic. Physically, they often look like the coils in Fig. 10-10. These inductors seldom have more than 5 millihenries (5 mH) of inductance.

Ferrite and Powdered-Iron Cores

The coils shown in Fig. 10-10 may also be iron-core inductors. In this case the core material would be ferrite or powdered iron. The correct symbol would then be the iron-core symbol of Fig. 10-6(*a*). Most of the inductors

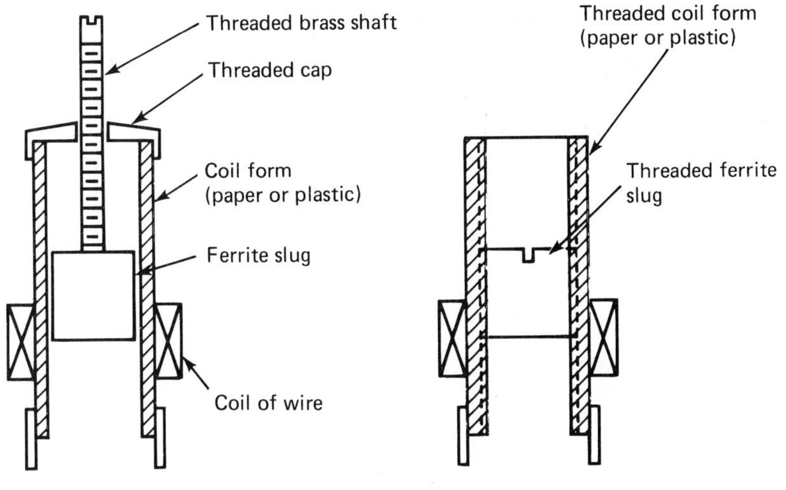

(a) Threaded shaft

(b) Threaded slug

Fig. 10-8 Methods of adjusting inductance.

Toroid core

Molded inductors

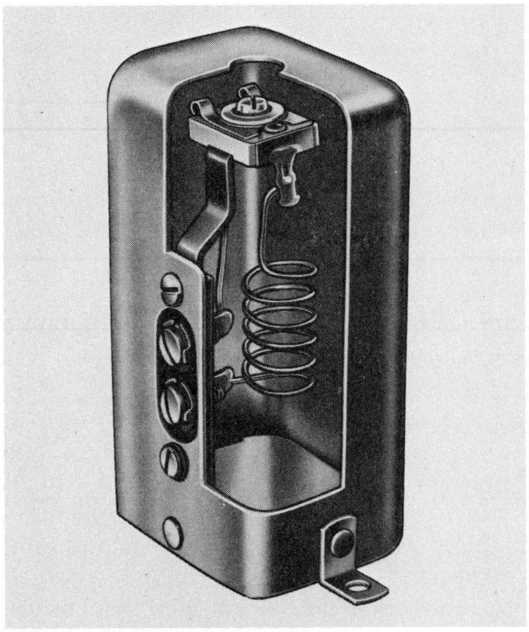

Fig. 10-9 Air-core inductor. (Courtesy of Bell Industries, J. W. Miller Division.)

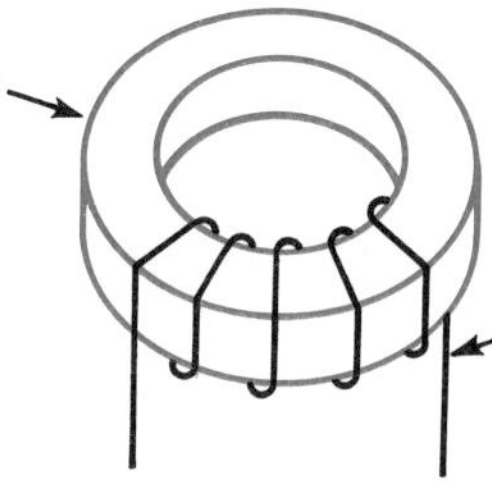

Fig. 10-11 Toroid core. All of the flux produced by the coil is concentrated in the low-reluctance core.

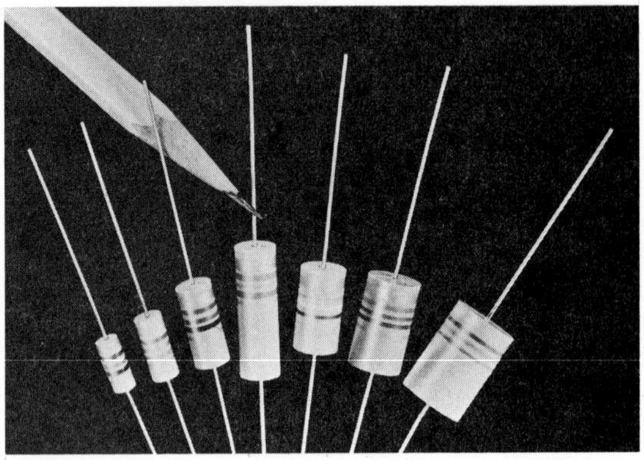

Fig. 10-12 Molded inductors. (Courtesy of Bell Industries, J. W. Miller Division.)

of this type have less than 200 mH of inductance. They are used primarily at frequencies above the audio (sound) range.

Toroid Cores

The cores of the inductors discussed so far have all been straight. The magnetic flux loops must extend through the air as well as the core material. With a toroid core the flux loops all exist within the core. Toroid cores (Fig. 10-11) are donut-shaped. Each turn of wire is threaded through the center of the core as shown in Fig. 10-11. Inductors made with toroid cores are called toroidal inductors. The toroid core is usually made from powdered iron or ferrite. Toroidal inductors can have high inductance values for their physical size.

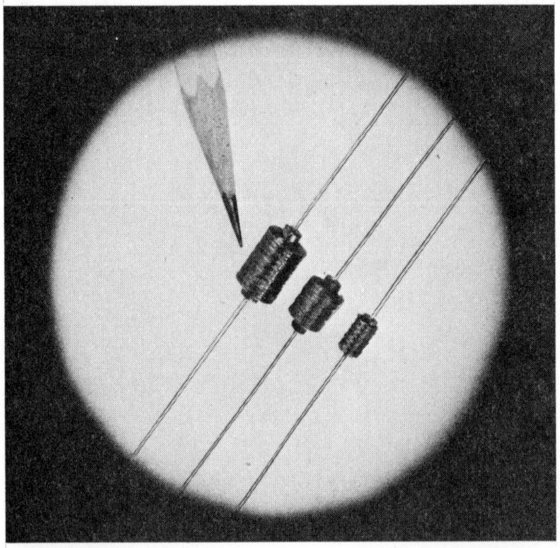

Fig. 10-10 Miniature high-frequency inductors. (Courtesy of Bell Industries, J. W. Miller Division.)

Molded Inductors

Physically, some inductors look like resistors (Fig. 10-12). These inductors have been enclosed in an insulating material to protect the inductor winding. Molded inductors can have cores of air, ferrite, or powdered iron.

Shielded Inductors

Inductors are often shielded to protect them from the influence of magnetic fields other than their own. The shield is made from a magnetic material. Figure 10-13 shows an exploded view of the parts of three shielded, adjustable coil forms. The coil windings are not shown. They would be wound on the cylindrical tubes or bobbins. The coil forms shown in Fig. 10-13 are the type that are used on printed circuit boards.

Some of the miniature chokes (inductors) like those in Fig. 10-12 are also shielded. Their shields are encased underneath the outside molding.

Laminated Iron Core

I and E laminations

Nearly all of the large inductors used at power frequencies (60 Hz, for example) use laminated iron cores. These inductors have inductances ranging from about 0.1 to 100 H.

The typical laminated core uses laminations like those in Fig. 10-14. From the illustration in Fig. 10-14 it is easy to see why these laminations are called I and E laminations. The E laminations are stacked together to the desired thickness. So are the I laminations. The winding is put on the center leg (Fig. 10-15) of the E lamination. The I lamination is then positioned across the open end of the E lamination.

As seen in Fig. 10-16 the I and E laminations form two parallel paths for flux. The center leg of the E lamination is twice as wide as the outside leg because it has to carry twice as much flux. For a given amount and rate of current change, the laminated-iron-core inductor creates more flux than other types of inductors. This changing flux, in turn, creates cemf. This is why laminated-iron-core inductors can provide large amounts of inductance.

Saturated

Filter choke

The inductance of an iron-core inductor is somewhat dependent on the amount of current flowing through it. The reason for this can be seen from Fig. 10-17 which illustrates the magnetic characteristics of the core material. The permeability of iron, and iron alloys, decreases as the magnetic field strength increases. The magnetic field strength of an inductor is a function of the amount of current flowing in the winding. Refer to Fig. 10-17. Suppose the current through the inductor changes from Point A to point B. The flux density would change from A′ to B′ and produce a certain amount of cemf. This amount of cemf would, of course, represent a certain amount of inductance. Now suppose the current in the inductor was greater and the current changed from C to D in Fig. 10-17. Although C to D is the same amount of change as A to B it produces a much smaller change

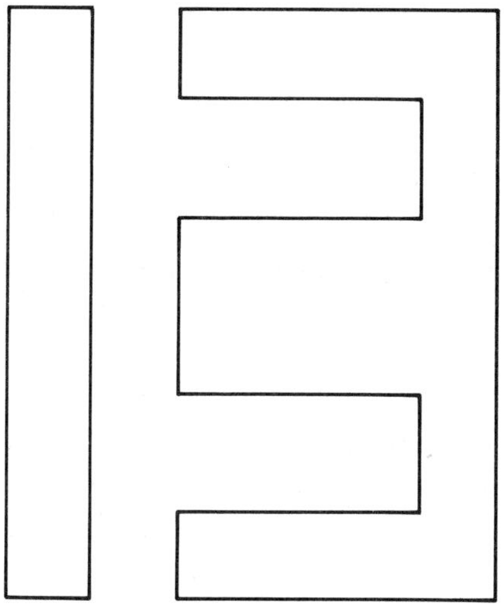

Fig. 10-14 I and E laminations. These lamina tions are stacked in various configurations to forn cores for electromagnetic devices.

Fig. 10-13 Shielded inductors. (Courtesy of Bell Industries, J. W. Miller Division.)

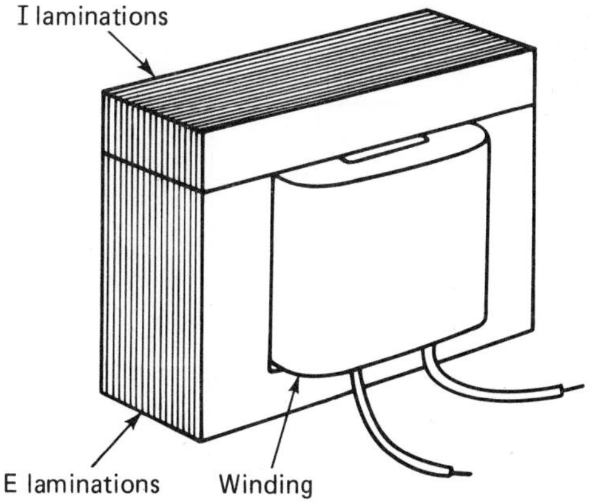

Fig. 10-15 Laminated iron-core inductor. The coil fits over the center leg of the E laminations.

Smoothing choke

Swinging choke

RF coil

Ohmic resistance

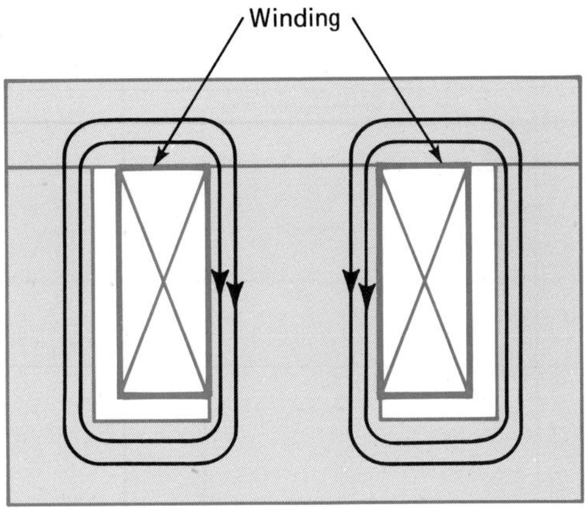

Fig. 10-16 Flux paths in a laminated core.

(C' to D') in flux. Thus, the inductor has less cemf and inductance at the higher current level. Points E and F in Fig. 10-17 show what happens when the core of an inductor is saturated. Since a change in current from E to F produces almost no change in flux, there would be very little inductance. Except for special applications, inductors are never operated in the saturation region of the permeability curve.

Filter Chokes

Laminated-iron-core inductors are often referred to as filter chokes. These chokes are used in the filter circuits of power supplies in a wide variety of electrical and electronic equipment. The power supply is often the part of the equipment which converts ac to dc. The filter circuit, which includes the inductor, smooths out the fluctuating or pulsating dc until it is nearly pure dc.

There are two types of filter chokes: the *smoothing* choke and the *swinging* choke. The swinging choke is one in which the I and E laminates are closely butted together so that there is a minimum air gap between them. This makes the amount of inductance vary with the amount of current (see Fig. 10-17). A typical swinging choke may be rated 20 H at 50 mA and 5 H at 200 mA.

The smoothing choke often has a small (0.1 mm) air gap between the I and E laminates. This makes the inductance less dependent on the amount of current because air doesn't saturate as easily as iron.

Radio Frequency (RF) Chokes or Coils

Inductors which are used at higher frequencies are often called RF chokes or RF coils. Since radio was one of the early popular uses of high-frequency inductors, they became known as radio frequency, or RF, coils. An RF coil or choke may have an air, powdered-iron, or ferrite core. It may be either fixed or variable.

10-5 RATINGS OF INDUCTORS

We have seen so far that one of the main ratings of an inductor is its inductance. Inductors are also rated for dc resistance, current, voltage, quality, and tolerance.

The dc resistance specifies the total resistance of the wire in the winding of the inductor. This is the resistance one would measure between the terminals of the inductor with an ohmmeter. Therefore this dc resistance is sometimes called the *ohmic resistance*.

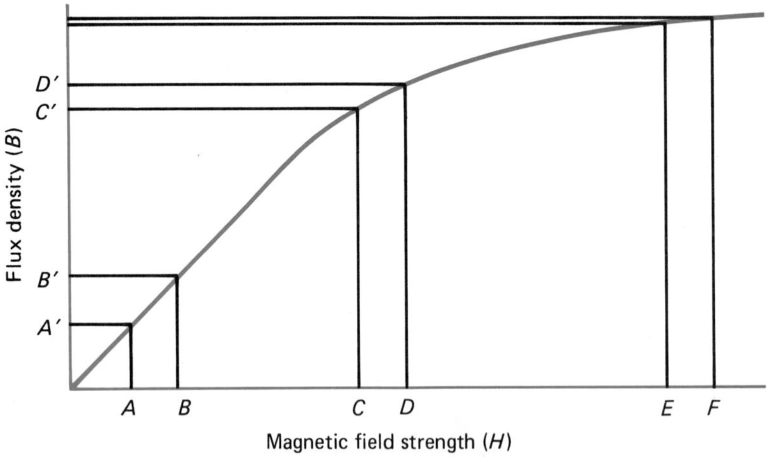

Fig. 10-17 Permeability curve. Permeability decreases as flux density and magnetic field strength increase.

The current rating of an inductor is important because it indicates how much current the inductor can continuously carry without overheating. With laminated-core inductors, the current rating also indicates the current level at which the inductance was measured. At lower current levels, the inductance will be greater than the specified value.

The voltage rating indicates how much voltage the insulation on the inductor winding can continuously withstand. Exceeding this voltage rating may not result in instantaneous breakdown of the insulation. However, it will shorten the life expectancy of the inductor's insulation and cause it to fail prematurely. Voltage ratings are used mostly with laminated-core inductors. With these inductors, the core is often physically and electrically connected to the chassis of an electrical device. However, the winding may be hundreds of volts positive or negative with respect to the chassis.

Quality

The quality of an inductor refers to the ratio of its reactance to its resistance. That is,

$$\text{Quality} = \frac{\text{reactance}}{\text{resistance}}$$

(Remember, reactance is a form of opposition which does not convert electrical energy to heat energy). Generally it is desirable to have a high-quality inductor. After all, inductors are used in circuits to provide reactance, not resistance. The quality of an inductor is not constant. It varies with the frequency at which the inductor is used. Therefore, the quality of an inductor is given for some specific frequency. All other factors being equal, the lower the dc resistance, the higher the quality of the inductor.

Like all other components, inductors have manufacturers' tolerances. Precision inductors can be obtained with a tolerance of less than ±1 percent. However, they are expensive. Typical inductors used in mass-produced electrical and electronic devices have tolerances of ±10 percent or more.

10-6 INDUCTORS IN DC CIRCUITS

In Chap. 6 we explored the behavior of coils in dc circuits with regard to electromagnetism. The behavior of an inductor in a pure dc circuit is contrasted to that of a resistor in Fig. 10-18. With a resistor [Fig. 10-18(*a*)] the current jumps to its maximum value almost the instant the switch is closed. When the switch is opened, it drops back to zero just as fast. An inductor in a dc circuit [Fig. 10-18(*b*)] forces the current to rise more slowly. This is due to the inductor's cemf. The time required for the current to reach its maximum value is dependent on the amount of inductance and resistance. With typical quality inductors the time will be much less than 1 s. Once the current reaches its peak value, the only opposition the inductor offers is its dc resistance. When the switch in Fig. 10-18(*b*) is opened, the cemf of the inductor prevents the current from instantaneously dropping to zero. It does this by ionizing the air between the switch contacts as

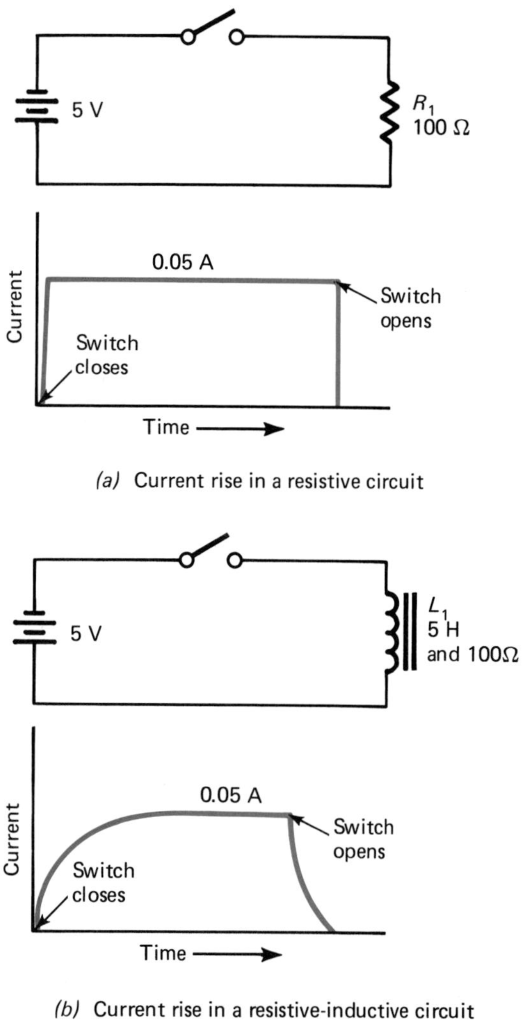

Fig. 10-18 Comparison of a resistive and an inductive dc circuit. The inductor opposes changes in current.

the switch opens. As the energy stored in the inductor's magnetic field is used up, the switch contacts de-ionize and current stops.

Inductive kick

When the switch in Fig. 10-18(*b*) is opened, the cemf of the inductor becomes much greater than the source voltage. The high voltage (cemf) generated when an inductive circuit is opened is known as an *inductive kick*. It is the voltage which ionizes the air between the switch contacts and causes the contacts to arc and burn. The inductive kick of an inductor is very high because the current drops very rapidly when the switch is opened. The difference between the source voltage and the cemf is dropped across the ionized air between the switch contacts. Kirchhoff's voltage law still applies. That is, the voltage across the switch plus the inductor's cemf still equals the source voltage. Notice the polarities in Fig. 10-19. The inductor's cemf and the voltage across the switch are series opposing. Thus, they both can be much greater than the battery voltage. The exact value of the cemf in Fig. 10-19 when the switch opens depends upon two factors. One factor is the amount of inductance and the other is the amount of current in the circuit before the switch is opened. The inductive kick of an inductor can be many thousands of volts. Inductive kick is the principle on which the ignition coil in an automobile operates.

The relationship between the current and voltage (cemf) in an inductor is illustrated in Fig. 10-20. The resistance of R in this figure is very high compared to the ohmic resistance of the inductor. Therefore, the voltage across the inductor will be almost zero once the current reaches its maximum value. Notice in Fig. 10-20 that the voltage across the inductor is maximum when the current through it is minimum. Also, the voltage is minimum when the current is maximum. Further, notice that the resistive current and voltage rise together.

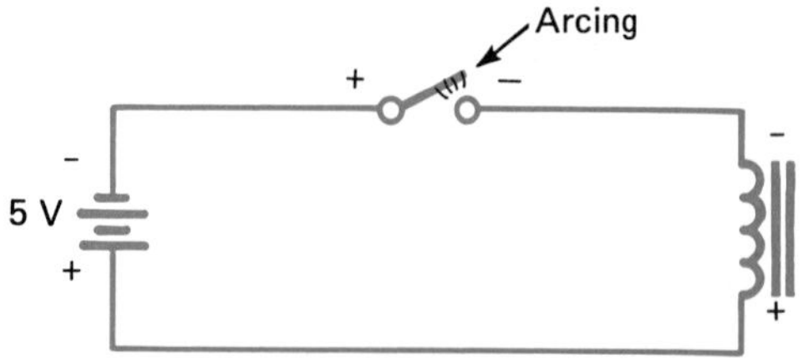

Fig. 10-19 Voltage polarities when an inductive circuit is opened.

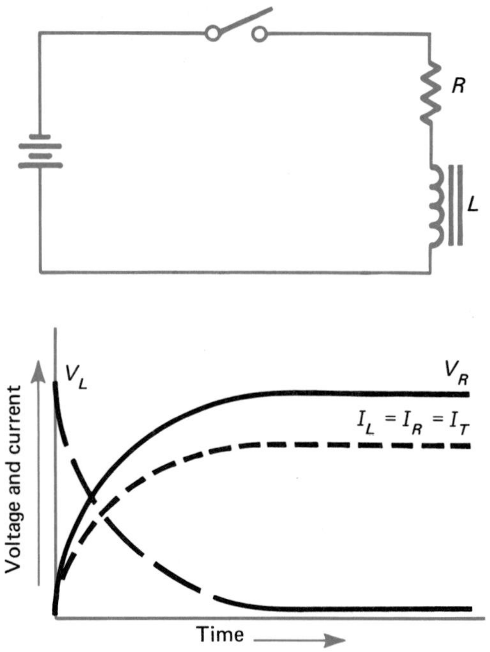

Fig. 10-20 Current and voltage relationships in an inductor. Maximum inductive voltage occurs before maximum current is reached.

Review Questions

17. Draw the symbols for:
    a. A fixed iron-core inductor
    b. A fixed air-core inductor
    c. A variable inductor
18. Should the air-core inductor symbol be used for an inductor that has a ceramic or plastic core?
19. Which symbol is used for an inductor that has a ferrite core?
20. What name is used to indicate the core material in a variable inductor?
21. Where would the brass core in a variable inductor be located to provide minimum inductance?
22. How is a ferrite core in a variable inductor positioned to give maximum inductance?
23. What is the most likely type of core material for a 50-mH inductor?
24. What type of core would be used in a 5-H inductor?
25. What style of core is shaped like a donut?
26. What type of core material would *not* be found in molded inductors?

27. Why is the center leg of an E lamination wider than the outside leg?
28. Why does an inductor have a current rating?
29. What is meant by the quality of an inductor?
30. Is the quality of an inductor dependent on the frequency?
31. What is an RF choke?
32. What type of core does a filter choke have?
33. What is meant by inductive kick?
34. An inductive circuit and a resistive circuit have equal currents. When the switch is opened in each circuit, which circuit produces the most arcing across the switch contacts?
35. When an inductive dc circuit is opened, is the cemf or the source voltage greatest?
36. Does the inductor's voltage (cemf) or its current reach its maximum value first in a dc inductive circuit?

10-7 IDEAL INDUCTORS IN AC CIRCUITS

Ideal inductor

An ideal inductor would be an inductor that had no resistance. It would not convert any electrical energy into heat energy. Ideal inductors are impossible to make because the wire used in the coil must have some resistance. However, air-core inductors have very low resistance. Therefore, in the discussions that follow, we will assume we have ideal inductors.

Reactance

Reactance

The opposition that inductance (and ideal inductors) offer to ac is known as *reactance*. The symbol for reactance is X. Since inductors are not the only components that have reactance, we use a subscript with the symbol X. Thus X_L is used to signify inductive reactance. That is, X_L means the reactance of an inductor. Reactance is like resistance in that it controls the amount of current in circuit. The base unit of reactance is also the ohm (Ω). Reactance is unlike resistance in that it does not convert electrical energy into heat energy. Therefore, it is incorrect to interchange the terms "resistance" and "reactance" even though both are expressed in ohms.

Inductive reactance (X_L) is the result of the cemf of the inductor. The inductor lets just enough alternating current flow to produce a cemf which is equal (and opposite) to the source voltage. This idea is illustrated in Fig. 10-21. During the positive half cycle of the source (an ac generator) the cemf of the inductor produces a positive half cycle of sinusoidal voltage. During the negative half cycle, it also produces a negative-going

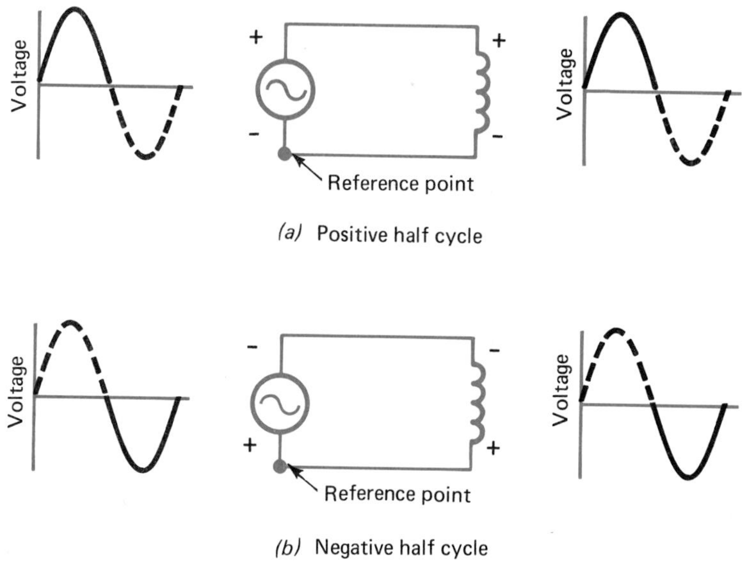

Fig. 10-21 Source and inductor ac voltage. The inductor's cemf opposes the source voltage.

$X_L = 6.28fL$

sinusoidal wave. At any instant of time the two voltages (source and cemf) are equal. The cemf of the inductor reacts against the source voltage.

The reactance of an inductor can be calculated with the following formula:

$$X_L = 2\pi fL$$

The inductive reactance is in ohms (Ω), when the frequency (f) is in hertz (Hz) and the inductance (L) is in henries (H). Pi (π) is a constant equal to approximately 3.14. Using this value for pi, the formula becomes

$$X_L = 6.28fL$$

From the above formula it can be seen that inductive reactance is directly proportional to both frequency and inductance. Doubling either will double the reactance. This direct, proportional relationship makes sense when one recalls two things:

1. The higher the frequency the more rapidly the current is changing. Thus more cemf and more reactance is produced.
2. The higher the inductance the more flux change per unit of current change. Again, more cemf and reactance is produced.

Example 10-1

What is the reactance of a 3-H inductor when the frequency is 120 Hz?

Given: Inductance (L) = 3 H
frequency (f) = 120 Hz

Find: Inductive reactance (X_L)

Known: $X_L = 6.28fL$

Solution: $X_L = 6.28 \times f \times L$
$= 6.28 \times 120 \text{ Hz} \times 3 \text{ H}$
$= 2260\ \Omega$

Answer: The inductive reactance is 2260 Ω.

Example 10-2

A 2.5 mH inductor is placed in a circuit where the frequency in 100 kHz. What is its inductive reactance?

Given: $L = 2.5$ mH
$f = 100$ kHz

Find: X_L

Known: $X_L = 6.28fL$

Solution: 2.5 mH = 0.0025 H
100 kHz = 100,000 Hz
$X_L = 6.28 \times 100{,}000 \text{ Hz} \times 0.0025 \text{ H}$
$= 1570\ \Omega$

Answer: The 2.5 millihenry inductor has 1570 ohms of reactance at 100 kilohertz.

The reactance of an inductor cannot be measured with an ohmmeter. The ohmmeter measures only the dc resistance of the inductor. Reactance must be calculated. It can be calculated using the reactance formula above or by using a modified form of Ohm's law. To use Ohm's law in an inductive circuit, replace the R with X_L. That is,

$$X_L = \frac{V_L}{I_L} \quad \text{or} \quad I = \frac{V_L}{X_L} \quad \text{or} \quad V_L = I_L X_L$$

Example 10-3

The ac voltage measured across an inductor is 40 V. The current measured through the inductor is 10 mA. What is its reactance?

Given: $V_L = 40$ V
$I_L = 10$ mA

Find: X_L

Known: $X_L = \frac{V_L}{I_L}$

Solution: $X_L = \frac{40 \text{ V}}{0.01 \text{ A}} = 4000\ \Omega$

Answer: The inductor has 4000 ohms of reactance.

Phase Relationships of *I* and *V*

The sinusoidal cemf of the inductor in Fig. 10-21 is produced by a sinusoidal current through the inductor. This current wave shown in Fig. 10-22 is 90° out of phase with the cemf and the source voltage. The current must be 90° out of phase because the cemf can be zero only when the current is not changing. The only instant that the current is not changing is when it is exactly at its peak value. That is, at the instant the current has just stopped rising and has not yet begun to fall, it is effectively a constant value. This is the instant at which zero cemf occurs. It can be

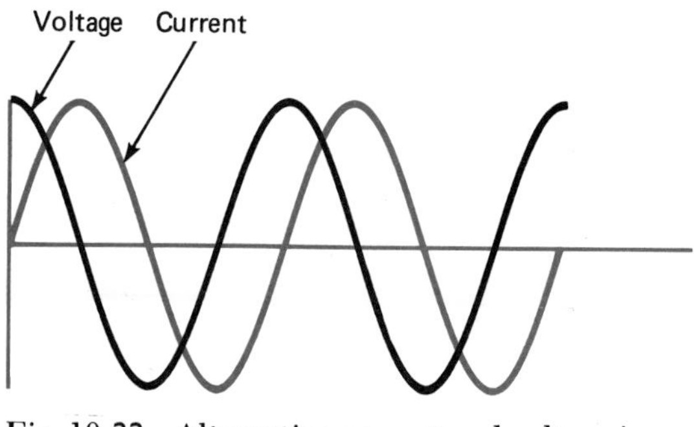

Fig. 10-22 Alternating current and voltage in an inductor. The voltage leads the current by 90°.

seen from Fig. 10-22 (and Fig. 10-20) that the *voltage leads the current* in an inductor circuit. More precisely, the voltage leads the current by exactly 90°.

Voltage leads the current

Power in an Inductor

Since the current and voltage are 90° out of phase, the inductor uses no power. No energy is converted to heat. (Remember, $P = IV \cos \theta$ and $\cos 90° = 0$.) Thus, in a pure inductance (ideal inductor) both current and voltage are present but no net energy is converted. The inductance controls the current from the source by creating a cemf which reacts against the source voltage.

From an energy point of view, one can say that energy is transferred back and forth between the source and the inductor. During the quarter cycle in which current is rising [Fig. 10-23(*a*)] energy is taken from the source (generator). The energy from the generator is converted into magnetic energy and stored in the inductor's field. During the next quarter cycle [Fig. 10-23(*b*)] when the current is decreasing, the field of the inductor collapses. Its stored energy is converted back to electrical energy and returned to the generator. As shown in Fig. 10-23(*c*), energy is again taken from the generator during the third-quarter cycle. This energy is again returned during the fourth-quarter cycle [Fig. 10-23(*d*)]. Notice in Fig. 10-23(*b*) and (*c*) that the polarity of the cemf remains the same. Yet the magnetic field changes from an expanding field to a collapsing field. This is because the polarity of the magnetic field also changed [between Fig. 10-23(*b*) and (*c*)] when the direction of the current reversed.

Review Questions

37. What is an ideal inductor?
38. Are ideal inductors common in electrical circuits?
39. What is reactance?
40. What is the symbol for inductive reactance?
41. What is the base unit for reactance?
42. What is the formula for calculating the reactance of an inductor?
43. When the frequency in an inductive circuit doubles, what happens to the reactance?

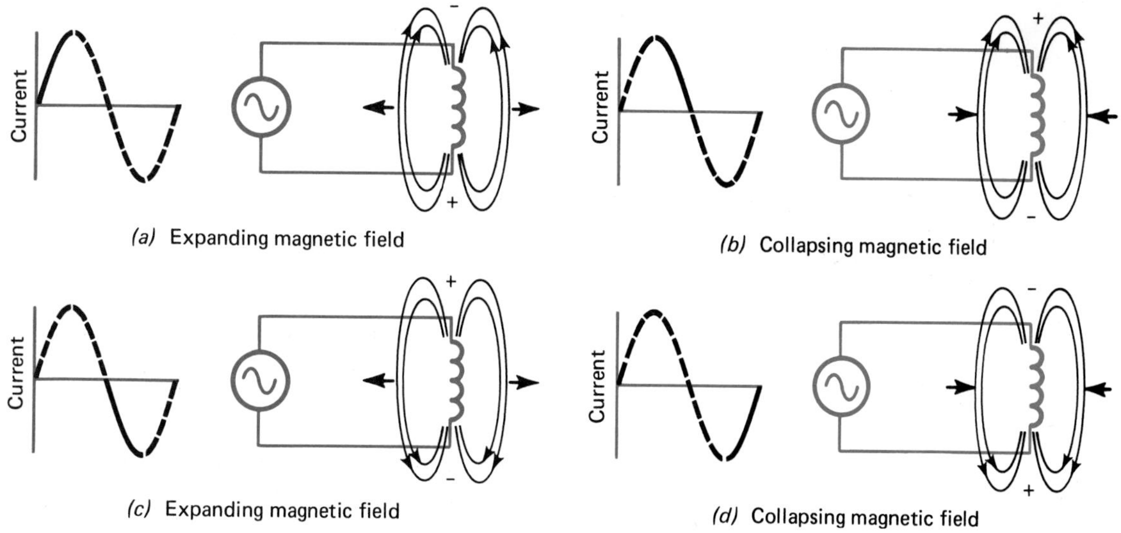

Fig. 10-23 Power in an inductor. Every other quarter cycle (*b* and *d*) the inductor returns its stored energy to the source.

44. What is the reactance of each of the following inductances at the specified frequencies?
   a. 2 H at 100 Hz
   b. 110 mH at 20 kHz
   c. 50 μH at 144 MHz
45. A reactor has 0.5 A of ac flowing through it. It drops 20 V of ac voltage. What is its reactance?
46. What is the phase relationship between inductive current and inductive voltage?
47. Why does inductance (or an ideal inductor) use no power?

$Q = \frac{X_L}{R}$

Impedance

Skin effect

Litz wire

10-8 REAL INDUCTORS IN AC CIRCUITS

All inductors possess resistance as well as reactance. Therefore, they all use some power and convert some electrical energy to heat.

Quality of Inductors

The quality of an inductor was defined in Sec. 10-5 as the ratio of resistance over reactance. Quality is often represented by the symbol Q. Mathematically, Q is defined as:

$$Q = \frac{X_L}{R}$$

Both X_L and R have units of ohms, and so Q is a pure number (it has no units). Since X_L (and also R to a lesser extent) changes with frequency, Q is frequency dependent. Its value can be given only for some specified frequency.

Example 10-4

What is the Q of a 10-mH coil (inductor) at 150 kHz if its resistance is 60 Ω?

Given: L = 10 mH
f = 150 kHz
R = 60 Ω

Find: Q

Known: $Q = \frac{X_L}{R}$ and $X_L = 6.28fL$

Solution: $X_L = 6.28 \times f \times L$
$= 6.28 \times 150{,}000 \text{ Hz} \times 0.01 \text{ H}$
$= 9420\ \Omega$

$$Q = \frac{9420\ \Omega}{60\ \Omega} = 157$$

Answer: The Q of the coil is 157.

The Q of iron-core inductors used at low frequencies is often less than 10. With air-core inductors operating at high frequencies, the Q can be over 200. Typical RF chokes have a Q ranging from 30 to 150. The higher the Q of the coil, the less power the inductor will use. Also, the higher the Q, the further the current and voltage are out of phase. Current and voltage are 90° out of phase only when there is no resistance.

The combined opposition offered by resistance and reactance is called *impedance*. Since an inductor has both resistance and reactance, it offers impedance to an ac current. To be technically correct we should specify the impedance of an inductor. However, the dominant form of opposition of an inductor is reactance. Therefore, we usually talk about its opposition in terms of reactance only. When the Q of the inductor is above 5, the difference between its reactance and its impedance is less than 2 percent. Thus, the use of reactance instead of impedance when calculating the current in an inductor is quite reasonable. It is especially reasonable, when you consider that the inductance may be 20 percent above or below its rated value.

Power Losses in Inductors

It has already been emphasized that real inductors use power because of their resistance. However, their resistance may actually be greater than the resistance measured by an ohmmeter. This higher-than-measured resistance is the result of the *skin effect*. Skin effect is caused by the tendency for electrons to travel close to the outer surface of a conductor (Fig. 10-24). The higher the frequency, the more pronounced the skin effect becomes. Because of the skin effect, the center of a conductor does not contribute to the current-carrying capacity of the conductor. Thus, the effective resistance of the conductor at high frequencies is greater than that which is measured by an ohmmeter.

The skin effect can be minimized by using *litz wire*. Litz wire is actually a multiple-conductor cable. Each conductor has a very thin insulation on it. The conductors are very small in diameter (about 44 gauge). These

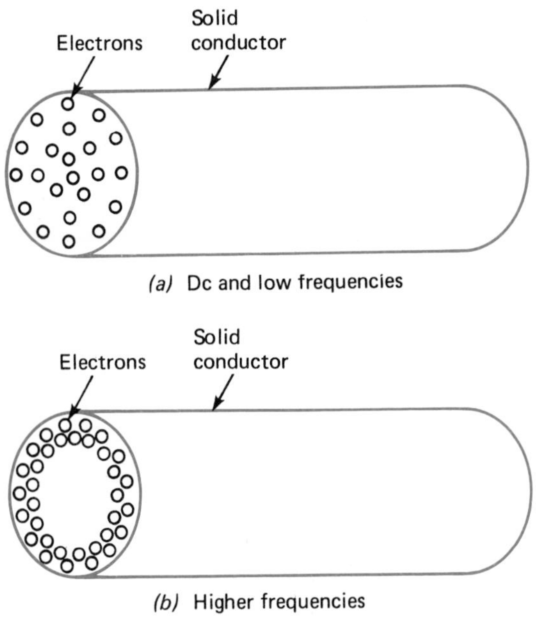

Fig. 10-24 Skin effect. At high frequencies, the current concentrates near the surface of the conductor.

small, insulated conductors are twisted together to form a very small cable. When using litz wire, the individual conductors are soldered together at the terminating ends of the coil. This ties all the conductors in parallel to effectively make a single wire out of the multiple-conductor cable. For a given overall diameter, litz wire provides more surface area than a single-strand conductor. Because of this greater surface area, litz wire has a lower resistance at high frequencies.

Iron-core inductors have power losses in their core material as well as in their winding. Two actions cause the core to convert electrical energy into heat energy. One action is that of the magnetic field of the winding inducing a voltage in the core material. The induced voltage causes small currents to flow in the core. This current produces heat in the core. The second action which causes power loss in the core is the periodic reversal of the magnetic field. Every time the polarity of the magnetic field reverses, it creates a small amount of heat in the core. Methods of minimizing these core losses are discussed in the chapter on transformers.

Review Questions

48. What is the symbol for the quality of an inductor?
49. What is the base unit for quality?
50. What is the quality of a 0.2-H inductor at 20 kHz if it has an effective resistance of 200 Ω?
51. What is impedance?
52. What causes the resistance of a conductor to be greater at high frequencies than at low frequencies?
53. What is litz wire?
54. What causes the core of a laminated-iron-core inductor to heat up?

10-9 INDUCTORS IN PARALLEL

Parallel inductors

Parallel inductors (with no mutual inductance) can be treated just like parallel resistors. The formulas used with resistors can be used with inductors by substituting L for R. The formulas are:

General method:

$$L_T = \frac{1}{\frac{1}{L_1} + \frac{1}{L_2} + \frac{1}{L_3} + \text{etc.}}$$

Two inductors in parallel:

$$L_T = \frac{L_1 \times L_2}{L_1 + L_2}$$

n equal inductors in parallel:

$$L_T = \frac{L}{n}$$

Example 10-5

What is the inductance of a 0.4-H inductor and a 600-mH inductor connected in parallel?

Given: $L_1 = 0.4\ \text{H}$
$L_2 = 600\ \text{mH} = 0.6\ \text{H}$

Find: L_T

Known: $L_T = \frac{L_1 \times L_2}{L_1 + L_2}$

Solution: $L_T = \frac{0.4\ \text{H} \times 0.6\ \text{H}}{0.4\ \text{H} + 0.6\ \text{H}} = \frac{0.24}{1.0} = 0.24\ \text{H}$

Answer: The total (or equivalent) inductance is 0.24 henry or 240 millihenries.

Total inductive reactance

Parallel reactances

Notice that the total inductance is less than the smallest of the parallel inductances. It always is with parallel inductors.

The total inductive reactance of parallel inductors can be found by either of two methods. The first method is to find the total inductance as in Example 10-5. Then find the total reactance by using the reactance formula. The second method is to determine the reactance of the individual inductors. Then, using the parallel formula, combine the individual reactances to find the total reactance. The formulas for combining parallel inductive reactances are similar to those used for parallel resistances and parallel inductances.

General method:

$$X_L = \frac{1}{\frac{1}{X_{L_1}} + \frac{1}{X_{L_2}} + \frac{1}{X_{L_3}}} + \text{etc.}$$

Two parallel inductive reactances:

$$X_L = \frac{X_{L_1} \times X_{L_2}}{X_{L_1} + X_{L_2}}$$

n equal inductive reactances in parallel:

$$X_L = \frac{X_L}{n}$$

Example 10-6

Using the first method above, find the total inductive reactance for Example 10-5 when the frequency is 20 kHz.

Given: $L_T = 0.24$ H
$f = 20$ kHz
Find: X_{LT}
Known: $X_{LT} = 6.28fL_T$
Solution: $X_{LT} = 6.28 \times f \times L_T$
$= 6.28 \times 20{,}000 \text{ Hz} \times 0.24 \text{ H}$
$= 30{,}144\ \Omega$
Answer: The total inductive reactance is 30,144 ohms or 30.144 kΩ.

Example 10-7

Using the second method above, find the total inductive reactance for Example 10-5 when the frequency is 20 kHz.

Given: $L_1 = 0.4$ H
$L_2 = 0.6$ H
$f = 20$ kHz
Find: X_{LT}
Known: $X_L = 6.28fL$

$$X_L = \frac{X_{L_1} \times X_{L_2}}{X_{L_1} + X_{L_2}}$$

Solution: $X_{L_1} = 6.28 \times 20{,}000 \times 0.4$
$= 50{,}240\ \Omega$

$X_{L_2} = 6.28 \times 20{,}000 \times 0.6$
$= 75{,}360\ \Omega$

$$X_{LT} = \frac{50{,}240 \times 75{,}360}{50{,}240 + 75{,}360}$$

$= 30{,}144\ \Omega$

The total inductive reactance is 30,144 ohms.

As you might expect, the two methods used in Examples 10-6 and 10-7 give exactly the same answer.

In parallel-inductor circuits the total current splits up in inverse proportion to the inductance of the individual inductors. The lowest inductance carries the highest current (Fig. 10-25). The exact value of the current in each branch of the circuit can be found by using Ohm's law. Just replace the R in Ohm's law with X_L. The currents recorded in Fig. 10-25 were calculated by using the reactances found in Example 10-7.

$$I_{L_1} = \frac{V_{L_1}}{X_{L_1}} = \frac{40 \text{ V}}{50{,}240\ \Omega} = 0.0008 \text{ A} = 0.8 \text{ mA}$$

$$I_{L_2} = \frac{V_{L_2}}{X_{L_2}} = \frac{40 \text{ V}}{75{,}360\ \Omega} = 0.00053 \text{ A} = 0.53 \text{ mA}$$

$$I_T = \frac{V_T}{X_{LT}} = \frac{40 \text{ V}}{30{,}144\ \Omega} = 0.00133 \text{ A} = 1.33 \text{ mA}$$

Of course, I_T can also be found by using Kirchhoff's current law. For the circuit of Fig. 10-25 we have:

$$I_T = I_{L_1} + I_{L_2} = 0.8 \text{ mA} + 0.53 \text{ mA} = 1.33 \text{ mA}$$

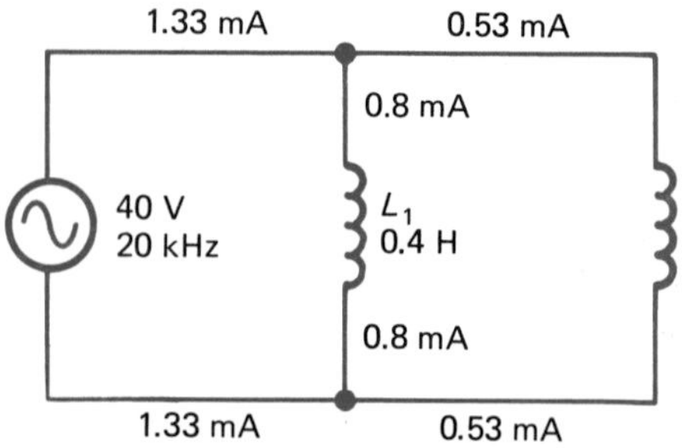

Fig. 10-25 Currents in parallel inductors. The smaller inductor carries the greater current.

10-10 INDUCTORS IN SERIES

Treat series inductances and series reactances the same as you would treat series resistances. The appropriate formula for series inductances is:

$$L_T = L_1 + L_2 + L_3 + \text{etc.}$$

For inductive reactance in series the formula is:

$$X_{L_T} = X_{L_1} + X_{L_2} + X_{L_3} + \text{etc.}$$

Again the total reactance can also be determined by the reactance formula if the total inductance is known. That is:

$$X_{L_T} = 6.28 f L_T$$

Example 10-8

Using the reactance formula, find the total reactance at 60 Hz of a 3-H choke and a 5-H choke connected in series.

Given: $L_1 = 3\text{ H}$
$L_2 = 5\text{ H}$
$f = 60\text{ Hz}$

Find: X_{LT}

Known: $X_L = 6.28 f L_T$
$L_T = L_1 + L_2$

Solution: $L_T = 3\text{ H} + 5\text{ H} = 8\text{ H}$

$X_{LT} = 6.28 \times 60\text{ Hz} \times 8\text{ H}$
$= 3014\ \Omega$

Answer: The total inductive reactance is 3014 ohms.

Example 10-9

Find the total reactance for Example 10-8 without first finding the total inductance.

Given: $L_1 = 3\text{ H}$
$L_2 = 5\text{ H}$
$f = 60\text{ Hz}$

Find: X_{LT}

Known: $X_{LT} = X_{L_1} + X_{L_2}$
$X_{L_1} = 6.28 f L_1$
$X_{L_2} = 6.28 f L_2$

Solution: $X_{L_1} = 6.28 \times 60\text{ Hz} \times 3\text{ H}$
$= 1130\ \Omega$

$X_{L_2} = 6.28 \times 60\text{ Hz} \times 5\text{ H}$
$= 1884\ \Omega$

$X_{LT} = 1130\ \Omega + 1884\ \Omega$
$= 3014\ \Omega$

Answer: The total reactance is 3014 ohms.

As shown in Fig. 10-26, the total voltage in a series inductor circuit splits in direct proportion to the individual inductances.

$$V_{L_1} = \frac{L_1}{L_T} \times V_T = \frac{3\text{ H}}{8\text{ H}} \times 20\text{ V} = 7.5\text{ V}$$

$$V_{L_2} = \frac{5\text{ H}}{8\text{ H}} \times 20\text{ V} = 12.5\text{ V}$$

The exact voltage values can also be found by using Ohm's law and the reactances of Example 10-9. Again, Kirchhoff's laws apply. The total voltage equals the sum of the individual circuit voltages.

10-11 PREVENTING MUTUAL INDUCTANCE

Mutual inductance can be reduced or prevented by the following methods:

1. Axis orientation
2. Physical separation
3. Shielding

Suppose the center axis of two coils are at 90° to each other as shown in Fig. 10-27(*a*). Under this condition, very little of the flux from one coil cuts the other coil. We say that very little *coupling* occurs between the coils (inductors). When the axis of the coils are lined up and physically close together as in Fig. 10-27(*b*), mutual inductance results.

Series inductances

Series reactances

Coupling

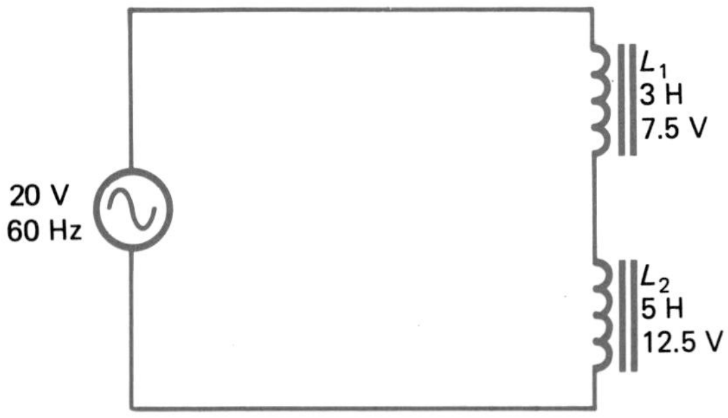

Fig. 10-26 Voltages in series inductors. The larger inductor drops the greater voltage.

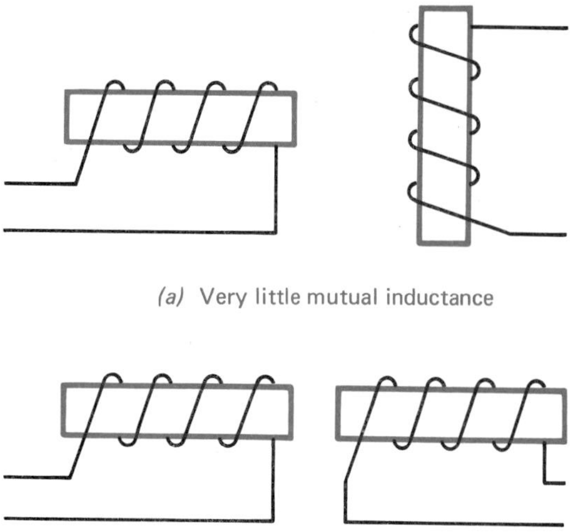

(a) Very little mutual inductance

(b) High mutual inductance

Fig. 10-27 Axis orientation and mutual inductance.

When inductors are physically separated, mutual inductance is reduced. The further apart the inductors are the less mutual inductance (coupling of flux) they will have.

An inductor which is enclosed in a magnetic shield will have very little mutual inductance with surrounding inductors. The flux from surrounding inductors will pass through the low reluctance (high permeability) of the shield rather than the inductor.

10-12 UNDESIRED INDUCTANCE

As mentioned earlier, all conductors possess inductance. The inductance of a single wire, although low, is significant at very high frequencies. Often this inductance is undesirable because of its effects on the overall electrical or electronic circuit. In high frequency circuits interconnecting leads are kept as short as possible to reduce the resulting inductance. Whenever possible the inductance of the leads is used as part of the required inductance of the circuit.

Noninductive resistors

Wire-wound and deposited-film resistors may have an appreciable amount of inductance. Their resistive elements are coils of conductive material wound on a nonmagnetic insulator form. In dc and low-frequency ac circuits this undesired inductance has very little reactance. Thus, it can be ignored. However, at higher frequencies the reactance becomes greater and the total opposition (impedance) of the resistor significantly exceeds its resistance. In many electronic circuits this is undesirable and unacceptable.

To minimize the above problem, special noninductive, wire-wound resistors are produced. In these resistors, half of the turns of wire are wound clockwise and half are wound counterclockwise. Thus the magnetic field of half of the turns cancels the field of the other half of the turns. These resistors are often used in high-power circuits when the load on the circuit must be independent of the frequency.

Review Questions

55. A 0.6-H and a 0.3-H inductor are connected in parallel. They are connected to a 10 V, 200 Hz source.
    a. What is their total or equivalent inductance?
    b. What is the total reactance?
    c. What is the total current?
    d. What is the current through the 0.6 H inductor?
56. Suppose the inductors in question 55 are now connected in series rather than parallel.
    a. What is the total inductance?
    b. What is the total reactance?
    c. What is the total current?
    d. What is the voltage across the 0.6-H inductor?
57. How can you minimize the mutual inductance between two coils?
58. How is the wire-wound, noninductive resistor constructed?

Summary

1. Inductance opposes changes in current.
2. Inductance results from induced voltage.
3. Inductors are physical devices which provide inductance. Chokes, coils, and reactors are other names for inductors.
4. The symbol for inductance is *L*.
5. The induced voltage in an inductor is known as counter electromotive force (cemf) or back electromotive force (bemf).
6. Lenz's law is concerned with the polarity of an induced voltage (cemf).
7. A cemf opposes the change that created it.

8. Inductors convert energy back and forth between the magnetic form and the electrical form.
9. The henry is the base unit of inductance. The abbreviation for henry is H.
10. Inductance is determined by (1) core material, (2) number of turns, (3) spacing of turns, and (4) diameter of turns.
11. Inductors are rated for inductance, dc resistance, current, voltage, quality, and tolerance.
12. Dc resistance of an inductor is also called ohmic resistance.
13. The quality of an inductor is the ratio of its reactance to its resistance. The symbol for quality is Q.

$$Q = \frac{X_L}{R}$$

Quality is a pure number. It does not have a base unit. For actual inductors it ranges from less than 10 to over 200.
14. Current in a dc inductive circuit rises more slowly than in a dc resistive circuit.
15. Voltage (cemf) reaches its peak value before the current does in a dc inductive circuit.
16. Inductive kick causes arcing in switch contacts when an inductive circuit is opened.
17. Inductive reactance is the opposition of an inductor to ac.
18. The symbol for reactance is X.
19. The base unit of reactance is the ohm.
20. The symbol for inductive reactance is X_L.

$$X_L = 6.28fL$$

21. Inductive reactance is directly proportional to both frequency and inductance.
22. Ohm's law can be used in inductive circuits by replacing R with X_L.
23. In an inductive circuit, current lags voltage by 90°.
24. Inductance uses no power or energy.
25. Real inductors have resistance; therefore, they do use some power.
26. Impedance is the combined opposition of reactance and resistance.
27. Skin effect increases the effective resistance of a conductor at high frequencies.
28. Litz wire is multistrand wire designed to reduce the skin effect.
29. Core losses are caused by induced currents in the core and by periodic reversal of the magnetic field.
30. Inductors (and inductive reactances) in parallel behave like resistors in parallel. The same formulas are used except that R is replaced by L or X_L.
31. Inductors (and inductive reactances) in series behave like resistances in series.
32. The lowest series inductance will drop the least voltage.
33. Mutual inductance can be reduced by axis orientation, physical separation, and shielding.
34. Undesired inductance occurs in conductors and resistors.

Chapter Review Questions

For questions 10-1 to 10-15, determine whether each statement is true or false.

10-1. A straight length of conductor has no inductance.

10-2. The induced voltage in an inductor is caused by mutual inductance between the turns of the inductor.

10-3. When the current in an inductor is increasing, its bemf opposes the source voltage.

10-4. A 2-H inductor would most likely have a powdered-iron core.

10-5. Magnetic shields for inductors are usually made from plastic.

10-6. The dc resistance of an inductor can be measured with an ohmmeter.

10-7. The core material in a variable inductor is often called a slug.

10-8. Maximum inductance of a variable inductor occurs when the ferrite slug is centered within the coil core.

10-9. When an inductive circuit is opened, the cemf exceeds the source voltage.

10-10. Inductive reactance is inversely proportional to frequency.

10-11. Current leads voltage by 90° in a circuit containing only inductance.

10-12. The iron core of an inductor converts some electrical energy into heat energy.

10-13. In a parallel inductor circuit, a low value of inductance carries more current than a high value.

10-14. The lowest inductance of series-connected inductors will produce the highest voltage drop.

10-15. The quality of a coil is independent of frequency.

For questions 10-16 to 10-20, choose the letter that best completes each statement.

10-16. The symbol for inductance is
(A) *H* (B) *I* (C) L (D) X

10-17. When the current in an inductance is increasing
(A) The cemf aids the source voltage (B) The bemf aids the source voltage (C) Electrical energy is being converted to heat energy (D) Electrical energy is being converted to magnetic energy

10-18. The base unit of inductance is the
(A) Coil (B) Ampere-turn (C) Weber (D) Henry

10-19. The symbol for reactance is
(A) *H* (B) *I* (C) L (D) X

10-20. The base unit of reactance is the
(A) Ohm (B) Henry (C) Ampere (D) Coil

Answer the following questions.

10-21. List three other names for inductors.

10-22. List four factors that determine the inductance of an inductor. Give an example of how to decrease inductance with each of these factors.

10-23. List all of the electrical ratings used to completely describe an inductor.

10-24. Does the center leg of a laminated core carry more, less, or the same amount of flux as does an outside leg?

10-25. A 500-mH inductor is connected to a 30-V, 10-kHz source. What is the reactance of the inductor? How much current flows in the circuit?

10-26. How much power is used in the above circuit if the inductor is an ideal inductor?

10-27. What is the quality of 100-mH inductor at 20 kHz if its resistance is 126 Ω?

10-28. What is the total inductance of a 2.5-H and a 5-H inductor connected in series? What is the total reactance if the frequency is 100 H?

10-29. If the inductors in the above question were connected in parallel, what would be the value of the total inductance?

10-30. List three ways to minimize or eliminate mutual inductance.

Answers to Review Questions

1. Inductance is the opposition to changes in current. It is caused by cemf.
2. An inductor is an electrical component which provides inductance.
3. Reactor, choke, coil
4. Yes
5. No, because the wire was not constructed specifically to provide inductance in a circuit.
6. Aid
7. Back electromotive force
8. Mutual inductance involves flux interaction between two electrically isolated inductors. Self-inductance refers to inductance of a single inductor.
9. Lenz's law
10. Electrical energy is converted to magnetic energy.
11. L
12. Henry
13. H
14. One henry is one volt of cemf for a current change of one ampere per second.
15. Type of core, number of turns, spacing between turns, and diameter of turns
16. Use a core with higher permeability or less resistance, increase the number of turns, put the turns closer together, increase the diameter of the turns.
17. *a*. See Fig. 10-6(*a*).
    b. See Fig. 10-6(*b*).
    c. See Fig. 10-7.
18. Yes
19. Iron-core inductor symbol
20. Slug
21. Centered under the coils of wire
22. Centered under the coils of wire
23. Powdered-iron or ferrite
24. Laminated-iron core
25. Toroidal core
26. Laminated-iron
27. Because it carries twice as much flux
28. Because too much current will cause the winding to overheat
29. The ratio of reactance to resistance
30. Yes
31. An inductor or coil designed to be used at radio frequencies
32. Laminated iron
33. The high value of cemf that occurs when an inductive circuit is opened
34. The inductive circuit
35. The cemf
36. The cemf
37. An inductor with no resistance
38. No, they do not exist.
39. A form of opposition to current that uses no energy
40. X_L
41. Ω
42. $X_L = 6.28fL$
43. It also doubles.
44. *a.* 1257 Ω
    b. 13,823 Ω
    c. 45,239 Ω
45. 40 Ω
46. The current lags the voltage by 90°, or the voltage leads the current by 90°.
47. Because I and V are 90° out of phase and thus the cos θ is 0
48. Q
49. It has none. It is a pure number.
50. Q = 126
51. The combined opposition of resistance and reactance
52. Skin effect
53. Multistrand wire used to reduce skin effect
54. Induced currents and magnetic polarity reversal
55. *a.* 0.2 H
    b. 251.3 Ω
    c. 39.8 mA
    d. 13.3 mA
56. *a.* 0.9 H
    b. 1131 Ω
    c. 8.8 mA
    d. 6.7 V
57. Orient the axis 90° to each other, separate the coils, and shield them.
58. Half of the turns are wound in the opposite direction to the other half.

11 Transformers

- **Transformers are multiple-winding inductors. They operate on the principle of mutual inductance. For a relatively simple device, they are extremely versatile. Without transformers, our present power distribution system could not exist.**

  When finished with this chapter, you will be able to use a transformer for impedance matching and isolation. You will be able to calculate transformer efficiency and transformer losses. You will be able to use three-phase transformers in delta and wye connections. You will know how to connect windings to obtain different voltages. Finally, you will understand transformer terminology and ratings.

11-1 TRANSFORMER FUNDAMENTALS

A transformer consists of two or more coils that are linked together by magnetic flux (see Fig. 11-1). The changing flux from one coil (the primary) induces a voltage in the other coil (the secondary). In other words, the coils are coupled, or linked, together by mutual inductance.

Without mutual inductance, there would be no such thing as a transformer. The amount of mutual inductance, like the amount of self-inductance, is specified in henries. One henry exists when one volt is induced in a coil by a current change of one ampere per second in another coil. Suppose the primary current in Fig. 11-1 changes at a rate of one ampere per second. Further, suppose the secondary voltage in Fig. 11-1 is three volts. Then, the transformer would have three henries of mutual inductance.

Symbols

The symbol for a transformer is basically two coils with their axis parallel to each other. As illustrated in Fig. 11-2, the basic symbol can be modified in many ways. These modifications are needed to more fully describe the various types of transformers.

The dashed lines in Fig. 11-2(c) represent the metal enclosure which houses the windings (coils). When this enclosure is made of aluminum or copper or other nonmagnetic materials, it is intended as a shield against electric fields rather than magnetic fields.

The symbols shown in Fig. 11-2 are not the only ones you may see. For example, iron-core transformers sometimes have shields. In this case, the symbol of Fig. 11-2(c) would also have lines which indicate an iron core. Not all tapped secondaries are center-tapped. An

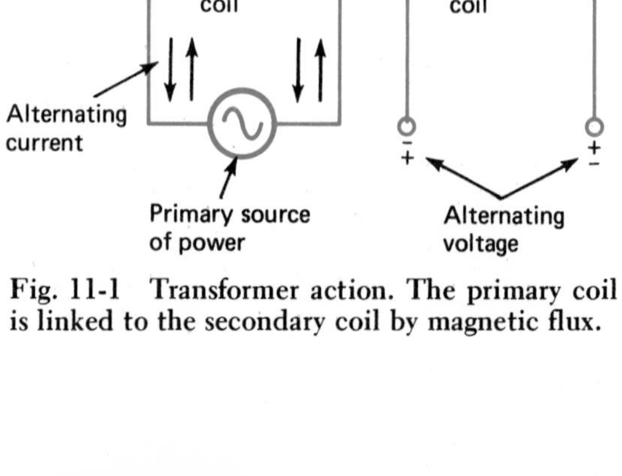

Fig. 11-1 Transformer action. The primary coil is linked to the secondary coil by magnetic flux.

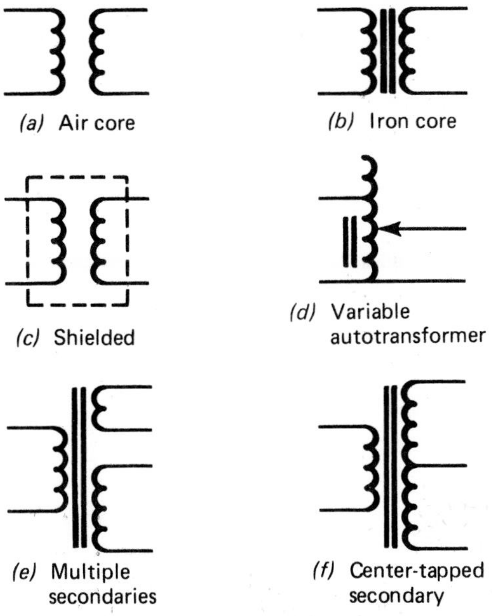

Fig. 11-2 Transformer symbols.

off-center-tapped winding would be represented by having more turns indicated on one side of the tap than on the other.

Primary and Secondary Windings

A transformer is a device which transfers electrical energy (or power) from a primary winding to a secondary winding. There is no electrical connection between the primary winding and the secondary winding. The primary converts the electrical energy into magnetic energy. The secondary converts the magnetic energy back into electrical energy. The primary and secondary are said to be electrically isolated from each other, but magnetically connected (coupled) together.

A transformer is designed to receive power from some primary power source such as the 23-kV output of a power company generator, or the 120-V outlet in your home. The transformer winding designed to receive power from the primary source is called the primary winding. Electrical power is taken from the secondary winding by a load (see Fig. 11-3). Therefore, the secondary winding becomes the power source for the load. Furthermore, the secondary voltage need not be the same as the primary voltage. In fact, in most cases the two voltages differ greatly. This is an important property of transformers. For example, the primary of the transformer in Fig. 11-3 is connected to a 120-V supply and the secondary delivers 20 V to the load.

Transformers are two-way devices. The winding that the manufacturer calls the secondary can be used as a primary. Of course, a primary power source of the correct voltage must be used. The winding designated as the primary will now serve as a secondary. It will, of course, provide power at the voltage for which it was originally designed. For example, the transformer in Fig. 11-3 can be used to provide a 120-V secondary source from a 20-V primary source as shown in Fig. 11-4. As you can see, labeling one coil "primary" and another "secondary" is somewhat arbitrary. It is based on the intended use of the transformer; that is, the intended primary power source.

Coefficient of Coupling

The portion of the flux from one coil that links the other coil is referred to as the *coefficient of coupling*. The coefficient of coupling can range from 0 to 1. When all the flux is coupled, the coefficient of coupling is 1. Sometimes the coefficient of coupling is expressed as a percentage. Thus, 100 percent coupling means a coefficient of coupling of 1.

Transformers with laminated-iron cores have percentages of coupling that are very close to 100 percent. This is because all of the flux is concentrated in the high-permeability core on which the coils are wound. The iron core provides a complete, low-reluctance path for the flux loops. Therefore, essentially none of the flux leaks into the surrounding air.

On the other hand, air-core transformers can have very low coefficients of coupling. The paths through the surrounding air

From page 202:
Primary

Secondary

Mutual inductance

On this page:
Coefficient of coupling

Transformer symbols

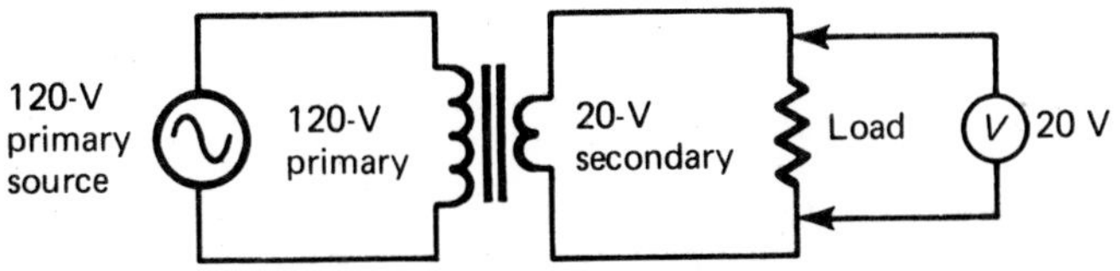

Fig. 11-3 Primary-secondary terminology. The primary winding receives power from a power source.

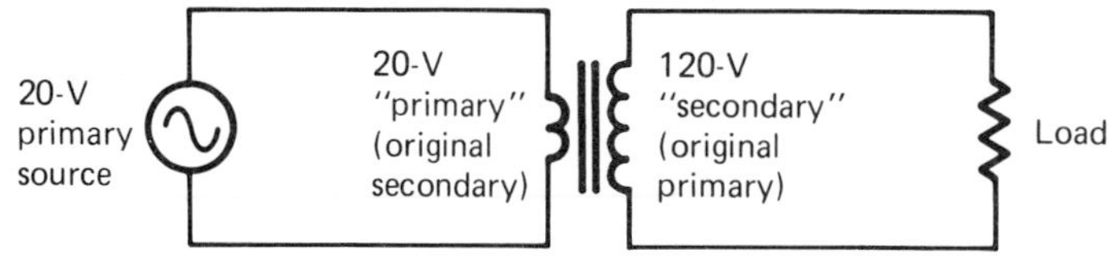

Fig. 11-4 **Reversing the primary-secondary functions.**

Step-up

Step-down

Turns ratio

Turns-per-volt ratio

around the inductors offer no more reluctance than the path through the cores. Therefore, much of the flux from one coil never links up with the other coil. It just "leaks" off into surrounding air paths. The amount of coupling in an air-core transformer can be controlled by the spacing between the coils. The further apart the coils are, the more flux leakage occurs, and the less the percentage of coupling. The percentage of coupling can also be controlled by the axis orientation of the coils. When the axis of coils are perpendicular to each other, the coefficient of coupling is close to zero.

Review Questions

1. What is a transformer?
2. What are the two coils of a transformer called?
3. What is the base unit of mutual inductance?
4. What is mutual inductance?
5. Draw the symbol for a shielded magnetic-core transformer.
6. What energy conversion processes are involved in a transformer?
7. Which winding of a transformer receives power or energy from another source?
8. Which winding of a transformer provides power or energy to the load?
9. Can a power source be connected to the winding that the manufacturer of a transformer calls the secondary?
10. What is a coefficient of coupling?
11. What is the maximum value of a coefficient of coupling?
12. Do laminated-iron-core transformers or air-core transformers have the greater coefficient of coupling?
13. How can the coefficient of coupling in air-core transformers be varied?

Changing Voltage Values

A transformer can either *step up* or *step down* a voltage. If the primary voltage is greater than the secondary voltage, the transformer is stepping the voltage down. Thus, the transformer in Fig. 11-3 would be called a step-down transformer. If the voltage in the secondary exceeds the voltage in the primary (Fig. 11-4) the transformer is a step-up transformer. Some transformers with multiple secondaries will have one or more step-up secondaries and one or more step-down secondaries.

Whether a secondary will be a step-up or step-down winding is determined by the primary-to-secondary *turns ratio*. When the primary turns exceed the secondary turns, *and the coupling is 100 percent*, the transformer will step-down the voltage. In fact, with 100 percent coupling, the *turns ratio* and the *voltage ratio* will be equal. Mathematically we can write

$$\frac{V_{pri}}{V_{sec}} = \frac{N_{pri}}{N_{sec}}$$

In this formula, N is the abbreviation for the number of turns. This formula can be rearranged to show that

$$\frac{N_{pri}}{V_{pri}} = \frac{N_{sec}}{V_{sec}}$$

In this new arrangement, the relationship between voltage and turns is very informative. It shows that the *turns-per-volt* ratio is the same for both the primary and the secondary. Furthermore, the turns-per-volt ratio of all secondary windings in a multiple-secondary transformer are equal. Thus, once you determine the turns-per-volt ratio of any winding you know the ratio of all other windings.

Referring to Fig. 11-5 will aid in understanding why all windings have the same turns-per-volt ratio. Remember that the cemf of a coil is always equal to the ac voltage applied to the coil. Therefore, the cemf in the primary in Fig. 11-5 must be 1 V. This 1 V of cemf is created by the flux in the circuit. The flux, therefore, must induce 0.25 V in each turn of the primary. Notice that all the flux that produced cemf in the primary turns also goes through every secondary. Therefore, each turn in the secondary windings will have 0.25 V induced in it. If it requires 4 turns in the primary for each volt, then 4 turns in a secondary will also develop 1 V.

The turns-per-volt ratio used for the windings of a laminated-iron-core transformer varies with the size of the transformer. Small transformers (less than 10-W rating) may have 7 or 8 turns-per-volt. Larger transformers (over 500-W) may have less than a turn-per-volt.

The concept of turns-per-volt is useful in modifying the secondary voltage of a transformer.

Example 11-1

A transformer is to be rewound to provide a voltage of 14 V. The transformer's present secondary is rated at 6.3 V and contains 20 turns. How many turns will be required for the new 14-V winding?

Given: Present secondary winding delivers 6.3 V with 20 turns.

Find: The number of turns necessary to provide 14 V

Known: $\dfrac{N_{pri}}{V_{pri}} = \dfrac{N_{sec}}{V_{sec}}$

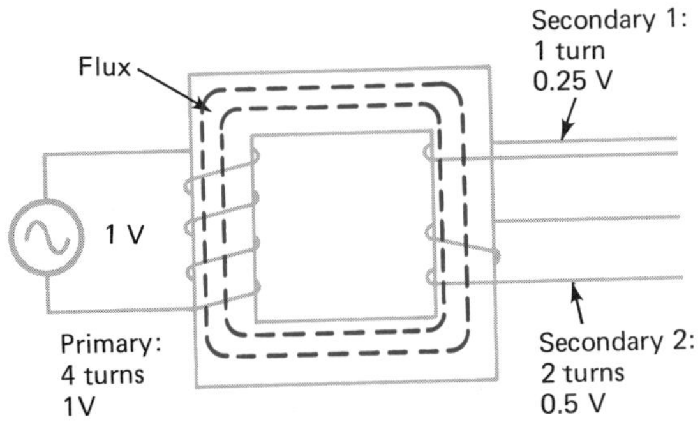

Fig. 11-5 Turns-per-volt. The turns-per-volt ratio is the same in all windings.

Solution: $\dfrac{N_{pri}}{V_{pri}} = \dfrac{20}{6.3} = 3.175$ turns per volt

The transformer has a 3.175 turns-per-volt ratio. Since a 14-V secondary is required

N(new secondary)
= 3.175 turns per volt × 14 V
= 44.5 turns or 45 turns

Answer: A 14-V winding would need 45 turns.

Example 11-2

The designer of a transformer has calculated that the 120-V primary will have 2.6 turns-per-volt. How many turns will be required for a 230-V secondary?

Given: Turns per volt = 2.6
$V_{pri} = 120$ V
$V_{sec} = 230$ V

Find: N_{sec}

Known: $\dfrac{N_{pri}}{V_{pri}} = \dfrac{N_{sec}}{V_{sec}}$

Solution: $\dfrac{N_{pri}}{V_{pri}} = 2.6$ turns per volt

$N_{sec} = 230$ V × 2.6 turns per volt
= 598 turns

Answer: The 230-V secondary will need 598 turns.

Notice in Example 11-2 that we did not need to calculate the number of turns in the primary. Also, we did not need to know the voltage of the primary. All we needed to know was the turns-per-volt ratio.

One of the chief uses of transformers is to change voltages from one value to another. Perhaps an example will show why it is so necessary to be able to transform voltage levels. It is usually necessary to transmit electrical power from the power plant where it is produced to the location where it is used. Often the distance between these two points is many hundreds of miles. Obviously it is desirable to use as small a conductor as possible in the power lines between the two points. The size of the conductor needed is directly dependent on the amount of current it must carry. So we need to keep the current as low as possible. Assuming a 100% PF, power equals current times voltage ($P = IV$). Thus, the lower the current the higher the voltage must be for a given amount of power. However, the generators at the power plant and the loads at the point of use have a limited operating voltage. The solution to this problem is illustrated in Fig. 11-6. If the 10 MW from the generator were directly connected to the loads, the transmission lines would have to carry 500 A. With a step-up transformer at the power plant and a step-down transformer at a substation near the load, the power lines carry only 25 A. (This illustration assumes that the power factor is 1, and that the transformer uses no power).

Efficiency

Figure 11-6 also illustrates another transformer relationship. When the voltage of a transformer is stepped-up, the current will be stepped-down and vice versa. The exact current ratio depends upon the power factor, the power consumed by the transformer, and the voltage ratio. For the ideal transformer (one with no losses) the primary power is approximately equal to the secondary power when the transformer has a full resistance load. Stating this as a formula we have

$$P_{pri} \approx P_{sec}$$

This leads to the formula for the current relationship:

$$\frac{V_{pri}}{V_{sec}} \approx \frac{I_{sec}}{I_{pri}}$$

This points out that the current varies inversely to the voltage. In other words, a step-up *voltage* transformer is, in effect, a step-down current transformer.

Review Questions

14. What is a step-up transformer?
15. What relationship exists between the turns ratio and the voltage ratio of a transformer with 100 percent coupling?
16. What relationship exists between the turns-per-volt ratio in the primary and secondary windings?
17. Does a large or small transformer have the higher turns-per-volt rating?
18. It is known that a transformer has three turns-per-volt on the primary. How many turns will a 190-V secondary have?
19. What is one of the most common uses of a transformer?
20. Why do power systems transmit power at as high a voltage as possible?
21. When a transformer steps-up voltage, what does it do to current?

11-2 EFFICIENCY OF TRANSFORMERS

The iron core and the copper coils of a transformer both convert some electrical energy into heat energy. This, of course, is why a transformer heats up when in operation. The purpose of a transformer is not to provide heat, but to transfer energy from the primary to the secondary. Therefore, any heat produced by the transformer represents inefficiency.

Since energy is equal to power times time, the efficiency of transformers is calculated in terms of power. The efficiency of a transformer (expressed in percentage) is calculated by the following formula:

$$\%\ \text{efficiency} = \frac{P_{sec}}{P_{pri}} \times 100$$

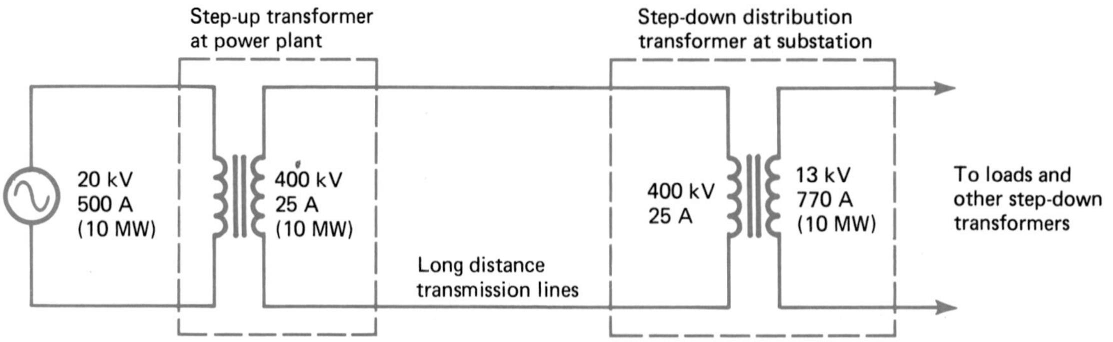

Fig. 11-6 Advantage of high-voltage power lines. The higher the voltage, the lower the current for a given amount of power.

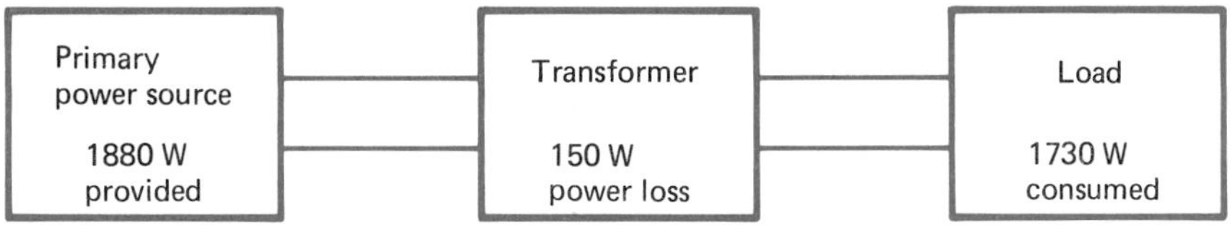

Fig. 11-7 Transformer efficiency. Power loss occurs because the transformer converts some electrical energy into heat energy.

Power loss

Hysteresis loss

Hysteresis loop

Eddy current

Example 11-3

What is the efficiency of a transformer that requires 1880 W of primary power to provide 1730 W of secondary power?

Given: $P_{pri} = 1880$ W
$P_{sec} = 1730$ W

Find: Efficiency

Known: $\% \text{ eff.} = \frac{P_{sec}}{P_{pri}} \times 100$

Solution: $\% \text{ efficiency} = \frac{1730 \text{ W}}{1880 \text{ W}} \times 100 = 92$

Answer: The transformer is 92% efficient.

In the above example, the 150-W difference between received power and delivered power was lost in the transformer. As indicated in Fig. 11-7, the power consumed by the transformer is referred to as a *power loss*. The power loss in a transformer is caused by:

1. Hysteresis loss
2. Eddy current loss
3. Copper (I^2R) loss

The first two of these losses occur in the transformer core material. The latter occurs in the windings. All three convert electrical energy to heat energy.

Hysteresis Loss

Hysteresis loss is caused by *residual magnetism*. That is, by the magnetism that remains in a material after the magnetizing force has been removed. The core of a transformer has to reverse its magnetic polarity every time the primary current reverses direction. Every time the magnetic polarity is reversed, the residual magnetism of the previous polarity has to be overcome. This produces heat. It requires energy from the primary to produce this heat. Hysteresis loss, then, refers to the energy required to reduce the residual magnetism to zero. This loss occurs once every half cycle just before the core is remagnetized in the opposite direction.

The hysteresis loop in Fig. 11-8 graphically illustrates hysteresis loss. The narrower the hysteresis loop, the smaller the hysteresis loss will be. Therefore, the core material for transformers, and other magnetic devices that operate on ac, should have a narrow hysteresis loop. Laminated-iron cores are not made from pure iron because iron has a high hysteresis loss. These cores are made from silicon steel. Silicon steel is an alloy which has a narrow hysteresis loop and still has high permeability.

Hysteresis loss will increase with an increase in the frequency of the primary current. This is one of the reasons that laminated-iron-core transformers are not used above the audio frequency range.

Eddy Current Loss

The current that flows in the core of a magnetic device is called *eddy current*. Eddy current results from the voltage that is induced into the core by the primary winding.

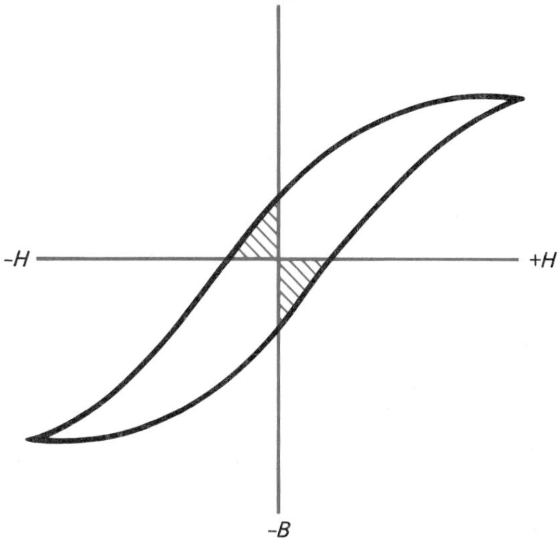

Fig. 11-8 Hysteresis loop. Magnetic energy is converted to heat energy in the shaded regions of the loop.

How voltage, and thus eddy current, is induced in the core is illustrated in Fig. 11-9. We know that the secondary coil receives induced voltage from the flux created by the primary coil. Notice that the secondary coil is wound around the core. Its axis is parallel to the flux. Now think of the core as a conductor (which it is). The varying flux in the primary cuts the conductors in the secondary, but it also cuts the core. The flux created by the primary current will certainly induce a voltage in the core as well as any other secondary winding.

Lamination

The energy taken from the primary to produce eddy current is converted to heat in the core. The amount of heat due to eddy current is dependent on the values of both the eddy current and the induced voltage ($P = IV$). There is nothing we can do to reduce the value of the induced voltage. Therefore, we must reduce eddy current loss by reducing the value of the eddy current produced by the induced voltage. This can be done by increasing the resistance of the path through which the eddy current must flow. Remember, $I = V/R$.) The resistance of the core, in the plane in which eddy current flows, is increased by laminating the core. Each lamination of the core is insulated with a thin layer of oxide [Fig. 11-10(*a*)]. The oxide has a much higher resistance than the rest of the silicon steel lamination. Notice in Fig. 11-10(*a*) that the eddy current would have to flow through the oxide layers as it circulates through the core. The equivalent circuit [Fig. 11-10(*b*)] for the core shows that the high resistance of the oxide on each lamination effectively blocks the flow of eddy currents. Thus laminating the core reduces the eddy current and its associated heat loss.

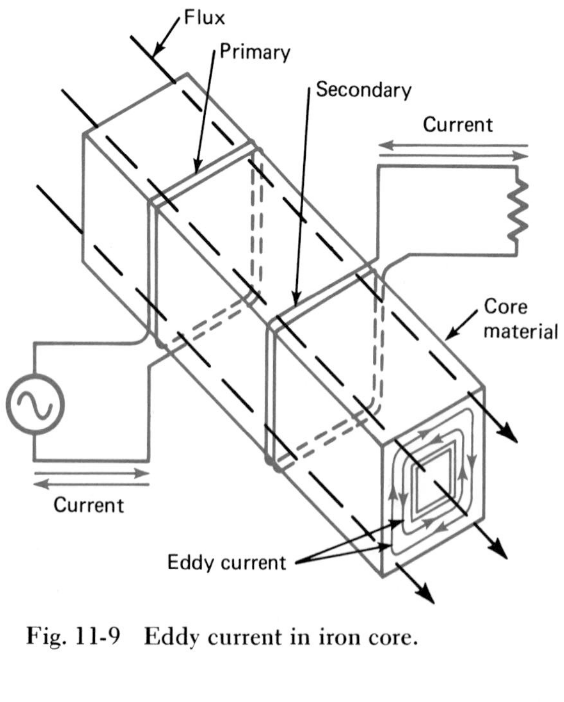

Fig. 11-9 Eddy current in iron core.

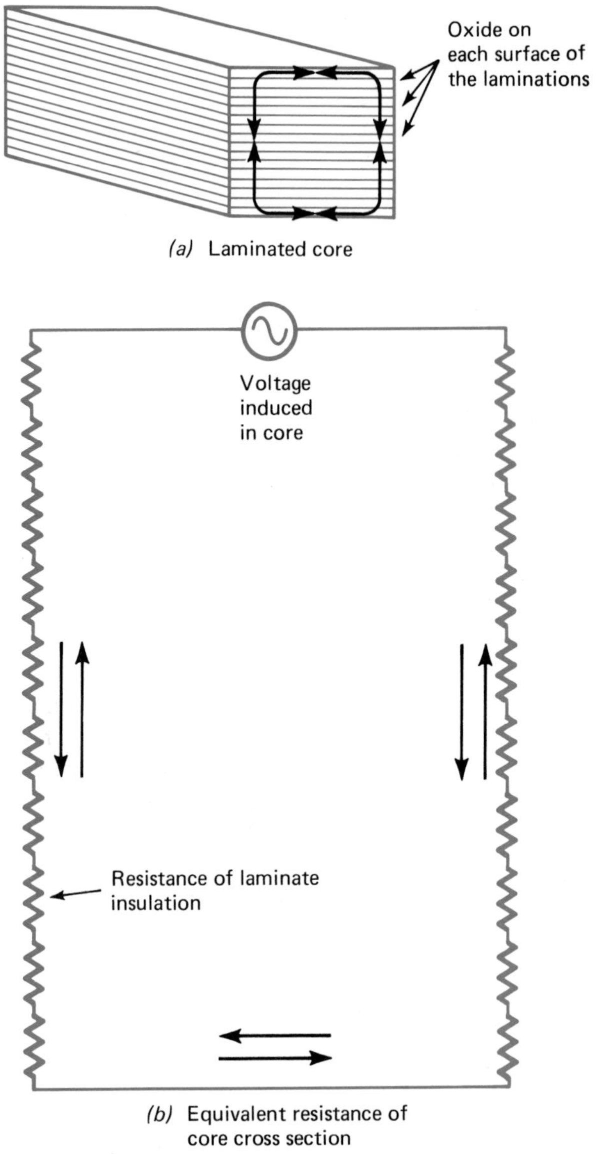

Fig. 11-10 Reducing eddy current loss.

The thinner the laminations, the more series resistance the core contains and the lower the eddy current will be. However, making the laminations thinner also increases the total amount of oxide in the core. The oxide has a lower permeability than does the silicon steel. Therefore, the core with thin laminations cannot carry as high a flux density as the one with thicker laminations. Flux density, of course, controls the amount of magnetic energy a given size of core can handle. This, in turn, controls the amount of power the transformer can handle. The transformer designer has to consider all of these factors, and others, as he designs a transformer core.

Copper Loss

Copper loss

Copper loss refers to the power dissipated in the windings of a transformer. Since this loss can be calculated by $P = I^2R$, it is called the I^2R loss. The R in the formula is the ohmic or dc resistance of the turns in the winding.

Obviously, copper loss is minimized by using as large a conductor as possible in the windings. However, conductor size is limited by the area of the windows (openings) in the core into which the winding must fit.

Load and Efficiency

Maximum efficiency is obtained from a transformer when it is fully loaded. For small transformers (less than 10 W) maximum efficiency may be less than 70 percent. With transformers larger than 1000 W it is more than 95 percent.

As the load is decreased the efficiency of the transformer also decreases. This is because current flow in a transformer primary does not decrease in direct proportion to decreases in the load. The primary current still causes substantial core losses and copper losses even when the secondary is lightly loaded.

Data which shows the changes in efficiency with changes in load is tabulated in Fig. 11-11. This data comes from tests conducted on a transformer rated at 75 W.

Primary power (W)	Secondary power (W)	Efficiency %
7	3.8	54.3
11	7.3	66.4
18	12.8	71.1
53	45.5	85.8
71	62.0	87.3

Fig. 11-11 Efficiency versus load. The transformer is most efficient when operating at its rated power.

Review Questions

22. What causes a transformer to be inefficient?
23. What is the efficiency of a transformer that uses 120 W to deliver 95 W?
24. Suppose a transformer with 90% efficiency delivers 400 W to its load. How much power will the transformer require from its source?
25. List three types of power losses in a transformer.
26. Which of the losses in question 25 occur in the core material?
27. What causes hysteresis loss?
28. How often must residual magnetism be overcome in the core of a transformer?
29. What is a hysteresis loop?
30. What are core laminations made of? Why?
31. Is a narrow or wide hysteresis loop desirable for the core of a transformer?
32. Will hysteresis loss be higher at 60 Hz or 100 Hz?
33. How are eddy current losses reduced?
34. For minimum eddy current loss should core laminations be thick or thin?
35. What is I^2R loss? How can it be minimized?
36. Do large or small transformers have the higher efficiency?
37. When does a transformer have maximum efficiency?

11-3 LOADED AND UNLOADED TRANSFORMERS

It has previously been shown that the fully loaded transformer has higher efficiency than an unloaded transformer. There are also many other differences associated with the amount of load on a transformer. These will become apparent in the discussion which follows.

An unloaded transformer acts like a simple inductor. It is a highly inductive load on the primary power source to which it is connected. The current in the primary winding is nearly 90° out of phase with the voltage [Fig.

Energizing current

Secondary magnetizing force

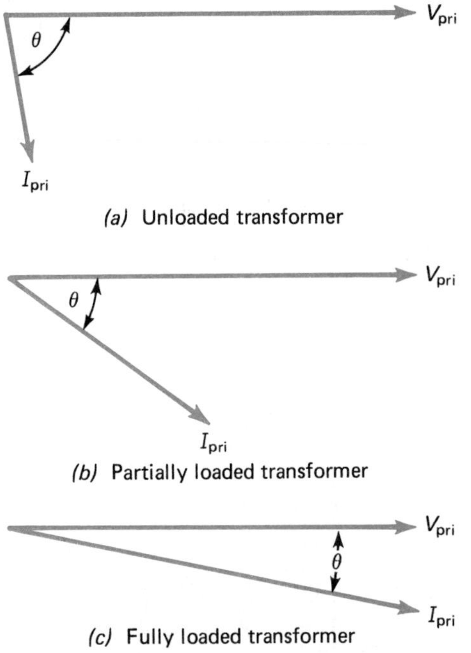

Fig. 11-12 Primary current and voltage phase relationships. When fully loaded a transformer appears to be close to a resistive load.

11-12(*a*)]. Only the copper and core losses keep the current from being a full 90° out of phase with the voltage. The current in the primary of an unloaded transformer is called the *energizing current*. It is the current needed to set up the flux in the core. The energizing current can be rather high and still not use much power. This is because the current and voltage are so far out of phase. The amount of energizing current is determined primarily by the inductive reactance of the primary winding.

When a load is connected to the secondary of a transformer, both the primary current and angle θ change. As indicated in Fig. 11-12(*b*) and 11-12(*c*), the amount of change depends upon how heavily the transformer is loaded. For this discussion the load on the secondary is assumed to be resistive. Therefore, the transformer is furnishing power to the load.

Figure 11-13 presents experimental data collected by testing a 75-VA transformer. Look at the secondary power and the angle θ columns. From them you can see that the transformer appears less inductive, and more resistive, as the load increases.

The last row of numbers in Fig. 11-13 indicates that the transformer is almost purely resistive at 72 VA. The apparent power and the true power of the primary are nearly equal; the cos θ (power factor) approaches 1. Thus, the primary is providing the power for the load connected to the secondary. In other words the load on the secondary is reflected back to the primary. The primary, in turn, draws the required power from the primary power source.

The secondary load is reflected back to the primary by the interaction of the primary and secondary magnetizing forces. Until now we have not mentioned a secondary magnetizing force. But, as soon as the secondary is loaded, current starts to flow in the secondary coil. This current in the secondary coil creates a magnetizing force which tries to produce a flux in the core material. The polarity of the magnetizing force of the secondary always opposes the magnetizing force of the primary (Fig. 11-14). Obviously the core cannot have flux flowing in opposite directions at the same time. Either the primary or the secondary magnetizing force must dominate. The primary magnetizing force always dominates. It must dominate because the flux created by the primary is what produces the secondary

Measured values				Calculated values		
V_{pri} (volts)	I_{pri} (amperes)	P_{pri} (watts)	P_{sec} (watts)	Apparent P_{pri} (voltamperes)	Cos θ	θ (degrees)
120	0.125	3	0	15	0.20	78
120	0.135	7	3.8	16.2	0.432	64
120	0.15	11	7.3	18	0.611	52
120	0.19	18	12.8	22.8	0.789	38
120	0.46	53	45.5	55.2	0.960	16
120	0.60	71	62	72	0.986	9.6

Fig. 11-13 Primary current and angle theta (θ) versus load.

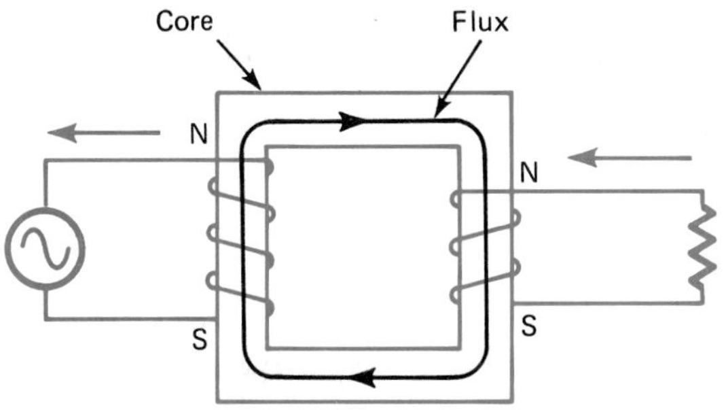

(a) One half cycle

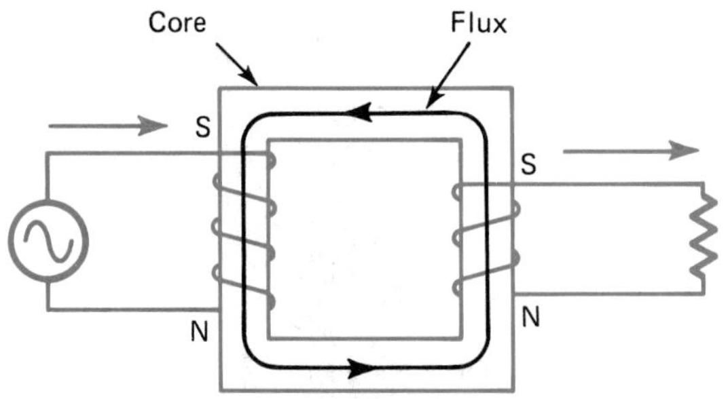

(b) Other half cycle

Fig. 11-14 Primary and secondary magnetizing forces oppose each other. However, notice that the primary force dominates.

voltage and current. Whenever the secondary magnetizing force increases, the primary current increases to provide a greater primary magnetizing force. This is why the primary current increases whenever the secondary load current increases. Any primary current caused by a load on the secondary will be an in-phase (resistive) current (Fig. 11-15). The total primary current (Fig. 11-15) is composed of this reflected resistive current and the energizing current (which is mostly inductive current). When the transformer is fully loaded the resistive current (caused by the secondary load) is predominate. Thus, the fully loaded transformer appears resistive to the primary power source.

Review Questions

38. Describe what happens to the following factors when a transformer's load is changed from no-load to full-load:
    a. Angle θ
    b. Cos θ (power factor)
    c. Type of load the transformer presents to the primary source
39. Define energizing current.
40. Is the energizing current mostly inductive or resistive?
41. What is meant by "the secondary load is reflected back to the primary"?
42. Does the magnetizing force of the secondary current aid or oppose the magnetizing force of the primary current?

11-4 TRANSFORMER CORES

I and E laminations

Transformers can be broadly grouped by the type of core materials they use. Like inductors, transformers can have either magnetic cores or air cores.

Physically, iron-core transformers look like iron-core inductors. They use the same type of I and E laminations in their cores. However, the I and E laminations for transformer cores are stacked differently than for inductors. With inductors all the I laminates are stacked together and all the E laminates are stacked together. This leaves a small continuous air gap where the I and E stacks butt together. This small air gap in the inductor core aids in keeping the inductance more constant for different amounts of current. (Sometimes this air gap is increased by putting one or more layers of paper between the I and E joint). With transformers, the I and E laminations are rotated 180° every few layers (see Fig. 11-16). This procedure breaks up the joint between the I and E laminations so that there is no continuous air gap. Thus flux leakage from the core is reduced to a minimum.

Laminated-iron-core transformers are used only at power and audio frequencies (frequencies up to 20 kHz). At frequencies above

(a) Unloaded condition

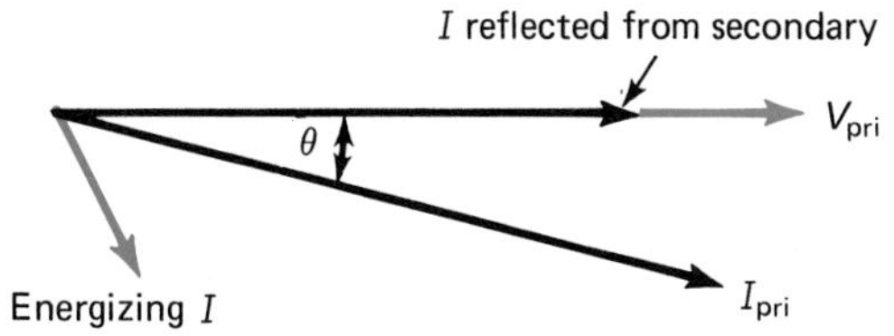

(b) Loaded condition

Fig. 11-15 Reflected load current.

Air-core transformer

Power transformer

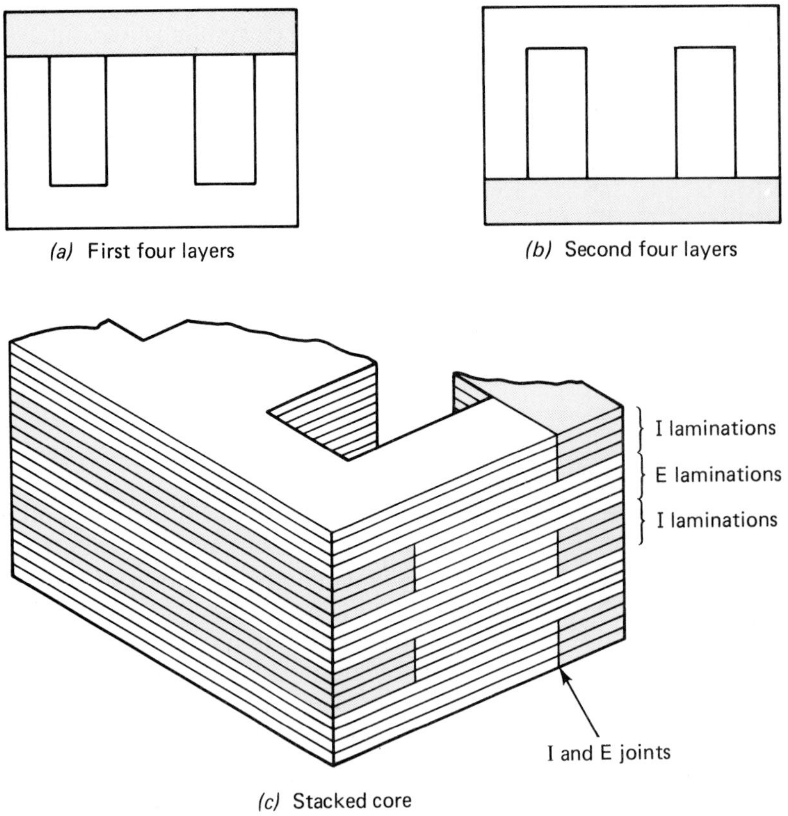

Fig. 11-16 Stacking of I and E laminations.

the audio range, their core losses become excessive.

Powdered-iron and ferrite are also used as core material for magnetic-core transformers. When used in the audio range the cores form a continuous path for the flux. Toroidal cores are often used at the higher audio frequencies. When used in the radio frequency range, the core is often just a slug. Notice in Fig. 11-17 that the coils are physically separated from each other. When greater coupling is desired, the spacing between the coils is reduced. Frequently one coil is wound on top of the other to obtain maximum coupling.

Air-core transformers are used exclusively at high (radio) frequencies. Often they are made from heavy enough wire that the individual coils are self-supporting. Some air-core transformers are made so that the coefficient of coupling is variable. This requires a mechanical arrangement that either varies the distance between the coils or the axis orientation of the coils.

11-5 TYPES OF TRANSFORMERS

Electrical and electronic parts catalogs list many different types of classifications of transformers. Usually these catalogs classify a transformer according to the application for which it was designed. Some of the more common types and their applications are discussed below.

Power transformers are those transformers that are designed to operate at power line frequencies and voltages (usually 60 Hz and from 115 V to several thousand volts). The larger transformers are for distribution and lighting. Smaller transformers are used for rectifier, filament, and plate circuits, usually in electronic systems. Rectifier transformers are used for providing low-voltage ac for rectification into dc. These transformers are also used for providing low-voltage ac for control circuits

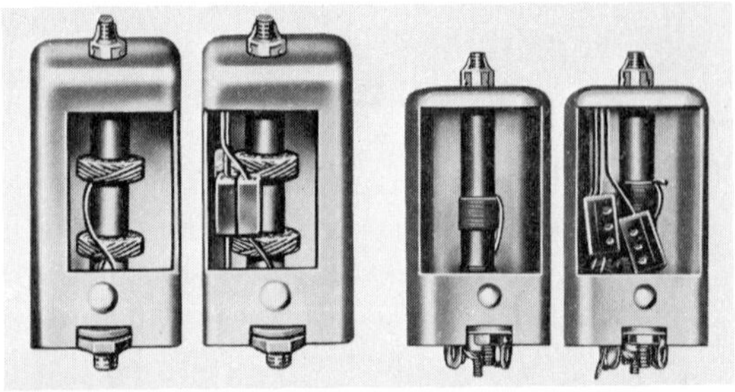

Fig. 11-17 Radio frequency transformers. (Courtesy of Bell Industries, J. W. Miller Division.)

(relays, solenoids, etc.). Therefore, they are also called *control transformers*.

Those transformers designed to operate at frequencies up to 20 kHz are often referred to as *audio transformers*. They are further categorized as *input*, *output* and *interstage* transformers. (These terms refer to audio amplifiers.) They are transformers to: (1) receive the input to the amplifier, (2) deliver the output from the amplifier, or (3) process the audio signal within the amplifier.

Radio frequency transformers perform functions similar to the audio transformers discussed above. Only these functions are performed at radio frequencies rather than audio frequencies. Radio frequency transformers may be either air core or magnetic core. They are often enclosed in a metal container (Fig. 11-17) which serves to shield against electric fields. These transformers are used in such devices as radio and television receivers and transmitters.

Some electrical and electronic equipment (such as processing equipment and computers) is very sensitive to voltage changes. Such equipment is often powered by *constant-voltage transformers* because regular, power-line voltage may vary too much. A constant-voltage transformer provides a stable secondary voltage even when the primary voltage is very unstable. Typically the primary voltage can vary from 95 to 130 V without causing more than 1 percent variation in the secondary voltage.

Isolation transformers have equal primary and secondary voltages. Their purpose is to electrically isolate a piece of electrical equipment from the power distribution system. An important use of isolation transformers is illustrated in Fig. 11-18. Many pieces of electronic

Control transformer

Audio transformer

Radio frequency transformer

Constant-voltage transformer

Isolation transformer

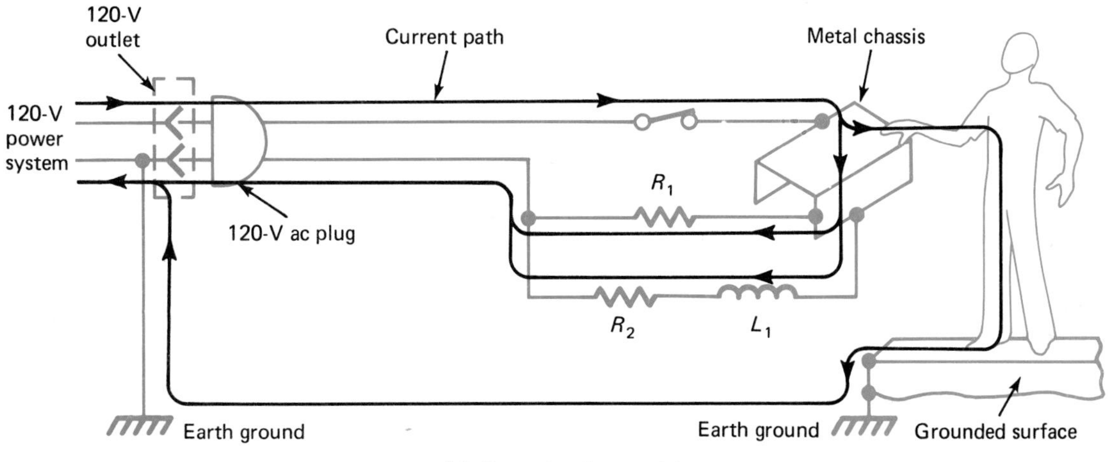

(a) Current path complete

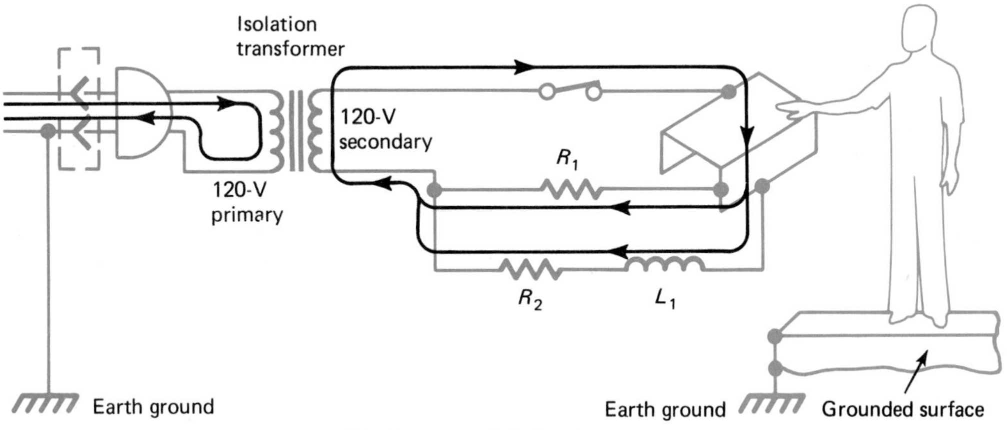

(b) Current path isolated

Fig. 11-18 Importance of isolation transformer.

equipment use the metal chassis on which components are mounted as a *common conductor.* In Fig. 11-18, the currents for both R_1 R_2, and L_1 flow through the chassis. Thus the chassis is part of the electrical circuit. Technicians servicing this equipment can accidentally come in contact with the chassis while power is applied to the equipment. If they do, they complete a circuit and receive a shock [see Fig. 11-18(*a*)]. The shock can be fatal when the resistance through ground is low. Inserting an isolation transformer [Fig. 11-18(*b*)] breaks the circuit that includes the technician. Current can no longer flow from the ungrounded side of the power source through the chassis and the technician to ground. Yet the other circuit containing R_1, R_2 and L_1 receives normal voltage, current, and power.

Autotransformer

An *autotransformer* is somewhat different than the other transformers we have studied. Its primary is also part of its secondary and vice versa. With a step-up autotransformer [Fig. 11-19(*a*)] the secondary consists of the primary plus some additional turns. These additional turns are wound so that their induced voltage is series aiding the cemf of the primary. The step-down autotransformer is shown in Fig. 11-19(*b*). Here the secondary is just a fraction of the primary. The cemf of that fraction of the primary provides the secondary voltage.

In most ways an autotransformer behaves like any other transformer. That is, a load on the secondary increases the primary current. Also, the turns-per-volt ratio of the primary and secondary is the same. The big difference between the autotransformer and other transformers is that the autotransformer does not provide electrical isolation.

Variable transformer

Variable transformers that operate at power frequencies are usually autotransformers with an adjustable secondary. One secondary lead is connected to a carbon brush that can be adjusted up and down on the turns of the winding. The secondary voltage of most variable transformers can be adjusted from 0 to about 110 percent of the primary voltage.

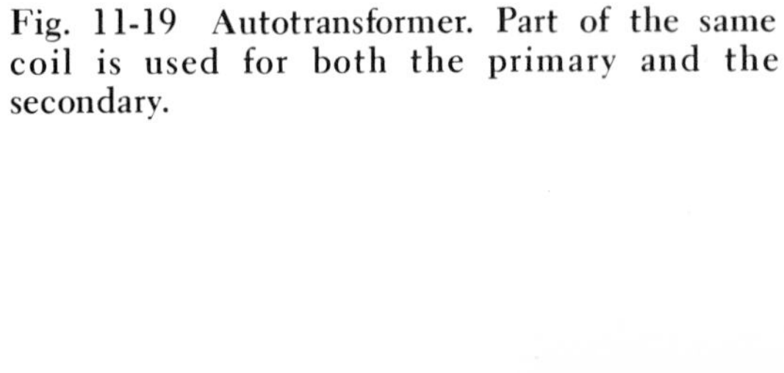

Fig. 11-19 Autotransformer. Part of the same coil is used for both the primary and the secondary.

Review Questions

43. Why are I and E laminations rotated every few layers?
44. What types of transformers use laminated iron cores?
45. List several types of power transformers.
46. Which voltage, primary or secondary, is stabilized by a constant-voltage transformer?
47. What is the purpose of an isolation transformer?
48. How does an autotransformer differ from other transformers?

11-6 IMPEDANCE MATCHING

Impedance matching

We have seen that transformers are used for changing voltage levels and for isolation. The third common use of a transformer is impedance matching.

Before we see how a transformer can match impedances, let us review why impedance matching is desirable. In Chap. 5 we discussed the advantages of matching the resistance of the source with that of the load. This achieved the maximum transfer of power in the dc circuit. Impedance matching, in effect, does the same thing for ac circuits. Suppose the internal impedance of a source is matched (equal) to the impedance of the load. Then, maximum power will be transferred from the source to the load. When the load impedance is either greater than or less than the source impedance, the load power will decrease. These ideas are numerically shown in Fig. 11-20. Although impedance includes both resistance and reactance, Fig. 11-20 uses only resistance. Using only resistance does not change the concept; it just keeps the calculations simpler. In Fig. 11-20(*a*) all of the source voltage is available at the terminals. No voltage drops across the internal resistance because no current flows. When a load is connected to the source [Fig. 11-20(*b*)] current

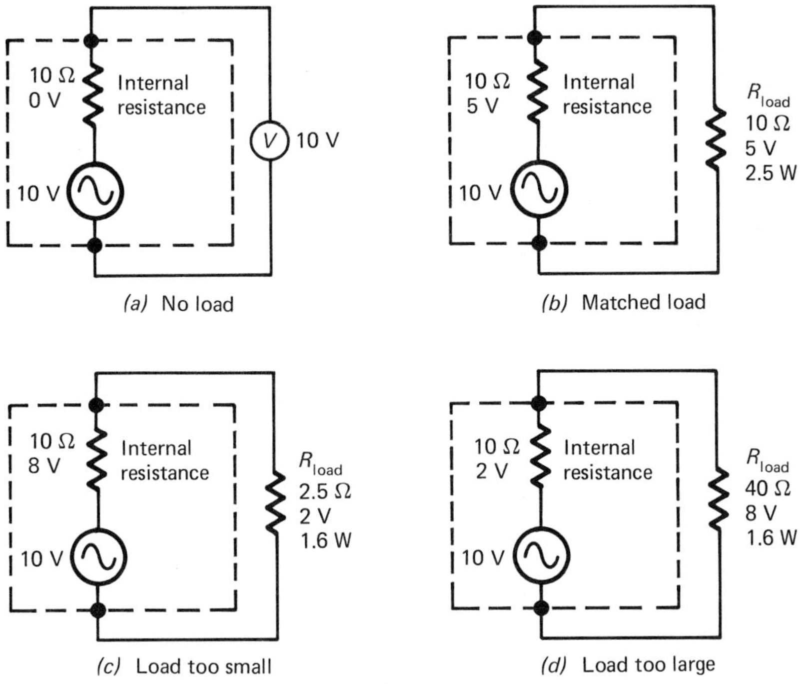

Fig. 11-20 Power transfer. Maximum transfer of power occurs when the source and load are matched.

flows and the internal resistance drops some voltage. The source resistance and the load resistance form a series circuit. Therefore, series rules and formulas can be used in analyzing the circuit. This is how the voltages and power for the various loads were calculated.

A transformer can make a load appear to the source to be either larger or smaller than its actual value. If the transformer is a step-up type, the load appears smaller. If it is a step-down transformer, the load appears larger. Figure 11-21 illustrates how a transformer changes the apparent value of a load. For purposes of calculating values, the transformer in Fig. 11-21(*a*) is assumed to be an ideal transformer. That is, it has 100% efficiency, 100% coupling, and a 0° phase shift when loaded. If the transformer has no power losses, then its primary power equals its secondary power. To obtain 0.1 W of secondary power, the primary must provide 0.1 W. This means that the primary has to draw 0.01 A of current from the source. Notice this is the same value of current as the 1000-Ω resistor in Fig. 11-21(*b*) draws. Thus you can see that the transformer makes the 10-Ω in the secondary resistor appear as 1000 Ω to the source. The source's output voltage, current, and power are identical in the two circuits [Figs. 11-21(*a*) and 11-21(*b*)]. The transformer has matched the impedance of a 1000-Ω source to a 10-Ω load.

Reflected impedance

Impedance ratio

The 1000-Ω impedance that the primary seemed to have is called *reflected impedance*. In other words, the 10 Ω in the above example was reflected back to the primary as 1000 Ω. We could say that the primary impedance of the transformer is 1000 Ω and the secondary impedance is 10 Ω. Therefore, the primary to secondary *impedance ratio* is 100 to 1. Notice in Fig. 11-21(*a*) that the turns ratio (and also the voltage ratio) is 10 to 1. Thus the turns ratio squared is equal to the impedance ratio. Also, the voltage ratio squared is equal to the impedance ratio. If we use the symbol Z for impedance, these relationships can be written:

$$\frac{Z_{pri}}{Z_{sec}} = \left(\frac{N_{pri}}{N_{sec}}\right)^2 \quad \text{or} \quad \frac{Z_{pri}}{Z_{sec}} = \left(\frac{V_{pri}}{V_{sec}}\right)^2$$

Example 11-4

An 8-Ω load is to be matched to a source that has 8000 Ω of internal impedance. What will be the turns ratio of the transformer?

Given: $Z_{sec} = 8\ \Omega$
$Z_{pri} = 8000\ \Omega$

Find: $\frac{N_{pri}}{N_{sec}}$ (that is, the turns ratio)

Known: $\frac{Z_{pri}}{Z_{sec}} = \left(\frac{N_{pri}}{N_{sec}}\right)^2$

Voltage rating

Current rating

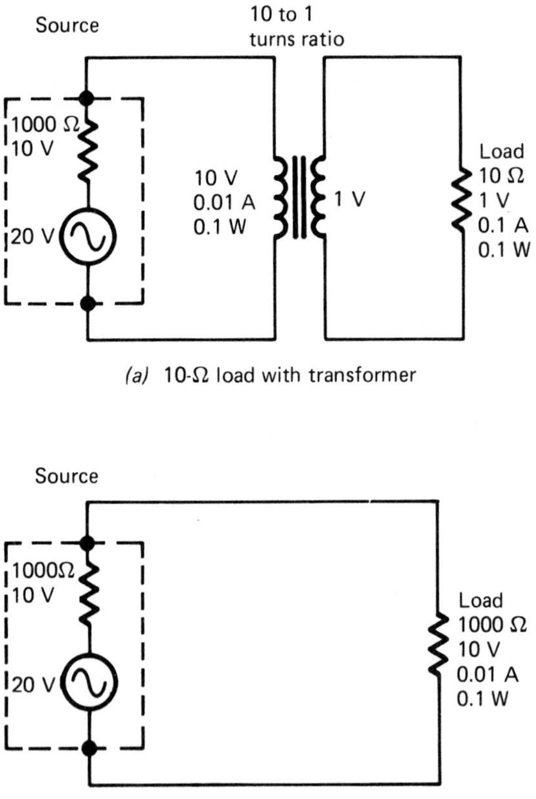

(a) 10-Ω load with transformer

(b) 1000-Ω load without transformer

Fig. 11-21 Impedance matching principle. The transformer makes the 10-Ω resistor appear to the source as a 100-Ω resistor.

Solution: $\dfrac{Z_{pri}}{Z_{sec}} = \dfrac{8000\ \Omega}{8\ \Omega} = 1000$

$$\left(\frac{N_{pri}}{N_{sec}}\right)^2 = 1000$$

$$\text{turns ratio} = \frac{N_{pri}}{N_{sec}} = \sqrt{1000}$$

$$= 31.6$$

Answer: The transformer must have a turns ratio of 31.6 to 1 (primary to secondary).

Review Questions

49. What does impedance matching mean?
50. What is the significance of impedance matching?
51. Does a load appear to the source as smaller or larger than its true value when a step-down transformer is used?
52. What does reflected impedance mean?
53. What is the impedance ratio of a transformer that steps down 230 V to 23 V?
54. The transformer in question 53 is connected to a 100-Ω load. What is the value of the impedance reflected back to the source?

11-7 TRANSFORMER RATINGS

To properly use a transformer one must know its voltage and current ratings. Of course, from these ratings the power rating of the transformer can be calculated. Most transformers are also specified by their voltampere (apparent power) ratings.

Voltage Rating

Manufacturers always specify the voltage rating of the primary and secondary windings. Operating the primary at higher than rated voltage will usually cause the transformer to overheat. More serious is the additional stress placed on the transformer insulation by the higher primary and secondary voltages. Operating the primary below rated voltage does no harm, but the secondary voltages will be lower than rated values.

The rated voltages of the secondaries are specified for full-load conditions with rated primary voltage. With no load, the secondary voltage will be slightly higher than rated voltage (usually 5 to 10 percent higher).

There is considerable variation in the way in which manufacturers specify center-tapped secondaries. For example, the secondary in Fig. 11-22 may be specified in any of the following ways:

1. 40 V C.T.
2. 20 V–0–20 V
3. 20 V each side of center

Current Rating

Manufacturers usually specify current ratings for secondary windings only. As long as the secondary current rating is not exceeded, the primary current-carrying capacity cannot be exceeded.

Exceeding the current rating of a secondary will cause its voltage to fall slightly below rated

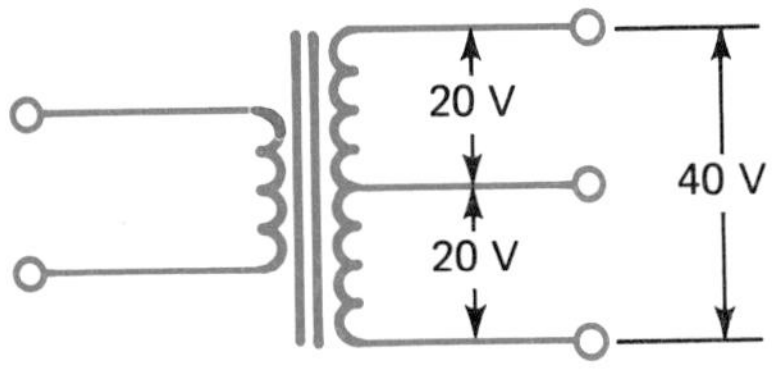

Fig. 11-22 Center-tapped secondary specifications. The secondary is classified as either 40 V center-tapped or 20 V each side of center.

value. More serious than decreased voltage, however, is the increase in I^2R loss in the secondary. The increased I^2R loss will cause the winding to overheat and eventually destroy the transformer.

Power Rating

Power rating

Some manufacturers specify a power rating (in watts) for their transformers. This is understood to be the power the transformer can deliver *into a resistive load*. Thus the power rating is merely the product of the current rating times the voltage rating of the secondary ($P = IV$). For multiple-secondary transformers, the power rating is the sum of the powers available from the individual secondaries ($P_T = P_1 + P_2 +$ etc.). The total power cannot be taken from a single-secondary on a multiple-secondary transformer. The current rating of the individual secondaries must not be exceeded.

Voltampere Rating

Voltampere rating

The voltampere rating of a transformer is an apparent power rating. It is applicable to any type of load—resistive, reactive, or combination (impedance). The voltampere rating, like the power rating, is given for the total transformer instead of individual secondaries. With a multiple-secondary transformer the total voltampere rating cannot be taken from a single secondary.

A transformer can be loaded to its full voltampere rating and be delivering only a fraction of its power rating. Refer to Fig. 11-23 for an example. Here the load is a motor which has a power factor (cos θ) of 0.6. It is connected to a transformer which is rated at 750 VA and has a secondary voltage of 120 V. The motor draws 6.25 A. The transformer is providing 750 VA (120 V × 6.25 A). However, the motor is drawing only 450 W ($P = IV \cos \theta = 6.25 \text{ A} \times 120 \text{ V} \times 0.6$).

As you can see manufacturers use many ways to rate transformers. You will never overload a transformer or exceed any of its ratings if you observe two rules:

1. Never apply more than the rated voltage to the primary.
2. Never draw more than the rated current from any secondary.

Review Questions

55. Is it possible to exceed the VA rating of a transformer, but still be operating within its rated power?
56. Can the power rating of a transformer be exceeded without exceeding either the primary voltage rating or the secondary current rating?
57. Can the VA rating of a transformer be exceeded without exceeding either the primary voltage rating or the secondary current rating?
58. Under what conditions are secondary voltages rated?

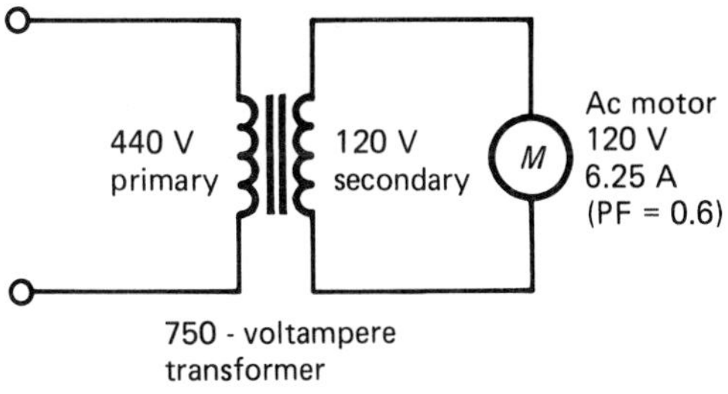

Fig. 11-23 Voltampere versus power rating. The limiting rating is the voltampere rating.

11-8 SERIES AND PARALLEL WINDINGS

Some transformers are made with more than one primary as well as more than one secondary. These multiple transformer windings can be connected in series or parallel to change their voltage or current capabilities. Either (or both) primary and secondary windings can be series or parallel connected.

Correct phasing

Series aiding

Bucking

Windings in Parallel

For parallel connections, windings should have identical ratings. They *must* have identical voltage ratings. Before windings are connected in parallel their *phasing* must be correct. Phasing refers to the instantaneous polarity of the windings. Correct phasing for parallel connection is illustrated in Fig. 11-24. The polarities shown are instantaneous. Notice that the negative ends of the two secondary windings are connected together as are the positive ends. With this phasing of the windings no secondary current flows when there is no load. Neither winding can force current through the other because their voltages are equal and opposing. The output voltage (Fig. 11-24) is still 10 V, but the current capability has doubled from 3 to 6 A.

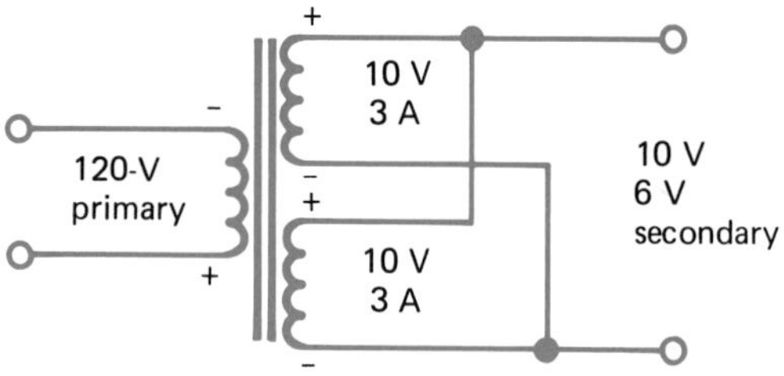

Fig. 11-24 Windings connected in parallel. The voltage remains the same, but the current capacity increases.

Some transformer manufacturers indicate the phasing of windings by terminal numbers or color code. However, when phasing is not indicated, the method shown in Fig. 11-25 can be used to determine the correct phasing. If points 2 and 3 of Fig. 11-25 are of the same instantaneous polarity, the voltmeter across points 1 and 4 will indicate 0 V. This means that points 1 and 4 are also of the same polarity. If the two windings are to be operated in parallel, points 1 and 4 can be connected together. Points 2 and 3 would form one terminal of the parallel winding and points 1 and 4 would be the other terminal.

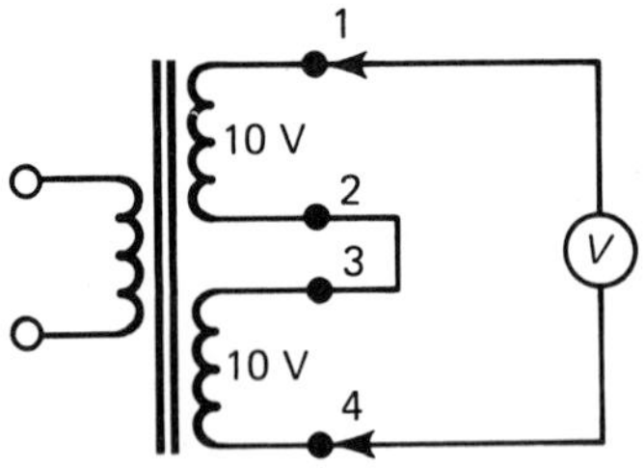

Fig. 11-25 Determining correct phasing for a parallel connection. If correctly phased, the meter will indicate zero volts.

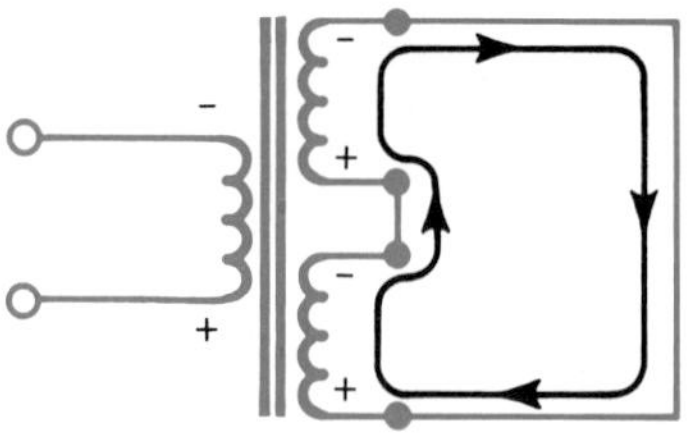

Fig. 11-26 Short circuit caused by incorrect phasing.

If, on the other hand, the meter in Fig. 11-25 indicates 20 V, points 2 and 3 would have opposite polarities. Then, for a parallel connection, points 1 and 3 would be connected together as would points 2 and 4.

Incorrect phasing (Fig. 11-26) causes a "dead" short on the secondaries of the transformer. The voltage of each winding aids the other winding in producing secondary current. The secondary current is limited only by the resistance of the secondaries themselves. Therefore, the current becomes very high. If not protected against overload, the transformer would soon burn out.

Some transformer primaries have two identical windings so that they can be powered from either of two voltages. When operated on the lower of the two voltages, the primaries are connected in parallel. Proper phasing of the primary windings can also be determined using the method of Fig. 11-25. Just use one of the secondaries as a temporary primary and apply an appropriate voltage source to it. The voltage applied to the secondary can be any value equal to *or less than* its rated value.

Windings in Series

Transformer windings can be connected in series so that they either aid or oppose each other as in Fig. 11-27. In this figure the polarity markings indicate the instantaneous polarities for one half cycle. In the series aiding configuration [Fig. 11-27(*a*)] the output voltage is the sum of the two secondary voltages. Notice, however, that the current capability is restricted to the lowest rating of the two windings. This is because all of the load current flows through both secondaries.

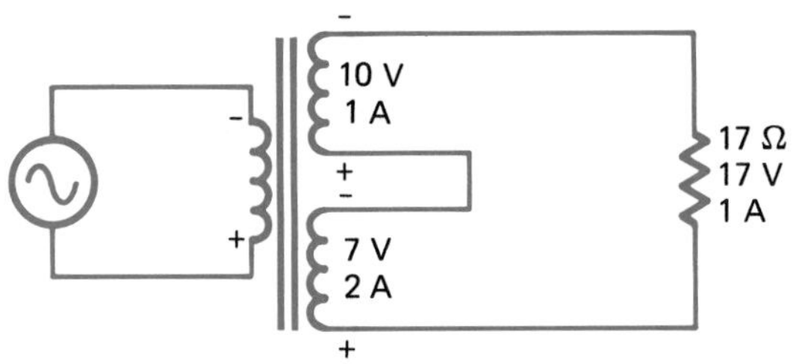

(a) Series-aiding connection

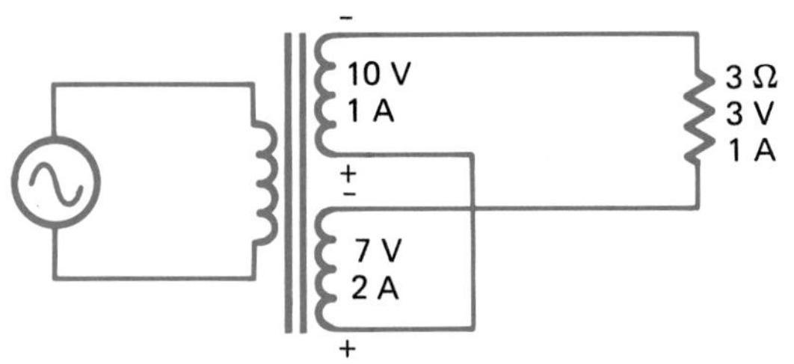

(b) Series-opposing connection

Fig. 11-27 Series aiding and opposing windings.

When connected in series opposing [Fig. 11-27(*b*)], the two windings produce a voltage equal to the difference between the two voltages. We say the two voltages are *bucking* one another. Again note that the current is limited to the lower current rating of the two windings.

In summary, it is possible to obtain four different voltages from two secondary windings. The four voltages are:

1. The voltage of secondary number 1
2. The voltage of secondary number 2
3. The sum of the voltages from secondaries 1 and 2
4. The difference between the voltages of secondaries 1 and 2

Dual primary windings are connected in series for operation on the higher voltage. The connection must be series aiding. Otherwise the magnetic field from one winding would cancel the field from the other. The primary current would be very high. The primary would burn out immediately unless it was properly protected against overloads. Phasing the primary windings for series-aiding connections can be done as described for parallel primary windings. The connection that yields the maximum voltmeter reading is the correct connection for series aiding primaries. Power is applied to the two terminals where the voltmeter was connected.

Review Questions

59. List two requirements for parallel-connected windings.
60. What does phasing refer to?
61. Two secondary windings are to be connected in series. One winding is rated for 15 V and 3 A. The other is rated for 10 V and 4 A.
 a. How much voltage and current will be available if one instantaneous + is connected to the other instantaneous + ?
 b. How much voltage and current will be available if + is connected to −?
62. What happens when parallel windings are incorrectly phased?

11-9 THREE-PHASE TRANSFORMERS

If you have not recently done so, you should review Sec. 8-8 before reading this section. The ideas developed in Sec. 8-8 are essential in understanding three-phase transformers.

Three-phase voltages can be transformed by either a single three-phase transformer or by three single-phase transformers. The end results are the same; all three of the phase voltages are changed.

The structure of a three-phase transformer is illustrated in Fig. 11-28. The flux in the phase-1 leg is equal to the phase-2 flux plus the phase-3 flux. Phase-2 flux equals phase-3 flux plus phase-1 flux, etc. This is because the flux, like the current, of each phase is displaced by 120°.

Figure 11-29 graphically presents the idea of how the flux splits up in a three-phase transformer core. In this figure, the flux of each phase is assumed to be in step with the current in that phase. (This assumption implies that the core has no hysteresis loss.) At instant A [Fig. 11-29(*a*)] the phase-1 current is 0. Therefore, the phase-1 flux is also 0. In Fig. 11-29(*b*) (at instant *B*) currents in phase 1 and phase 3 are both positive. This produces flux in the direction shown in both the phase-1 and phase-3 legs of the core. At this same instant (instant *B*) phase-2 current is negative and of twice the value of either phase-1 or phase-3 current. Thus, the phase-2 leg of the core has twice as much flux as either phase 1 or phase

Series opposing

Three-phase transformer

Flux splits up

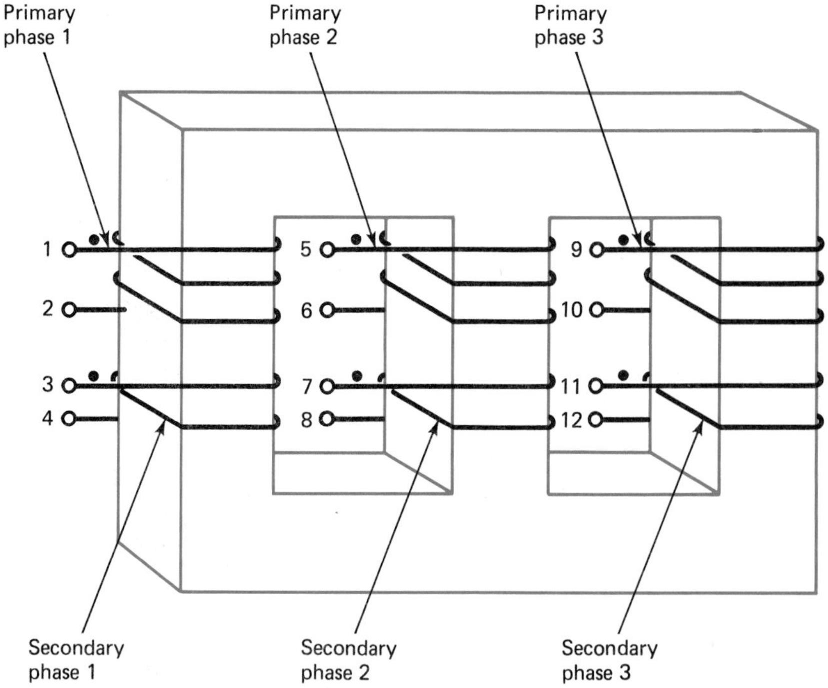

Fig. 11-28 Three-phase transformer. One phase is wound on each leg of the transformer.

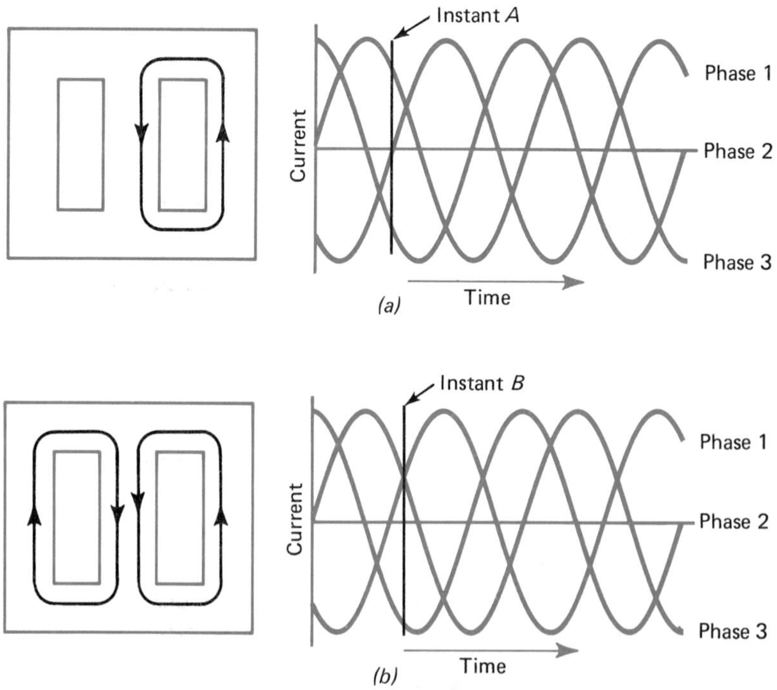

Fig. 11-29 Core flux in a three-phase transformer core.

3. Also the flux in phase-2 is in the opposite direction to the flux in the other two phases. Close inspection of Fig. 11-29(*c*) and 11-29(*d*) will show how the flux continues to shift around in the core.

Delta-connected

Wye connection

The primary and secondary windings of a three-phase transformer may be either wye- or delta-connected. The secondary does not have to have the same configuration (wye or delta) as the primary. Figure 11-30 shows four possible ways to connect a three-phase transformer. The dots on one end of each winding (Fig. 11-30) indicates the physical beginning of each winding. Refer back to Fig. 11-28. Notice that all windings are wound in the same direction (counterclockwise) when you start at the dotted end of the winding. Identifying the start of the windings is necessary before they can be properly phased.

The diagrams in Fig. 11-30 show one way of connecting the windings to obtain correct phasing. With a wye connection, correct phasing can also be obtained by connecting all the dotted ends to the star point. In the delta connection all three windings can be reversed; just be sure that two dotted ends are not connected together. On physical transformers, the dotted (start) end of a winding is identified by the manufacturer. The identification may be made in several different ways. It may be a colored strip wrapped around the insulation of the start lead. It may be identified by a number on a diagram mounted on the transformer.

Incorrect phasing of the primary, in either the wye or delta configuration, causes excessively high primary current. If not protected against overload, the incorrectly phased primary can be destroyed by the excess current.

Improper phasing of a delta-connected secondary also causes excess destructive currents. A quick check for correct phasing of a delta-connected secondary is shown in Fig. 11-31. If the voltmeter indicates 0 V, the windings are properly phased. The ends of the windings to which the meter is connected can be connected together to complete the delta. If the meter indicates a high voltage (twice the phase voltage), incorrect phasing exists. Reverse the lead connections on one winding at a time until the meter indicates 0 V. With some wye-delta-connected transformers the meter in Fig. 11-31 may not indicate 0 V when properly phased. In these cases the meter will indicate the lowest reading when the phasing is correct.

A wye-connected secondary provides equal line voltages when correctly phased. If the line voltages are unequal, reverse one winding at a time until the line voltages are balanced.

The relationships between phase and line

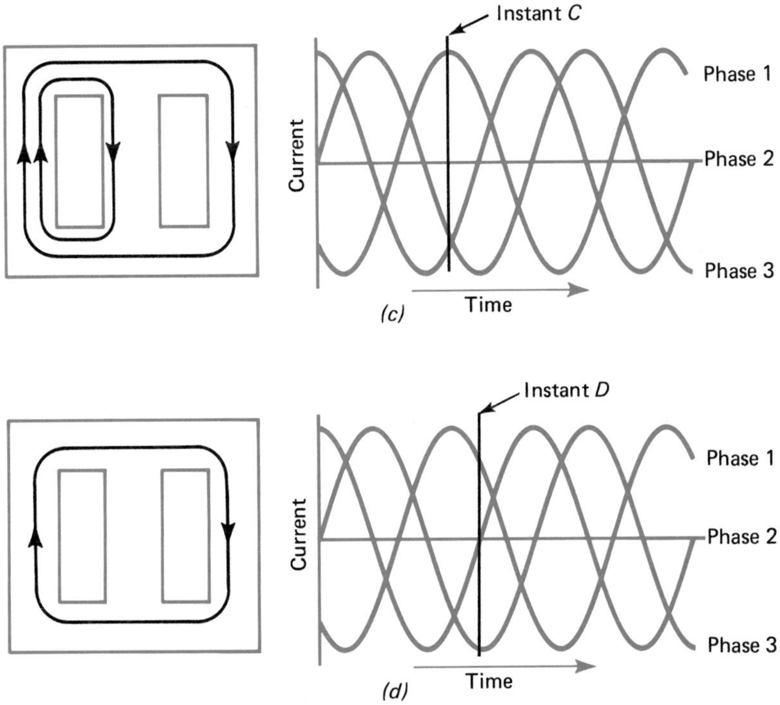

Fig. 11-29 (*Cont.*)

voltages and phase and line currents were explained in Sec. 8-8. These relationships apply to three-phase transformers as well as three-phase generators.

Review Questions

Wye connection

63. List two configurations in which three-phase transformers can be connected.
64. In which configuration can all the start ends of the windings be connected together?
65. What is the correct phasing for a delta connection?
66. Must the primary and secondary of a three-phase transformer be connected in the same configuration?

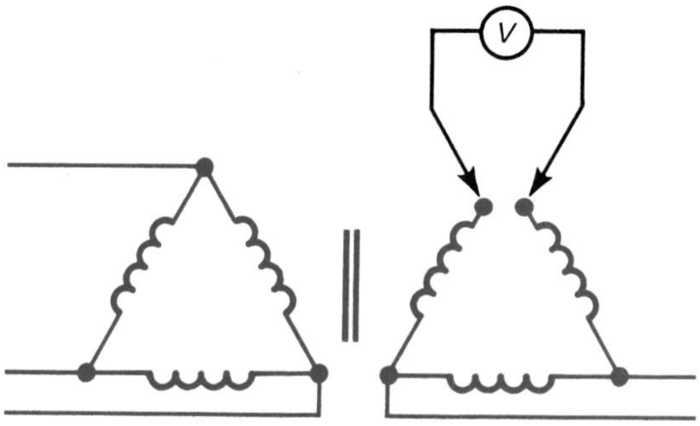

Fig. 11-31 Checking delta phasing. When properly phased, the meter will indicate zero volts.

67. What happens when a delta winding is incorrectly phased?
68. Refer to Fig. 11-28. Which terminals would be connected together to form a wye-connected primary?

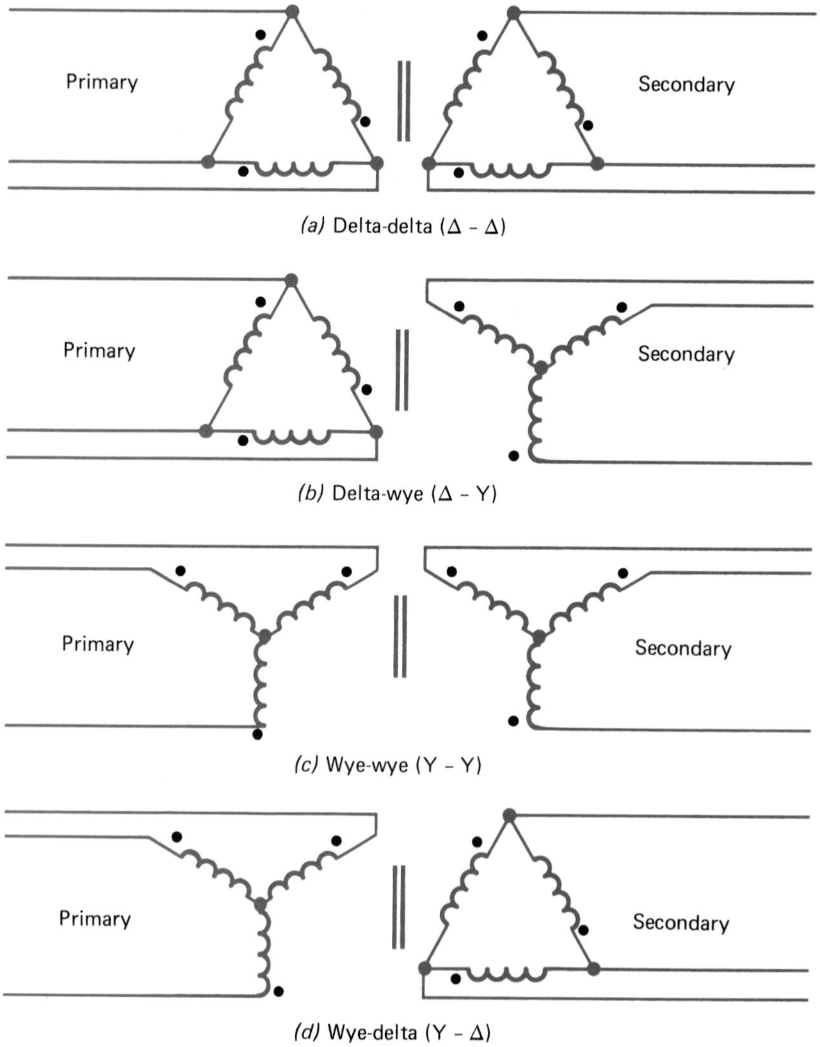

Fig. 11-30 Delta and wye connections.

Summary

1. Transformers operate on the principle of mutual inductance.
2. Mutual inductance is measured in henries.
3. Magnetic flux links or couples the two coils of a transformer.
4. Primaries receive power. Secondaries deliver power.
5. Primaries and secondaries, except on autotransformers, are electrically isolated.
6. Primaries and secondaries are reversible.
7. The coefficient of coupling specifies the portion of the primary flux which links with the secondary.
8. The coefficient of coupling ranges from 0 to 1. It is greatest (almost 1) with iron-core transformers.
9. Flux leakage refers to primary flux that does not couple to the secondary.
10. Turns ratio and voltage ratio are the same when the coupling is 100 percent.
11. The turns-per-volt ratio is the same in all windings of a transformer.
12. One of the major uses of transformers is to step-up and step-down voltages in a power transmission and distribution system.
13. When voltage is stepped-up, current is stepped-down.
14. When voltage is stepped down, current is stepped up.
15. Transformer losses occur in both the core and the coils.
16. $\% \text{ eff.} = \frac{P_{sec}}{P_{pri}} \times 100$
17. Transformer loss consists of hysteresis, eddy current, and copper loss.
18. Copper loss is called I^2R loss.
19. Core loss consists of hysteresis and eddy current loss.
20. Hysteresis loss results from residual magnetism.
21. Hysteresis loss increases with increased frequency.
22. Eddy currents are currents induced in the core by primary flux.
23. Eddy currents are reduced by using laminations and oxidizing or coating the surface of the lamination with an insulating material.
24. An unloaded transformer behaves just like an inductor. The energizing current and supply voltage are nearly 90° out of phase. In the unloaded condition, the input power is dissipated in the form of copper and core losses.
25. Energizing current is the current drawn by the primary when the transformer is unloaded. Its magnitude is controlled by the reactance of the primary.
26. A fully loaded transformer appears to be almost entirely resistive to the source.
27. Power factor (cos θ) approaches one with a fully loaded transformer.
28. As secondary current increases so does primary current.
29. Laminated-core transformers are used at power and audio frequencies.
30. Powdered-iron and ferrite cores are used in the audio and lower radio frequency range.
31. Air-core transformers are used only in the radio frequency range. Their coupling can be controlled by the spacing and axis orientation of the coils.
32. Constant-voltage transformers provide a stable secondary voltage.
33. Isolation transformers have equal primary and secondary voltages.
34. Isolation transformers help protect the service technician from receiving a shock through the chassis of electrical equipment.
35. Autotransformers use a common primary-secondary winding. They are often used as variable transformers at power frequencies.
36. Matched impedances provide maximum power transfer.
37. Transformers may have voltage, current, power, and voltampere ratings.
38. Transformer power ratings apply to resistive loads only.
39. Power and voltampere ratings refer to the total of all secondaries.
40. Connecting windings in parallel increases the available current, but does not change the voltage rating.
41. Connecting windings in series either increases or decreases the available voltage, but doesn't increase the current rating.
42. Parallel windings must be properly phased and have identical voltages.
43. Three single-phase transformers can be used to transform three-phase voltages.
44. Three-phase transformer windings can be connected in either delta or wye configurations.
45. The primary and secondary of a three-phase transformer need not be connected in the same configurations.

Chapter Review Questions

For questions 11-1 to 11-16, determine whether each statement is true or false.

11-1. Symbols for transformers indicate whether the core is a magnetic or nonmagnetic material.

11-2. The winding of a transformer designated as the primary should never be used as a secondary.

11-3. Secondaries of a transformer convert electrical energy to magnetic energy.

11-4. In an autotransformer part or all of a coil will be used as both the primary and the secondary.

11-5. The coefficient of coupling of an air-core transformer is greater than that of an iron-core transformer.

11-6. If a transformer steps up the voltage, it also steps up the current.

11-7. Transformers operate on the principle of mutual inductance.

11-8. When the coupling is 100 percent, the turns ratio is numerically equal to the voltage ratio.

11-9. In a step-up transformer, the turns-per-volt ratio in the primary is smaller than in the secondary.

11-10. Primary flux that does not link up with the secondary is known as leakage flux.

11-11. Large transformers usually have a higher turns-per-volt ratio than do small transformers.

11-12. A transformer is most efficient when the secondary is not loaded.

11-13. Power factor approaches 1 when a transformer is fully loaded.

11-14. The primary current of a transformer increases whenever the secondary current increases.

11-15. The primary and secondary of a three-phase transformer must be connected in the same configuration.

11-16. Three single-phase transformers can be used to transform a three-phase voltage.

For questions 11-17 to 11-19, choose the letter that best completes each statement.

11-17. A transformer has two secondary windings. One is rated 14 V, 2 A and the other is rated 6 V, 5 A. The transformer can be connected to provide
(A) 20 V, 5 A *(B)* 20 V, 7 A *(C)* 8 V, 5 A *(D)* 8 V, 2 A

11-18. Primary current and voltage are furthest out of phase when the
(A) Primary draws only energizing current *(B)* Primary draws its maximum current *(C)* Secondary draws its maximum current *(D)* Secondary furnishes its maximum power

11-19. A step-down transformer causes the
(A) Primary current to exceed the secondary current *(B)* Primary

power to exceed the secondary power (*C*) Secondary impedance to appear to be larger than its true value (*D*) Secondary current and voltage to be out of phase

Answer the following questions.

11-20. A transformer primary has 4 turns per volt. How many turns are needed for a 40-V secondary?

11-21. List three types of transformer loss that cause a transformer to be less than 100% efficient. Tell how each type of loss is minimized.

11-22. List three common uses of transformers.

11-23. The secondary of a transformer provides 50 V across a 10-Ω resistor. The primary draws 300 W. What is the percentage of efficiency of the transformer?

11-24. What type of material is used for I and E laminations?

11-25. An 1800-W transformer has a single secondary rated at 240 V. What is its VA rating? Could it be used to power a 240-V motor that draws 10 A at a power factor of 60 percent?

11-26. Explain how to connect a three-phase winding in a wye configuration.

Answers to Review Questions

1. Two inductors coupled together by mutual inductance. Or, a device that transfers power between electrically isolated circuits.
2. The primary and the secondary
3. The henry
4. The magnetic linking together of two coils
5. 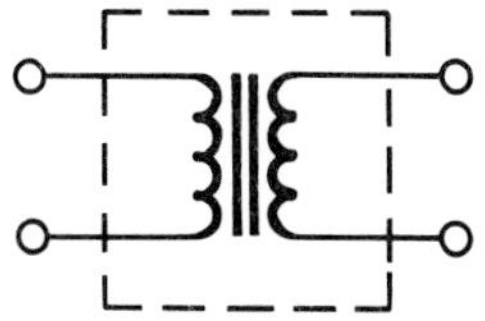
6. The primary converts electrical energy to magnetic energy. The secondary converts magnetic energy to electrical energy.
7. The primary
8. The secondary
9. Yes
10. It is a number that indicates the portion of the primary flux that links the secondary coil.
11. 1 or 100 percent
12. Laminated-iron-core transformers
13. By varying the distance between primary and secondary or by changing the axis orientation between the primary and secondary coils
14. A transformer with a secondary voltage that is higher than the primary voltage
15. They are equal.
16. They are equal.
17. The smaller
18. 570 turns
19. Changing voltage levels
20. To keep current low and thus wire size small
21. It steps current down.
22. Heat losses in the core and the windings
23. 79.2%
24. 444.4 W
25. Copper loss, hysteresis loss, and eddy current loss
26. Hysteresis and eddy current
27. The energy required to overcome residual magnetism
28. Twice each cycle
29. A curve on a graph produced by plotting magnetizing force against flux density
30. Silicon steel, because it has high permeability and low hysteresis loss
31. Narrow
32. 100 Hz
33. By using laminations in the core that have oxidized or have insulation-coated surfaces
34. Thin
35. Power lost in the transformer windings. I^2R loss is minimized by

using large diameter conductors in the coils.
36. Large
37. When it is loaded to full-capacity
38. *a.* Decreases from nearly 90 to nearly 0°
b. Increases from nearly 0 to nearly 1
c. Changes from inductive to resistive
39. The current drawn by an unloaded transformer
40. Resistive
41. It means that the primary draws in-phase current to supply the secondary-load power requirements.
42. Oppose
43. To break up the air gap and thereby reduce flux leakage
44. Power and audio frequency transformers
45. Distribution, lighting, rectifier, plate, filament and control
46. Secondary
47. It isolates a piece of equipment from the power line. This reduces the shock hazard.
48. Its primary and secondary use the same winding. It provides no electrical isolation.
49. Impedance matching means making the load impedance equal to the source impedance.
50. It provides maximum power transfer from source to load.
51. Larger
52. It means the impedance of the secondary load is reflected back to the source through the transformer.
53. 100 to 1 (primary to secondary)
54. 10,000 Ω
55. Yes
56. No
57. No
58. Full load and rated primary voltage
59. They must have equal voltages and correct phasing.
60. The momentary polarity of transformer windings
61. *a.* 5 V, 3 A
b. 25 V, 3 A
62. The windings are short circuited. The current in the winding goes to a very high value.
63. Delta and wye
64. Wye
65. The start-end of one winding must be connected to the finish-end of the next winding.
66. No
67. The winding has excessive current flowing in it.
68. 1, 5, and 9; or 2, 6, and 10

Capacitance

- Capacitors, the physical devices which possess capacitance, are almost as common as resistors in electrical and electronic circuits. They are used in electric motors, automobile ignition systems, electronic photo-flashes, and fluorescent lamp starters to name just a few applications.

  In this chapter you will learn how capacitors behave in ac and dc circuits. After completing the chapter, you will be able to determine the values of reactance, voltage, and current in capacitive circuits. You will be able to write the specifications needed to order capacitors for various uses. You will be able to test capacitors for opens and shorts. And, you will be able to determine how long it takes a capacitor to charge in a resistor-capacitor circuit.

12-1 TERMINOLOGY

The ability to store energy in the form of electrical charge is called capacitance. The symbol used to represent capacitance is C. A physical device designed to possess capacitance is called a capacitor. In its simplest form a capacitor is nothing more than two conductors separated by an insulator. Figure 12-1 illustrates a capacitor consisting of two pieces of aluminum (a conductor) separated by a sheet of polyester plastic (an insulator). In capacitors the conductors are called *plates*, and the insulator is called a *dielectric*.

12-2 BASIC CAPACITOR ACTION

A capacitor stores energy when an electric charge is forced onto its plates by some other energy source such as a battery. When a battery is connected to the capacitor of Fig. 12-1, a current flows until the capacitor becomes charged (Fig. 12-2.). Electrons from the negative terminal of the battery move through the connecting lead and pile up on one of the plates. At the same instant electrons from the other plate move through the connecting lead to the positive terminal of the battery. The net result is that one plate of the capacitor ends up with an excess of electrons (negative charge). The other plate ends up with a deficiency of electrons (positive charge). Note that in the process of charging the capacitor no electrons move from one plate through the dielectric to the other plate. Further, once the capacitor is charged all current stops, and the capacitor behaves like an open switch.

Because of opposing voltages, the current stops once the capacitor is charged. The opposite charges on the plates of the capacitor create a new energy source. The energy stored in the capacitor produces a voltage equal to that of the battery (the original energy source). Since the capacitor's voltage is equal to, and in opposition to, the battery voltage, a

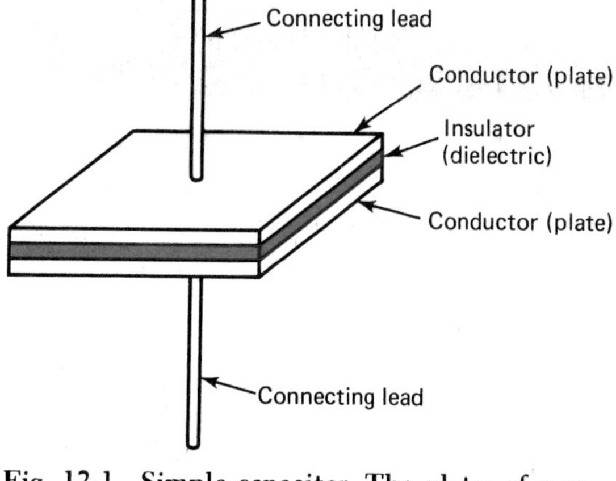

Fig. 12-1 Simple capacitor. The plates of a capacitor are always insulated from each other.

From page 227:
Plates

Dielectric

On this page:
Charged
capacitor

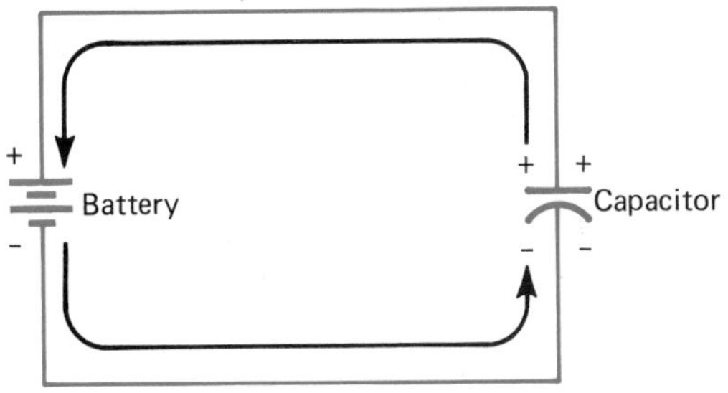

Fig. 12-2 Capacitor being charged. The current stops when the capacitor is fully charged.

state of equilibrium exists. No current can flow in either direction.

A charged capacitor can be disconnected from the original energy source (the battery) and used as a new energy source. If a voltmeter is connected to the capacitor it will register a voltage. If a resistor is connected to the capacitor (Fig. 12-3), current will flow through the resistor. However, a capacitor has limited use as a primary energy source for two reasons: (1) For its weight and size, the amount of energy it can store is small compared to a battery. (2) The voltage available from the capacitor rapidly diminishes as energy is removed from the capacitor.

In a dc circuit, a capacitor will act as an open circuit after its initial charge. In ac circuits, capacitors periodically store and return energy. In other words, they charge and discharge as the ac source voltage changes.

Although the amount of energy stored in a capacitor is small, a capacitor can deliver a shock. The shock can be very severe (even fatal) if the capacitor is large and charged to a high voltage. Treat charged capacitors as you would any other electrical energy source.

When a capacitor is charged, an electric field is established between the two charged plates. This electric field can be represented by lines between the charged plates as shown

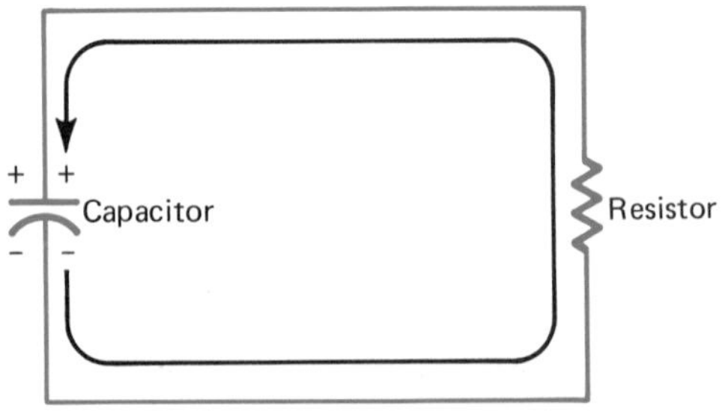

Fig. 12-3 Capacitor as an energy source. The capacitor forces current through the resistor until the capacitor is discharged.

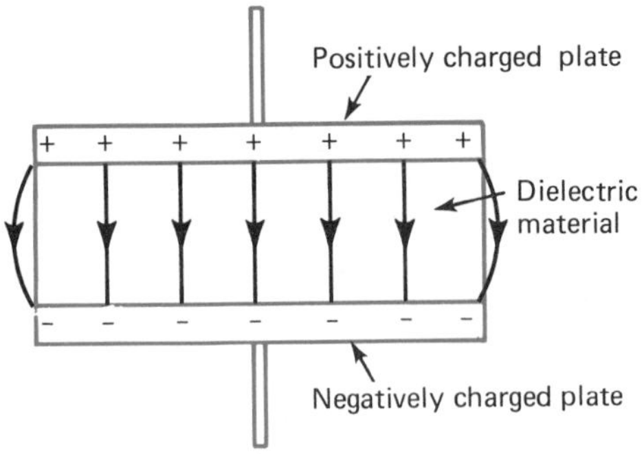

Fig. 12-4 Electric field of a charged capacitor. The field stresses the dielectric material.

in Fig. 12-4. These lines represent an electric field in the same way that lines between magnetic poles represent a magnetic field. The electric field exerts its force on any charge within the field. A negatively charged particle, such as an electron, will be forced toward the positive plate of the capacitor. The dielectric material between the plates of a capacitor is under stress because of the force of the electric field. The force of the field reacts against the charged particles (electrons) contained within the dielectric material. One way to visualize this stress is illustrated in Fig. 12-5. In Fig. 12-5 the orbital path of an electron around its nucleus is elongated. The electric field forces the electron toward the positive plate and away from the negative plate. The modified orbital path of the electron raises the energy level of the electron. This is the way the capacitor stores energy.

A dielectric material, such as that illustrated in Fig. 12-5, can only withstand a certain

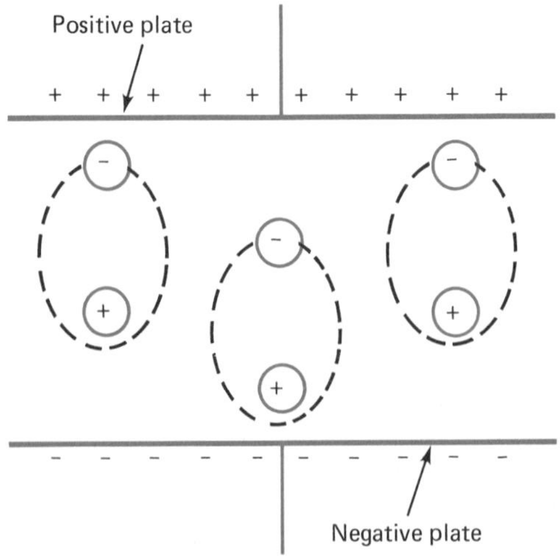

Fig. 12-5 Stress on the dielectric material. Energy is stored in the stressed dielectric material.

amount of stress. Then the electrons in the material break free of the parent atoms and become free electrons available as current carriers. Of course, when this happens the dielectric is no longer an insulator. The capacitor's negative plate discharges through the dielectric to the positive plate. So much energy is released in the discharge that a hole is burned in the dielectric material in the region where the capacitor discharged. This then becomes a path for further electron flow between the plates, and the capacitor is no longer usable.

12-3 VOLTAGE RATING

The strength of the electric field in a capacitor is related to the distance between plates and the voltage across the plates. Therefore, a specific capacitor can withstand being charged only to a given voltage before its dielectric breaks down, allowing current to flow between the plates. The maximum voltage to which a capacitor should be charged is determined by the manufacturer. It is listed as part of the specifications of the capacitor. The manufacturer's voltage rating is usually listed as the "Direct Current Working Voltage" (DCWV). Suppose a capacitor is rated at 200 DCWV. It can be operated at 200 V for long periods of time with no deterioration of the dielectric. If it is subjected to 250 or 300 V dc, it may not immediately cause the dielectric to break down. However, its life expectancy will be greatly shortened.

Review Questions

1. What is a capacitor?
2. What is the symbol for capacitance?
3. Name the parts of a capacitor.
4. Describe what happens when a capacitor is charged.
5. When a capacitor is being charged, do electrons move through the dielectric?
6. Can a charged capacitor store energy once it is disconnected from a battery?
7. Where is the energy in a capacitor stored?
8. What does the abbreviation DCWV stand for?
9. Why do capacitors have voltage ratings?
10. What two factors determine the electric field strength in a capacitor?

12-4 UNIT OF CAPACITANCE

Voltage rating

Working voltage

Farad

$C = \frac{Q}{V}$

The base unit of capacitance is the *farad*. The abbreviation for farad is F. A farad is that amount of capacitance which will store one coulomb (C) of charge when the capacitor is charged to one volt. In other words, a farad is a coulomb per volt (C/V).

As noted in Sec. 2-20, a letter is often used to denote two different quantities. In that section it was pointed out that the letter *W* is used as a symbol for energy and W is the abbreviation for the base unit for power, the watt. In a similar manner, *Q* is used as the symbol for charge, and *Q* is used as the symbol for coil quality. Now, notice that *C* is used as the symbol for capacitance, and C is used as the abbreviation for the base unit of charge, the coulomb.

Farad is the unit of capacitance (C), coulomb is the unit of charge (*Q*), and volt is the unit of voltage (V). Therefore, capacitance can be mathematically expressed as:

$$C = \frac{Q}{V}$$

This relationship will be very useful in understanding voltage distribution in series capacitor circuits.

Example 12-1

What is the capacitance of a capacitor that requires 0.5 coulombs (C) to charge it to 25 V?

Given: Charge (*Q*) = 0.5 C
Voltage (V) = 25 V

Find: Capacitance (C)

Known: $C = \frac{Q}{V}$

Solution: $C = \frac{0.5\text{ C}}{25\text{ V}} = 0.02\text{ F}$

Answer: The capacitance is 0.02 farads.

The 0.02 F calculated above is a very high value of capacitance. In most circuits the capacitances used are much lower. In fact, they are usually so much lower that the base unit of farad is too high to conveniently express their values. The microfarad (μF), which is 1/1,000,000 or 1×10^{-6} farad, and the picofarad (pF), which is 1/1,000,000,000,000 or 1×10^{-12} farad, are more convenient units. Both are used extensively in specifying values of capacitors in electronic circuits. Often it is necessary to convert from one unit to another. For example, a manufacturer may specify a 1000 pF capacitor in his parts list. The technician repairing the equipment may not have a capacitor marked 1000 pF. However, an equivalent capacitor marked in μF may be available. The technician must convert pF into μF by multiplying pF by 10^{-6}. The net result is:

$$1000 \text{ pF} = 1000 \times 10^{-6} \ \mu\text{F} = 0.001 \ \mu\text{F}$$

To convert from μF to pF, the reverse process is used. Divide the μF value by 10^{-6} to obtain the equivalent pF value. Since dividing by 10^{-6} is the same as multiplying by 10^6, the conversion proceeds as follows:

$$\begin{aligned} 0.001 \ \mu\text{F} &= 0.001 \times 10^6 \text{ pF} \\ &= 1000 \text{ pF} \end{aligned}$$

which agrees with the results above.

12-5 DETERMINING CAPACITANCE

The capacitance of a capacitor is determined by four factors:

1. Area of the plates
2. Distance between the plates
3. Type of dielectric material
4. Temperature

Temperature coefficient

The temperature of the capacitor is the least significant of the four factors. It need not be considered for many general purpose applications of capacitors. However, in more critical applications (such as oscillator circuits) the temperature characteristic of a capacitor is very important. The temperature characteristic of a capacitor is determined primarily by the type of dielectric material used in constructing the capacitor. Some dielectric materials will cause an increase in capacitance as the temperature increases. This is referred to as a positive temperature coefficient. Other dielectric materials have negative temperature coefficients; that is, the capacitance increases as the temperature decreases. Still other dielectric materials have zero temperature coefficients. Their capacitance is independent of temperature. The temperature coefficient of a capacitor is specified by the manufacturer in "parts per million per degree celsius." Negative temperature coefficient is abbreviated N, positive temperature coefficient is P, and zero temperature coefficient is NPO. The ambient, reference temperature at which the capacitor is rated is 25°C. (Note that the letter C is used here as the abbreviation for celsius. With only 26 letters in the alphabet and thousands of abbreviations and symbols, it is no wonder that some letters are used more than once.)

Plate area

The capacitance of a capacitor is directly proportional to the area of its plate (or the area of its dielectric). All other factors remaining the same, doubling the plate area will double the capacitance [Fig. 12-6(*a*)]. This is so, because doubling the plate area also doubles the area of dielectric material. For a given voltage across the capacitor, the strength of the electric field is independent of the plate area. Thus, when the dielectric area is increased, the amount of energy stored in the dielectric is increased and the capacitance is also increased. (Remember, capacitance was defined as *the ability to store energy*.)

Distance between plates

Other factors being equal, the amount of capacitance is inversely proportional to the distance between the plates [Fig. 12-6(*b*)]. As the plates are separated, the strength of the electric field between the plates decreases. The force on the electrons within the dielectric decreases accordingly. Again, the amount of energy stored in the capacitor, for a given voltage applied to the capacitor, would decrease. Thus, the capacitance decreases.

Dielectric constant

The capacitance of a capacitor is also controlled by the type of dielectric material used [Fig. 12-6(*c*)]. When subjected to the same electric field some materials will undergo greater molecular distortion than others. In general, those materials which undergo the greatest distortion will store the most capacitance. The ability of a dielectric material to distort and store energy is indicated by its *dielectric constant* (K). The dielectric constant of a material is a pure number. It compares the material's ability to distort and store energy when in an electric field to the ability

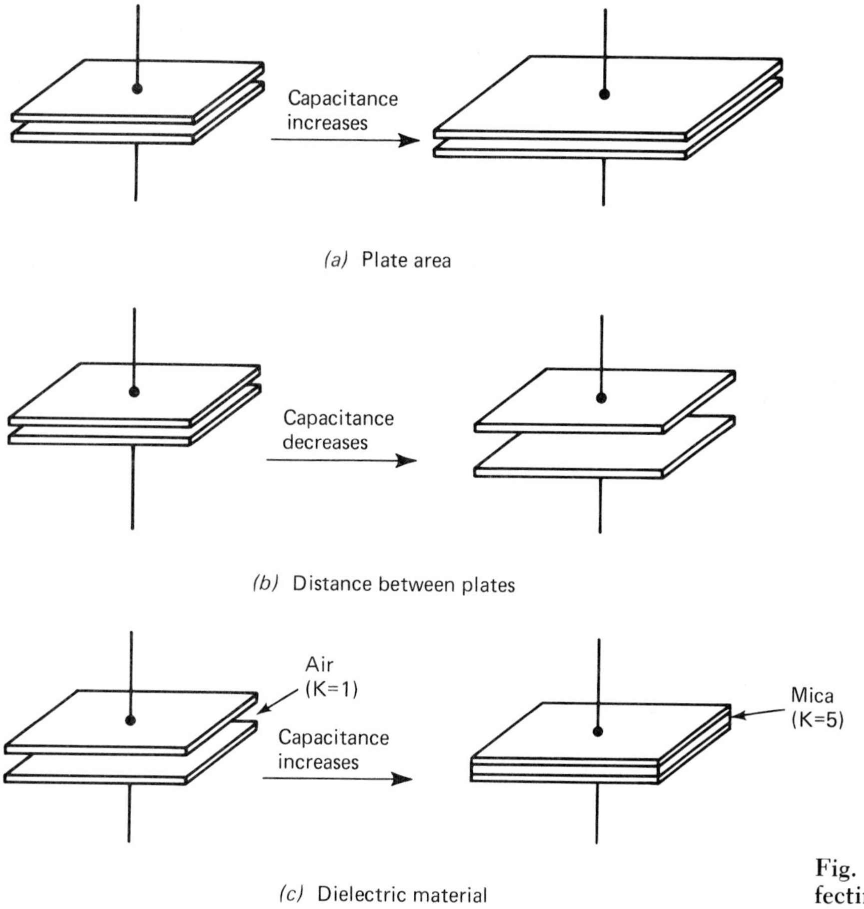

Fig. 12-6 Major factors affecting capacitance.

of air to do the same. Since air is used as the reference, it has been given a K equal to 1. Mica, often used as a dielectric, has a dielectric constant approximately 5 times that of air. Therefore, for mica, K = 5 (approximately). Suppose all other factors (plate area, distance between plates, and temperature) are the same. Then a capacitor with a mica dielectric will have 5 times as much capacitance as one using air as its dielectric. Dielectric constants for materials commonly used for dielectrics range from 1 for air to over 4000 for some types of ceramics.

The capacitance of a two-plate capacitor can be calculated by using the formula

$$C = \frac{0.885 \times 10^{-6} \times K \times A}{D}$$

In this formula K is the dielectric constant. The area A is of *either* of the plates and is in square centimeters. The distance between the plates D is in millimeters. The permittivity of air or a vacuum is the constant 0.885×10^{-6}. *Permittivity* is a factor indicating the ability to concentrate electric flux or field. The capacitance C will be in microfarads.

Review Questions

11. What is the base unit of capacitance?
12. What is the abbreviation for the base unit of capacitance?
13. How many pF are there in 0.003 μF?
14. How many μF are there in 2200 pF?
15. Define a farad in terms of other base units.
16. List four factors that determine the capacitance of a capacitor.
17. Which of the above four factors in question 16 has the least effect on capacitance?
18. When the plate area of a capacitor is reduced by half, what happens to its capacitance?

19. Suppose the distance between the plates of a capacitor is reduced by half. What happens to the capacitance?

20. What does the dielectric constant of a material indicate?

21. Does air have a smaller or larger dielectric constant than other common dielectric materials used in making capacitors?

12-6 TYPES OF CAPACITORS

Many different types and styles of capacitors (Fig. 12-7) are manufactured to satisfy the needs of the electronics industry. Capacitors may be named to indicate their dielectric material, their enclosure, the process used in their construction, or their intended use.

Electrolytic Capacitors

Electrolytic capacitor

Electrolytic capacitors [Fig. 12-7(*c*) and (*e*)] provide more capacitance for their size and weight than any other type of capacitors. This is their primary advantage over other capacitors.

A common electrolytic capacitor consists of two aluminum foil plates separated by a layer of fine gauze or other absorbent material. The foil plates and separators are long, narrow strips. These strips are rolled up and inserted in an aluminum container (Fig. 12-8). One plate (the negative plate) is usually electrically connected to the aluminum can (container). Electrical terminals from the foil plates are brought out one end of the container. Then the container is sealed.

One of the aluminum plates in Fig. 12-8 is oxidized. Since aluminum oxide is an in-

(a) Disc Ceramic

(b) Tuning (variable)

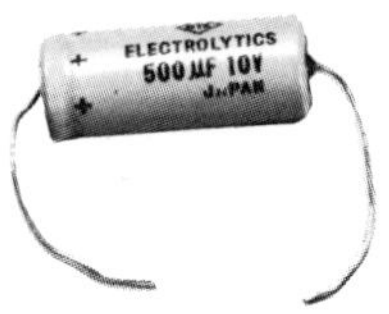

(c) Electrolytic (axial lead)

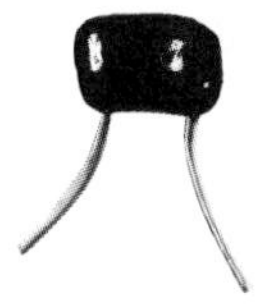

(d) Dipped Mica

(e) Electrolytic (radial lead)

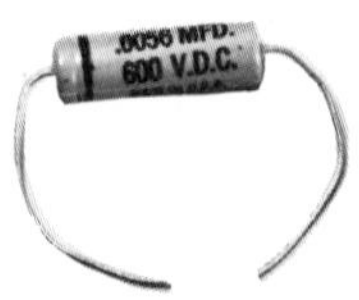

(f) Paper (molded)

Fig. 12-7 Types of capacitors.

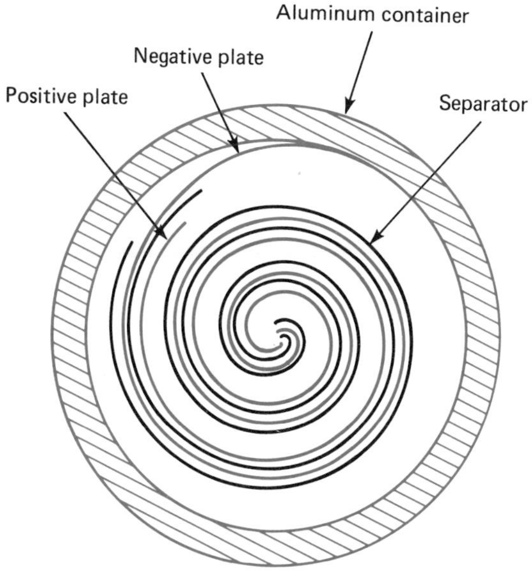

Fig. 12-8 Electrolytic capacitor. The container is electrically connected to the negative plates.

sulator, it is used for the dielectric. The separators are saturated with a chemical solution (such as borax) which is called the *electrolyte*. The electrolyte, having relatively good conductivity, serves as part of one of the plates of the capacitor (Fig. 12-9). It is in direct contact with both the dielectric and the pure aluminum foil plate. The electrolyte is also necessary in forming and maintaining the oxide on one of the plates.

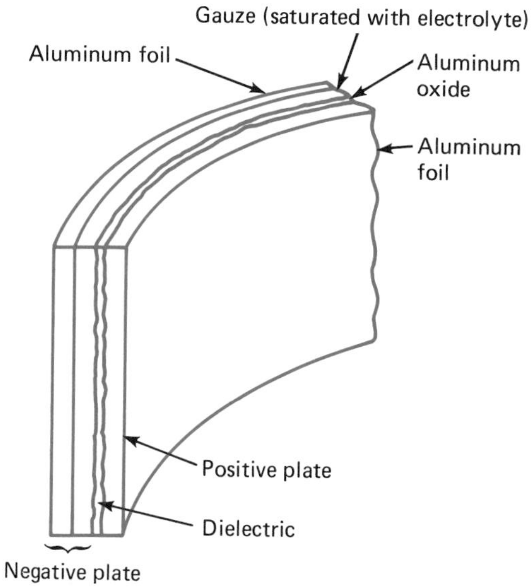

Fig. 12-9 Section of an electrolytic capacitor.

The method used to produce electrolytic capacitors results in the plates being polarized. When using these capacitors, always keep the same voltage polarity as was used in manufacturing them. This polarity is always marked on the body of the capacitor. Again, the aluminum container (which may be encased in an insulating tube) is usually connected to the negative plate. It is never connected to the positive plate. Reverse polarity on an electrolytic capacitor causes excessively high current in the capacitor. It causes the capacitor to heat up, and it can cause the capacitor to explode. Thus, the common electrolytic capacitor is limited to use in dc circuits.

Electrolyte

Polarized

Special electrolytic capacitors are manufactured for use in ac circuits. These capacitors are usually listed in parts catalogs as "nonpolarized" or "ac" electrolytic capacitors. An ac electrolytic capacitor is really two electrolytic capacitors packaged in a single container (see Fig. 12-10). The two internal capacitors are in series with their positive ends connected together. Regardless of the polarity on the leads of the ac electrolytic capacitor, one of the two internal capacitors will be correctly polarized. The correctly polarized capacitor will limit the current flowing through the reverse polarized capacitor.

Many nonpolarized (ac) electrolytic capacitors are designed for intermittent use only. These capacitors will overheat (and often rupture) if used for extended periods of time.

The oxide on the positive plate of an electrolytic capacitor (Fig. 12-9) is very thin (approximately 0.25 micrometers) and rough. Its thinness keeps the plates very close together. Its roughness effectively increases the plate area. These two factors make the capacitance of an electrolytic capacitor very high for its weight and volume.

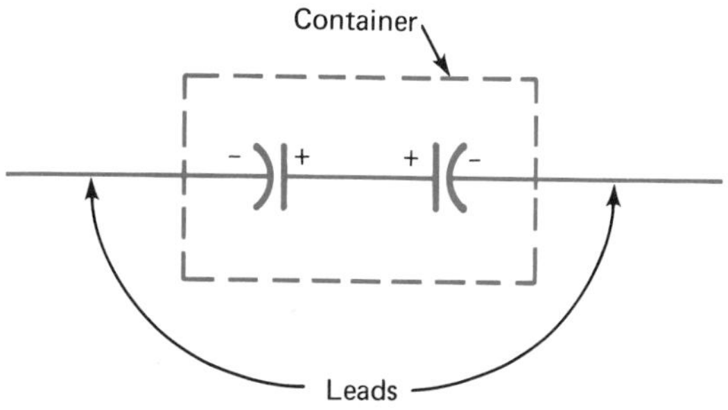

Fig. 12-10 Nonpolarized electrolytic capacitor. Regardless of lead polarity, one of the internal capacitors will be correctly polarized.

Tantalum

Tantalum, a conductive metallic element, is also used for the foil plates of electrolytic capacitors. Tantalum electrolytic capacitors are smaller, more stable and more reliable than aluminum electrolytic capacitors. In an actual capacitor, some electrons do find their way from one plate to the other. This produces a *leakage current*. Tantalum electrolytic capacitors have less leakage current (higher dielectric resistance) than do aluminum electrolytic capacitors. However, tantalum electrolytic capacitors are more expensive than their aluminum counterparts.

Molded capacitor

Dipped capacitor

Tubular capacitor

Metallized film capacitor

Film and Paper Capacitors

Paper capacitors and film capacitors are constructed using a rolled-foil technique illustrated in Fig. 12-11. Notice in this figure that the two foil plates are offset so that one plate extends out each end. The leads for the capacitor are then connected to the ends of the completed roll. After the leads are attached, the complete assembly (plates, lead connections, and dielectric) is covered with a protective coating of insulation. Sometimes the insulation is molded around the assembly to produce a *molded capacitor*. Sometimes it is dipped in a plastic insulating material producing a *dipped capacitor*. Sometimes it is placed in an insulated tube and the ends of the tube are then sealed. This is referred to as a *tubular capacitor*.

The paper-dielectric capacitor is usually bulkier, less stable, and has less dielectric resistance than the film capacitors. The paper dielectric may absorb moisture from the atmosphere if there is the slightest flaw in the end seals. Of course, changing the moisture content of the paper changes the dielectric constant and the insulation resistance of the paper. Because of these limitations of paper as a dielectric material, it is rapidly being replaced by plastic film materials. Such materials as polystyrene, polyester (Mylar) and polycarbonate are popular replacements.

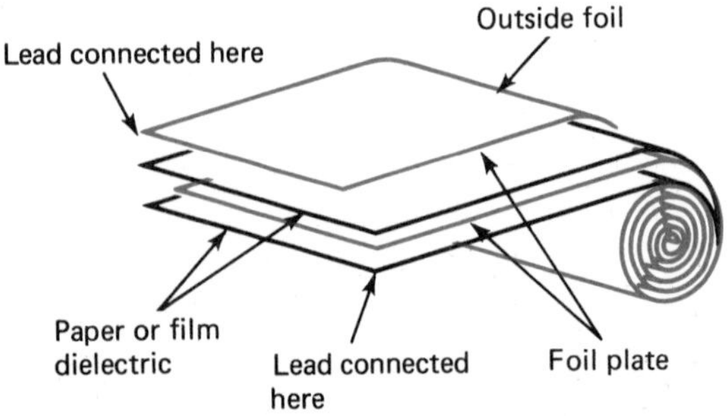

Fig. 12-11 Paper or film capacitor construction.

Many of the plastic film capacitors use metallized plates. With these metallized film capacitors, the metal plate is deposited directly on the film. This keeps the distance between plates as small as possible and produces a small compact capacitor.

Film and paper capacitors used in industrial applications can be as large as several hundred microfarads. They are often rated in terms of VA as well as DCWV. These large values are usually enclosed in a metal container which is filled with a special insulation oil. However, in most electronic circuits paper and film capacitors are less than one microfarad in value.

Ceramic Capacitors

For low value capacitors (less than 0.1 μF) ceramic is a popular dielectric material. The most common style of ceramic capacitor is the disc ceramic [see Fig. 12-7(*a*)].

The structure of a typical ceramic capacitor is detailed in Fig. 12-12. As indicated in this figure, these capacitors have a simple, strong structure. This makes them a tough, reliable, general-purpose capacitor. They are used in a wide variety of applications where low values of capacitance are needed.

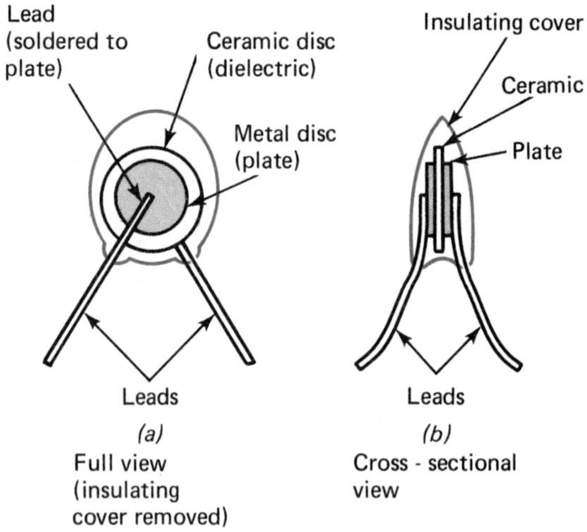

Fig. 12-12 Ceramic capacitor construction.

Mica Capacitors

Mica capacitors are limited to even lower values than are ceramic capacitors. This is be-

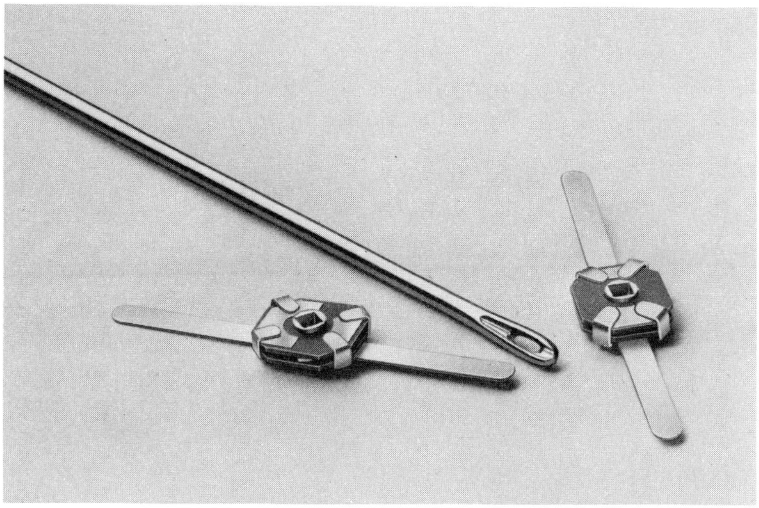

Fig. 12-13 Trimmer capacitors. (Courtesy of Johanson Manufacturing Corp.)

Variable capacitor

Feed-through capacitor

Stand-off capacitor

Trimmer capacitor

cause mica has a lower dielectric constant than ceramic. However, it is easier to control production tolerances with mica dielectrics, and mica has good high-temperature characteristics.

It takes less space to provide a given value of capacitance with ceramic than with mica. Also it is easier to construct a ceramic capacitor. Therefore, mica capacitors are not as common as ceramic capacitors.

Specific-Use Capacitors

So far we have classified capacitors by the type of dielectric materials used in their manufacture. Capacitors can also be classified by their function or by the way they are connected in a circuit. Some of the more common names in this classification are feed-through, filter, padder, trimmer, tuning, and stand-off.

Padder, trimmer, and tuning capacitors are all variable capacitors (Fig. 12-7). They are used in circuits which allow one frequency to pass. Circuits of this type are used for tuning radio and television sets to a particular station. The trimmer and padder are adjusted by the electronics technician at the factory. The tuning capacitor is varied by the user of the receiver as he tunes in a station. These variable capacitors generally use either air or one of the plastics as a dielectric. Examples of miniature trimmer capacitors are pictured in Fig. 12-13.

One of the plates of a feed-through capacitor is the tubular, metal case which encloses the capacitor (Fig. 12-14). The two leads extending from the ends of the capacitor are both connected to the other plate. In use, the capacitor protrudes through a hole in an electrical (metal) shield which separates two parts of an electronic system. The case (one plate) is electrically connected to the shield through which it protrudes. The case may be soldered as in Fig. 12-14, or it may be threaded and held in place with a nut. This arrangement allows a dc current to feed through the shield while radio frequencies are shorted through the shield to ground. The radio frequencies are bypassed from either lead.

Stand-off capacitors (Fig. 12-15) are also used to bypass high frequencies to the ground of an electronic system. These capacitors have one plate connected to a threaded terminal which bolts the capacitor to the grounded

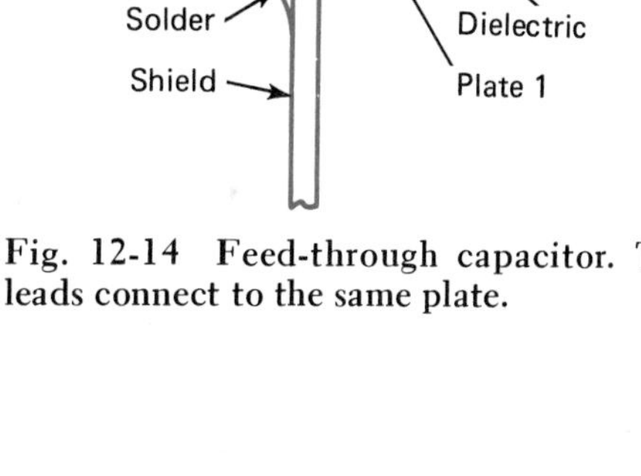

Fig. 12-14 Feed-through capacitor. The two leads connect to the same plate.

Filter capacitor

Energy storage capacitor

$W = 0.5CV^2$

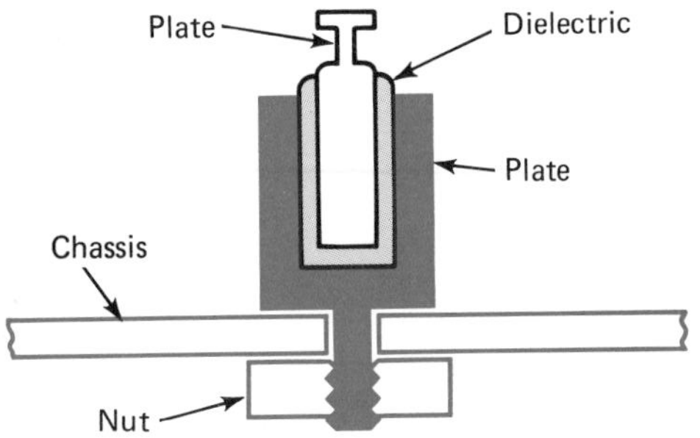

Fig. 12-15 Stand-off capacitor.

chassis. The other plate is connected to a terminal which serves as a tie point for other electronic components. Radio frequencies going through the components connected to the stand-off capacitors are bypassed to the chassis and effectively eliminated.

Filter capacitors are high-value capacitors used to filter (flatten) out pulsations in dc voltage. For example, a filter capacitor can change the pulsating dc in Fig. 12-16(*a*) to that shown in Fig. 12-16(*b*). Many filter capacitors are electrolytic capacitors which are polarized. They must be used in circuits where the polarity of the voltage applied to them will never be reversed.

Most filter capacitors have a value somewhere between 1 μF and 5000 μF. The voltage rating may be anywhere from a few volts to thousands of volts.

Energy Storage Capacitors

Capacitors designed to release their stored energy (discharge) during a short time interval are sometimes called *energy storage capacitors*. They may be either electrolytic or nonelectrolytic capacitors. They are used in such devices as electronic flash units (used in photography) and capacitor discharge ignition systems for automobiles. In the electronic flash unit a charged capacitor is discharged through a xenon flash tube. In the ignition system the capacitor is discharged through the ignition coil. In both of these examples the limitations of a capacitor as an energy source are not important. Neither circuit requires much energy, and constant voltage is not necessary. However, because of the short time interval involved, both devices achieve high instantaneous power levels. During the rapid discharge of an energy storage capacitor, the terminal leads and plates must carry many amperes of current. The dielectric material, plates, and enclosure must be able to withstand extreme stress created by such rapid discharge.

Large, high-voltage capacitors like those shown in Fig. 12-17 can be charged to many thousands of volts. They can store several thousand joules of energy. These capacitors are used in industrial applications such as capacitor discharge welding. They can also be used for less demanding applications such as filtering.

In addition to the standard voltage and capacitance rating, most energy storage capacitors have two additional ratings. One is a peak current rating and the other is an energy rating. The energy rating specifies how many joules (watt-seconds) of energy the capacitor will store when charged to its rated voltage. The peak current rating allows one to calculate the minimum resistance through which the fully charged capacitor can be discharged. Exceeding the peak current rating one time will not immediately destroy the capacitor. However, repeated discharging at excess current levels will cause premature failure of a capacitor.

The amount of energy a capacitor can store depends upon its capacitance and voltage rating. We can calculate the amount of stored energy with the formula

$$W = 0.5CV^2$$

The calculated energy (W) will be in joules when capacitance (C) is in farads and voltage (V) is in volts.

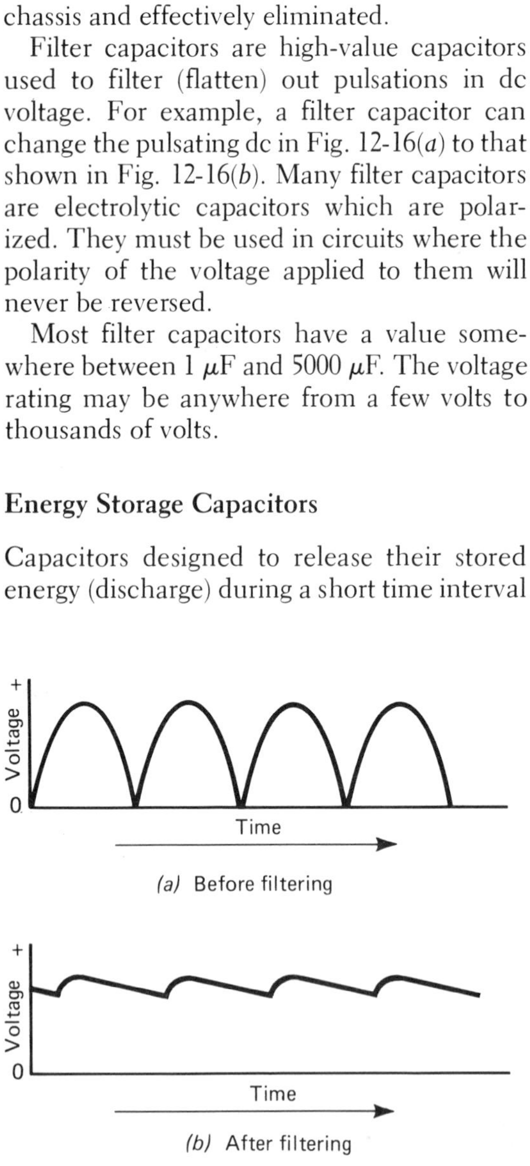

Fig. 12-16 Filtering pulsating dc.

Fig. 12-17 Energy storage capacitors. (Courtesy of Aerovox Industries, Inc.)

Example 12-2

An energy storage capacitor is rated at 300 μF and 450 V. How much energy can it store?

Given: $C = 300\ \mu F$
$V = 450\ V$

Find: W

Known: $W = 0.5CV^2$

Solution: $W = 0.5 \times 300 \times 10^{-6} \times 450 \times 450$
$= 30.4\ J$

Answer: The capacitor stores 30.4 joules of energy.

The amount of energy in the above example is not very great. A 40-W lamp uses more energy every second than the capacitor stored. However, the capacitor can provide high power if it is discharged rapidly. For instance, if the capacitor in Example 12-2 is discharged in two milliseconds it will produce the power calculated below.

$$P = \frac{W}{t} = \frac{30.4\ J}{0.002\ s} = 15{,}200\ W$$

Review Questions

22. What is the major advantage of electrolytic capacitors?
23. What material is used as the dielectric in an electrolytic capacitor?
24. What purpose does the electrolyte serve in an electrolytic capacitor?
25. Is the negative or positive plate of an electrolytic capacitor connected to the aluminum container?
26. List two metals used as plate foil in electrolytic capacitors.
27. What happens when an electrolytic capacitor has reverse-polarity voltage connected to it?
28. List four types of capacitors that are named after the dielectric material used in them.
29. What are the advantages of film capacitors over paper capacitors?
30. What is a metallized capacitor?
31. Why are mica capacitors not as popular as ceramic capacitors?
32. List three types of variable capacitors often used in radio receivers.
33. What are stand-off and feed-through capacitors used for?
34. What is a filter capacitor?
35. Where are energy storage capacitors used?
36. How much energy will a capacitor store if it is rated at 150 μF and 500 V?

12-7 SCHEMATIC SYMBOLS

Capacitor symbols

Capacitors can be broadly classified as either fixed or variable and as either polarized or nonpolarized. All variable capacitors are nonpolarized. Schematic symbols for capacitors are shown in Fig. 12-18. The curved line rep-

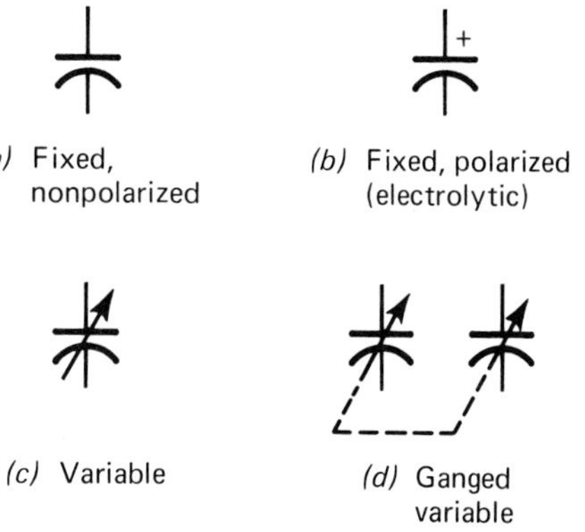

Fig. 12-18 Schematic symbols for capacitors.

Outside foil

Capacitors in series

resenting one of the plates in the symbols of Fig. 12-18 has no particular significance for some types of capacitors. For other types it indicates a specific lead of the capacitor.

For a polarized capacitor, the curved plate of the symbol indicates the negative terminal of the capacitor. With variable capacitors, the curved line identifies the plate which is electrically connected to the frame or adjusting mechanism. The curved line identifies the outside foil (Fig. 12-11) on a rolled-foil (film or paper) capacitor. On the actual capacitor, the lead connecting to the outside foil has to be identified. This is done with a color band around the end of the capacitor from which this lead protrudes.

12-8 CAPACITORS IN DC CIRCUITS

When the switch of a dc circuit containing only capacitance is closed, a surge of current charges the capacitor. Once this surge is over, no more current flows (Fig. 12-19). The capacitor, once charged, appears to the battery just like an open switch. For practical purposes it has infinite impedance.

The magnitude of the surge of current in Fig. 12-19 is controlled by the resistance in the circuit. In this circuit the resistance is composed of the lead resistance and the internal resistances of the battery and the capacitor. These resistances are usually very low; therefore, the surge current can be very high. However, this surge current only lasts for an extremely short period of time.

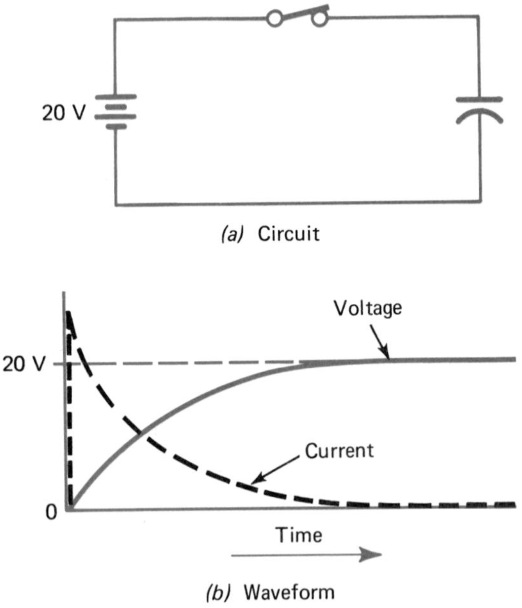

Fig. 12-19 Capacitor charging in dc circuit. The current reaches its maximum value almost the instant the switch is closed. The voltage developed across the capacitor increases more slowly.

Notice in Fig. 12-19(*b*) that the current is at peak value while the voltage is just leaving zero. By the time the voltage is at maximum, the current is at zero. This is a characteristic of capacitors in either dc or ac circuits.

Voltage Distribution

The dc source voltage will divide among series capacitors in inverse proportion to their capacitance (see Fig. 12-20). The voltmeter across the 1 μF capacitor indicates twice as much voltage as the one across the 2 μF capacitor. This happens because capacitors in series must receive the same quantity of charge. (There is only one path for the charging current.) The same amount of charge produces less voltage in a large capacitor than it does in a small capacitor. Remember that $C = Q/V$; therefore, $V = Q/C$. If Q is the same for both capacitors, then the higher the capacitance is, the lower the voltage will be.

Sometimes capacitors are connected in series because the voltage rating of either capacitor is lower than the source voltage. In this case the capacitors should have as close to equal capacitances as possible. Then the source voltage will divide equally between the two capacitors.

Review Questions

37. Which plate of an electrolytic capacitor is

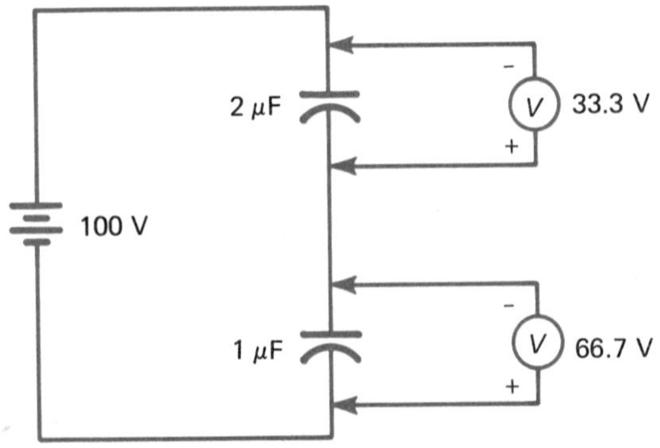

Fig. 12-20 Dc voltage distribution with series capacitors. The smallest capacitor develops the most voltage.

represented by the curved line of the capacitor symbol?

38. How is the outside-foil lead identified on a paper or film capacitor?

39. Does a capacitor allow dc to flow through it continuously?

40. A 10 μF and 5 μF capacitor are connected in series to a dc source. Which capacitor will develop the higher voltage? Why?

12-9 CAPACITORS IN AC CIRCUITS

A capacitor controls the current in an ac circuit. It controls current by storing energy which produces a voltage in the capacitor. The voltage produced by the capacitor's stored energy is always in opposition to the source voltage.

Careful study of Fig. 12-21 will show exactly how a capacitor controls circuit current. The voltage waveform in Fig. 12-21 represents the source (generator) voltage applied across the capacitor. The current waveform represents the current flowing through the ammeter. Assume the source voltage in Fig. 12-21(*a*) has just changed polarity and is increasing in the positive direction. Also assume the capacitor is completely discharged at the instant the source voltage changes polarity. Now, as the generator (source) voltage builds up, the capacitor *must charge* to keep up with the increasing generator voltage. The current in the circuit will be maximum at this time because the generator's voltage is changing most rapidly at this time. (Remember that the faster the capacitor's voltage must rise, the more charge per second it requires. And, more charge per second means more current.) As the first quarter cycle progresses [Fig. 12-21(*a*)] the generator's voltage increases less rapidly. Consequently, the current in the circuit decreases. At the end of the first quarter cycle there is an instant when the generator's voltage is stable (neither increasing nor decreasing). At that instant the circuit current is zero. The capacitor's voltage is equal and opposite to the source voltage.

Let us note four things from observing the action during this first quarter cycle [Fig. 12-21(*a*)]:

1. The generator is charging the capacitor. Thus electrons (current) leave the negative terminal and enter the positive terminal of the generator.
2. The generator is transferring energy to the capacitor as it charges the capacitor.
3. The capacitor allows just enough current to charge it to the value of the source voltage.
4. The current and voltage are 90° out of phase with each other. Specifically the *current leads the voltage* by 90°.

Current leads voltage by 90°

Transferring energy

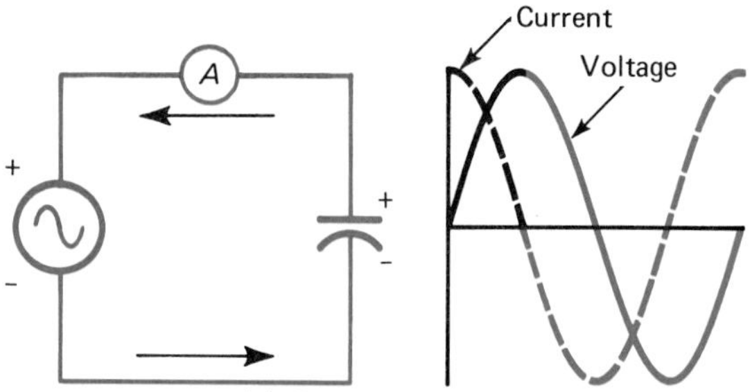

(a) First quarter cycle

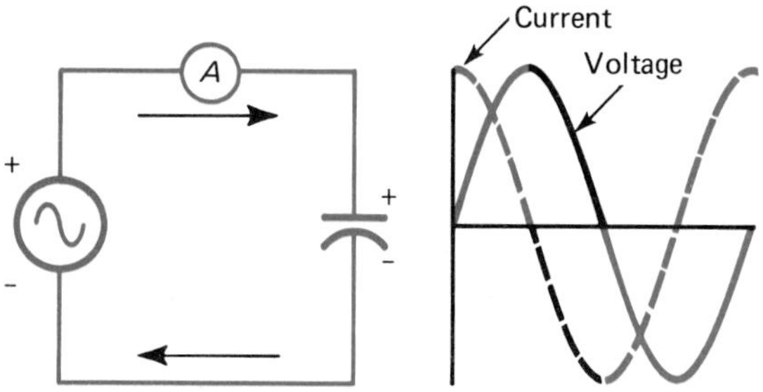

(b) Second quarter cycle

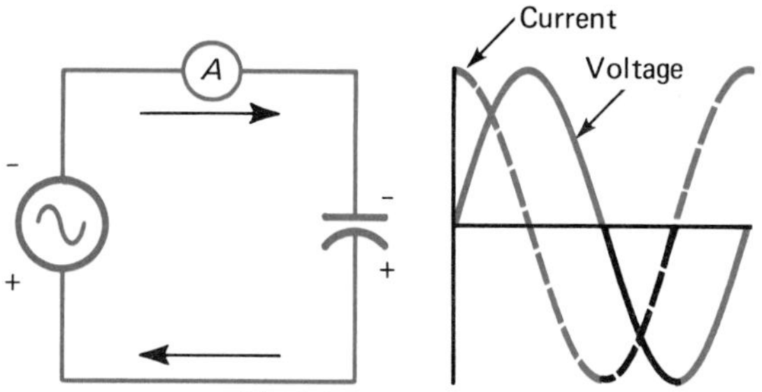

(c) Third quarter cycle

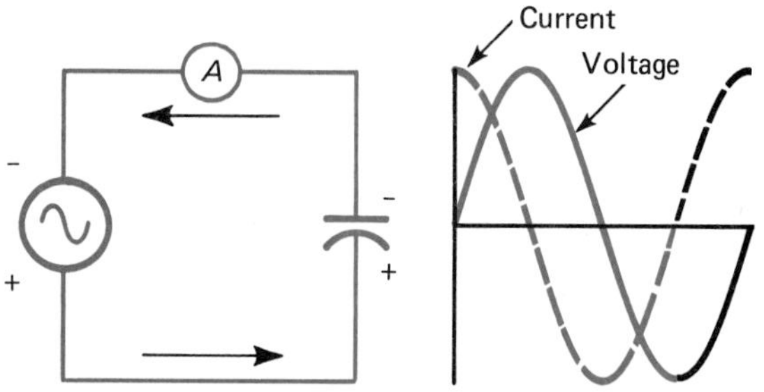

(d) Fourth quarter cycle

Fig. 12-21 Energy transfer in a capacitor circuit. Every other quarter cycle the capacitor becomes the energy source.

Now we will look at the second quarter cycle [Fig. 12-21(*b*)]. During this part of the cycle, the generator's voltage is decreasing. Since the capacitor is connected directly across the generator, its voltage must also decrease. For the capacitor to lose voltage, it must discharge. This means that electrons must leave its negative terminal and enter its positive terminal. To discharge, the capacitor has to force electrons through the generator against (in opposition to) the generator's voltage. Thus, the capacitor is now returning its stored energy back to the generator. Figure 12-21(*b*) shows that the decrease in the generator's voltage causes an increase in the current in the circuit in the opposite direction. The current is maximum when the generator voltage is zero. The current in this case, however, is a result of the capacitor discharging. Again, let us summarize the major action during the second quarter cycle:

Resistance loss

Dielectric loss

Equivalent series resistance

Returning stored energy

1. The capacitor is discharging through the generator.
2. The capacitor is transferring energy back to the source (generator) as it discharges.
3. The capacitor furnishes just enough current to discharge it to the value of the source voltage.
4. The current still leads the voltage by 90°.

Inspection of the third quarter cycle [Fig. 12-21(*c*)] will show that it is essentially the same as the first quarter cycle. The only difference is that the voltage and current have reversed. The generator again takes control and furnishes the energy needed to charge the capacitor. The capacitor stores energy during this third quarter of the cycle.

The fourth quarter cycle [Fig. 12-21(*d*)] is like the second quarter cycle. The capacitor now discharges through the generator and returns its stored energy to the source. As in the previous three quarters of the cycle, the current and voltage remain 90° out of phase.

Power in Capacitor Circuits

As you can gather from the previous discussion and illustration (Fig. 12-21) pure capacitance can use no energy. It merely stores energy for one quarter of a cycle and returns it the next quarter of a cycle. Since capacitance uses no energy, it cannot use power either (power = energy/time). This idea is illustrated in Fig. 12-22. This figure shows power (or energy) taken from the source as positive power and power (or energy) returned to the source as negative energy. The negative power exactly cancels the positive power. You have seen the waveform of Fig. 12-22 before. It is the power waveform obtained in Chap. 9 for a current and voltage that are 90° out of phase. Therefore, we can conclude two things about a circuit that contains only capacitance.

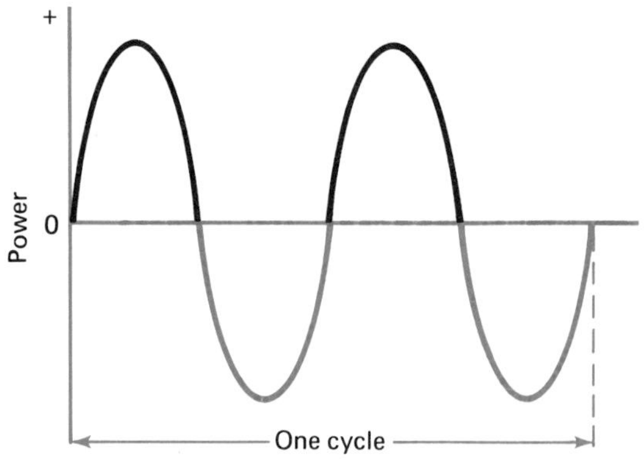

Fig. 12-22 Power waveform for capacitance. Capacitance uses no power.

1. Current and voltage are 90° out of phase.
2. The circuit uses no net energy or power.

Notice that the term used was *capacitance*, not *capacitor*. Capacitance was used because capacitors, like actual inductors, do have some energy losses.

Energy losses in capacitors can be classified as resistance losses and dielectric loss. The resistance losses are caused primarily by current flowing through the resistance of the capacitor leads and plates. There is also a very minute current flowing through the high resistance of the dielectric material. The dielectric resistance appears as a parallel resistance, whereas the plate and lead resistance appears as a series resistance (Fig. 12-23). Dielectric loss results from continually reversing the polarity of the electric field through the dielectric. Each time the molecular structure of the dielectric has to adjust to a new polarity, a little electrical energy is converted into heat energy. This dielectric loss appears as if it were a small resistance in series with the capacitor. The effects of all these resistances (Fig. 12-23) are often combined as the equivalent series resistance (ESR). A capacitor with a small ESR will have a small energy loss.

The relative amount of energy loss in a capacitor can be indicated by any of three terms. The terms are dissipation factor (DF), power factor (PF), and quality (Q). All of these terms give about the same information. *Power factor* is the ratio of power (P) to apparent power (P_{app}). In a capacitor, power represents the energy-per-second lost, and apparent power represents the energy-per-second stored. If a capacitor has very little energy loss, its power factor will be low (less than 0.01 or 1 percent). *Quality* (*Q*) was defined as a ratio of reactance (X) to resistance (R). If a capacitor has very little energy loss, R will be low and Q will be high (more than 100). *Dissipation factor* is the reciprocal of quality. In other words, dissipation factor is a ratio of resistance (R) to reactance (X). Obviously, if Q is typically a large number, dissipation factor will be a small number or percentage. For all but the lowest quality capacitors, power factor and dissipation factor are essentially equal to each other. In summary, remember that capacitors with small energy losses will have a high Q and low dissipation and power factors.

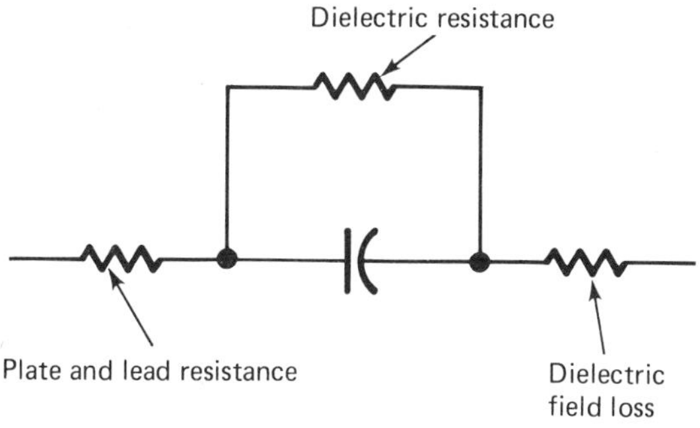

Fig. 12-23 Equivalent resistances in a capacitor. These resistances convert electrical energy into heat energy.

With lower quality electrolytic capacitors, energy losses can be considerable (1 to 10 percent of the stored energy). However, with high quality nonelectrolytic capacitors the losses are so small they can be ignored in most applications. In this text, we will assume capacitors have negligible energy loss. In terms of phase shift, this means we will treat nonelectrolytic capacitors as ideal capacitors (Fig. 12-24). We will assume the current and voltage are a full 90° out of phase.

Capacitive Reactance

Capacitance, like inductance, controls circuit current without converting electrical energy

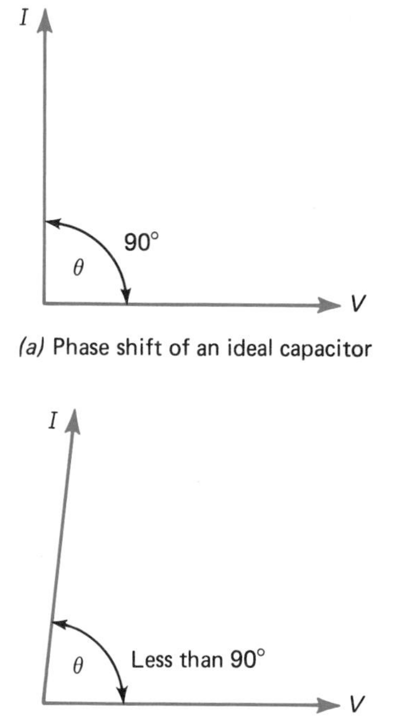

(a) Phase shift of an ideal capacitor

(b) Phase shift of a low-quality capacitor

Fig. 12-24 Phase shift in capacitors.

into heat energy. Therefore, its opposition, like the opposition of inductance, is called reactance (X). To separate the two types of reactances, we use the symbol X_C for *capacitive reactance*. Since X_C is a form of opposition it can be used in Ohm's law to determine circuit current or voltage. The Ohm's law formula is

$$I = \frac{V}{X_C}$$

Capacitive reactance is controlled by two factors. One, the frequency of the current (and voltage); and two, the amount of capacitance. Further, the reactance is inversely proportional to both of these factors. If either frequency or capacitance is doubled, the reactance is halved. If either frequency or capacitance is halved, the reactance is doubled. Why reactance is inversely proportional to both frequency and capacitance can be explained by the example which follows.

Figure 12-25 shows a capacitor circuit in which frequency, capacitance, and voltage are indicated. We know that the current in the circuit is controlled by the capacitive reactance. If the reactance increases the current will decrease and vice versa. Therefore, if we determine what happens to the current in the circuit, we will know what happened to the

Dissipation factor

Power factor

Quality

Capacitive reactance

$X_C = \frac{1}{6.28fC}$

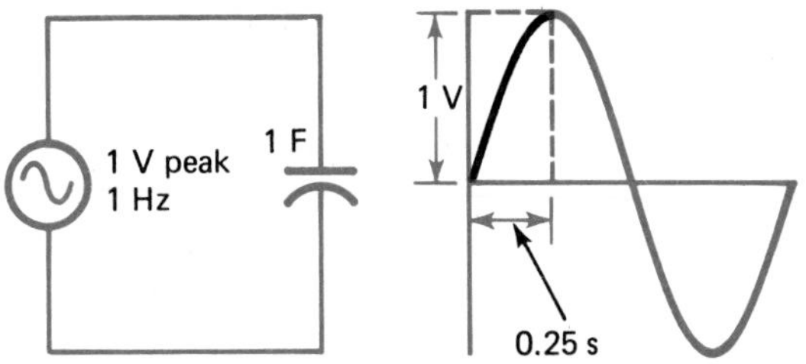

Fig. 12-25 Determining charge and average current. In 0.25 s the capacitor must receive one coulomb of charge.

reactance. First let us determine the circuit current for the frequency, voltage and capacitance given in Fig. 12-25. Then we can change either frequency or capacitance and see what happens to the current. To determine the current, we will first determine the amount of charge needed to charge the capacitor to the source voltage. This can be done with the basic formula for capacitance:

$$C = \frac{Q}{V}$$

Rearranging yields

$$Q = CV$$

In Fig. 12-25 then, the charge will be $Q = 1\text{ F} \times 1\text{ V} = 1$ coulomb. Next we can determine the amount of time available to transfer the charge from the source to the capacitor. The time available is the time required for one-quarter of a cycle. The period of 1 Hz is 1 s. Thus the capacitor has 0.25 s in which to charge. Finally, we know that

$$\text{Current } (I) = \frac{\text{charge } (Q)}{\text{time } (t)} = \frac{1\text{ C}}{0.25\text{ s}} = 4\text{ C/s} = 4\text{ A}$$

A coulomb per second is an ampere, so the average current for the first quarter cycle in Fig. 12-25 is 4 A. The remaining quarters of the cycle handle the same charge in the same amount of time as the first quarter. Therefore, the average current in the circuit of Fig. 12-25 is 4 A.

Now suppose we double the frequency to 2 Hz. The new period is 0.5 s, and the time for one-quarter cycle is 0.125 s. Since the voltage and capacitance are the same, the charge will be the same. The current will now increase to:

$$I = \frac{Q}{t} = \frac{1\text{ C}}{0.125\text{ s}} = 8\text{ C/s} = 8\text{ A}$$

Since doubling the frequency doubled the current, the reactance of the circuit must have halved.

Now suppose we leave the frequency at 1 Hz and double the capacitance to 2 F. The new charge will be:

$$Q = CV = 2\text{ F} \times 1\text{ V} = 2\text{ C}$$

The time for one-quarter cycle at 1 Hz is still 0.25 s. Therefore, the new current is:

$$I = \frac{Q}{t} = \frac{2\text{ C}}{0.25\text{ s}} = 8\text{ C/s} = 8\text{ A}$$

Again, the current doubled so the reactance must be half as great as it was with 1 F.

The exact value of capacitive reactance (X_C) can be calculated with the formula

$$X_C = \frac{1}{2\pi fC} = \frac{1}{6.28fC}$$

Capacitive reactance (X_C) will be in ohms when frequency (f) is in hertz (Hz) and capacitance (C) is in farads (F).

Example 12-3

What is the reactance of a 0.01 μF capacitor at 400 Hz?

Given: $C = 0.01\ \mu\text{F}$, $f = 400$ Hz

Find: X_C

Known: $X_C = \frac{1}{6.28fC}$

Solution: $X_C = \frac{1}{6.28 \times 400 \times 0.01 \times 10^{-6}} = 39{,}800\ \Omega$

Answer: The capacitive reactance equals 39,800 ohms or about 40 kΩ.

Notice in the above example the microfarad was converted to farad by multiplying by 10^{-6}.

From the above example we can quickly estimate the reactance of a 0.005 μF capacitor at 400 Hz. Since 0.005 μF is half as great as 0.01, the reactance will be twice as great. Thus, the 0.005 μF capacitor at 400 Hz will have approximately 80 kΩ of capacitive reactance.

Example 12-4

How much current flows in the circuit shown in Fig. 12-26?

Given: $f = 2$ MHz
$C = 500$ pF
$V = 20$ V

Find: I

Known: $I = \dfrac{V}{X_C}$ and $X_C = \dfrac{1}{6.28fC}$

Solution: $X_C = \dfrac{1}{6.28 \times 2 \times 10^6 \times 500 \times 10^{-12}}$

$= 159\Omega$

$I = \dfrac{20\text{ V}}{159\ \Omega} = 0.126\text{ A} = 126\text{ mA}$

Answer: The current in the circuit is 126 milliamperes.

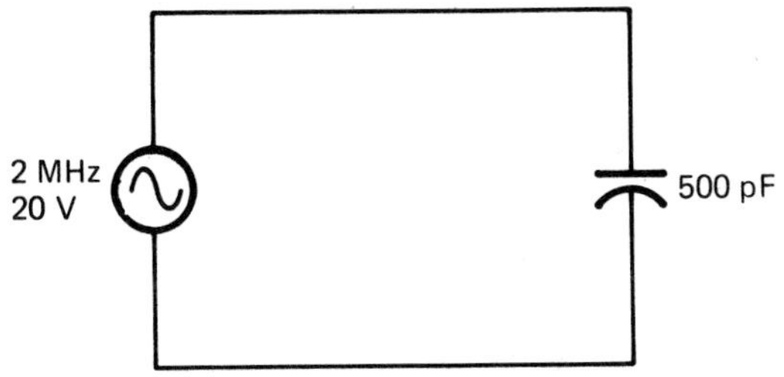

Fig. 12-26 Circuit for Example 12-4.

Review Questions

41. How does a capacitor control current in an ac circuit?
42. Does capacitance use power?
43. What is the phase relationship between current and voltage in an ideal capacitor?
44. How often does a capacitor return energy to the primary source in an ac circuit?
45. List three terms used to indicate relative energy loss in a capacitor.
46. What is the opposition to current of a capacitor called?
47. What is the symbol for a capacitor's opposition?
48. What type of relationship exists between capacitive reactance, frequency, and capacitance?
49. Calculate the capacitive reactance of a
    a. 2-μF capacitor at 300 Hz.
    b. 100-pF capacitor at 4 MHz.
    c. 0.003-μF capacitor at 250 kHz.

Series capacitors

12-10 CAPACITORS IN SERIES

When capacitors are connected in series the total capacitance decreases. In fact, the total capacitance of series capacitors is always less than the capacitance of the smallest capacitor. The logic of this statement can be seen by referring to Fig. 12-27 which shows two capacitors connected in series. Assume these capacitors are made from flat metal plates separated by an insulator. Suppose the lead connecting the two capacitors together is shortened until the plates of the capacitors touch each other. The result would be as indicated in Fig. 12-27(*b*) where the two plates combine to make one plate. The combined plates, isolated between two insulators, serve no useful purpose. If they are removed, as in Fig. 12-27(*c*), the result is a single, equivalent capacitor. Look carefully at the plates of the equivalent capacitor. They are spaced further apart than were the plates of either of the original capacitors. Also, the effective plate area is limited by the smaller plate. Therefore, the equivalent capacitor has less capacitance than either of the two series capacitors.

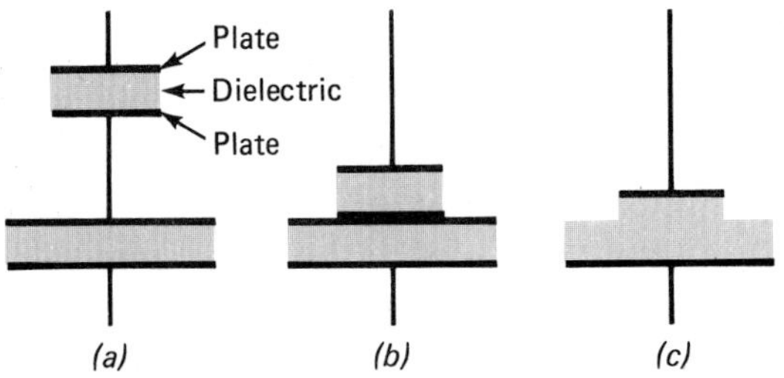

Fig. 12-27 Capacitors in series. The equivalent capacitance in (*c*) is less than the lower of the two capacitances in (*a*).

The total (or equivalent) capacitance of series capacitors can be calculated with any of these formulas:

General formula for series capacitors:

$$C_T = \frac{1}{\frac{1}{C_1} + \frac{1}{C_2} + \frac{1}{C_3} + \text{etc.}}$$

Total capacitance

Two capacitors in series:

$$C_T = \frac{C_1 \times C_2}{C_1 + C_2}$$

n equal capacitors in series:

$$C_T = \frac{C}{n}$$

Notice that the general form of these formulas is the same as the formulas used for parallel resistances and inductances. The formulas are indeed the same; just substitute *C* for *R*.

Example 12-5

What is the total capacitance of a 0.001-μF capacitor in series with a 2000-pF capacitor?

Given: $C_1 = 0.001\ \mu F$
$C_2 = 2000$ pF

Find: C_T

Known: $C_T = \frac{C_1 \times C_2}{C_1 + C_2}$

Solution: 2000 pF = 0.002 μF

$$C_T = \frac{0.001\ \mu F \times 0.002\ \mu F}{0.001\ \mu F + 0.002\ \mu F}$$

$$= \frac{0.000002}{0.003}$$

$$= 0.00067\ \mu F$$

Answer: The total series capacitance is 0.00067 μF.

Notice in the above example that the capacitors were converted into the same units (μF) before using the formula. This way the total (equivalent) capacitance comes out in the same unit as the two original capacitors.

Total reactance

The reactances of series capacitors are additive. The total reactance can be computed by the formula:

$$X_{C_T} = X_{C_1} + X_{C_2} + X_{C_3}$$

Example 12-6

Determine the total reactance of the circuit shown in Fig. 12-28.

Given: $C_1 = 2\ \mu F$
$C_2 = 4\ \mu F$
$f = 100$ Hz

Find: X_{CT}

Known: $X_{CT} = X_{C_1} + X_{C_2}$ and $X_C = \frac{1}{6.28 fC}$

Solution:

$$X_{C_1} = \frac{1}{6.28 \times 100 \times 2 \times 10^{-6}} = 796\ \Omega$$

$$X_{C_2} = \frac{1}{6.28 \times 100 \times 4 \times 10^{-6}} = 398\ \Omega$$

$$X_{CT} = 796\ \Omega + 398\ \Omega = 1194\ \Omega$$

Answer: The total reactance of the circuit is 1194 ohms.

The answer to Example 12-6 could also be found by first determining the total capacitance and then finding the total reactance. For example:

$$C_T = \frac{C_1 \times C_2}{C_1 + C_2} = \frac{2 \times 4}{2 + 4} = \frac{8}{6} = 1.33\ \mu F$$

$$X_{CT} = \frac{1}{6.28 f C_T}$$

$$= \frac{1}{6.28 \times 100 \times 1.33 \times 10^{-6}}$$

$$= 1194\ \Omega$$

Ohm's law is also applicable in capacitor circuits. Just replace *R* with X_C.

Example 12-7

Determine the circuit current and the voltage across each capacitor in Fig. 12-28.

Given: From the previous example: $X_{C_1} = 796\ \Omega$, $X_{C_2} = 398\ \Omega$ and $X_{CT} = 1194\ \Omega$. From Fig. 12-28: $V_T = 40$ V

Find: I_T, V_{C_1}, and V_{C_2}

Known: $I_T = \frac{V_T}{X_{CT}}$
$I_T = I_{C_1} = I_{C_2}$
$V = IX_C$

Fig. 12-28 Circuit for Examples 12-6 and 12-7.

Solution: $I_T = \frac{40\text{ V}}{1194\ \Omega} = 0.0335\text{ A}$
$= 33.5\text{ mA}$

$V_{C_1} = 0.0335\text{ A} \times 796\ \Omega$
$= 26.7\text{ V}$

$V_{C_2} = 0.0335\text{ A} \times 398\ \Omega$
$= 13.3\text{ V}$

Answer: The total current is 33.5 milliamperes. The voltage across C_1 is 26.7 volts, and across C_2 the voltage is 13.3 volts.

As a check on the computations done in the above example, we can apply Kirchhoff's voltage law:

$$V_T = V_{C_1} + V_{C_2} = 26.7\text{ V} + 13.3\text{ V} = 40\text{ V}$$

Notice in the above example that the largest capacitor dropped the least amount of voltage. This is because capacitance and capacitive reactance are inversely proportional.

In electronic circuits, capacitors can be used as ac voltage dividers. The advantage of using capacitors rather than resistors as an ac voltage divider is that the capacitors use no power.

12-11 CAPACITORS IN PARALLEL

Parallel capacitors

The total capacitance of parallel capacitors is found by adding the individual capacitance:

$$C_T = C_1 + C_2 + C_3 + \text{etc.}$$

Figure 12-29 shows that connecting capacitors in parallel effectively increases the plate area. Suppose the two capacitors shown in cross section in Fig. 12-29(*a*) are moved together until the ends of their plates touch. Then the two connecting conductors can be removed. The result is an equivalent capacitor [Fig. 12-29(*b*)] with a larger plate area. A larger plate area, of course, means more capacitance.

Total reactance

The reactances of capacitors in parallel can be treated like resistances or inductive reactances in parallel. The formulas are:

General formula for parallel capacitors:

$$X_{C_T} = \frac{1}{\frac{1}{X_{C_1}} + \frac{1}{X_{C_2}} + \frac{1}{X_{C_3}} + \text{etc.}}$$

Two capacitors in parallel:

$$X_{C_T} = \frac{X_{C_1} \times X_{C_2}}{X_{C_1} + X_{C_2}}$$

n equal capacitors in parallel:

$$X_{C_T} = \frac{X_C}{n}$$

Total reactance of parallel capacitors can also be obtained by either the reactance formula or Ohm's law. The formulas are:

$$X_{C_T} = \frac{1}{6.28fC_T} \quad \text{and} \quad X_C = \frac{V_T}{I_T}$$

Of course, total capacitance must be determined before applying the reactance formula.

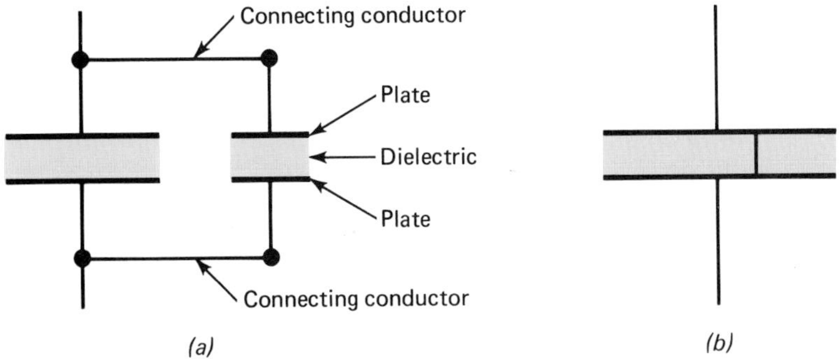

Fig. 12-29 Capacitance in parallel. The equivalent capacitance in (*b*) is equal to the sum of the capacitances in (*a*).

Example 12-8

Refer to Fig. 12-30. Determine the total current, the current through C_1, the total capacitance, and the total reactance.

Given: $C_1 = 0.2\ \mu F$
$C_2 = 0.3\ \mu F$
$f = 1000$ Hz
$V = 50$ V

Find: I_T, I_{C_1}, C_T, and X_C

Known: $I_T = \dfrac{V_T}{X_C}$

$I_{C_1} = \dfrac{V_{C_1}}{X_{C_1}}$

$C_T = C_1 + C_2$

$X_{CT} = \dfrac{1}{6.28fC}$

$V_T = V_{C_1} = V_{C_2}$

Solution: $C_T = 0.2\ \mu F + 0.3\ \mu F = 0.5\ \mu F$

$$X_{CT} = \frac{1}{6.28 \times 1 \times 10^3 \times 0.5 \times 10^{-6}} = 318\ \Omega$$

$$I_T = \frac{50\text{ V}}{318\ \Omega} = 0.157\text{ A} = 157\text{ mA}$$

$$X_{C_1} = \frac{1}{6.28 \times 1 \times 10^3 \times 0.2 \times 10^{-6}} = 796\ \Omega$$

$$I_{C_1} = \frac{50\text{ V}}{796\ \Omega} = 0.0628\text{ A} = 62.8\text{ mA}$$

In the above example, total reactance was determined without using a parallel reactance formula or determining the value of X_{C_2}. Let us check the answers to Example 12-8 by calculating X_{C_2} and then recalculating X_{CT}.

$$X_{C_2} = \frac{1}{6.28 \times 1 \times 10^3 \times 0.3 \times 10^{-6}} = 531\ \Omega$$

$$X_{CT} = \frac{796 \times 531}{796 + 531} = 318\ \Omega$$

Since the value obtained for X_{CT} is the same in both cases, we have made no mathematical errors. If we wanted to know the current through C_2 in Fig. 12-30, we could use Kirchhoff's current law. (We have already determined the value of I_T and I_{C_1}.)

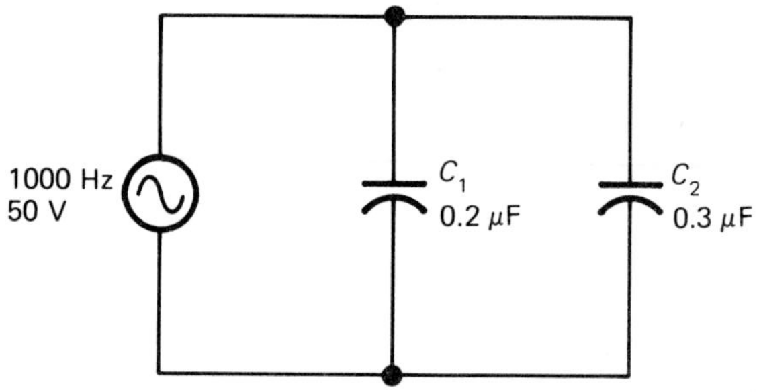

Fig. 12-30 Circuit for Example 12-8.

$$I_T = I_{C_1} + I_{C_2}$$

$$I_{C_2} = I_T - I_{C_1} = 157\text{ mA} - 62.8\text{ mA} = 94.2\text{ mA}$$

We could also determine I_{C_2} by using Ohm's law because we know X_{C_2} and V_{C_2}.

$$I_{C_2} = \frac{V_{C_2}}{X_{C_2}} = \frac{50\text{ V}}{531\ \Omega} = 0.0942\text{ A} = 94.2\text{ mA}$$

Review Questions

50. The equivalent capacitance of a 0.01-μF capacitor and a 0.04-μF capacitor is 0.05 μF. Are the capacitors connected in series or parallel?
51. The equivalent capacitance of a 1000-pF capacitor and a 0.001-μF capacitor is 0.0005 μF. Are the capacitors connected in series or parallel?
52. Two capacitors, 0.3 μF and 0.5 μF, are connected in parallel. What is the total reactance at 600 Hz?
53. If the two capacitors in question 52 were connected in series, what would the total reactance be?
54. A current of 0.02 A flows through a capacitor that has 3000 Ω of reactance. How much voltage is across the capacitor?
55. For the circuit of Fig. 12-31 find the following:
    a. X_{C_2} *c.* V_{C_1}
    b. C_T *d.* I_T
56. For the circuit of Fig. 12-32 find the following:
    a. X_{CT}

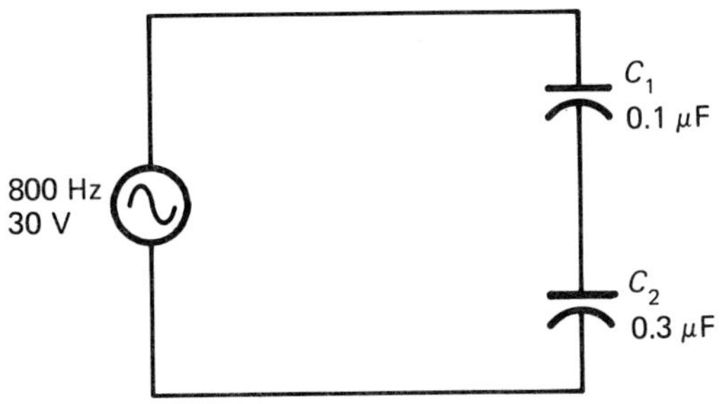

Fig. 12-31 Circuit for question 55.

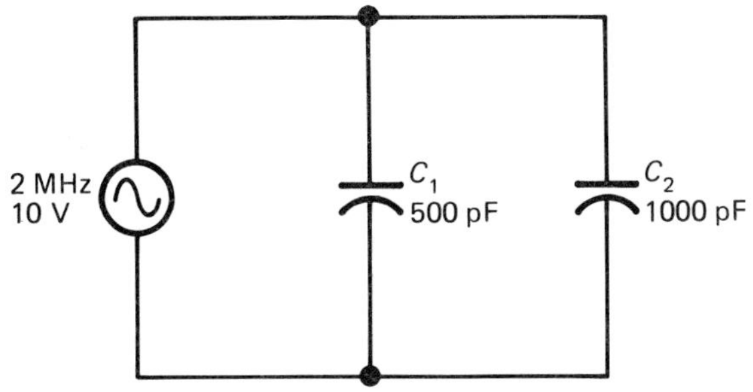

Fig. 12-32 Circuit for questions 56 and 57.

b. I_T
c. C_T

57. Will C_1 or C_2 in Fig. 12-32 carry the most current?

12-12 TIME CONSTANTS

How rapidly a capacitor charges or discharges is determined by the amount of resistance and capacitance in the circuit. The exact rate at which the capacitor charges or discharges can be determined by multiplying resistance times capacitance. The multiplication of a specific amount of resistance times a specific amount of capacitance yields a specific amount of time. This time is called the *time constant* for that specific value of R and C. Written in equation form, the time constant T is

$$\text{Time constant } (T) = RC$$

In this formula T will be in seconds when R is in ohms and C is in farads. During one time constant, a capacitor will charge or discharge, through a resistor, to 63.2 percent of the available voltage. The *available voltage* when charging is the difference between the capacitor's voltage and the source voltage. When discharging, the available voltage is the voltage left on the capacitor. The curves in Fig. 12-33 graphically demonstrate the meaning of a time constant. Figure 12-33(*a*) shows that a capacitor charges from 0 to 63.2 percent of the source voltage in one time constant. During the next time constant it charges another 63.2 percent of the remaining available voltage. Thus, at the end of two time constants the capacitor is 86.5 percent charged [63.2 + (0.632 × 36.8) = 86.5]. The percentage of charge at the end of additional time constants is tabulated in Fig. 12-33(*a*).

Time constant

$T = RC$

The actual amount of time required to charge to 86.5 percent of the source voltage is strictly a function of resistance and capacitance. We can determine the exact time by calculating the time for a time constant. For example, the time constant for the resistance-capacitance circuit in Fig. 12-34 would be:

$$T = RC = 2 \times 10^6\ \Omega \times 4 \times 10^{-6}\ \text{F}$$
$$= 8\ \text{s}$$

The capacitor requires two time constants to charge to 86.5 percent of the source voltage. Therefore, it takes 16 s for the capacitor in Fig. 12-34 to charge to 86.5 percent of the battery voltage.

The exact amount of voltage on the capacitor after two time constants is dependent on the value of the source voltage. If the source voltage in Fig. 12-34 is 10 V, it will be 8.65 V. If the source is 100 V, it will be 86.5 V.

The discharge of a capacitor is illustrated in Fig. 12-35. Notice the values of the resistance and the capacitance. They are the same as used in Fig. 12-34; therefore, the time constant is still 8 s. When the switch is closed in Fig. 12-35 the capacitor will be charged to 200 V. When the switch is opened, the capacitor will discharge through the resistor. After 8 s (one time constant) the capacitor will have discharged 63.2 percent of its voltage. The voltage across the capacitance will, therefore, decrease by 126.4 V. It will still have 73.6 V left. During the next time constant (Fig. 12-35) it will discharge 63.2 percent of the remaining 73.6 V. Thus, it will discharge 46.5 V (0.632 × 73.6 = 46.5). As shown in Fig. 12-35, there will be about 27 V left on the capacitor at the end of two time constants (16 s).

Inspect the curves in Fig. 12-33 again. Notice that a capacitor is essentially charged

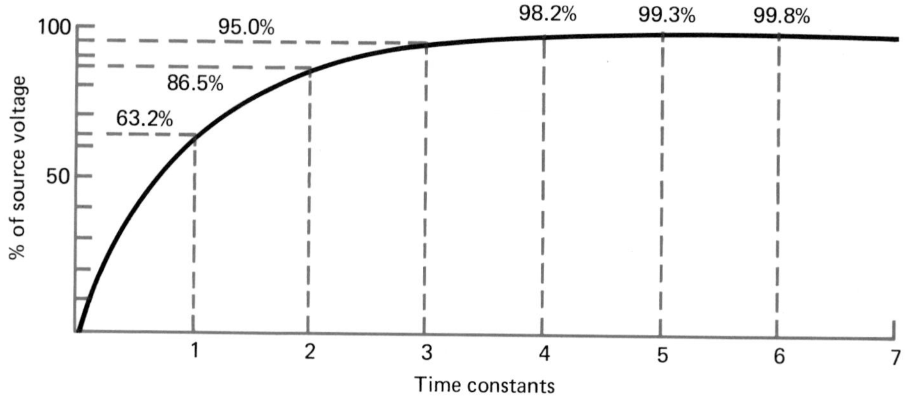

(a) Charge curve

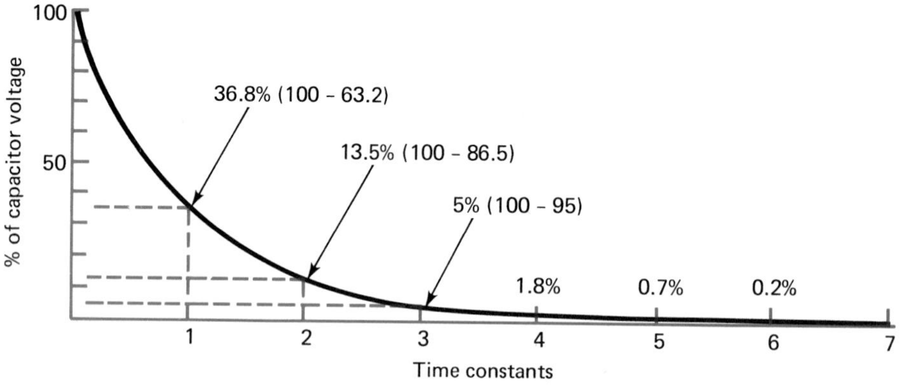

(b) Discharge curve

Fig. 12-33 Universal charge-discharge curves. After five or six time constants, the capacitor is considered to be fully charged or discharged.

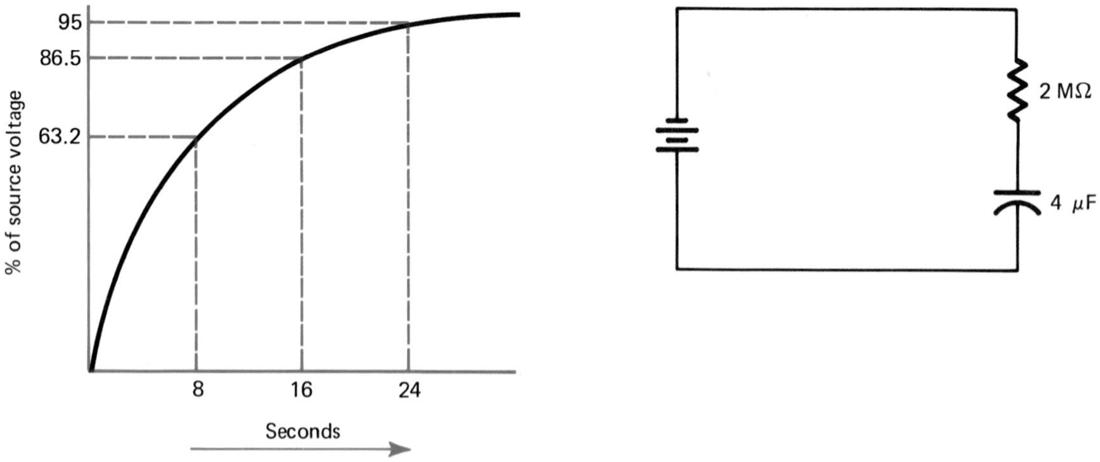

Fig. 12-34 Charging time for a capacitor. After 24 s (three time constants) the capacitor is 95 percent charged.

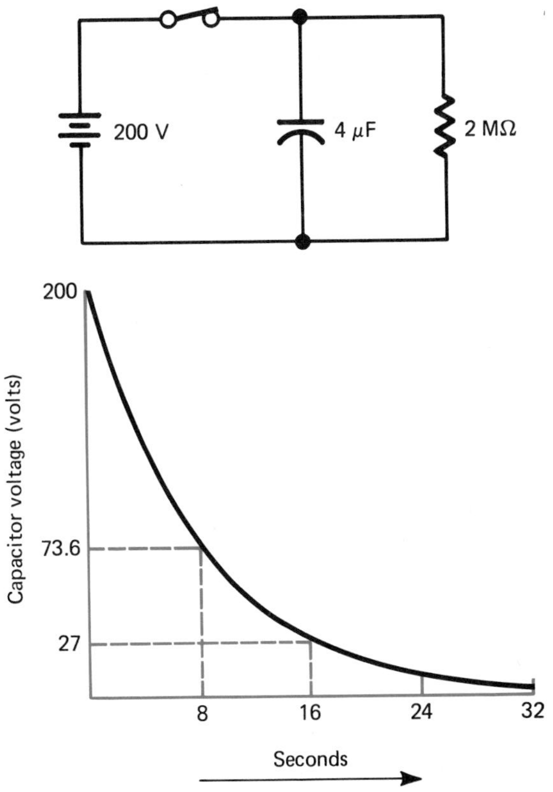

Fig. 12-35 Discharging time for a capacitor.

or discharged by the end of six time constants. For analyzing and designing circuits many technicians consider a capacitor to be charged (or discharged) after only five time constants.

Resistance-capacitance circuits, and their associated time constants, are used extensively in electronic circuits. For example, they are used to control the frequency of many nonsinusoidal waveforms. When you adjust the vertical-hold control on a television receiver you are adjusting a time constant.

12-13 DETECTING FAULTY CAPACITORS

An ohmmeter can be used to check the condition of a capacitor. When an ohmmeter is connected across a completely discharged capacitor the capacitor will charge. The capacitor will charge up to a voltage equal to the voltage of the battery used in the ohmmeter. How long it will take the capacitor to charge depends on two things. One, the size of the capacitor, and two, the internal resistance of the ohmmeter. In other words, it depends on the time constant. When checking capacitors, the ohmmeter should be on a high range to provide a large time constant so you can "see" the capacitor charge.

High value capacitor

With high values of capacitance, the ohmmeter pointer will deflect toward zero ohms and then return to a high reading. If the capacitor is nonelectrolytic, the final indication on the meter should be as great as the meter can indicate. (After the capacitor has charged, the ohmmeter is just measuring the resistance of the dielectric material.) With low quality electrolytic capacitors the final reading may be as low as 100 kΩ. This is because the dielectric material is an oxide which is not as good an insulator as plastics or ceramic.

Low value capacitor

With low value capacitors the charging current may not last long enough to cause a deflection on the ohmmeter. In this case, the meter just indicates a very high resistance. How large a capacitor must be to show a deflection depends on the type of meter. For a particular meter this can be determined by experimentation.

So far we have dealt with the ohmmeter action when a capacitor is not defective. How does it react to a defective capacitor? If the capacitor is open (a lead does not make contact with a plate) the meter won't deflect. If a capacitor is shorted, the pointer will never return to a high resistance reading. This means that current is flowing through the dielectric or the plates are touching each other. If the ohmmeter indicates zero ohms, we say the capacitor is "dead shorted" or completely shorted. This usually means the plates are making direct contact. If the dielectric material has become contaminated or deteriorated it may indicate a few thousand ohms. We would say the capacitor is partially shorted, has a high-resistance short, or is leaky.

Stray capacitance

12-14 UNDESIRED OR STRAY CAPACITANCE

Any two conductive surfaces separated by insulation possess capacitance. Thus, every electrical and electronic circuit has some undesired capacitance. For example, capacitance exists between:

1. An insulated conductor and a metal chassis
2. Two conductors in an electrical cable
3. The turns of an inductor
4. The primary and secondary of a transformer

5. Between the input and output leads of a transistor

The complete list of places where stray capacitance is found would be almost endless.

Whether or not undesired capacitance affects a circuit depends on the amount of capacitance and the frequency of the circuit. For example, the capacitance between two parallel, insulated conductors 5 cm in length is very low. In dc and most ac power circuits, it would have no effect. However, at the frequency used in a television tuner, it would have a very pronounced effect. In fact, if the conductors were 6 rather than 5 cm long, the tuner might not function properly.

Specifications

12-15 CAPACITOR SPECIFICATIONS

When ordering capacitors one has to include the specifications needed to insure receiving the desired capacitors. The minimum specifications for ordering capacitors for general use would be:

1. Capacitance
2. Tolerance
3. Direct current working voltage
4. Type (ceramic disk, polystyrene, variable, etc.)
5. Lead configuration (axial or radial)

If the capacitors are for circuits where the capacitance value is critical, then the temperature coefficient should also be specified. For extreme environmental conditions, the operating temperature range should be included in the specifications. When available space is limited, the physical size and shape should also be specified.

Relaxation oscillator

12-16 USES OF CAPACITORS

Capacitors are used with other components to accomplish a wide variety of jobs. For example, capacitors are used:

1. In power supply filters which smooth pulsating dc into pure dc
2. In oscillators which convert dc into either sinusoidal or nonsinusoidal ac

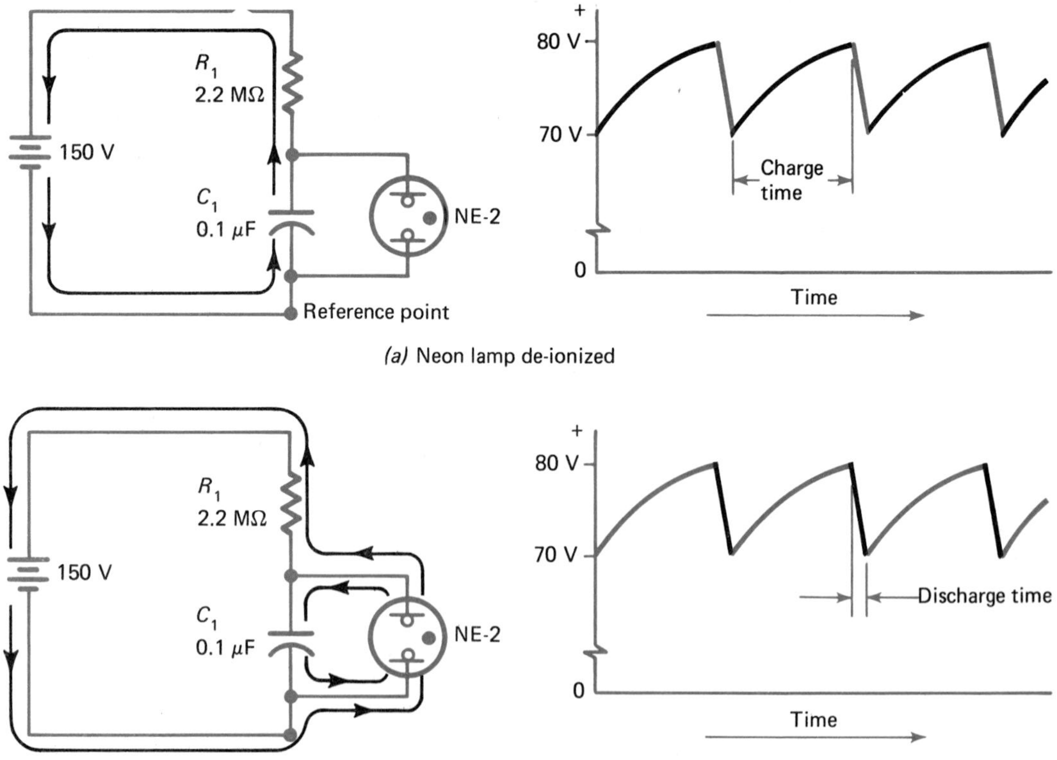

Fig. 12-36 Relaxation oscillator circuit. The capacitor discharges during the time that the lamp is ionized.

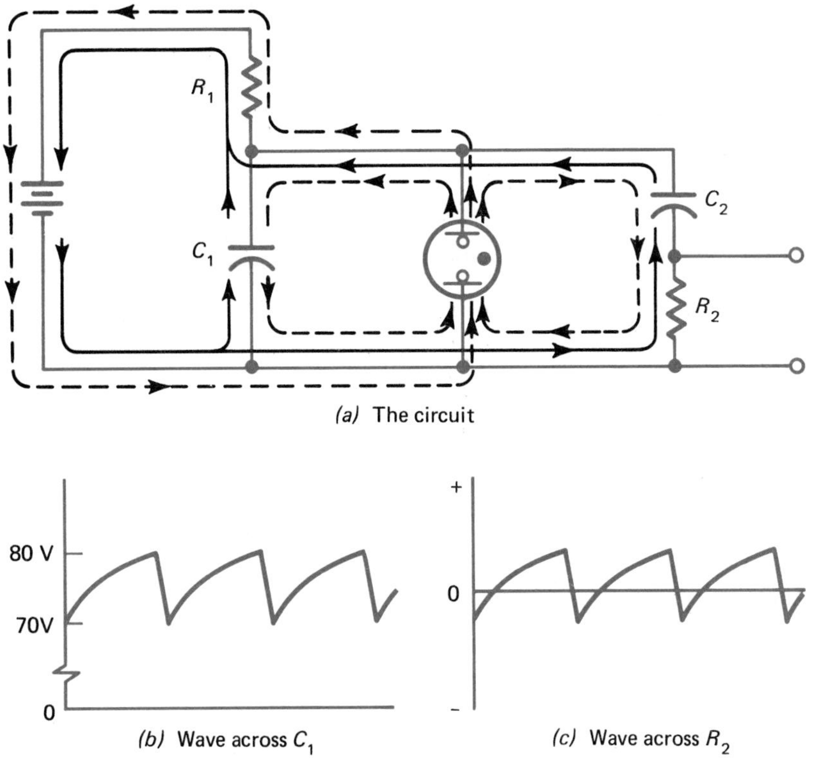

Fig. 12-37 Coupling capacitor. The coupling capacitor separates the ac from the dc.

3. In filters which separate low-frequency ac from high-frequency ac
4. For coupling amplifiers together. Here they separate ac from dc.
5. For power factor correction. Here their leading current compensates for the lagging current of an inductive load.

The circuit in Fig. 12-36 will illustrate how a capacitor works in a complete functional system. This circuit is called a *relaxation oscillator.* Its purpose is to produce a sawtooth voltage from a pure dc voltage. The neon lamp in the circuit fires (ionizes) at 80 V and de-ionizes at 70 V. When fired its resistance is extremely low; when de-ionized its resistance is high.

Figure 12-36(*a*) shows the current flow when the lamp is de-ionized; at this time the capacitor is charging. Since C_1 charges through R_1, the time constant is relatively long as indicated in the waveform graph. Once the capacitor has charged to 80 V the lamp fires [Fig. 12-36(*b*)]. The internal resistance of the lamp drops to a low value when it fires. This allows the capacitor to quickly discharge through the low resistance. When the capacitor has discharged to the lamp's critical voltage (about 70 V), the lamp de-ionizes. The capacitor then stops discharging [as in Fig. 12-36(*b*)] and again starts to recharge [as in Fig. 12-36(*a*)].

The above charging and discharging process repeats itself endlessly as long as the battery energy lasts. It produces a continuous sawtooth waveform with a 10-V peak-to-peak amplitude. Notice in Fig. 12-36 that the waveform is a fluctuating dc voltage. It varies from 70 to 80 V positive with respect to the reference point [Fig. 12-36(*a*)].

Suppose we wished to separate the ac sawtooth part of the waveform from the dc part (the +70 V). We can do this with another capacitor and resistor combination like C_2 and R_2 in Fig. 12-37. In this figure C_2 couples the sawtooth (ac) to R_2 while blocking the dc voltage. This is what is being referred to by the statement, "A capacitor blocks dc and passes ac." The broken black lines in Fig. 12-37 show the paths of current flow when the lamp is fired. When fired, the lamp carries discharge current for both C_1 and C_2 as well as current for R_1. The solid line shows the paths when the lamp is de-ionized. Notice that the current through R_2 is ac; it periodically reverses itself.

Review Questions

58. What is a time constant?
59. What is the time constant of a 270-kΩ resistor and a 4-μF capacitor?
60. How many time constants do technicians figure it takes to charge a capacitor?
61. How much resistance should an ohmmeter indicate across the terminals of a good ceramic capacitor?
62. What indication does an open capacitor give when its resistance is checked with an ohmmeter?
63. What is another name for undesired capacitance?
64. List the minimum specifications for ordering capacitors.

Summary

1. Capacitance is the ability to store electrical energy.
2. The symbol for capacitance is *C*.
3. Capacitors are devices constructed to provide specific amounts of capacitance.
4. Capacitors have two plates and a dielectric.
5. A charged capacitor stores energy and creates a voltage between its plates.
6. A charged capacitor can provide energy for other parts of an electrical circuit.
7. Current does not flow through the dielectric material.
8. The farad is the base unit of capacitance.
9. The abbreviation for farad is F.
10. Capacitance is determined by plate area, distance between plates, dielectric material, and temperature.
11. Capacitors are named after their dielectric material, enclosure, construction process, or intended use.
12. Electrolytic capacitors are usually polarized. They have maximum capacitance for their size and weight.
13. Typical paper and film capacitors are less than 1 μF.
14. Ceramic and mica capacitors are limited to low values.
15. Energy storage capacitors can provide very high power levels.
16. The curved line on a capacitor symbol identifies a specific plate on variable, polarized, paper, and film capacitors.
17. Except for a charging surge, capacitors block dc.
18. The smallest of two series capacitors develops the most voltage.
19. Capacitors control current flow in an ac circuit.
20. Current leads voltage by 90° in an ideal capacitor.
21. Capacitance uses no power or energy.
22. Energy loss in a capacitor results from dielectric and resistance losses.
23. Dissipation factor, power factor, and quality all are ways of rating a capacitor's relative energy loss.
24. Capacitive reactance is the opposition of a capacitor to sinusoidal ac.
25. The symbol for capacitive reactance is X_C.
26. Capacitive reactance is inversely proportional to both frequency and capacitance.
27. The total capacitance of series capacitors is less than the smallest capacitance.
28. The total reactance of series capacitors is the sum of the individual reactances.
29. The total capacitance of parallel capacitors is the sum of the individual capacitors.
30. Reactances of parallel capacitors are added reciprocally.
31. Ohm's law and Kirchhoff's laws apply to capacitor circuits.
32. A time constant defines the rate at which a capacitor charges or discharges through a resistor.
33. The capacitance formulas are:

$$C = \frac{Q}{V}$$

$$W = 0.5\ CV^2$$

$$X_C = \frac{1}{6.28fC}$$

$$T = RC$$

For series capacitors

$$C_T = \frac{1}{\frac{1}{C_1} + \frac{1}{C_2} + \frac{1}{C_3}}$$

$$X_{C_T} = X_{C_1} + X_{C_2} + X_{C_3}$$

For parallel capacitors

$$C_T = C_1 + C_2 + C_3$$

$$X_{CT} = \frac{1}{\frac{1}{X_{C_1}} + \frac{1}{X_{C_2}} + \frac{1}{X_{C_3}} + \text{etc.}}$$

Chapter Review Questions

For questions 12-1 to 12-15, determine whether each statement is true or false.

12-1. The dielectric material in a capacitor is an insulator.

12-2. A capacitor charges and discharges twice during each ac cycle.

12-3. When a capacitor is being charged, electrons travel from one plate through the dielectric to the next plate.

12-4. The DCWV of a capacitor is specified to keep the capacitor from overheating.

12-5. The base unit of capacitance is the picofarad.

12-6. A 2000-pF capacitor has the same capacitance as a 2.0-μF capacitor.

12-7. Electrolytic capacitors are smaller and lighter than other types of capacitors that have the same amount of capacitance.

12-8. Energy storage capacitors rarely produce a peak current of more than 0.5 A while being discharged.

12-9. The color-banded end of a paper or foil capacitor indicates the negative plate of the capacitor.

12-10. The curved plate of a capacitor symbol indicates the negative end of an electrolytic capacitor.

12-11. Except for the charging current, a capacitor blocks dc.

12-12. Two capacitors are connected in series. The smallest capacitor will drop the most voltage when either ac or dc is applied to the circuit.

12-13. The final indication of an ohmmeter connected to a "good" capacitor should be less than 100 kΩ.

12-14. The symbol for capacitance is *F*.

12-15. A farad is a coulomb per volt.

Answer the following questions.

12-16. List four factors that determine the capacitance of a capacitor. Also indicate how the capacitance can be increased by changing each factor.

12-17. Where is the energy stored in a capacitor? In what form (chemical, heat, etc.) is it stored?

12-18. List two factors that determine the strength of the electric field between the plates of a capacitor.

12-19. What type of capacitors are usually polarized?

12-20. A 500-μF capacitor is charged to 100 V. How much energy does it store?

12-21. List the specifications needed for ordering capacitors.

12-22. What is the time constant of a 470-kΩ resistor and an 8-μF capacitor connected to a 50-V source?

12-23. What is the reactance of a 0.02-μF capacitor at 3000 Hz?

12-24. A 4-μF capacitor and a 6-μF capacitor are connected in series across a 100-V, 100-Hz source. Determine the
a. Total capacitance
b. Total reactance
c. Total current
d. Voltage across the 6-μF capacitor

12-25. If the capacitors in question 12-24 were connected in parallel, what would be the value of the
a. Total capacitance?
b. Total reactance?
c. Current through the 4-μF capacitor?

12-26. A capacitor and a resistor are series connected to a 100-V dc source. What will the voltage across the capacitor be after two time constants?

Answers to Review Questions

1. A capacitor is an electrical device which stores energy directly as electrical energy.
2. *C*
3. Plates and dielectric
4. Energy is transferred from a primary source (like a battery) to the capacitor.
5. No
6. Yes
7. In the dielectric material
8. Direct current working voltage
9. Because too much electric field strength between the plates will destroy the dielectric material
10. The distance between the plates and the voltage to which it is charged
11. Farad
12. F
13. 3000 pF
14. 0.0022 μF
15. A farad equals a coulomb per volt.
16. Plate area, distance between plates, dielectric material, and temperature.
17. Temperature
18. It is halved also.
19. It doubles.
20. The material's ability to store energy when used as a dielectric
21. Smaller
22. The large amount of capacitance per unit of volume and weight
23. Aluminum or tantalum oxide
24. It helps form and maintain the oxide and it acts as part of a plate.
25. Negative
26. Aluminum and tantalum
27. It draws excessively high current from the source and overheats. It may explode.
28. Mica, ceramic, paper and film (plastics)
29. Smaller, more stable, and higher dielectric resistance
30. A capacitor in which the plate is deposited directly onto the film (or paper) dielectric
31. Because they are harder to construct and they are larger per unit of capacitance
32. Trimmer, padder and tuning capacitors
33. To remove high (radio) frequencies from dc circuits
34. A capacitor used to remove pulsations from dc voltage
35. In applications that require very rapid discharge of a capacitor
36. 18.75 J
37. The negative plate
38. By a color band on the end of the capacitor
39. No
40. The 5 μF, because $V = Q/C$ and both capacitors receive the same amount of Q
41. By producing a voltage which opposes the source voltage
42. No
43. Current leads voltage by 90°

44. Twice each cycle
45. Dissipation factor, power factor, and quality
46. Capacitive reactance
47. X_C
48. Capacitive reactance is inversely proportional to both frequency and capacitance.
49. *a.* 265.4 Ω
    b. 398.1 Ω
    c. 212.3 Ω
50. Parallel
51. Series
52. 331.7 Ω
53. 1415.4 Ω
54. 60 V
55. *a.* 663.5 Ω
    b. 0.075 μF
    c. 22.5 V
    d. 11.3 mA
56. *a.* 53.0 Ω
    b. 0.1884 A
    c. 1500 pF
57. C_1
58. The time required to charge or discharge a capacitor by 63.2% of the available voltage
59. 1.08 s
60. Five
61. Nearly infinite
62. The meter pointer does not momentarily deflect toward zero.
63. Stray capacitance
64. Capacitance, tolerance, type, lead arrangement, and voltage rating

R, C, and L Circuits

- Previous chapters have dealt with resistance, inductance and capacitance used individually in ac circuits. This chapter deals with the results of combining two or more of these quantities in a single circuit.

  When you have finished this chapter you will be able to calculate impedance, resonance, and phase angle. You will be able to add, both graphically and mathematically, out-of-phase voltages or currents. Also, you will understand the purpose and operation of basic filter circuits.

13-1 IMPEDANCE

The combined opposition to current of resistance and reactance is called *impedance*. The symbol for impedance is Z. Like resistance and reactance, impedance has base units of ohms.

In a circuit that contains resistance and reactance, the current and voltage cannot be in phase. Neither can they be a full 90° out of phase. If the circuit contains resistance and capacitance, the current leads the voltage. If it contains resistance and inductance, the current lags the voltage. When the circuit contains all three (R, C, and L), the phase relationship depends upon the relative sizes of L and C.

Circuits containing resistance and reactance will have more apparent power than true power. Their power factor will be less than one (less than 100 percent).

Many common electrical and electronic loads possess impedance consisting of more than resistance or reactance alone. Such loads include speakers, microphones, motors, and earphones.

Review Questions

1. Can the voltage and current be in phase in a circuit which contains capacitance and resistance?
2. What is impedance?
3. What is the symbol for impedance?
4. What is the base unit of impedance?

13-2 ADDING PHASORS

Suppose we wish to add the two phasors (Y_1 and Y_2) shown in Fig. 13-1(*a*). We can add them either graphically or mathematically. First, we will do it graphically and then we will develop the mathematical method.

The graphical addition of the phasors in Fig. 13-1(*a*) is illustrated in Fig. 13-1(*b*). The process involves the construction of a rectangle. First, line 1 is drawn from the tip of the Y_1 phasor parallel to the Y_2 phasor. Then, line 2 is drawn from the tip of the Y_2 phasor parallel to the Y_1 phasor. The point at which line 1 and line 2 cross is the tip of the resultant Y_T phasor.

The three phasors of Fig. 13-1(*b*) can be rearranged into the configuration shown in Fig. 13-1(*c*). This rearrangement merely replaces line 2 of Fig. 13-1(*b*) with phasor Y_1. This is permissible because line 2 and Y_1 are parallel and of the same length.

The rearrangement of phasors in Fig. 13-1(*c*) produces a right triangle. This right triangle is redrawn in Fig. 13-2 with the sides labeled. You are already familiar with the trigonometric functions listed in this figure.

From page 256:
Impedance

Graphical addition

On this page:
Pythagorean theorem

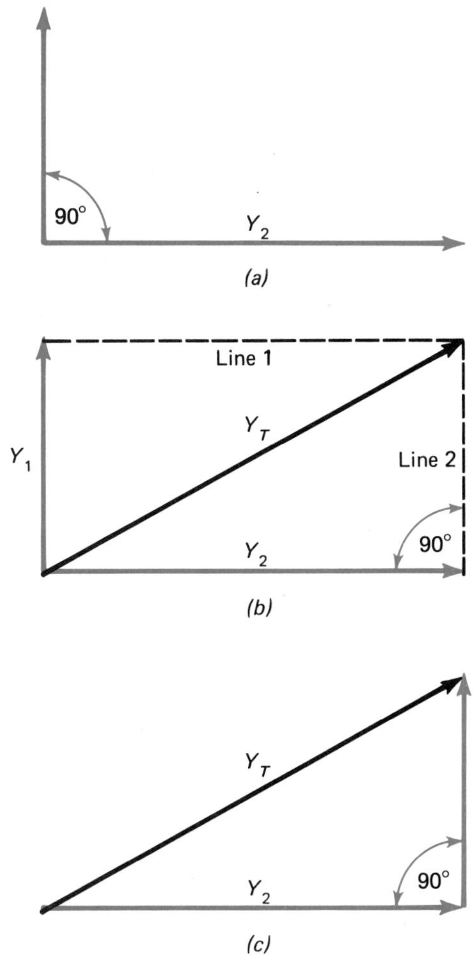

Fig. 13-1 Graphical addition of phasors. Adding Y_1 and Y_2 in (*a*) yields the resultant phasor (Y_T) in (*b*) and (*c*).

They are listed here merely as a review and easy reference for this chapter. Below the trigonometric functions in Fig. 13-2 is listed a very important relationship of the sides of a right triangle. The relationship states that the sum of the squares of the lengths of two sides of a right triangle is equal to the square of the length of the hypotenuse. This can be written as a mathematical formula:

$$(\text{Hypotenuse})^2 = (\text{opposite})^2 + (\text{adjacent})^2$$

This formula applies to all right triangles regardless of size. Thus, knowing the length of two sides of a triangle, you can calculate the length of the third side. This relationship is known as the Pythagorean theorem, in honor of the Greek mathematician Pythagorus, who first derived this relationship.

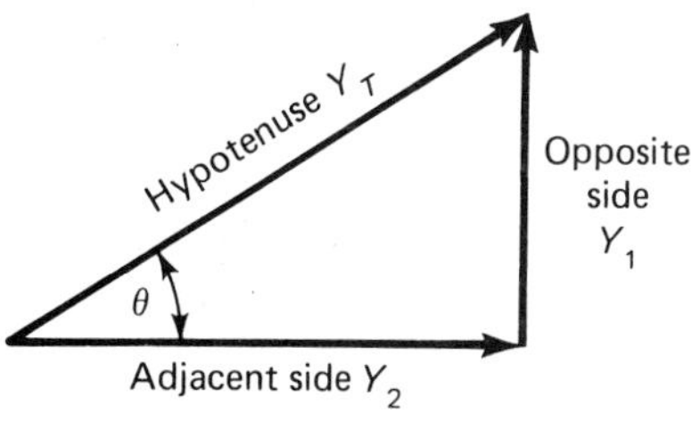

$$\operatorname{Tan}\theta = \frac{\text{opposite}}{\text{adjacent}} = \frac{Y_1}{Y_2}$$

$$\operatorname{Sin}\theta = \frac{\text{opposite}}{\text{hypotenuse}} = \frac{Y_1}{Y_T}$$

$$\operatorname{Cos}\theta = \frac{\text{adjacent}}{\text{hypotenuse}} = \frac{Y_2}{Y_T}$$

$$(\text{Hypotenuse})^2 = (\text{adjacent})^2 + (\text{opposite})^2$$

$$Y_T^2 = Y_1^2 + Y_2^2$$

$$Y_T = \sqrt{Y_1^2 + Y_2^2}$$

Fig. 13-2 Properties of a right triangle.

Example 13-1

Refer to Fig. 13-2. Suppose Y_1 represents 6 A and Y_2 represents 8 A. What is the value of Y_T which represents the total current?

Given: $I_1 = 6$ A
$I_2 = 8$ A

Find: I_T

Known: $Y_T = \sqrt{Y_1^2 + Y_2^2}$

Solution: $I_T = \sqrt{I_1^2 + I_2^2}$
$= \sqrt{6^2 + 8^2} = \sqrt{36 + 64}$
$= \sqrt{100} = 10$ A

Answer: The total current is 10 amperes.

Figure 13-3 illustrates how to add two phasors (Y_1 and Y_2) when the phasors are 180° apart. The resultant phasor (Y_X) is equal in length to the difference between the two phasors. The direction of the resultant phasor is the same direction as the longest phasor. As an example, assume Y_1 is 14 Ω and Y_2 is 10 Ω. Then Y_X would be 4 Ω in the direction of Y_1.

Suppose three phasors [Fig. 13-4(*a*)], all displaced 90° from each other, are to be added. The first step is to add the two phasors that are 180° apart. Adding Y_1 and Y_3 together reduces the problem to the two phasors shown in Fig.

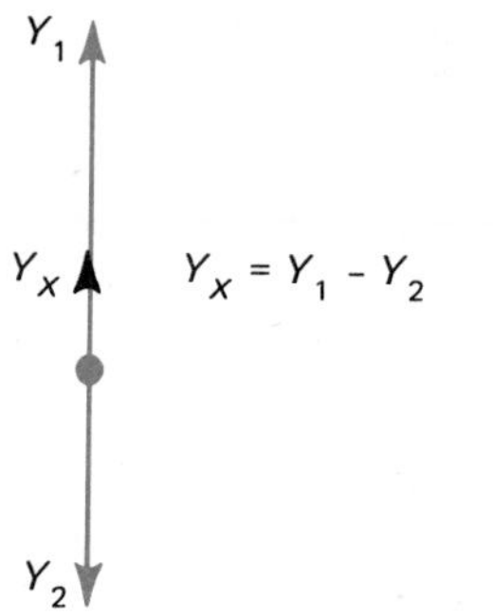

Fig. 13-3 Adding phasors separated by 180°.

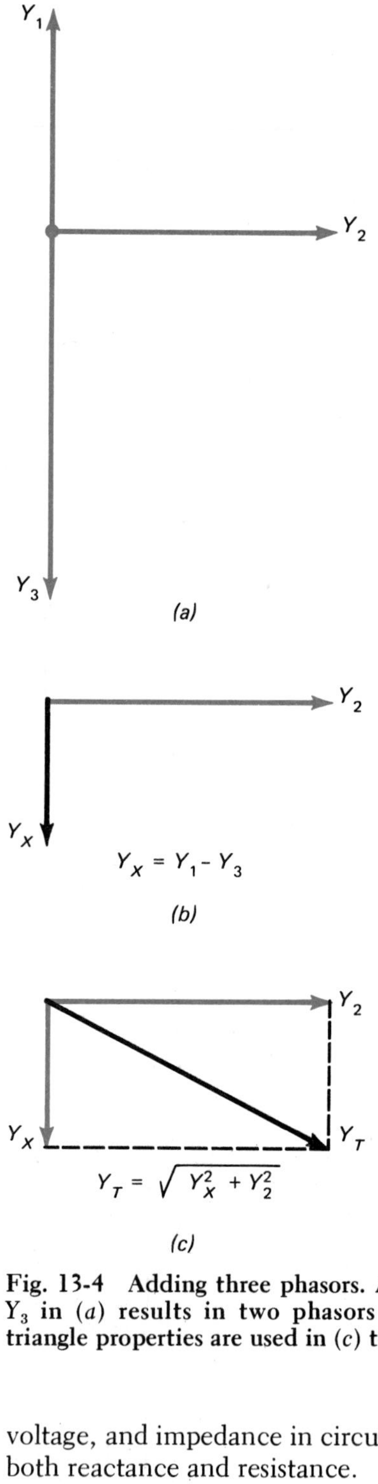

(c)

Fig. 13-4 Adding three phasors. Adding Y_1 and Y_3 in (*a*) results in two phasors in (*b*). Right triangle properties are used in (*c*) to find Y_T.

13-4(*b*). The second, and last, step involves adding together the two phasors of Fig. 13-4(*b*). This is accomplished by using the Pythagorean theorem. The result [Fig. 13-4(*c*)] is a single phasor (Y_T) which is equivalent to the original three phasors of Fig. 13-4(*a*). The formula given in Fig. 13-4(*c*) can be expanded so that it will add all three phasors in one step. It can be expanded by substituting $Y_1 - Y_3$ for Y_X. The result is the general formula:

$$Y_T = \sqrt{(Y_1 - Y_3)^2 + Y_2^2}$$

This formula can be used to combine any three phasors that are displaced 90° from each other. The three phasors may represent voltage, current, or impedance (and its components, resistance and reactance). The only requirement is that all the phasors have the same units, that is, volts, amperes, or ohms.

Example 13-2

Find the total current represented by the three current phasors in Fig. 13-5.

Given: $I_1 = 5$ A
$I_2 = 4$ A
$I_3 = 8$ A

Find: I_T

Known: $I_T = \sqrt{(I_1 - I_3)^2 + (I_2)^2}$

Solution: $I_T = \sqrt{(5-8)^2 + (4)^2}$
$= \sqrt{(-3)^2 + (4)^2} = \sqrt{9 + 16}$
$= \sqrt{25} = 5$ A

Answer: The three phasors represent a total current of 5 amperes.

Now that we know how to add phasors, we can return to problems that involve impedance. That is, we can now calculate current, voltage, and impedance in circuits containing both reactance and resistance.

Review Questions

5. Refer to Fig. 13-1(*a*). Assume $Y_1 = 3$ m, and $Y_2 = 5$ m. What is the length of the resultant?
6. Refer to Fig. 13-4(*a*). Assume $Y_1 = 20$ N,

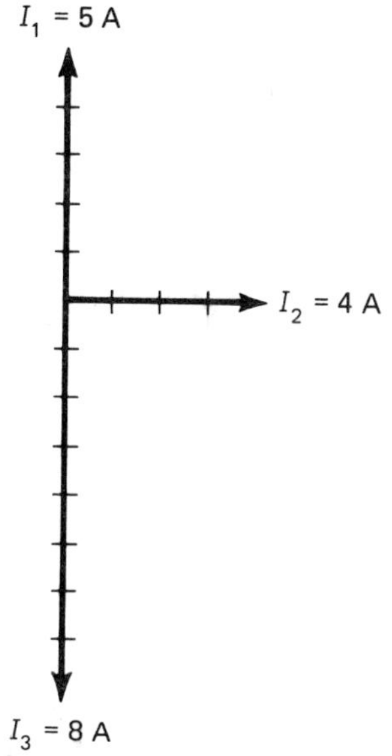

Fig. 13-5 Diagram for Example 13-2.

$Y_2 = 30$ N and $Y_3 = 40$ N. Find the total force.

7. What is the Pythagorean theorem?

13-3 SOLVING *RC* CIRCUITS

Circuits containing resistors and capacitors are very common in electronics. Unless stated otherwise, we will assume the components we use are ideal devices. That is, the resistors contain only resistance and the capacitors contain only capacitance. This assumption is acceptable when developing the basic relationship. Of course, actual resistors usually have both inductance and capacitance. Actual capacitors and inductors have resistance, capacitance, and inductance.

Series *RC* Circuits

Figure 13-6 shows a series *RC* circuit and the phasor diagrams that represent its voltages and impedance. In a series circuit the current through all parts of the circuit has to be the same. There is but one path for current; the capacitive current has to be the same as the resistive current. Therefore, the current phasor is often used as the reference phasor in series circuits [see Fig. 13-6(*b*)].

In any circuit we know that the resistive voltage and the resistive current are in phase. This means that the resistive voltage phasor [V_R in Fig. 13-6(*b*)] will be in the same direction as I_T.

We also know that the current through a capacitor leads the voltage across the capacitor by 90°. Another way of looking at this is to say the voltage lags the current by 90°. Of course, in a series circuit the value of the capacitive current is the same as the total current. They are one and the same. Thus, the capacitive voltage phasor [V_C in Fig. 13-6(*b*)] is plotted 90° behind the current phasor. (Recall that phasors are assumed to rotate in a counterclockwise direction.)

Series *RC* circuit

Now that we have V_R and V_C, they can be added together to obtain V_T. Phasor V_T establishes angle θ. Angle θ shows how much the source current (I_T) leads the source voltage (V_T). As you can see in Fig. 13-6, V_C is greater than V_R. This makes angle θ greater than 45°.

Also from Fig. 13-6(*b*) we can conclude that

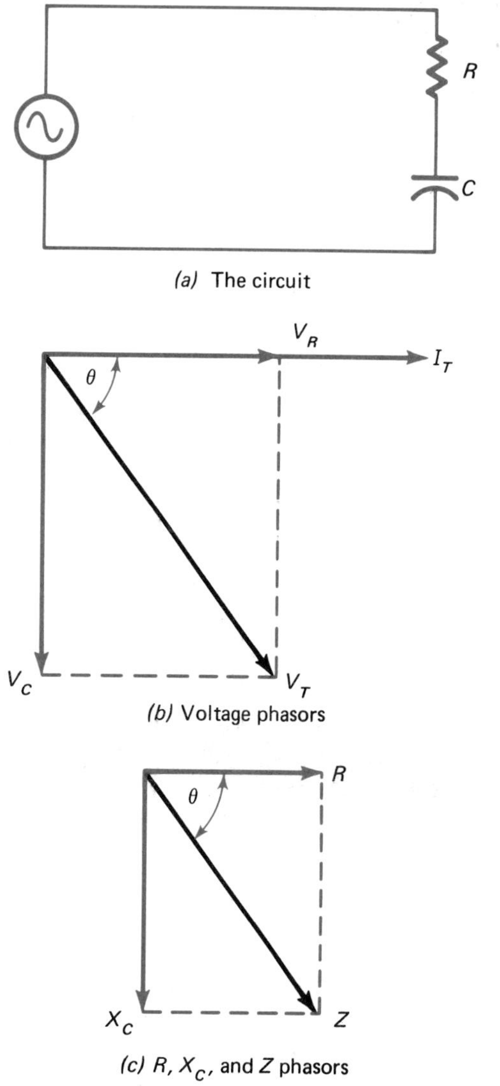

(*a*) The circuit

(*b*) Voltage phasors

(*c*) *R*, X_C, and *Z* phasors

Fig. 13-6 Series *RC* c[illegible]t. The voltages and the oppositions are show[illegible] for this circuit.

$V_T = \sqrt{V_R^2 + V_C^2}$

$Z = \sqrt{R^2 + X_C^2}$

$\text{Cos } \theta = \frac{V_R}{V_T} = \frac{R}{Z}$

X_C must be greater than R. We make this conclusion because of the following facts:

1. $I_R = I_C$
2. Ohm's law applies individually to both the resistor and the capacitor.
3. Ohm's law states the voltage and resistance or reactance are directly proportional if current is held constant.

Thus, if V_C is greater than V_R then X_C must also be greater than R.

Figure 13-6(*c*) shows that the resistance and capacitive reactance are also 90° out of phase. This is to be expected. The resistor and capacitor currents are in phase, but their voltages are out of phase. Therefore, X_C and R must also be 90° out of phase.

The phasors of Fig. 13-6(*b*) and (*c*) can be redrawn into triangles as shown in Fig. 13-7. From Fig. 13-7 it can be seen that the Pythagorean theorem can be used to calculate V_T and Z. The formulas are:

$$V_T = \sqrt{V_R^2 + V_C^2} \quad \text{and} \quad Z = \sqrt{R^2 + X_C^2}$$

The ratio of R over X_C [in Fig. 13-6(*c*)] must be equal to the ratio of V_R over V_C [in Fig. 13-6(*b*)]. This statement is easily seen by looking at Ohm's law:

$$I_T = \frac{V_R}{R} \quad \text{and} \quad I_T = \frac{V_C}{X_C}$$

Therefore,

$$I_T = \frac{V_R}{R} = \frac{V_C}{X_C}$$

Transposing terms:

$$I_T = \frac{V_R}{V_C} = \frac{R}{X_C}$$

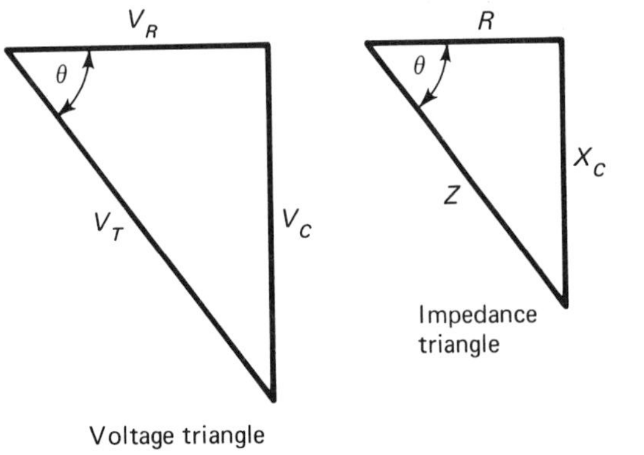

Fig. 13-7 Voltage and impedance triangles. Angle θ is the [illegible] triangles.

This means that angle θ can be figured equally well with voltages as well as impedances. In other words:

$$\cos \theta = \frac{\text{adjacent}}{\text{hypotenuse}} = \frac{V_R}{V_T} = \frac{R}{Z}$$

Now that we know the relationships between impedances and voltages in series RC circuits, let us use them.

Example 13-3

Find the impedance of the circuit in Fig. 13-8.

Given: $C = 1\ \mu F$
$R = 1000\ \Omega$

Find: Z

Known: $Z = \sqrt{R^2 + X_C^2}$

$X_C = \frac{1}{6.28fC}$

Solution: $X_C = \frac{1}{6.28 \times 100 \times 1 \times 10^{-6}}$

$= 1592\ \Omega$

$Z = \sqrt{1000^2 + 1592^2} = 1880\ \Omega$

The impedance of the circuits is 1880 ohms.

We can check the reasonableness of our answer very easily. Look again at Fig. 13-7 and notice that Z must be greater than either R or X_C. Also notice that Z must be less than the arithmetic sum of $R + X_C$. Since 1880 is greater than 1592 but less than 2592, it seems to be a reasonable answer.

Impedance can be used in Ohm's law just like resistance and reactance. For instance, the impedance and source voltage can be used to find the total current. Then the current can be used to calculate individual voltages.

Example 13-4

Find the current (I_T), the resistive voltage (V_R), and the capacitive voltage (V_C) for Fig. 13-8.

Given: $V_T = 40$ V
$R = 1000\ \Omega$
$X_C = 1592\ \Omega$ (from Example 13-3)
$Z = 1880\ \Omega$ (from Example 13-3)

Find: I_T, V_C, and V_R

Known: Ohm's law

Solution: $I_T = \frac{40\text{ V}}{1880\ \Omega} = 0.0213$ A

$= 21.3$ mA

$V_R = 0.0213\text{ A} \times 1000\ \Omega$
$= 21.3\text{ V}$

$V_C = 0.0213\text{ A} \times 1592\ \Omega$
$= 33.9\text{ V}$

Answer: The circuit current is 21.3 milliamperes. The resistive voltage is 21.3 volts and the capacitive voltage is 33.9 volts.

We can verify our answer by checking whether the phasor sum of V_R and V_C equals the source voltage (V_T).

$$V_T = \sqrt{21.3^2 + 33.9^2} = 40\text{ V}$$

Since this calculated value of V_T agrees with the value given, we have not made any errors in our calculation.

Now we can continue our study of the circuit in Fig. 13-8 by determining θ and power.

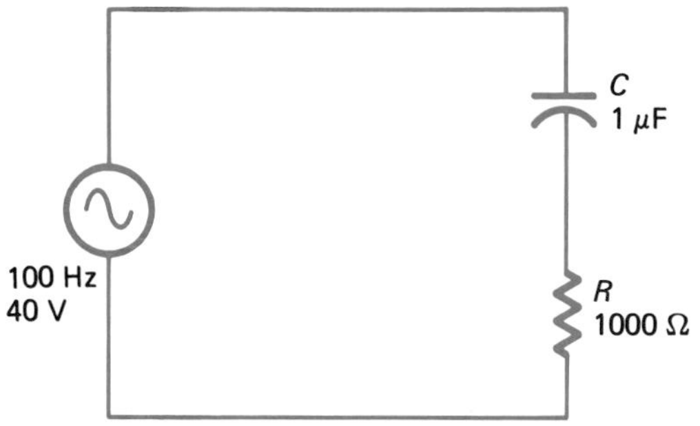

Fig. 13-8 Circuit for Examples 13-3, 13-4, and 13-5.

Example 13-5

For the circuit of Fig. 13-8 find angle θ and the total true power.

Given: $R = 1000\ \Omega$
$V_T = 40\text{ V}$
$Z = 1880\ \Omega$
$I_T = 21.3\text{ mA}$

Find: θ and P

Known: $\cos\theta = \frac{R}{Z}$, power formulas

Solution: $\cos\theta = \frac{1000\ \Omega}{1880\ \Omega} = 0.532$

$\theta = 58°$ (from App. 6)

$P_T = IV\cos\theta$
$= 0.0213 \times 40 \times 0.532 = 0.453\text{ W}$

Answer: The current leads the voltage by 58° and the power in the circuit is 0.453 watts.

In the above example we could have found the total power without first calculating θ. Recall that only the resistance in a circuit can use power. Therefore, the power used by the resistor (P_R) must equal the total power. Thus,

$$P_T = P_R = I_R \times V_R = 0.0213\text{ A} \times 21.3\text{ V}$$
$$= 0.454\text{ W}$$

This answer agrees (within round-off error) with the answer in Example 13-5. Remember $P_T = P_R$ for all circuits.

$P_T = P_R$

Also, in Example 13-5 we could have found $\cos\theta$ using voltage ratios. This would give us:

$$\cos\theta = \frac{V_R}{V_T} = \frac{21.3\text{ V}}{40\text{ V}} = 0.533$$

Again, the answer agrees (within round-off error) with that obtained in Example 13-5.

Suppose the frequency of the source in Fig. 13-8 were decreased. What would happen to angle θ, the impedance and voltage distributions? The answer to this question is illustrated in Fig. 13-9(*a*) and (*b*). In Fig. 13-9(*a*) the phasor diagrams for the circuit, when operated at 100 Hz, are shown. Figure 13-9(*b*)

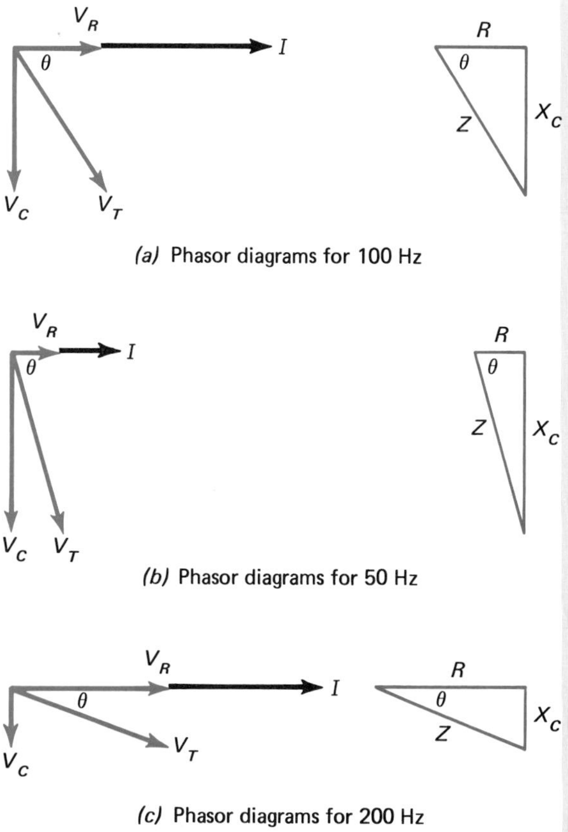

Fig. 13-9 Effects of *f* on Z, V, and θ.

shows what happens when the frequency is decreased to 50 Hz. When the frequency decreases, the reactance of the capacitor increases. This causes an increase in the impedance and a corresponding decrease in the current. Less current means less voltage across the resistor. However, the capacitive voltage increases because its reactance increased proportionately more than the current decreased. The net result is that the circuit is more capacitive than before. Angle θ has increased and the power in the circuit has decreased.

When the frequency increases, the results [Fig. 13-9(*c*)] are just the opposite from when it decreases. The circuit becomes more resistive; θ decreases and the power increases.

Figure 13-10 illustrates what happens when the capacitor in Fig. 13-8 is changed. The diagrams in Fig. 13-10(*a*) show the situation with the original capacitor (1 μF). When the capacitance is decreased [Fig. 13-10(*b*)], the effects are the same as when the frequency was decreased. This is because capacitive reactance is inversely proportional to both capacitance and frequency. Decreasing capacitance causes the capacitive reactance to increase; that is, θ becomes larger. Increasing the capacitance, of course, decreases the capacitive reactance and causes the circuit to be more resistive [see Fig. 13-10(*c*)].

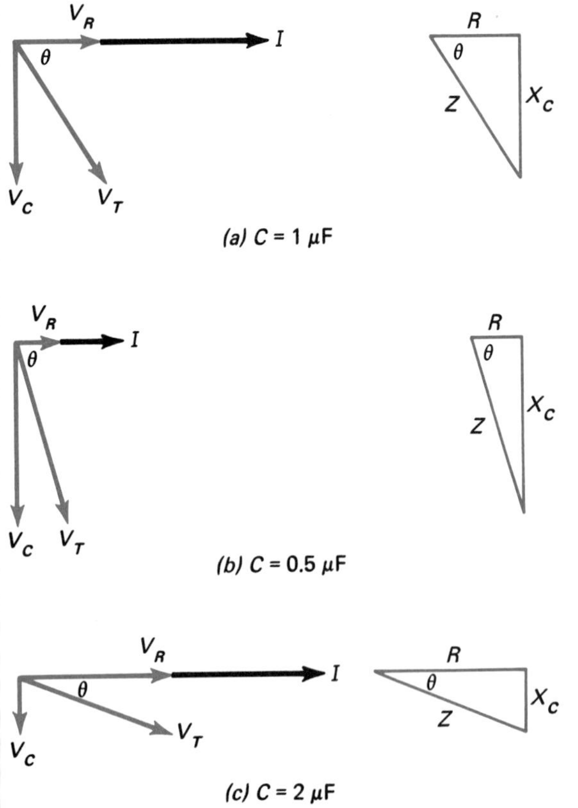

Fig. 13-10 Effects of C on Z, V, and θ.

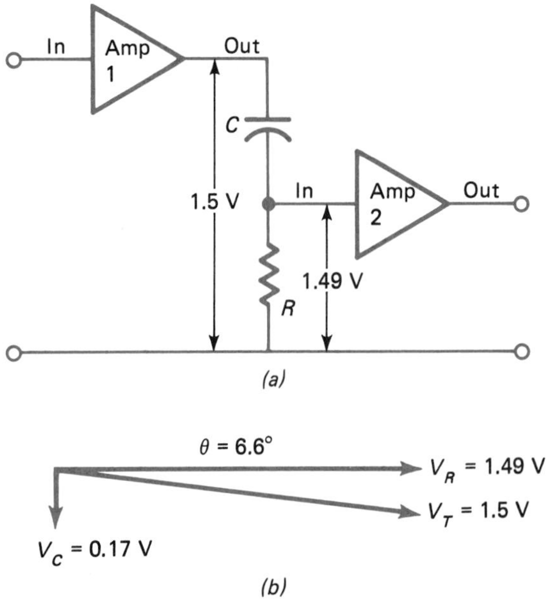

Fig. 13-11 Use of an *RC* circuit. The capacitor couples the amplifiers together with minimum voltage loss or phase shift.

An example of how an *RC* circuit is used is shown in Fig. 13-11. Amplifier 1 replaces the generator in Fig. 13-8. In other words, amplifier 1 is the source which provides the voltage across the *RC* circuit. Whatever voltage develops across *R* becomes the input voltage to the second amplifier. It is desirable to get as much as possible of the voltage from amplifier 1 to the input of amplifier 2. Therefore, the reactance of the capacitor must be low compared to the resistance of the resistor. The resistance should be about 10 times as high as the reactance. This makes the resistive voltage [Fig. 13-11(*b*)] nearly equal to the source voltage. Also, Fig. 13-11(*b*) shows that very little phase shift occurs when *R* is high compared to X_C. This means that the voltage input to amplifier 2 is nearly in phase with the output of amplifier 1.

Parallel *RC* circuit

Parallel *RC* Circuits

Reference phasor

In a parallel circuit [Fig. 13-12(*a*)], the voltage is the same across all components. Obviously V_R, V_C, and V_T have to be in phase. Therefore, the voltage phasor is used as the reference phasor [Fig. 13-12(*b*)] in solving parallel *RC* circuits. The capacitive current still must lead the capacitive voltage by 90°. Thus, the I_C phasor is drawn 90° ahead of the voltage phasor [Fig. 13-12(*b*)]. As is always the case, the resistive current is in phase with the resistive voltage. The I_R phasor has to be in line with the V_T phasor. Adding phasors I_C and I_R yields the I_T phasor. Since the circuit has capacitance, the source current (I_T) will lead the source voltage (V_T).

The impedance of the circuit shown in Fig. 13-12(*a*) can be found by using Ohm's law with V_T and I_T. That is:

$$Z = \frac{V_T}{I_T}$$

Of course, one must calculate I_T before Z is determined. The value of I_T can be calculated by using the Pythagorean theorem. Just express the sides of the triangle in terms of currents. The formula is:

$$I_T = \sqrt{I_C^2 + I_R^2}$$

Example 13-6

Find the impedance of the circuit in Fig. 13-12.

Given: $V_T = 60$ V
$f = 400$ Hz
$C = 0.02\ \mu F$
$R = 27\ k\Omega$

Find: Z

Known: Ohm's law

$$X_C = \frac{1}{6.28fC}$$

$$I_T = \sqrt{I_C^2 + I_R^2}$$

Solution:

$$X_C = \frac{1}{6.28 \times 400 \times 0.02 \times 10^{-6}} = 19{,}900\ \Omega = 19.9\ k\Omega$$

$$I_C = \frac{60\ V}{19{,}900\ \Omega} = 3.0\ mA$$

$$I_R = \frac{60\ V}{27{,}000\ \Omega} = 0.00222\ A = 2.22\ mA$$

$$I_T = \sqrt{(0.003)^2 + (0.00222)^2} = 0.00373\ A = 3.73\ mA$$

$$Z = \frac{60\ V}{0.00373\ A} = 16{,}100\ \Omega = 16.1\ k\Omega$$

Answer: The impedance of the circuit is 16.1 kilohms.

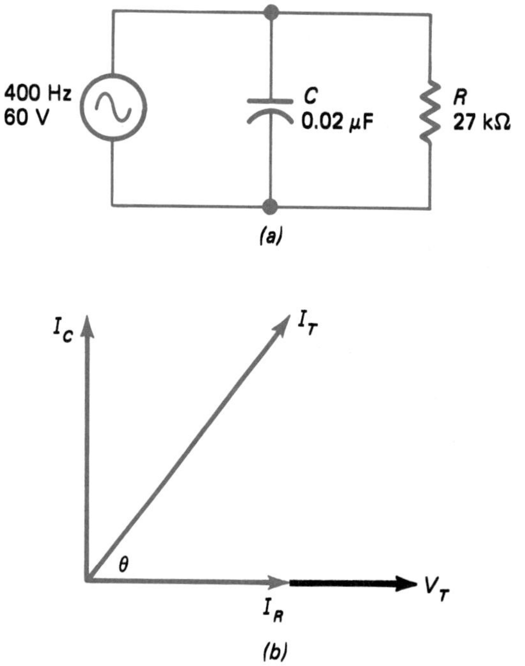

Fig. 13-12 Parallel *RC* circuit. The circuit phasor currents are related. The voltage phasor is the reference (zero degrees) phasor.

Notice in this example that the impedance is less than the capacitive reactance or the resistance. This is because the reactance and the resistance are in parallel as well as 90° out of phase.

The impedance of a parallel *RC* circuit can also be calculated from the resistance and reactance. The formula is:

$$Z = \frac{X_C \times R}{\sqrt{X_C^2 + R^2}}$$

This formula yields the same results as the Ohm's law method. Therefore, we can use it to check the answer arrived at in Example 13-6.

$$Z = \frac{19{,}900\ \Omega \times 27{,}000\ \Omega}{\sqrt{(19{,}900\ \Omega)^2 + (27{,}000\ \Omega)^2}} = 16{,}000\ \Omega = 16.0\ k\Omega$$

Since the two methods provide the same answer (within round-off error), we can assume no calculation errors were made.

The effects of changing *f* or C in a parallel *RC* circuit are illustrated in Fig. 13-13. Notice that the effect on angle θ is the exact opposite of the effect in a series *RC* circuit. Increasing *C* or *f* causes a decrease in reactance which results in less impedance, more current, and a

$$I_T = \sqrt{I_C^2 + I_R^2}$$

$$Z = \frac{X_C \times R}{\sqrt{X_C^2 + R^2}}$$

$\text{Cos}\ \theta = \frac{I_R}{I_T}$

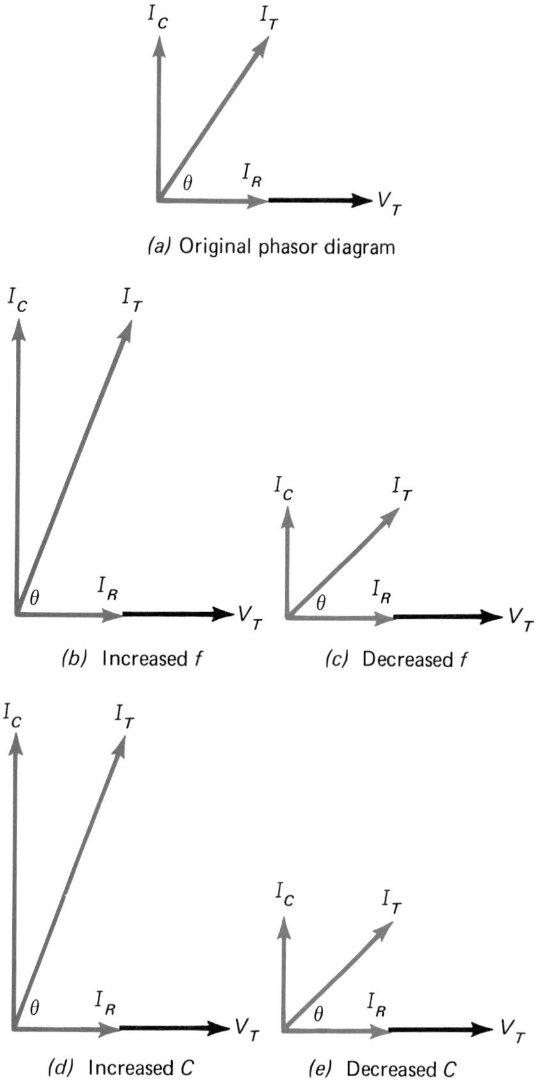

Fig. 13-13 Effects of *f* and *C* on *I* and θ.

larger angle θ. Of course, decreasing C or *f* has the opposite effect.

The value of cos θ can be calculated from either the currents or the powers. The formulas are:

$$\text{Cos}\ \theta = \frac{P}{P_{\text{app}}} \quad \text{or} \quad \cos\theta = \frac{I_R}{I_T}$$

These are also the power factor formulas. Once the cos θ is determined, angle θ can be determined from a table of trigonometric functions. We can find the power factor of the circuit in Example 13-6 using the values for I_R and I_T:

$$\text{Cos}\ \theta = \text{PF} = \frac{I_R}{I_T} = \frac{0.00222}{0.00373}$$
$$= 0.595 \quad \text{or } 59.5\%$$

Since the total current is leading the voltage, the power factor of the circuit is 59.5%, leading.

Review Questions

8. In a series *RC* circuit does the current lead or lag the voltage?
9. In a parallel *RC* circuit does the current lead or lag the voltage?
10. Which phasor is usually the reference phasor in a parallel *RC* circuit?
11. When V_R is greater than V_C in a series *RC* circuit, is θ less than or more than 45°?
12. For the circuit specified in question 11, will X_C be greater than or less than *R*?
13. What is the phase relationship between resistance and reactance?
14. Write two formulas for finding the impedance of a series *RC* circuit.
15. Write the formula for combining the resistive and reactive voltages in a series *RC* circuit.
16. Write three formulas for finding the cosine of θ in a series *RC* circuit.
17. Which of the formulas in question 16 can also be used in a parallel *RC* circuit?
18. What other formula (besides the one listed in question 17) can be used to find cos θ in a parallel *RC* circuit?
19. If the frequency of a parallel *RC* circuit increases what happens to
    a. The impedance?
    b. The resistive current?
    c. The power?
    d. Cos θ?
20. If the capacitance of a series *RC* circuit decreases what happens to
    a. The impedance?
    b. The resistive voltage?
    c. The power?
    d. Cos θ?
21. Refer to Fig. 13-12. Change the value of R to 10 kΩ. Now calculate the following:
    a. Z
    b. P
    c. θ
    d. I_C

22. Refer to Fig. 13-8. Change R to 2200 Ω. Now calculate the following:
    a. I_T
    b. V_R
    c. P
    d. Z

13-4 SOLVING *RL* CIRCUITS

Although *RL* circuits are not as common as *RC* circuits in electronics they are important. One must know something of their characteristics to fully understand such devices as ac motors and high-frequency amplifiers.

Series *RL* Circuits

The series *RL* circuit [Fig. 13-14(*a*)] causes the source voltage to lead the source current. This is shown in Fig. 13-14(*b*) where V_T leads I_T (the reference phasor). The same general formulas used for series *RC* circuits can also be used for series *RL* circuits; merely replace the C with L and the X_C with X_L. The important formulas for series *RL* circuits are:

Series *RL* circuit

$Z = \sqrt{R^2 + X_L^2}$

$V_T = \sqrt{V_R^2 + V_L^2}$

$$Z = \sqrt{R^2 + X_L^2}$$

$$V_T = \sqrt{V_R^2 + V_L^2}$$

$$\text{Cos } \theta = \frac{V_R}{V_T} = \frac{R}{Z} = \frac{P}{P_{\text{app}}}$$

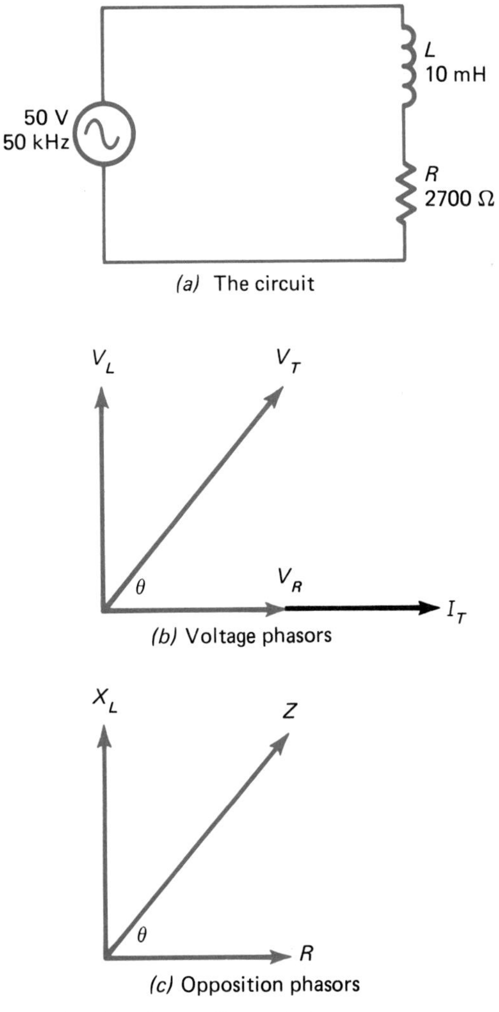

Fig. 13-14 **Series *RL* circuit.**

Example 13-7

Determine the impedance and voltage across the inductance for the circuit in Fig. 13-14.

Given: $V_T = 50$ V
$f = 50$ kHz
$L = 10$ mH
$R = 2700$ Ω

Find: Z and V_L

Known: $Z = \sqrt{R^2 + X_L^2}$
$X_L = 6.28fL$
Ohm's law

Solution: $X_L = 6.28 \times 50 \times 10^3 \times 10 \times 10^{-3}$
$= 3140\ \Omega$

$Z = \sqrt{(2700\ \Omega)^2 + (3140\ \Omega)^2}$
$= 4141\ \Omega$

$I_T = \dfrac{50\text{ V}}{4141\ \Omega} = 0.012\text{ A} = 12\text{ mA}$

$V_L = 0.012\text{ A} \times 3140\ \Omega = 37.7\text{ V}$

Answer: The impedance of the circuit is 4141 ohms, and the voltage across the inductor is 37.7 volts.

The series *RL* and series *RC* circuits respond to changes in frequency in exactly opposite fashions. This is because inductive reactance is directly proportional to frequency while capacitive reactance is inversely proportional to frequency. Increasing the frequency of a series *RL* circuit causes the following to occur: X_L increases, Z increases, I_T decreases, V_R decreases, V_L increases, and the angle θ increases. Inductive reactance is also directly proportional to inductance. Therefore, increasing the inductance has the same effect as increasing the frequency.

Power and angle θ in the series *RL* and series *RC* circuits are determined in exactly the same way. The difference in the two types of

Ac motor

Parallel *RL* circuit

circuits is that the current lags the voltage in an *RL* circuit while the current leads the voltage in an *RC* circuit (compare Fig. 13-6 and 13-14).

Many small, single-phase, ac motors are started by the phase shift between two *RL* circuits. These motors have two separate windings in them. The run winding uses large diameter wire and, therefore, has only a small amount of resistance. The run winding can be represented by a small resistor in series with an ideal inductor [see Fig. 13-15(*a*)]. Since the run winding has a low resistance and a high reactance, its current lags far behind the voltage [Fig. 13-15(*b*)]. The start winding is wound with small diameter wire; it also has fewer turns of wire than the run winding. Therefore, the start winding has more resistance and less inductance than the run winding. This causes the start-winding current to lag the voltage by fewer degrees than does the run-winding current. Thus, the start-winding current is out of phase with the run-winding current [see Fig. 13-15(*b*)]. These out of phase currents produce the special magnetic field needed to start the motor. Once the motor has started, a switch [Fig. 13-15(*a*)] disconnects the start winding from the circuit and the motor continues to run on the run winding alone.

Parallel *RL* Circuits

A parallel *RL* circuit and its phasor diagram are shown in Fig. 13-16. Parallel *RL* circuits can be treated like parallel *RC* circuits. There are only these two differences between the two types of circuits:

1. The current in the *RL* circuit lags the voltage, whereas the current leads the voltage in the *RC* circuit. [Compare Figs. 13-12(*b*) and 13-16(*b*).]
2. They respond in an opposite manner to changes in frequency and inductance or capacitance. When either f or L is increased in the parallel *RL* circuit, the impedance increases. This results in the inductive current and the total current decreasing and angle θ decreasing.

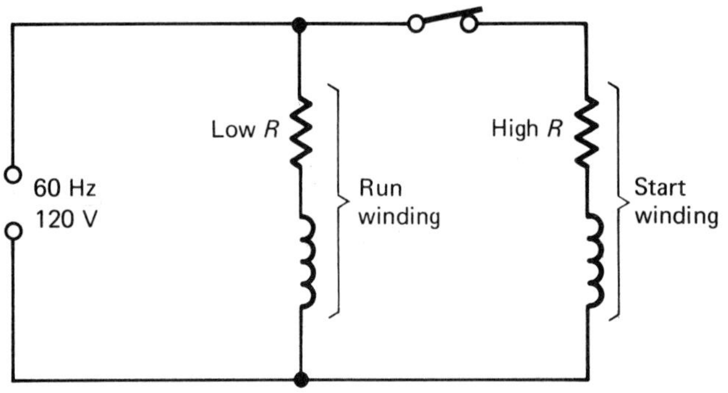

(*a*) Equivalent motor circuit

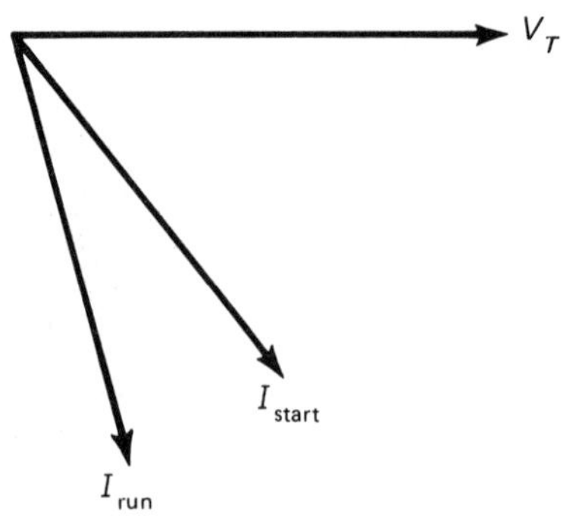

(*b*) Phasor diagram

Fig. 13-15 Ac split-phase motor circuit. (*a*) Equivalent circuit of a split-phase motor under starting conditions. (*b*) Phasor diagram showing the current in the run winding out of phase with the current in the start winding.

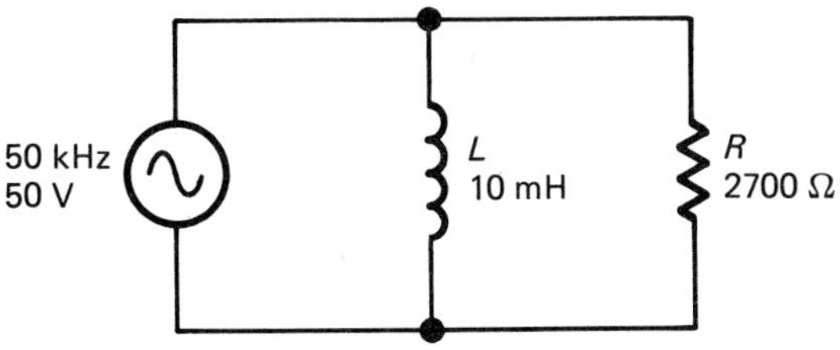

(a) Parallel *RL* circuit

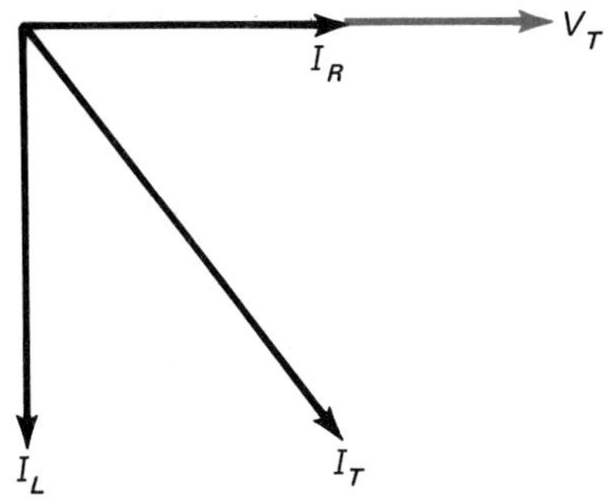

(b) Phasor diagram

Fig. 13-16 Parallel *RL* circuits. The total current lags the total voltage.

The formulas listed below are the major ones needed to work with parallel *RL* circuits.

$$I_T = \sqrt{I_L^2 + I_R^2}$$

$$Z = \frac{V_T}{I_T} = \frac{X_L \times R}{\sqrt{X_L^2 + R^2}}$$

$$\text{Cos } \theta = \frac{P}{P_{\text{app}}} = \frac{V_R}{V_T} = \frac{R}{X_L}$$

Notice these formulas are the same as the formulas used for parallel *RC* circuits except *L* is substituted for *C*.

Example 13-8

For the circuit of Fig. 13-16, calculate the total current and the impedance.

Given: $V_T = 50$
$f = 50$ kHz
$L = 10$ mH
$R = 2700\ \Omega$

Find: I_T and Z

Known: Ohm's law
$X_L = 6.28fL$
$I_T = \sqrt{I_L^2 + I_R^2}$

Solution: $I_R = \dfrac{50\text{ V}}{2700\ \Omega} = 0.0185\text{ A} = 18.5\text{ mA}$

$$X_L = 6.28 \times 50 \times 10^3 \times 10 \times 10^{-3} = 3140\ \Omega$$

$$I_L = \frac{50\text{ V}}{3140\ \Omega} = 0.0159\text{ A} = 15.9\text{ mA}$$

$$I_T = \sqrt{(0.0185)^2 + (0.0159)^2} = 0.0244\text{ A} = 24.4\text{ mA}$$

$$Z = \frac{50\text{ V}}{0.0244\text{ A}} = 2049\ \Omega$$

Answer: The current is 24.4 milliamperes and the impedance is 2049 ohms.

$I_T = \sqrt{I_L^2 + I_R^2}$

$Z = \dfrac{X_L \times R}{\sqrt{X_L^2 + R^2}}$

Review Questions

23. In *RL* circuits does the source current lead or lag the source voltage?
24. Suppose the inductance of a series *RL* circuit is reduced. What happens to I_T, V_L, and Z?
25. When the frequency of a parallel *RL* circuit is increased, what happens to Z and θ?
26. What is the impedance at 50 kHz of a 2-mH inductor and a 1000 Ω resistor when connected in series?
27. How much current would flow in the circuit of question 26 if the source voltage were 36 V?
28. Change L in Fig. 13-16 to 5 mH and solve for I_T and Z.

13-5 SOLVING *RCL* CIRCUITS

The techniques used to solve *RC* and *RL* circuits can be combined to solve *RCL* circuits [Fig. 13-17(*a*)]. However, *RCL* circuits have some unusual characteristics. For instance, a reactive voltage or current can be higher than the source voltage or current.

Series *RCL* circuit

Net reactive voltage

Series *RCL* Circuits

Refer to the phasors in Fig. 13-17(*b*). Notice that the inductive and capacitive voltage are 180° out of phase with each other. Adding these two voltage phasors results in V_X which is the net reactive voltage. V_X is then added to V_R to determine the total or source voltage. The reactance and resistance phasors are added in the same sequence to determine the impedance.

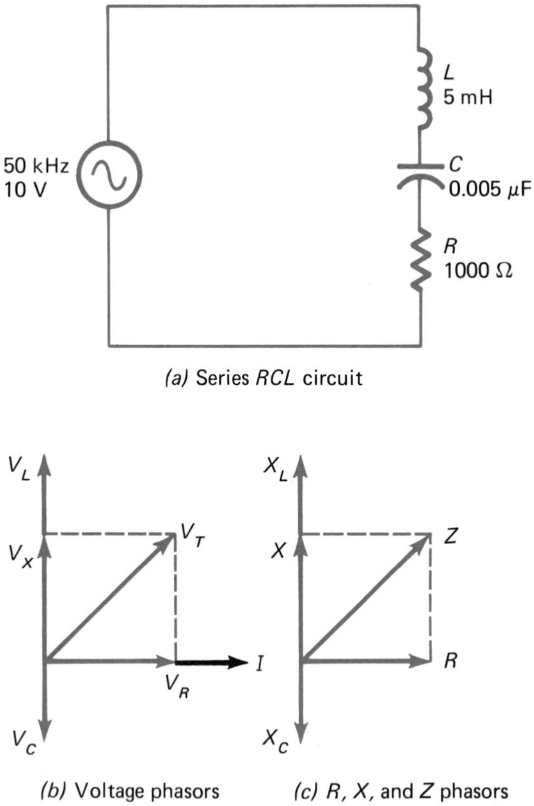

Fig. 13-17 Series *RCL* circuit.

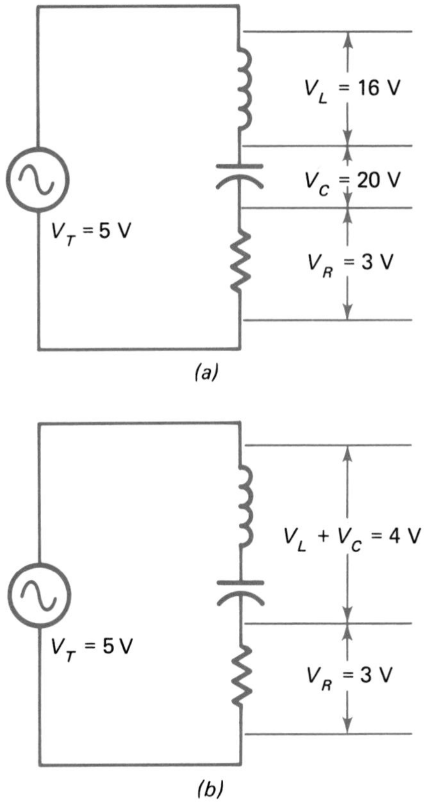

Fig. 13-18 Voltmeter readings in a series *RCL* circuit. Notice that both the inductive and the capacitive voltage in (*a*) exceed the source voltage.

In Fig. 13-17(*b*) you can see that V_L is greater than V_T. At first glance this may look like an exception to Kirchhoff's voltage law; however, it is not. At any instant the sum of the instantaneous voltages of the resistor, capacitor, and inductor will equal the source voltage. In other words, the three voltages are never of the same polarity at the same instant. This idea is illustrated in Fig. 13-18. Notice the voltage drops across each component (*L*, *C*, and *R*) in Fig. 13-18(*a*). These voltage drops are a result of the series current passing through each component. The voltage across *L*, however, is 180° out of phase with the voltage across *C*. This means the two voltages are directly opposing one another. The result is the net reactive voltage shown in Fig. 13-18(*b*) across the combination of *L* and *C*.

Notice in Fig. 13-18(*a*) that both the inductive and the capacitive voltages are greater than the source voltage. This is common in series *RCL* circuits. If the individual voltages exceed the source voltage, then the individual reactances will also exceed the impedance.

Impedance and total voltage in a series circuit can be calculated with the aid of the following formulas:

$$Z = \sqrt{(X_L - X_C)^2 + R^2}$$

$$Z = \frac{V_T}{I_T}$$

$$V_T = \sqrt{(V_L - V_C)^2 + V_R^2}$$

In these formulas, $V_L - V_C$ is the same as V_X in Fig. 13-17(*b*), and $X_L - X_C$ is the same as *X*.

The power factor (cos θ), and angle θ of the series *RCL* circuits can be found using the same formulas used for series *RL* and *RC* circuits.

Example 13-9

Find the impedance and the voltage across the resistance for the circuit in Fig. 13-17(*a*).

Given: $V_T = 10$ V
$f = 50$ kHz
$L = 5$ mH
$C = 0.005\ \mu$F
$R = 1000\ \Omega$

Find: Z and V_R

Known: Ohm's law

$$Z = \sqrt{(X_L - X_C)^2 + R^2}$$

$$X_L = 6.28fL$$

$$X_C = \frac{1}{6.28fC}$$

Solution: $X_L = 6.28 \times 50 \times 10^3 \times 5 \times 10^{-3}$

$= 1570\ \Omega$

$$X_C = \frac{1}{6.28 \times 50 \times 10^3 \times 0.005 \times 10^{-6}}$$

$= 637\ \Omega$

$$Z = \sqrt{(1570 - 637)^2 + (1000)^2}$$

$= 1368\ \Omega$

$$I_T = \frac{10\ \text{V}}{1368\ \Omega} = 0.0073\ \text{A} = 7.3\ \text{mA}$$

$$V_R = 0.0073\ \text{A} \times 1000\ \Omega = 7.3\ \text{V}$$

Answer: The impedance is 1368 ohms and the voltage across the resistance is 7.3 volts.

Parallel *RCL* Circuits

Figure 13-19 shows a parallel *RCL* circuit and its current phasor diagram. From the phasor diagram it is obvious that a branch current (I_C) can exceed the total current. (This is because the inductive current and the capacitive currents are 180° out of phase.) In fact, both the capacitive and inductive currents can exceed the total current. For example, if L in Fig. 13-19(*a*) were halved, the inductive current would double. I_X and I_T would both decrease; then I_L and I_C would each exceed I_T.

The formulas for working with parallel *RCL* circuits are:

$$I_T = \sqrt{(I_L - I_C)^2 + I_R^2}$$

$$\cos\theta = \frac{I_R}{I_T} = \frac{P}{P_{\text{app}}}$$

The impedance can also be calculated directly from the values of R, C, and L (see App. 2).

Parallel *RCL* circuit

Example 13-10

Calculate the value of I_T and Z for the circuit in Fig. 13-19. (Note: X_L and X_C will be the same value as in Example 13-9.)

Given: $V_T = 10$ V, $X_L = 1570\ \Omega$, $R = 1000\ \Omega$, $X_C = 637\ \Omega$

Known: Ohm's law

$$I_T = \sqrt{(I_C - I_L)^2 + I_R^2}$$

Solution:

$$I_C = \frac{10\ \text{V}}{637\ \Omega} = 0.0157\ \text{A} = 15.7\ \text{mA}$$

$$I_L = \frac{10\ \Omega}{1570\ \Omega} = 0.0064\ \text{A} = 6.4\ \text{mA}$$

$$I_R = \frac{10\ \text{V}}{1000\ \Omega} = 0.0010\ \text{A} = 10\ \text{mA}$$

$$I_T = \sqrt{(0.0064 - 0.0157)^2 + (0.01)^2}$$

$= 0.0137\ \text{A} = 13.7\ \text{mA}$

$$Z = \frac{10\ \text{V}}{0.0137\ \text{A}} = 730\ \Omega$$

Answer: The impedance is 730 ohms and the current is 13.7 milliamperes.

The parallel *RCL* circuit of Fig. 13-19 and the series *RCL* circuit of Fig. 13-17 have the same component values. They also have the same source voltage and frequency. As expected, the impedance of the parallel circuit is less than the series circuit and draws more current than the series circuit. Inspection of the phasor diagrams for the two circuits shows another difference between series and parallel *RCL* circuits. One of them is inductive and the other is capacitive. Notice in Fig. 13-17(*b*) that the source current lags behind the source

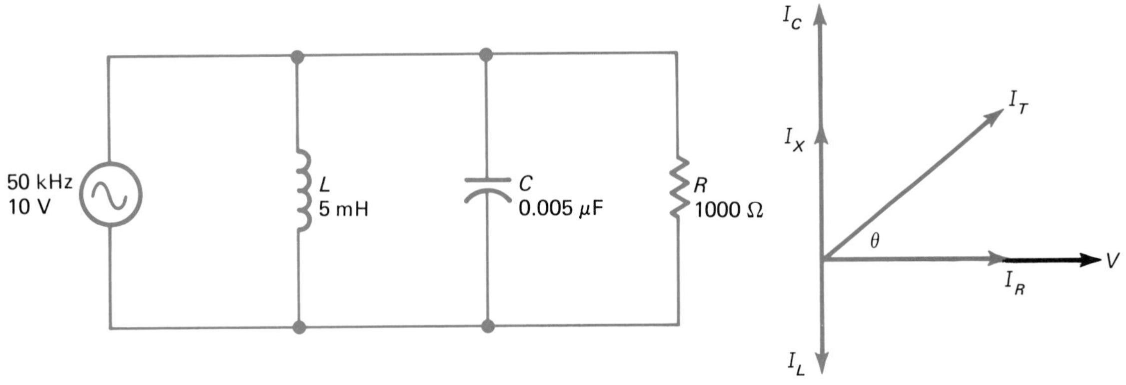

(*a*) The circuit (*b*) Current phasors

Fig. 13-19 Parallel *RCL* circuit.

voltage. Thus, this series *RCL* circuit is inductive. In Fig. 13-19(*b*), the source current leads the source voltage. Therefore, this parallel circuit (with the same components as the series circuit) is capacitive. In Figs. 13-17 and 13-19, X_L is greater than X_C. If X_C were greater than X_L, then the series circuit would be capacitive and the parallel circuit would be inductive (see Fig. 13-20).

Resonant circuit

Review Questions

29. Can the reactance in a series *RCL* circuit exceed the impedance?
30. Can the resistance in a series *RCL* circuit exceed the impedance?
31. Can the reactive current in a parallel *RCL* circuit exceed the total current?
32. Write two formulas for finding the impedance of a series *RCL* circuit.
33. Write the formula for adding the branch currents in a parallel *RCL* circuit.
34. Two *RCL* circuits (one series and the other parallel) are connected to the same frequency. If X_L is greater than X_C, which circuit will
    a. Be inductive?
    b. Be capacitive?
    c. Draw the higher source current?
35. A 7-mH inductor, a 0.003-μF capacitor, and a 1000-Ω resistor are series connected to a 20-V, 50-kHz source. Find the Z, I_T, and V_L.
36. Assume the components of question 35 were connected in parallel. Determine the impedance and the current through the capacitor.

13-6 RESONANCE

In the preceding section we worked with *RCL* circuits in which X_L was either greater or less than X_C. Now we will work with circuits in which X_L and X_C are equal. Circuits in which $X_L = X_C$ are called *resonant circuits*. They can

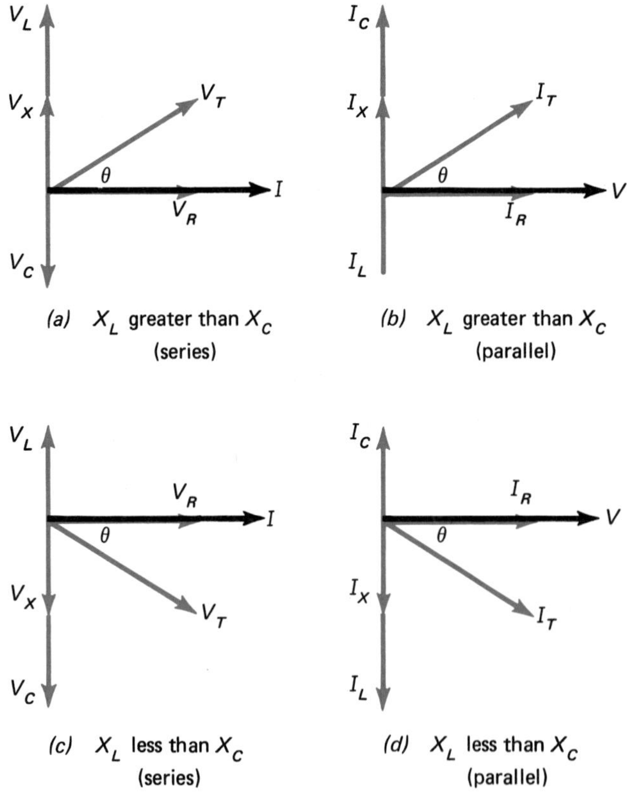

Fig. 13-20 Capacitive and inductive *RCL* circuits. Diagrams (*b*) and (*c*) are for capacitive circuits.

be either series or parallel circuits. They can be either *RCL* circuits or just *LC* circuits. Most often, resonant circuits are *LC* circuits; the only resistance in these circuits is that of the actual inductor and capacitor used.

Resonant Frequency

Resonant frequency

For a given value of L and C there is only one frequency at which X_L will equal X_C. This frequency, called the resonant frequency, can be calculated with the following formula:

$$f_r = \frac{1}{6.28\sqrt{LC}}$$

In this formula, the resonant frequency f_r will be in hertz (Hz) if inductance (L) is in henries (H) and capacitance (C) is in farads (F). The resonant frequency formula is derived from the two reactance formulas. By definition, the resonant frequency is that frequency at which $X_L = X_C$. Therefore:

$$6.28fL = \frac{1}{6.28fC}$$

Solving this equation for f yields the formula for f_r given above.

Figure 13-21 further explains the concept of resonant frequency. This figure is a plot of reactance against frequency. The reactance (X_{L_1}) of a specific value of inductance (L_1), when plotted against frequency, produces a straight line. The reactance (X_{C_1}) of a specific value of capacitance (C_1) produces a curved line. The resonant frequency of L_1C_1 is the frequency at which the X_{L_1} line and the X_{C_1} line cross.

We have seen that L_1C_1 can be resonant only at one specific frequency. However, any number of other values of L and C can also be resonant at the same frequency as L_1C_1. Suppose a higher inductance and a lower capacitance, such as L_2 and C_2 of Fig. 13-21, are used. They are resonant at the same frequency as L_1C_1. One could also use a lower value of L and a higher value of C to resonate at the same frequency as L_1C_1 and L_2C_2.

Notice two other points in Fig. 13-21. If the inductance decreases to L_2 and the capacitance remains at C_1, the resonant frequency drops to point A. On the other hand, if the inductance remains at L_1 and the capacitance

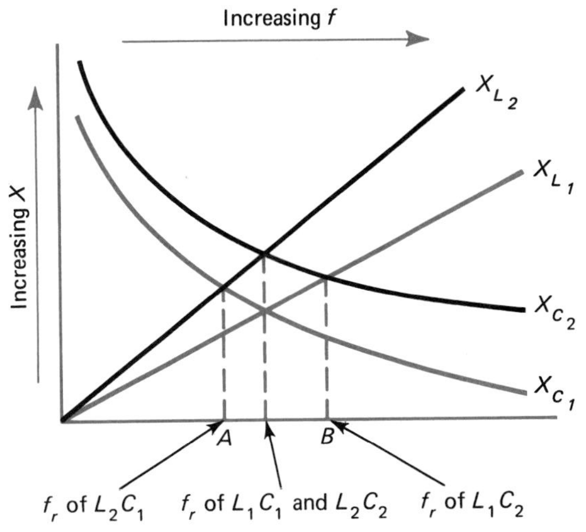

Fig. 13-21 Plot of X_L and X_C versus f. At resonance, $X_L = X_C$.

is decreased (that is, X_C is increased) to C_2, then the resonant frequency will increase to point *B*.

The relationship between change in L or C and the resultant change in f_r is also evident from the resonant frequency formula. From the formula, it can be seen that increasing either L or C will decrease f_r. Conversely, decreasing either L or C will increase f_r.

Example 13-11

What is the resonant frequency (f_r) of a 10-mH inductor and a 0.005-μF capacitor?

Given: $L = 10$ mH
$C = 0.005$ μF

Find: f_r

Known: $f_r = \frac{1}{6.28\sqrt{LC}}$

Solution: $f_r = \frac{1}{6.28\sqrt{10 \times 10^{-3} \times 0.005 \times 10^{-6}}}$
$= 22{,}500 \text{ Hz} = 22.5 \text{ kHz}$

Answer: The resonant frequency is 22.5 kilohertz.

The capacitor and inductor in the above example could be connected in either series or parallel. The resonant frequency is the same in either case.

Parallel Resonant Circuits

The vector diagram in Fig. 13-22 reveals the major characteristics of the parallel resonant

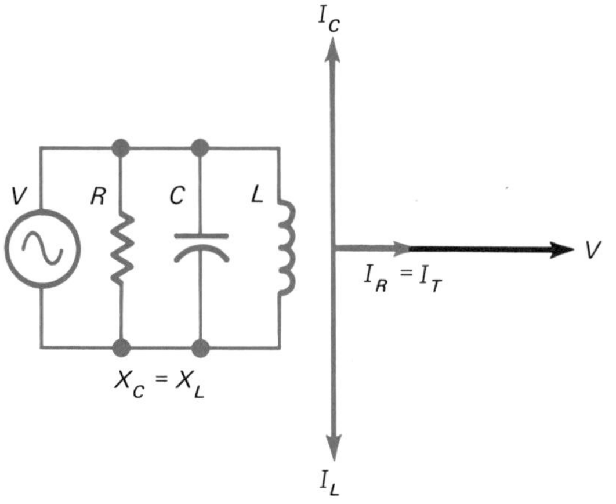

Fig. 13-22 Resonant parallel *RCL* circuit. Notice that $I_C = I_L$.

circuit. In this circuit, the inductive current exactly cancels the capacitive current. Therefore, the total current and the resistive current are the same. This means that the source current and voltage are in phase; angle θ is 0, and the power factor ($\cos\theta$) is 1. Thus, the parallel resonant circuit is purely resistive.

The only current drawn from the power source in Fig. 13-22 is the current which flows through the resistor. Yet, the capacitive and the inductive branches have large currents flowing in them. This is because the inductor and the capacitor are just transferring energy back and forth between themselves. This idea is graphically represented in Fig. 13-23 which shows the capacitor and inductor removed from the rest of the circuit. Suppose the inductor and capacitor part of the circuit are removed at the instant the capacitor is fully charged [Fig. 13-23(*a*)]. The capacitor will then be the energy source and start discharging through the inductor. This will cause a magnetic field to build up in the inductor and store the energy being transferred from the capacitor. Once the capacitor has discharged, the inductor's field will start to collapse as in Fig. 13-23(*b*). Now the inductor is the source. It provides the energy to recharge the capacitor in a reverse polarity as shown in Fig. 13-23(*b*). After the inductor's field has collapsed and recharged the capacitor, the capacitor again takes over as the source. It discharges [Fig. 13-23(*c*)] and again transfers energy to the inductor's magnetic field. The cycle is completed, as shown in Fig. 13-23(*d*), when the inductor's collapsing field recharges the capacitor. Then the cycle starts over again with the conditions shown in Fig. 13-23(*a*).

Tank circuit

The parallel *LC* circuit discussed above (and shown in Fig. 13-23) is often called a tank circuit. Tank circuit is a descriptive name because the circuit stores energy like a tank stores liquid. A tank circuit produces a sine wave, as shown in Fig. 13-24, as the capacitor charges and discharges repeatedly. If both the capacitor and inductor were ideal components (had no resistance) the tank circuit would produce a sine wave forever. Once the capacitor was given an initial charge and connected across the inductor, the cycling would continue indefinitely. However, all capacitors and inductors have some resistance. Therefore, each time the capacitor charges and discharges some energy is converted to heat. Thus, a real tank circuit produces a damped waveform like that shown in Fig. 13-24(*b*). How many cycles it takes for the waveform to completely dampen out depends on the quality (*Q*) of the circuit. (Recall that *Q* is a ratio X_L/R.) The frequency of the sine wave is

Damped waveform

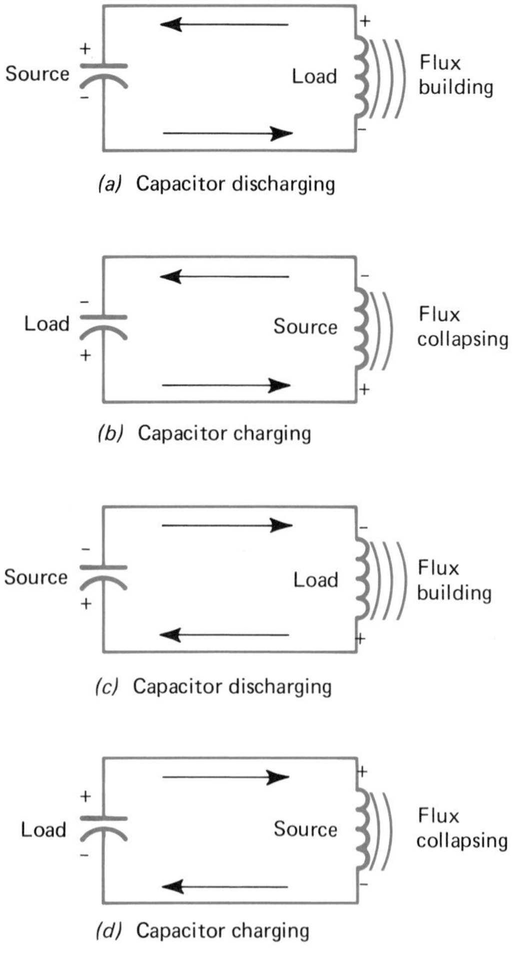

Fig. 13-23 Resonant tank circuit.

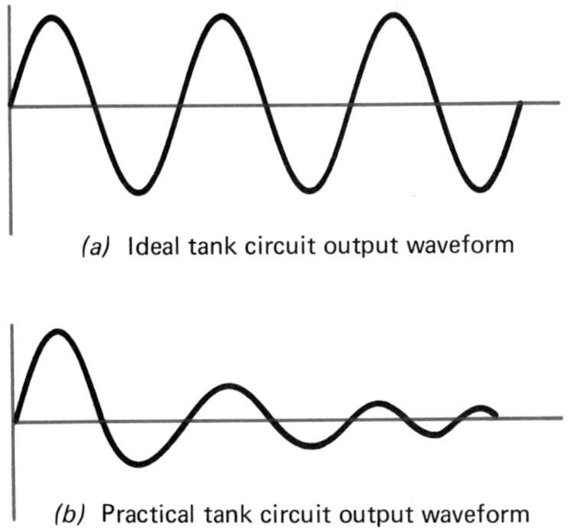

Fig. 13-24 Tank circuit waveforms.

determined by the value of L and C. It can be calculated with the resonant frequency formula.

Suppose the LC circuit of Fig. 13-23 is connected to an ac source [Fig. 13-25(*a*)]. Further, let the source frequency be equal to the LC circuit's resonant frequency. If L and C were ideal components, no current would be required from the source. If no source current flows, then the impedance of the circuit would be infinite. Using real components, a small source current will flow. This small current, together with the source voltage, furnishes the power used by the very low resistance of the inductor and capacitor. In the phasor diagrams of Fig. 13-25, this low current is labeled I_R since it is caused by the resistance of L and C. With high quality components, the source current is very low, and the impedance is still very high.

Suppose the frequency of the source is below the resonant frequency of the LC circuit as in Fig. 13-25(*b*). Then X_L will be less than X_C and I_L will exceed I_C. The difference between I_L and I_C becomes part of the current required from the source. The circuit is inductive and its impedance has decreased from the value it had at its resonant frequency. Its impedance could be calculated by dividing the source voltage by the source current.

Figure 13-25(*c*) shows what happens when the applied frequency is greater than the

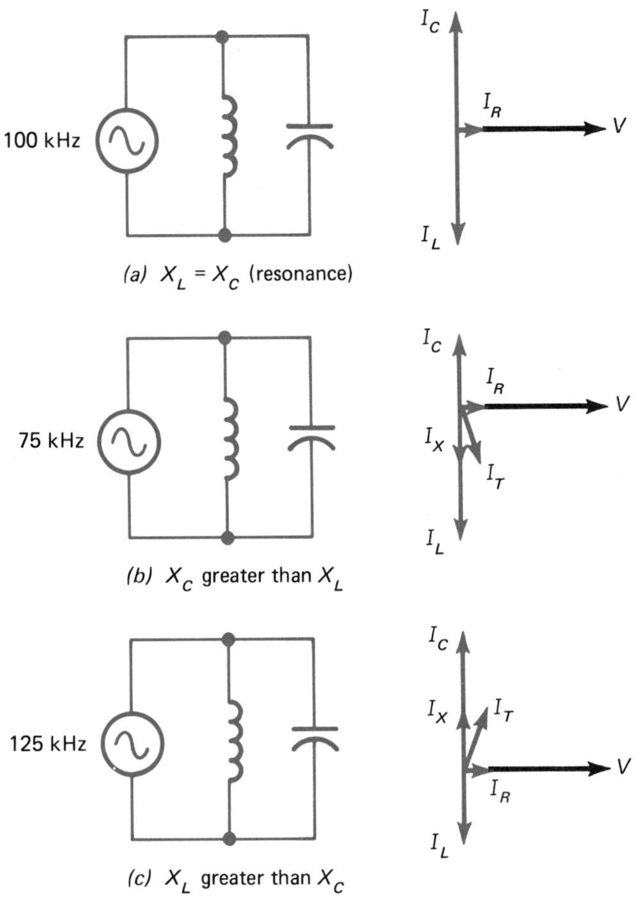

Fig. 13-25 Parallel LC circuit. The impedance (voltage divided by total current) is greatest at resonance.

resonant frequency. The circuit is capacitive, and, again, its impedance has decreased from the resonant frequency value.

In summary, a parallel resonant *LC* circuit has the following major characteristics:

1. Maximum impedance
2. Minimum source current
3. Phase angle almost 0°
4. High circulating inductive and capacitive currents

Series resonant circuit

Series Resonant Circuits

In the series *RCL* circuit in Fig. 13-26 the source voltage is at the resonant frequency for *L* and *C*. This means $X_L = X_C$ and $V_L = V_C$. Since V_L and V_C are 180° apart, they will cancel each other. The net voltage drop will be that across the resistance. The source voltage and current will be in phase and the circuit will have a power factor of 1. Since X_L and X_C cancel each other, the current in the circuit is limited only by the resistance.

Suppose the resistor in Fig. 13-26 is shorted out to produce an *LC* circuit like the one in Fig. 13-27(*a*). If *L* and *C* were ideal components, X_L would cancel X_C, and this series resonant circuit would have zero impedance. It would be a short circuit across the source. The current would be limited only by the internal resistance of the source. The voltages across the inductor and capacitor would be extremely high. However, real inductors and capacitors have some resistance. This resistance prevents the series resonant circuit from being a dead short. It drops an in-phase voltage (V_R in Fig. 13-27). This voltage, and the

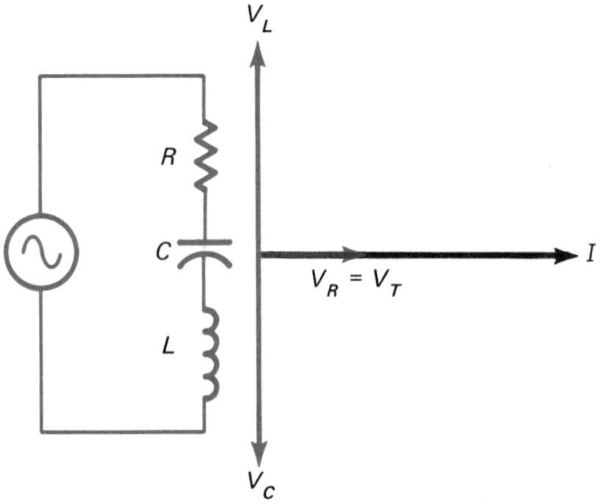

Fig. 13-26 Resonant series *RCL* circuit.

line current, represent the power used by the inductor and capacitor. The great majority of this power is used by the inductor.

The voltages across the capacitor and the inductor are many times greater than the source voltage in a series resonant circuit. Why this is so can be best explained by a practical example. The resistance of the typical inductor and capacitor in Fig. 13-27(*a*) would be about 20 Ω. Therefore, the current in the circuit would be:

$$I_T = \frac{V_T}{Z} = \frac{5\ \text{V}}{20\ \Omega} = 0.25\ \text{A}$$

This 0.25 A is also the current through the inductor and the capacitor. The reactance of the inductor (which, in a resonant circuit, is the same as the reactance of the capacitor) is:

$$\begin{aligned} X_L &= 6.28fL \\ &= 6.28 \times 107 \times 10^3\ \text{Hz} \times 1 \times 10^{-3}\ \text{H} \\ &= 672\ \Omega \end{aligned}$$

The voltage across the inductor can now be calculated:

$$V_L = I_L \times X_L = 0.25\ \text{A} \times 672\ \Omega = 168\ \text{V}$$

Thus, the voltage across the inductor (and across the capacitor) is more than 33 times (168 V/5 V) higher than the source voltage.

Figures 13-27(*b*) and (*c*) show the results of operating a series *LC* circuit below and above its resonant frequency. When operating below its resonant frequency, the series circuit is capacitive because V_C is greater than V_L. Again, notice that parallel and series circuits are opposites in many ways. The parallel *LC* circuit operating below its resonant frequency [Fig. 13-25(*b*)] is inductive.

The major characteristics of the series resonant *LC* circuit can be summarized as follows:

1. Minimum impedance
2. Maximum source current
3. Phase angle almost 0°
4. High inductive and capacitive voltages

Notice that the only characteristic shared by the series and parallel resonant circuits is the almost 0° phase angle.

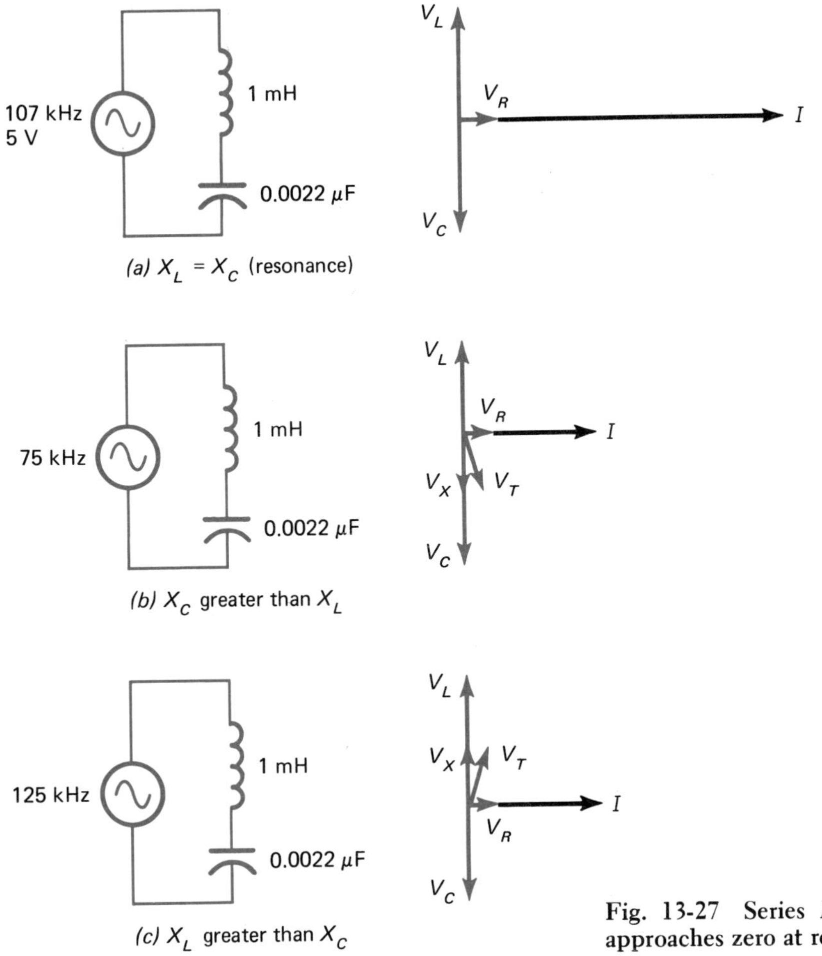

Fig. 13-27 Series *LC* circuit. The impedance approaches zero at resonance.

Response curves

Bandwidth

Selectivity

Response Curves, Bandwidth, and Selectivity

So far, only phasor diagrams have been used to show current, voltage, and impedance of *LC* circuits at several frequencies. A more complete picture of the behavior of *LC* circuits can be shown with the response curves of Fig. 13-28. These curves are made by either calculating or measuring the current, voltage, or impedance at numerous frequencies above and below resonance. The measured or calculated values are then plotted as points on a graph. Finally, the points are connected together to form the curve.

Compare the curves in Fig. 13-28(*a*) and (*b*). They clearly show the major differences between the series and parallel *LC* circuits.

The *bandwidth* (BW) of an *LC* circuit is expressed in hertz. It is the range of frequencies to which the circuit provides 70.7 percent or more of its maximum response. Suppose the parallel *LC* circuit of Fig. 13-28(*a*) is resonant at 500 kHz and has a resonant impedance of 100 kΩ. Further, suppose its impedance drops to 70.7 kΩ at 495 kHz [f_{low} in Fig. 13-28(*a*)] and at 505 kHz [f_{high} in Fig. 13-28(*a*)]. Then the circuit's bandwidth would be 10 kHz (505 − 495). The bandwidth of a series circuit can be determined from its current-versus-frequency curve [Fig. 13-28(*b*)]. The edges of the bandwidth (f_{low} and f_{high}) are those frequencies at which the circuit current is 70.7 percent of the current at resonance.

The bandwidth BW of a circuit determines the *selectivity* of the circuit. Selectivity refers to the ability of a circuit to select one frequency (the desired frequency) out of a group of frequencies. For example, the antenna of a radio receiver receives a signal from all of the local radio stations. Each station's signal is at a different frequency. It is the job of the resonant circuits in the receiver to select the frequency of one station and reject all the other stations. (When you tune a receiver, you are

Q determines bandwidth

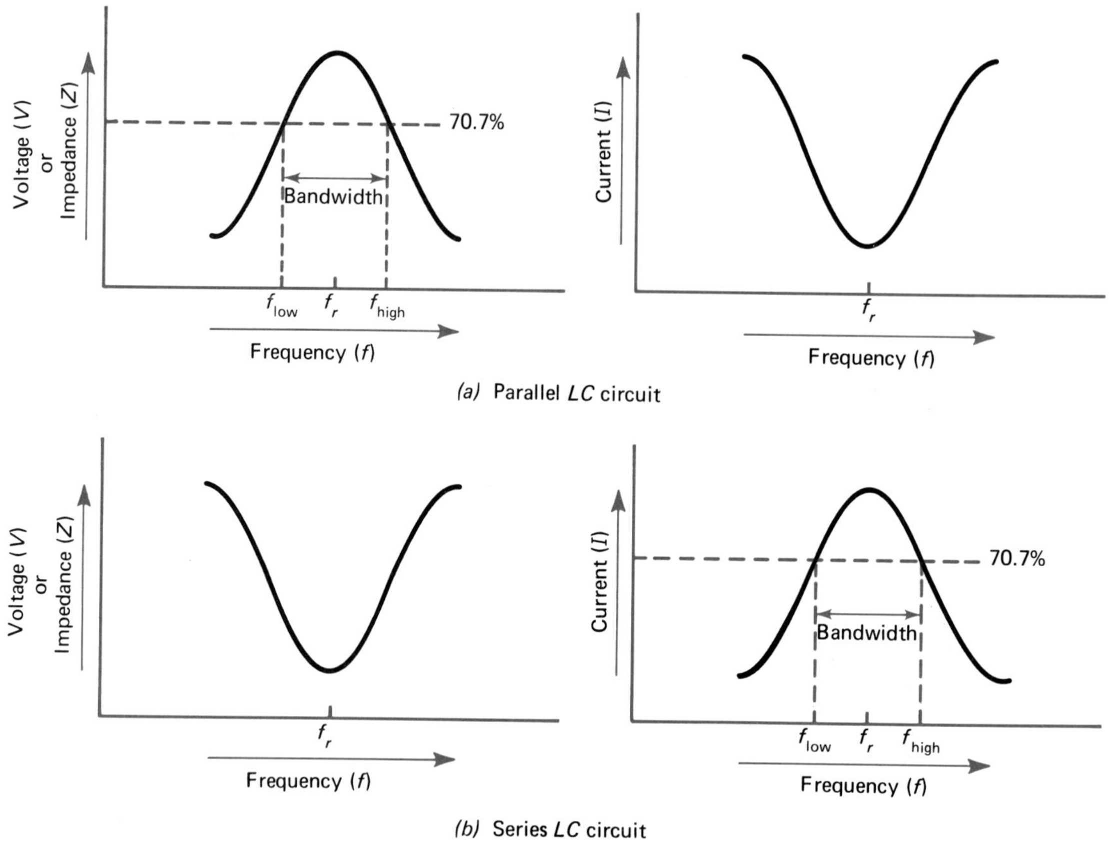

Fig. 13-28 Response curves of resonant circuits.

changing capacitor or inductor values to adjust the resonant frequency of the receiver circuits.) If the bandwidths of the receiver's circuits are too wide, two or more stations will be heard simultaneously. In this case we would say that the receiver has poor selectivity.

Q of Resonant Circuits

The term "quality" is also used as a rating of a resonant circuit. It is the ratio of the reactance of the inductor or capacitor to the equivalent series resistance of both components. Recall that the inductor possesses nearly all of the resistance of a resonant circuit. Therefore, the Q of the inductor is, for all practical purposes, also the Q of the resonant circuit.

The Q of the resonant circuit is important for two reasons:

1. The Q determines the minimum Z of the series resonant circuit and the maximum Z of the parallel resonant circuit. These ideas are illustrated in Fig. 13-29. For the series circuit, the impedance is:

$$Z = \frac{V_R}{I}$$

and, for the parallel circuit the impedance is:

$$Z = \frac{V}{I_R}$$

From Fig. 13-29(*a*) you can see that the high-Q circuit yields the lowest Z for the series circuit. The high-Q parallel circuit yields the highest Z [Fig. 13-29(*b*)].

2. The Q also determines the bandwidth of the resonant circuit.

The higher the Q, the narrower the bandwidth will be. A narrow bandwidth (high Q) provides a larger change in impedance for a given change in frequency than does a wide bandwidth. Figure 13-30 shows why a high-Q circuit provides a greater change in impedance than does a low-Q circuit. The phasor diagrams in Fig. 13-30(*a*) are for two parallel *LC* circuits operating at the same resonant frequency. The low-Q circuit has twice as much resistive current as the high-Q circuit.

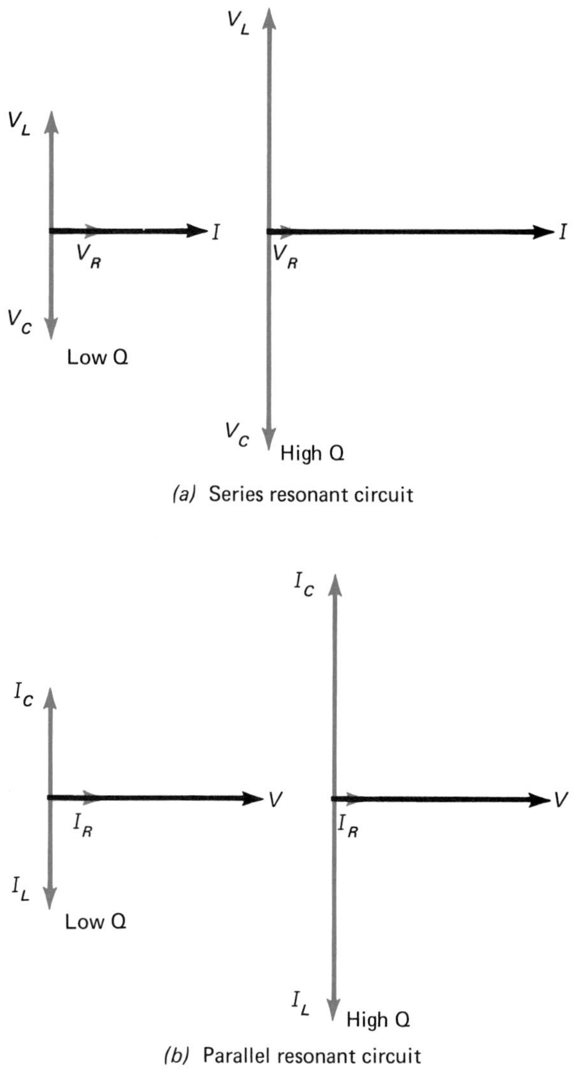

Fig. 13-29 Effects of Q on the resonant Z.

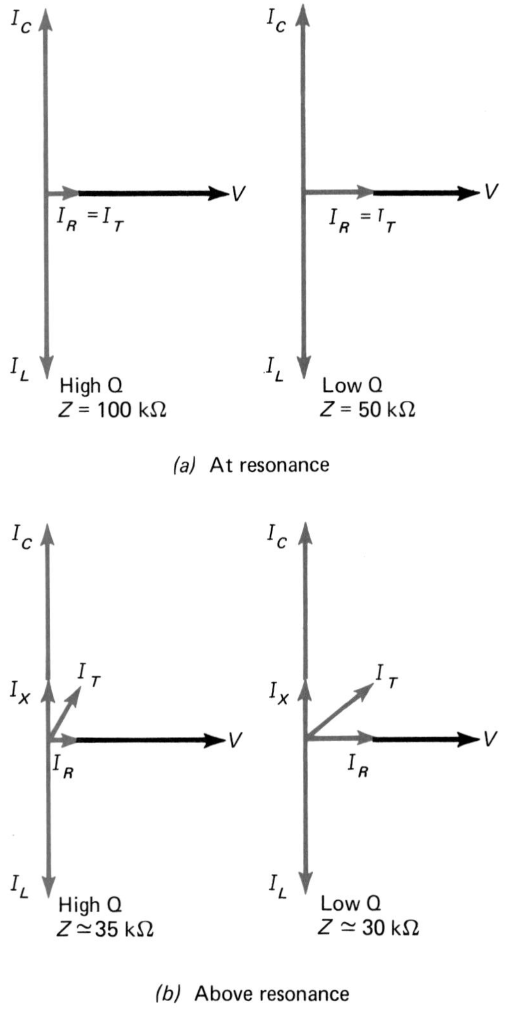

Fig. 13-30 Effects of Q on nonresonant Z.

$$BW = \frac{f_r}{Q}$$

Since $Z = V/I$, the low-Q circuit will also have only half as much impedance as the high-Q circuit. Figure 13-30(*b*) shows the results of increasing the source frequency the same amount for each circuit. Notice that the total current in the low-Q circuit is only slightly higher than the total current in the high-Q circuit. Therefore, the impedance of the high-Q circuit is now only slightly greater than the impedance of the low-Q circuit. Thus, the high-Q circuit changed impedance by a larger percentage than did the low-Q circuit.

Figure 13-30 illustrates why a high-Q circuit provides a proportionately greater Z change than does a low-Q circuit. The result of this greater Z change for the high-Q circuit is shown in Fig. 13-31. This figure shows that the high-Q circuit yields a narrower bandwidth for either series or parallel *LC* circuits. From Fig. 13-31 you can see that a high-Q circuit will be more selective than a low-Q circuit.

Resonant frequency f_r, quality Q, and bandwidth BW are all interrelated. Their relationship can be expressed by the following formula:

$$\text{Bandwidth (BW)} = \frac{f_r}{Q}$$

Although this formula is actually only an approximation, it is accurate enough for predicting circuit behavior. If the circuit Q is 10 or more, the error will be less than 1 percent.

Example 13-12

What is the Q of a circuit which is resonant at 100 kHz and has a bandwidth of 4 kHz?

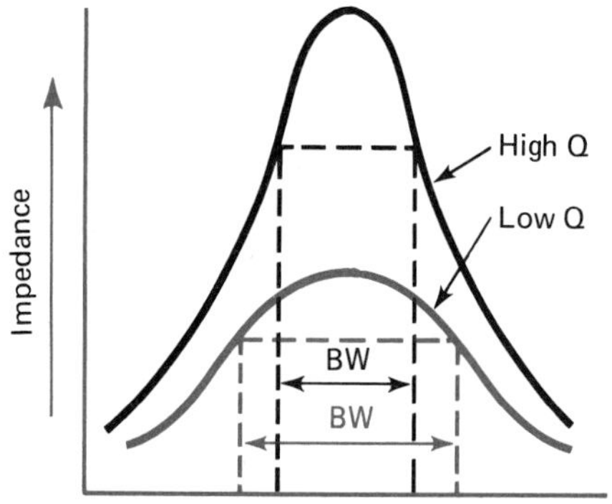

(a) Parallel *LC* circuits

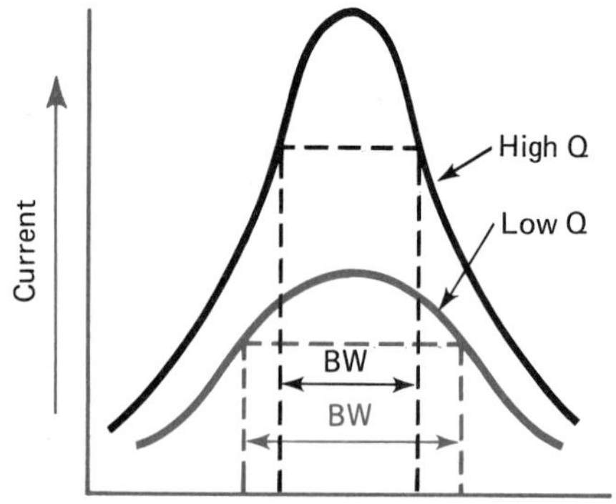

(b) Series *LC* circuits

Fig. 13-31 Response curves for low- and high-Q circuits.

Given: $f_r = 100$ kHz
BW = 4 kHz

Find: Q

Known: $\text{BW} = \frac{f_r}{Q}$

Solution: $Q = \frac{f_r}{\text{BW}} = \frac{100 \text{ kHz}}{4 \text{ kHz}}$
$= 25$

Answer: The quality Q of the circuit is 25.

The Q of a series resonant circuit can also be found from the following formula:

$$Q = \frac{V_L}{V_T}$$

In this formula the voltages (V_L and V_T) are the voltages at resonance. You can understand where this formula comes from by referring to Fig. 13-26 or Fig. 13-27(*a*). At resonance the resistive voltage and the source voltage are equal. Also, the resistance and the reactance have the same current flowing through them. Therefore, the ratio of reactance to resistance has to be equal to the ratio of reactive voltage to resistive voltage.

Review Questions

37. What is a resonant circuit?
38. Write the formula for calculating resonance.
39. Can a specific value of capacitance and inductance be resonant at more than one frequency?
40. Can more than one combination of values of *L* and *C* be resonant at a specified frequency?
41. What is a tank circuit?
42. What is the resonant frequency of a 0.002-μF capacitor and a 2-mH inductor?
43. If the capacitance of an *LC* circuit is increased, what happens to the resonant frequency?
44. Compare the series and parallel resonant circuits in terms of
    a. Impedance
    b. Source current requirements
    c. Phase angle
45. Will a parallel *LC* circuit operating above its resonant frequency be inductive or capacitive?
46. For high selectivity should a resonant circuit have a wide or a narrow bandwidth?
47. A series *LC* circuit allows 10 mA of source current to flow at its resonant frequency. How much source current will flow at a frequency which marks the lower edge of its bandwidth?
48. Will a low-Q or a high-Q circuit have the wider bandwidth?
49. Suppose the inductor in question 42 has 20 Ω of resistance.
    a. What is the Q of the circuit?
    b. What is the bandwidth of the circuit?

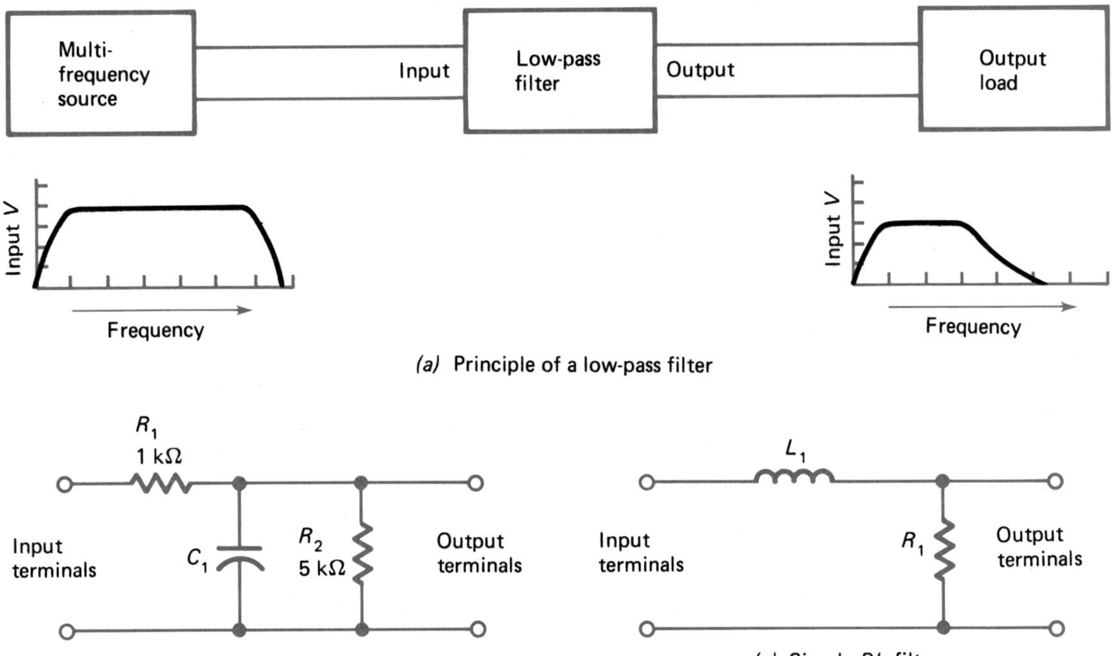

(a) Principle of a low-pass filter

(b) Simple RC filter

(c) Simple RL filter

Fig. 13-32 Low-pass filters.

Filtering

Low-pass filter

13-7 FILTERS

One of the major uses of *RC*, *RL*, *LC*, and *RCL* circuits is for filtering. *Filtering* refers to separating one group of frequencies from another group of frequencies. There are four general classes of filters: band-pass, band-reject, high-pass, and low-pass.

Low-Pass Filters

Low-pass filters offer very little opposition to low frequency signals [see Fig. 13-32(*a*)]. Most of the input signal appears at the output terminals. As the signal frequency increases, the filter provides more opposition. The filter drops more of the signal voltage and leaves less signal available at the output. As the graph of the output voltage shows [Fig. 13-32(*a*)], essentially none of the high frequency signals appears at the output terminal. Although low-pass filters can be constructed with either *RL* or *RC* circuits, the *RC* circuit is the most common.

Inspection of Fig. 13-32(b) will show why the low-pass filter behaves as described in the

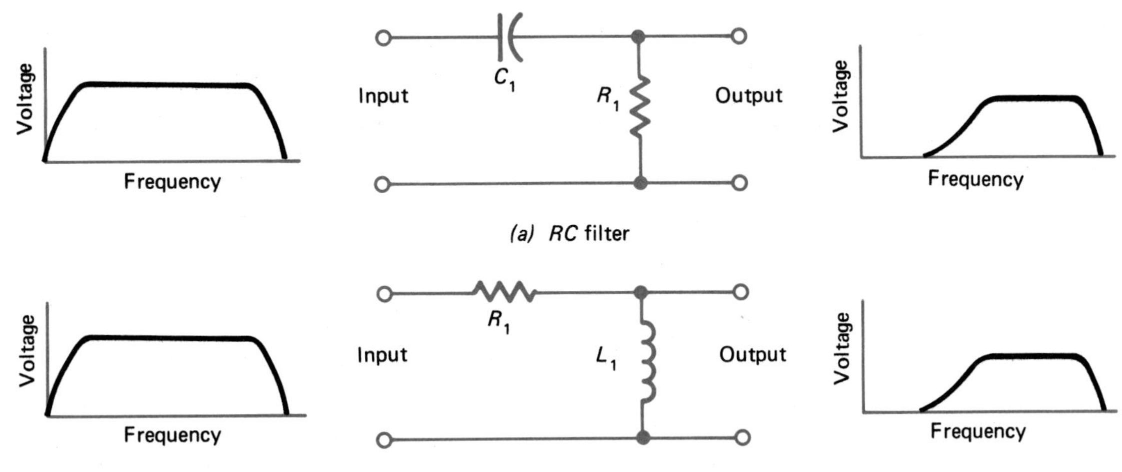

(a) RC filter

(b) RL filter

Fig. 13-33 High-pass filters.

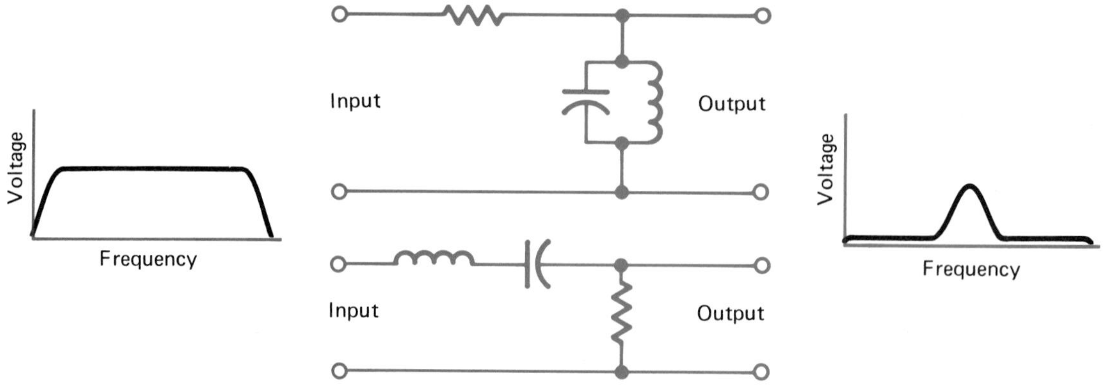

(a) Band-pass filters

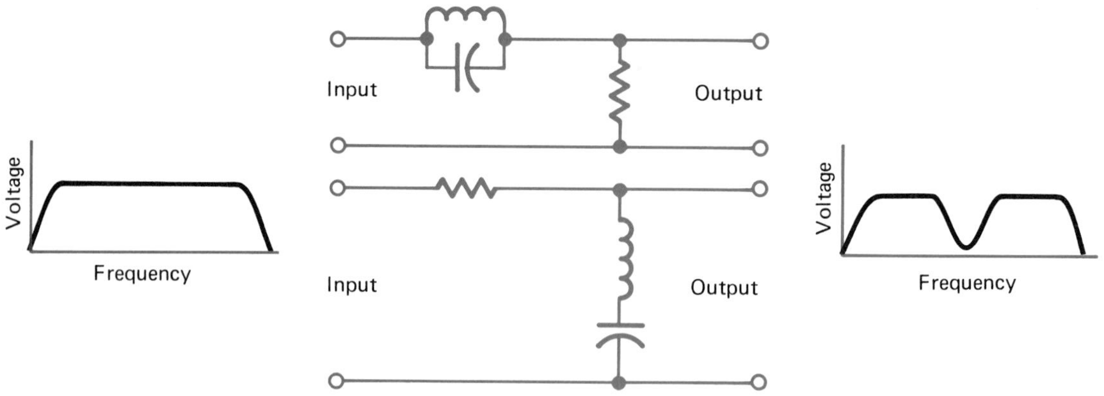

(b) Band-reject filters

Fig. 13-34 Band-pass and band-reject filters.

paragraph above. At low frequencies, the reactance of C_1 will be very high compared to R_2. The impedance of R_2C_1 will be essentially 5 kΩ. Thus, five-sixths of the input voltage will appear at the output terminals. At very high frequencies the reactance of C_1 will be very low compared to R_1. Now the impedance of R_2C_1 will be very low (essentially equal to X_{C_1}). Therefore, nearly all of the input signal will drop across R_1.

The reactance of L_1 in Fig. 13-32(*c*) is low (compared to R_1) at low frequencies. Thus, low frequencies pass through the filter. However, at high frequencies, the inductive reactance of L_1 is high and most of the signal drops across L_1. Very little of the high-frequency signal appears at the output terminals.

High-Pass Filters

High-pass filter

Several simple high-pass filters are shown in Fig. 13-33. With the *RC* filter of Fig. 13-33(*a*) the capacitor drops nearly all of the low frequency signals. At high frequencies, the capacitive reactance is very low. Therefore, R_1 (which provides the output) drops nearly all the high-frequency voltage. In other words the circuit passes high-frequency signals through it and blocks low frequencies.

Recall that $X_L = 6.28fL$. Therefore, L_1 in Fig. 13-33(*b*) drops most of the high-frequency voltage, but practically none of the low-frequency voltage. Thus, the series *RL* circuit can serve as a high-pass filter.

Band-Pass and Band-Reject Filters

You are already familiar with band-pass and band-reject filters. They are just series *LC* and parallel *LC* circuits. Frequencies within their response curve range are either passed or rejected (see Fig. 13-34).

Notice in Fig. 13-34 that a parallel *LC* circuit can serve as either a band-pass or band-reject filter. The same thing can be said for the series *LC* circuit. The determining factor is where the input and output terminals are connected to circuits.

Review Questions

50. List four classifications of filter circuits.

51. Which of the types listed in question 50 use resonant circuits?

52. If you wanted to eliminate high-frequency noise from a stereo record, what type of filter would you use?

Summary

1. Impedance is a combination of resistance and reactance.
2. Impedance causes phase shift.
3. Impedance may cause either a leading or lagging current.
4. The symbol for impedance is Z.
5. The ohm is the base unit of impedance.
6. In *RC* circuits the current will lead the voltage.
7. For series *RC* circuits the following generalizations can be made:
   a. The total impedance is higher than either R or X_C.
   b. The arithmetic sum of R and X_C is greater than Z.
   c. Decreasing either *f* or *C* causes Z to increase, *I* to decrease, *P* to decrease, and θ to increase.
8. The formulas that are applicable to all circuits involving combinations of *R*, *C*, and *L* are:

$$Z = \frac{V_T}{I_T}$$

$$f_r = \frac{1}{6.28\sqrt{LC}}$$

$$\text{Cos } \theta = \frac{P}{P_{\text{app}}} = \frac{P}{I_T V_T}$$

9. The formulas that are applicable to all series circuits using combinations of *R*, *C*, and *L* are:

$$\text{Cos } \theta = \frac{R}{Z} = \frac{V_R}{V_T}$$

10. The formula that applies to all parallel circuits using combinations of *R*, *C*, and *L* is:

$$\text{Cos } \theta = \frac{I_R}{I_T}$$

11. The following are characteristics of parallel *RC* circuits:
    a. The total impedance is less than R or X_C.
    b. Decreasing either *f* or C causes Z to increase, *I* to decrease, and θ to decrease. The power remains the same.
12. The current phasor is the reference phasor for series circuits.
13. The voltage phasor is the reference phasor for parallel circuits.
14. Resistance and reactance are 90° out of phase.
15. In all *RL* circuits the current lags the voltage.
16. A series *RL* circuit has more Z than either R or X_L.
17. A series *RL* circuit has more V_T than either V_R or V_L.
18. A parallel *RL* circuit has less Z than either R or X_L.
19. A parallel *RL* circuit has more I_T than either I_R or I_L.
20. In a series *RCL* circuit, Z may be either less or more than X_L or X_C.
21. In a resonant *RCL* circuit, Z is equal to R.
22. In *RCL* circuits, reactive voltages and currents can be greater than the total current or voltage.
23. In a series *RCL* circuit with X_L greater than X_C, the circuit will be inductive.
24. In a parallel *RCL* circuit with X_L greater than X_C, the circuit will be capacitive.
25. The formulas used only for series *RCL* circuits are:

$$Z = \sqrt{(X_L - X_C)^2 + R^2}$$

$$V_T = \sqrt{(V_L - V_C)^2 + V_R^2}$$

26. The formula used only for parallel *RCL* circuits is:

$$I_T = \sqrt{(I_L - I_C)^2 + I_R^2}$$

27. Resonance occurs when $X_L = X_C$.
28. Increasing either L or C will decrease the resonant frequency of an *LC* circuit.

29. A given L and C can produce only one resonant frequency.
30. For practical purposes, the Q of a resonant LC circuit is determined by the Q of the inductor.
31. All resonant circuits have a power factor (cos θ) of 1.
32. The impedance of a series resonant circuit approaches 0.
33. The impedance of a parallel resonant circuit approaches infinity.
34. A parallel LC circuit operating above its resonant frequency will be capacitive.
35. A series LC circuit operating above its resonant frequency will be inductive.
36. The higher the Q of a circuit the narrower its bandwidth.
37. The narrower the bandwidth of a circuit the more selective the circuit will be.
38. The bandwidth of a circuit is equal to f_r/Q.
39. Either RC or RL circuits can be used for high-pass and low-pass filters.

Chapter Review Questions

For questions 13-1 to 13-16, determine whether each statement is true or false.

13-1. In a circuit containing both resistance and capacitance, the source current will lead the source voltage.

13-2. The Pythagorean theorem is used to determine the resonant frequency of a circuit.

13-3. If the capacitive reactance in a series RC circuit is greater than the resistance, angle θ will be less than 45°.

13-4. Reactance is 180° out of phase with resistance.

13-5. In a parallel RC circuit, increasing the frequency will cause angle θ to increase.

13-6. In a parallel RL circuit, the source current leads the source voltage.

13-7. Reducing the inductance in a series RL circuit will cause the total power to increase.

13-8. In a series RL circuit the inductive voltage can be greater than the source voltage.

13-9. The resistance in a series RCL circuit can exceed the impedance.

13-10. In a parallel RCL circuit operating above its resonant frequency, the source current leads the source voltage.

13-11. There is only one value of inductance that will resonate with a 0.001-μF capacitor at 10,000 Hz.

13-12. At resonance, a series LC circuit provides its maximum impedance to the source.

13-13. A high-Q circuit is more selective than a low-Q circuit.

13-14. The Q of a resonant circuit is primarily determined by the quality of the capacitor.

13-15. In a resonant RCL circuit, the impedance is equal to the resistance.

13-16. Increasing either f or L in a parallel RL circuit causes angle θ to decrease.

For questions 13-17 to 13-19, choose the letter that best completes each statement.

13-17. The symbol for impedance is
(A) X (B) Z (C) Y (D) E

13-18. The resonant frequency of a tank circuit can be increased by
(A) Adding more turns to the inductor (B) Increasing the inductance (C) Increasing the capacitance (D) Decreasing the capacitance

13-19. At resonance the phase angle of a circuit is
(A) Approximately 0° (B) Approximately 90° (C) Approximately 180° (D) 180° for a series circuit and 0° for a parallel circuit

Answer the following questions.

13-20. Calculate the bandwidth of a circuit which is resonant at 500 kHz and has a Q of 40.

13-21. List four types (classifications) of filter circuits.

13-22. Which of the above filter circuits use resonant circuits?

13-23. A 1200-Ω resistor and a 1-μF capacitor are series connected to a 50-V, 180-Hz source. Calculate the following:
a. Impedance
b. Voltage across the resistor
c. Power
d. Cos θ

13-24. Assume the resistor and capacitor in the above circuit are connected in parallel instead of series. Calculate the following:
a. Impedance
b. Total current
c. Power

13-25. At 30 kHz, what is the impedance of a 5-mH inductor connected in series with a 1500-Ω resistor?

13-26. What is the resonant frequency of a 2.5-mH inductor connected in parallel with a 2000-pF capacitor?

13-27. A 0.002-μF capacitor, a 3-mH inductor and a 1000-Ω resistor are series connected to a 40-V, 50-kHz source. Determine the following:
a. Impedance
b. Total current
c. Voltage across the capacitor

Answers to Review Questions

1. No
2. It is the combined opposition of reactance and resistance
3. *Z*
4. Ohm
5. 5.83 m
6. 36 N
7. A law which states the relationship between the length of the three sides of a right triangle

8. Lead
9. Lead
10. The voltage phasor
11. Less than
12. Less than
13. They are separated by 90°
14. $Z = \sqrt{R^2 + X_C^2}$ and $Z = \dfrac{V_T}{I_T}$
15. $V_T = \sqrt{V_R^2 + V_C^2}$
16. $\cos\theta = \dfrac{R}{Z} = \dfrac{V_R}{V_T} = \dfrac{P}{P_{app}}$
17. $\cos\theta = \dfrac{P}{P_{app}}$
18. $\cos\theta = \dfrac{I_R}{I_T}$
19. *a.* Decreases
    b. Stays the same
    c. Stays the same
    d. Increases
20. *a.* Increases
    b. Decreases
    c. Decreases
    d. Increases
21. *a.* 8.94 kΩ
    b. 0.36 W
    c. 27°
    d. 3.0 mA
22. *a.* 14.73 mA
    b. 32.4 V
    c. 0.477 W
    d. 2716 Ω
23. Lag
24. I_T increases, V_L decreases, and Z decreases
25. Z increases and θ decreases
26. 1181 Ω
27. 30.5 mA
28. I_T is 36.84 mA, and Z is 1357 Ω
29. Yes
30. No
31. Yes
32. $Z = \dfrac{V_R}{V_T}$ and

    $Z = \sqrt{(X_L - X_C)^2 + R^2}$
33. $I_T = \sqrt{(I_L - I_C)^2 + I_R^2}$
34. *a.* The series circuit
    b. The parallel circuit
    c. The parallel circuit will draw the higher current
35. $Z = 1513.4\ \Omega$, $I_T =$ 13.2 mA and $V_L = 29$ V
36. $Z = 897.5\ \Omega$ and $I_C = 18.83$ mA
37. A circuit in which $X_L = X_C$
38. $f_r = \dfrac{1}{6.28\sqrt{LC}}$
39. No
40. Yes
41. A parallel resonant LC circuit
42. 79.6 kHz
43. Decreases
44. *a.* Series Z is very low, parallel Z is very high
    b. Series I is maximum, parallel I is minimum
    c. Both series and parallel are 0°
45. Capacitive
46. Narrow
47. 7.07 mA
48. High-Q
49. *a.* 50
    b. 1.6 kHz
50. Band-pass, band-reject, high-pass, and low-pass
51. Band-pass and band-reject
52. A low-pass filter

Ac Measurement

- **Many techniques and instruments are used to measure alternating current and voltage, and ac power. This chapter will cover the basic principle of some of the more common ac measurement techniques and instruments.**

  Once you have completed this chapter, you will be able to measure frequency, reactance, and impedance. You will also be able to use an ac bridge to measure capacitance and inductance. You will have developed the capability of measuring power in balanced and unbalanced, three-phase circuits.

14-1 AMMETERS AND VOLTMETERS

In Chap. 7, you learned about the basic meters used for measuring dc quantities. In this chapter, you will learn about ac measuring instruments. There are a number of types of meters for measuring ac current. Of course, any meter than can measure ac current can also measure voltage. All that is needed to convert a current meter to a voltmeter is a multiplier resistor.

Rectifier-type Meters

The rectifier-type meter consists of a rectifier connected in series with a d'Arsonval meter movement. The principle of operation of the d'Arsonval meter movement is illustrated in Fig. 14-1. The d'Arsonval movement is a dc instrument because current must flow in the correct direction to cause clockwise rotation. This means that ac must be rectified before the d'Arsonval movement can respond to it.

The process of rectification is shown in Fig. 14-2. A *rectifier* is a device that allows current to flow in only one direction. In other words, it has almost infinite resistance to current trying to flow in one direction and practically no resistance to current flowing in the opposite direction. Thus, the rectifier of Fig. 14-2 allows only one-half of an ac cycle to pass. It converts ac to pulsating dc. The pulsating dc provides the electromagnetic field of constant polarity which is needed to operate a d'Arsonval meter movement.

The combining of the d'Arsonval movement and a rectifier produces a rectifier-type of ac meter. The rectifier-type meter can then be combined with a shunt or a multiplier to produce appropriate current or voltage

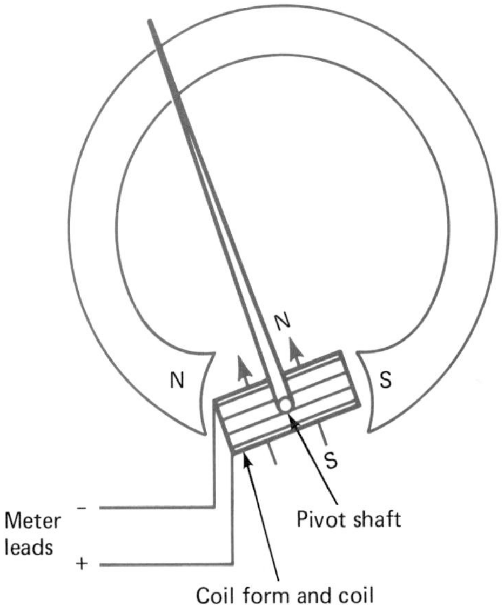

Fig. 14-1 Principle of the d'Arsonval meter movement. Notice that the movement is polarized.

From page 285:
d'Arsonval meter

On this page:
Sensitivity

Nearly linear

Iron-vane meter

Stationary vane

Moving vane

Extremely nonlinear

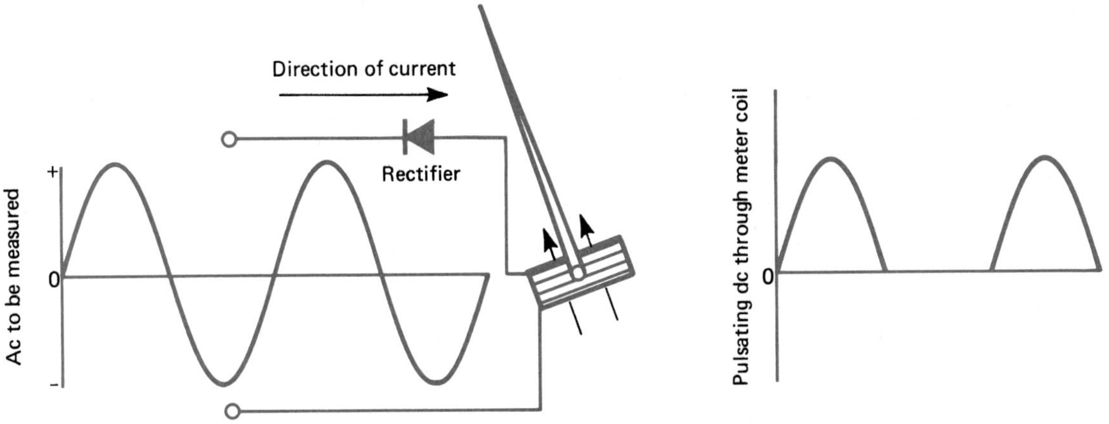

Fig. 14-2 Principle of rectification. The rectified current flows only in one direction.

ranges. Multirange ac meters (Fig. 14-3) can be constructed in the same way that multirange dc meters are constructed. Merely treat the movement-rectifier combination as a basic meter movement with a given internal resistance and full-scale current rating.

Rectifier-type meters are very popular analog meters, especially with multimeters that must measure both ac and dc. They have several advantages over other types of ac meters. First, they are very sensitive. (Recall from Chap. 7 that sensitivity relates to the amount of current required for full-scale deflection.) Second, they can have scales that are nearly linear. In fact, they are so close to linear that the same scales are sometimes used for both ac and dc.

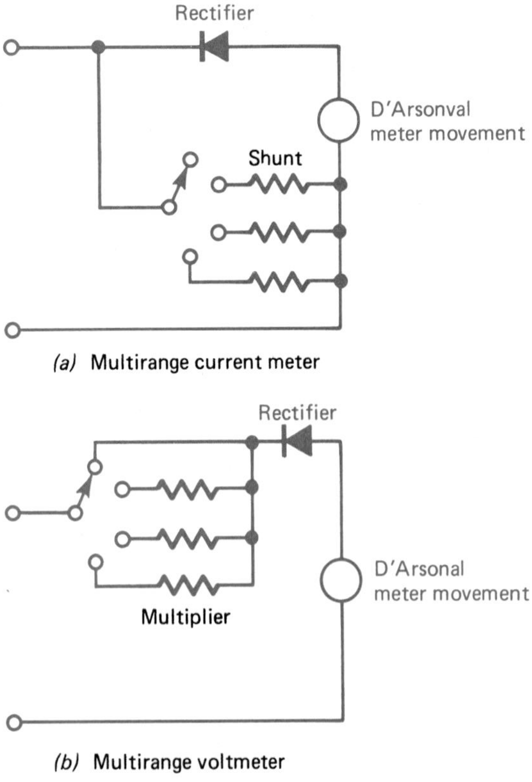

Fig. 14-3 Rectifier-type meters.

Iron-Vane Meter

In an *iron-vane* meter the coil which receives the current to be measured is stationary. The field set up by the coil magnetizes two iron vanes which become temporary magnets (see Fig. 14-4). Since the same field magnetizes both vanes, both vanes will have the same magnetic polarity. Consequently, there will be a force of repulsion between the two vanes. One of the vanes (the stationary vane) is attached to the coil form. The other vane (the moving vane) is mounted on the pivot shaft to which the meter pointer is attached. Thus, the magnetic force of repulsion pushes the moving vane away from the stationary vane. Of course, this force is offset by the countertorque of the spiral springs which are attached to the pivot shaft. The greater the current through the coil in Fig. 14-4, the stronger the magnetic repelling force. Thus, the further the moving vane rotates and the more current the pointer indicates.

The iron-vane meter can operate on either ac or dc. When the alternating current reverses, the magnetic polarity of both vanes reverses simultaneously. Therefore, a force of repulsion is maintained throughout the complete cycle.

The iron-vane meter has two shortcomings. It has an extremely nonlinear scale which is

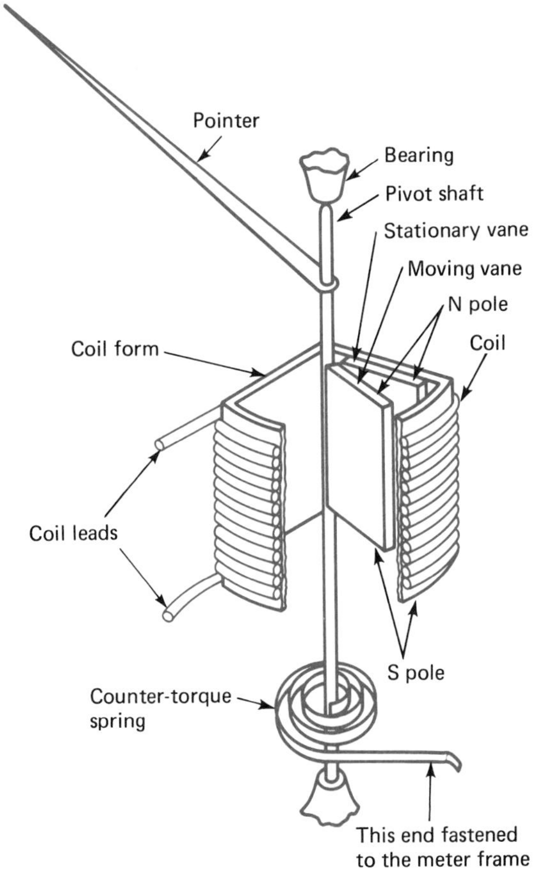

CUTAWAY VIEW

Fig. 14-4 Iron-vane meter movement. Since the movement uses no permanent magnet it responds to ac as well as dc.

very crowded at the low end of the scale. It requires considerable current for full scale deflection (low sensitivity).

Low sensitivity

The advantages of the iron-vane meter are its ease of construction, ruggedness, and dependability.

The iron-vane meter can also be used with shunts and multipliers to extend its range. However, its greatest use is as an ac ammeter.

Thermocouple Meter

Radio-frequency currents are hard to measure with the meters discussed thus far. At these high frequencies the inductive reactance of the meter coil is high and the capacitive reactance of the capacitance between the turns of the coil is low. The combination of these two reactances make some meters useless at higher frequencies. However, the thermocouple meter avoids the reactance problem by isolating the basic meter movement from the RF currents.

Thermocouple meter

Avoids reactance problem

The thermocouple meter, shown in Fig. 14-5, uses a d'Arsonval meter movement connected to a thermocouple. When current flows through the meter, the resistive element heats up and increases the temperature of the thermocouple junction. Recall that a thermocouple is a device that converts heat energy into electrical energy. Therefore, the heated thermocouple produces a current in the meter movement circuit. The higher the current through the resistive element, the hotter the thermocouple gets. The hotter the thermocouple, the higher the voltage it produces and the greater the current through the d'Arsonval meter movement.

Since the current in the thermocouple meter flows through the resistive element only, the reactance of the meter movement coil is not important. This is why the thermocouple meter can accurately respond to very high-frequency currents.

Electrodynamometer

Electro-dynamometer

Two electromagnetic fields

The electrodynamometer (Fig. 14-6) uses two electromagnetic fields in its operation. One field is created by current flowing through a pair of series-connected stationary coils. The other field is caused by current flowing through a movable coil which is attached to the pivot shaft. If the currents in the coils are in the correct directions, the pointer will rotate clockwise. The rotational torque on the movable coil is caused by the opposing magnetic forces [Fig. 14-6(*b*)] of the three coils.

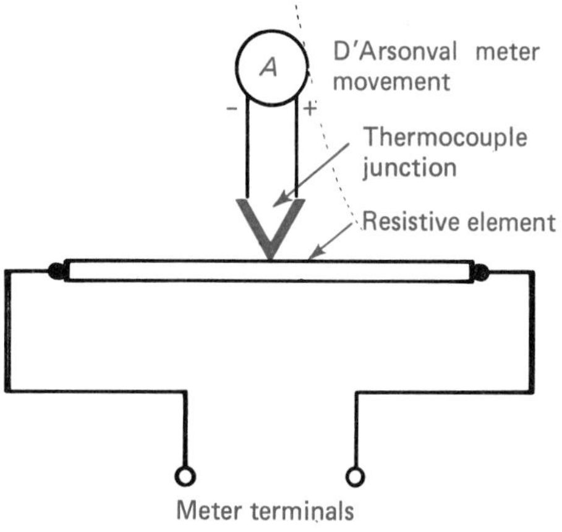

Fig. 14-5 Thermocouple meter. The current being measured does not flow through the meter movement.

The direction of the magnetic fields shown in Fig. 14-6(*b*) were arbitrarily chosen; all of the arrowheads could just as well be reversed. It is only the polarity of one coil relative to the other two that is important. (The two stationary coils are permanently connected so that their fields are always aiding each other.) The electrodynamometer can operate with alternating current because the alternating current reverses direction simultaneously in all three coils. The electrodynamometer can also operate on dc.

When used to measure current or voltage, the movable coil of the electrodynamometer is connected in series with the stationary coils. Shunts and multipliers can be used to extend the range of the basic movement. Although the electrodynamometer is used for measuring current and voltage, its most common use is for measuring power.

Electrodynamometer movements have low sensitivity compared to d'Arsonval movements. However, they can be very accurate, and they are quite stable. The accuracy of the electrodynamometer movement does not depend on the strength of a permanent magnet or the permeability of the iron parts.

Toroid-core transformer

Current Transformer

When measuring large values of direct current, we use an external shunt and a meter movement. The same general approach is used when measuring large values of alternating current. However, with ac we use a current transformer (Fig 14-7) instead of a shunt. The current transformer is essentially a toroid-core transformer that has no primary. When a current-carrying conductor is placed in the center of the current transformer, the conductor becomes the primary. An equivalent electrical circuit is shown in Fig. 14-8. The current in the conductor is stepped down by transformer action. Or, saying it another

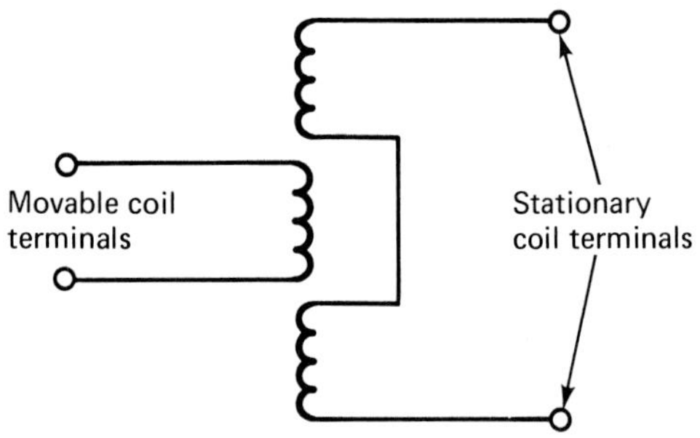

(a) Electrical connections

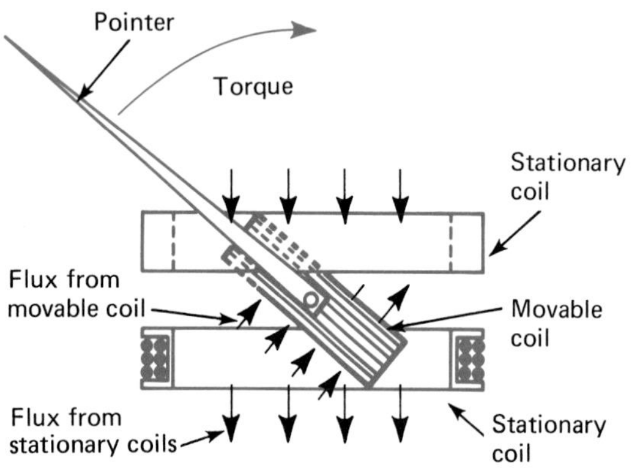

(b) Top view (front coil sectioned)

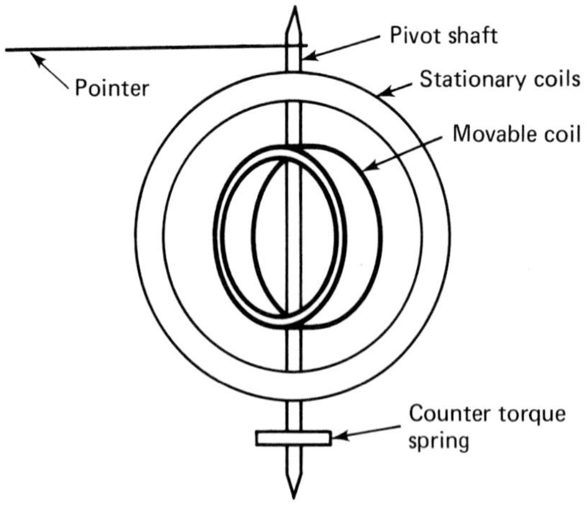

(c) Front view

Fig. 14-6 Electrodynamometer. This meter uses two electromagnetic fields and can respond to either ac or dc.

Fig. 14-7 Current transformer. (Courtesy of F. W. Bell, Inc.)

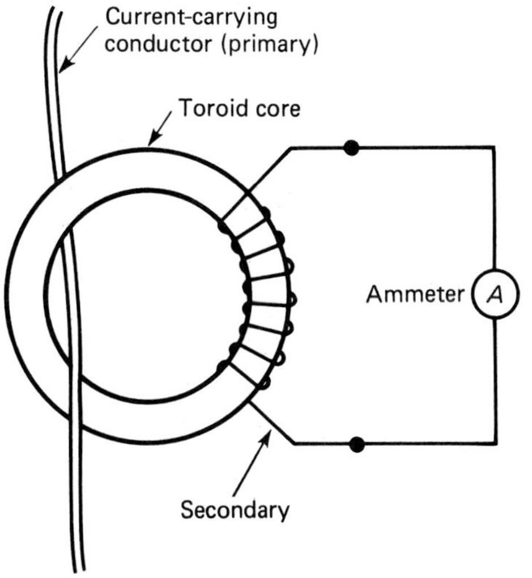

Fig. 14-8 Current transformer principle. The conductor in which current is being measured becomes a single-turn primary.

way, the voltage of a one-turn primary is stepped up by a multiturn secondary.

Measuring the current in a conductor with a current transformer requires that one end of the conductor be temporarily disconnected. The free end of the conductor is run through the large hole in the center of the current transformer (see Fig. 14-7). Then the conductor is reconnected to the circuit. Finally, a standard ammeter is connected to a pair of the terminals on the top of the current transformer. Of course, the ammeter's full-scale deflection voltage must equal the voltage output of the transformer under maximum current conditions. (Recall that ammeters have full-scale voltage ratings as well as full-scale current ratings.)

A current transformer may have several different secondaries. This allows it to be used with a variety of ammeters having different ratings.

Clamp-on Ammeter

Clamp-on ammeter

A clamp-on ammeter (also called a clamp-around ammeter) is pictured in Fig. 14-9. This meter uses the same principle as the current transformer-ammeter circuit of Fig. 14-8. However, with the clamp-on meter, the toroid core is made in two halves. The halves are hinged at one end and held together by a spring mechanism. The other end of the two halves can be temporarily separated to insert the conductor in which current is to be measured.

The clamp-on meter shown in Fig. 14-9 is a multirange unit. It can also measure ac voltage. When used to measure voltage, the clamp-on toroid core is not involved. Instead, the meter movement is converted to a voltmeter and the voltage is applied through a pair of test leads.

Meter Loading and Rating

All that you have learned about meter loading in dc circuits (Chap. 7) applies equally well to ac circuits. In general, ac meters tend to be even more prone to load a circuit than do dc meters. This is because ac voltmeters generally have less input resistance than do dc voltmeters. Likewise, ac ammeters tend to have more internal resistance than do dc ammeters with the same full-scale current rating.

Internal reactance

Meter loading in ac circuits is also caused by the internal reactances of meter movement, rectifier, leads, etc. Of course, these reactances vary with frequency. Therefore, the interaction between ac meters and the circuits in which they are used can be stated only at specified frequencies.

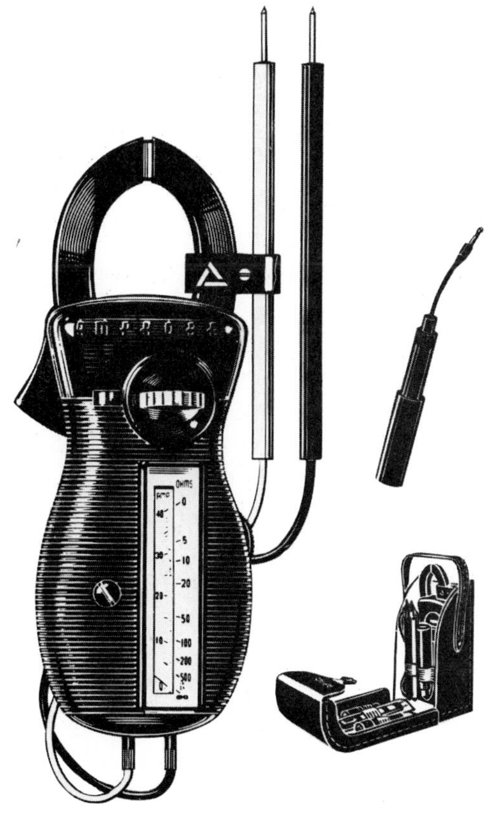

Fig. 14-9 Clamp-on meter. (Courtesy of Amprobe Instrument.)

Meters designed for use in ac circuits have a frequency rating. This rating tells the range of frequency over which the meter is accurate. It is important that ac meters be used only at their rated frequency.

Frequency rating

Review Questions

1. What type of meter movement does the rectifier-type meter use?
2. What is a rectifier?
3. Does a d'Arsonval meter movement respond to ac, dc, or both ac and dc?
4. Does an iron-vane movement respond to ac, dc, or both ac and dc?
5. Which of the following meter movements is the most sensitive? The iron-vane, the d'Arsonval, or the electrodynamometer?
6. Is the d'Arsonval or the iron-vane movement the most linear?
7. Does the iron-vane movement use a permanent magnet?
8. What type of meter movement is used in a thermocouple meter?
9. What are thermocouple meters used for?
10. Does the electrodynamometer movement respond to ac, dc, or both ac and dc?
11. What causes the rotary torque in an electrodynamometer movement?
12. How many coils create the stationary magnetic field in an electrodynamometer?
13. How many coils create the movable field in an electrodynamometer?
14. Name three meter movements that can be used to make a multirange ac meter.
15. What is a current transformer?
16. Does use of a current transformer require interruption of the circuit in which it is used?
17. Does use of a clamp-on meter require interruption of the circuit in which it is used?
18. What principle is used in the clamp-on ammeter?
19. Is meter loading as much of a problem with ac meters as it is with dc meters?
20. Is meter loading in ac circuits dependent on frequency?

14-2 WATTMETERS

Wattmeter

Voltage coil

Current coil

An electrodynamometer movement (Fig. 14-6) makes a perfect wattmeter. The moving coil (called the *voltage coil*) is used to detect the magnitude of the circuit voltage. The stationary coils are referred to as the *current coils*. The circuit current is detected by the current coils which are connected in series with the load (Fig. 14-10).

The stationary (current) coils are wound with large diameter wire. This keeps the resistance in series with the load as low as possible. The movable (voltage) coil is wound with thin wire to keep it as light as possible. Since the movable coil responds to voltage, it has a multiplier (see Fig. 14-10) in series with it.

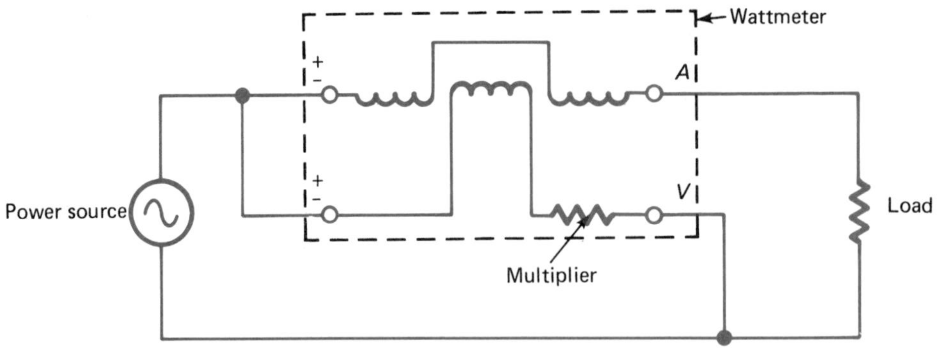

Fig. 14-10 Wattmeter connections. For clockwise torque, the instantaneous currents in the current coils and the voltage coil must be in the same direction.

Although the wattmeter in Fig. 14-10 is measuring ac power, relative polarity must still be observed. Connecting the plus and minus (±) terminals of both the voltage coil and the current coils to the power source will provide correct polarity. The other end of the current coils (marked A) goes to the load. Thus, all of the load current flows through the current coils. The other end of the voltage coil (marked V) returns to the power source. This puts the full source voltage across the voltage coil-multiplier circuit.

Correct polarity can also be achieved by connecting the ± end of the voltage coil to the A end of the current coils. Referring to Fig. 14-10, you can see that either connection gives the same current direction in the voltage coil. The differences in the two connections are minor. For instance, the current for the voltage coil has to flow through the current coil when the ± and A ends are connected together. When connected ± to ±, the voltage coil responds to the source voltage rather than to the load voltage.

The wattmeter indicates true power under all conditions. It automatically compensates for phase differences between the circuit voltage and current. How this happens can be understood by referring to Fig. 14-11. In Fig. 14-11(*a*) the voltage and current are in phase, that is, PF = 1. Therefore, the currents through the voltage coil and current coil will change at the same instance. Thus, the magnetic fields of the two coils are opposing and creating a clockwise torque during the complete cycle. However, when the current and voltage are out of phase [Fig. 14-11(*b*)], the situation changes. Now, for a portion of the cycle, the magnetic fields of the two coils will be aiding and will develop a counterclockwise torque. This will reduce the net clockwise torque and thus the power indicated by the meter. When current and voltage are 90° out of phase [Fig. 14-11(*c*)], the clockwise and counterclockwise torques are exactly equal. The net torque is zero, and the wattmeter indicates zero power.

True power

Three-phase power

Balanced load

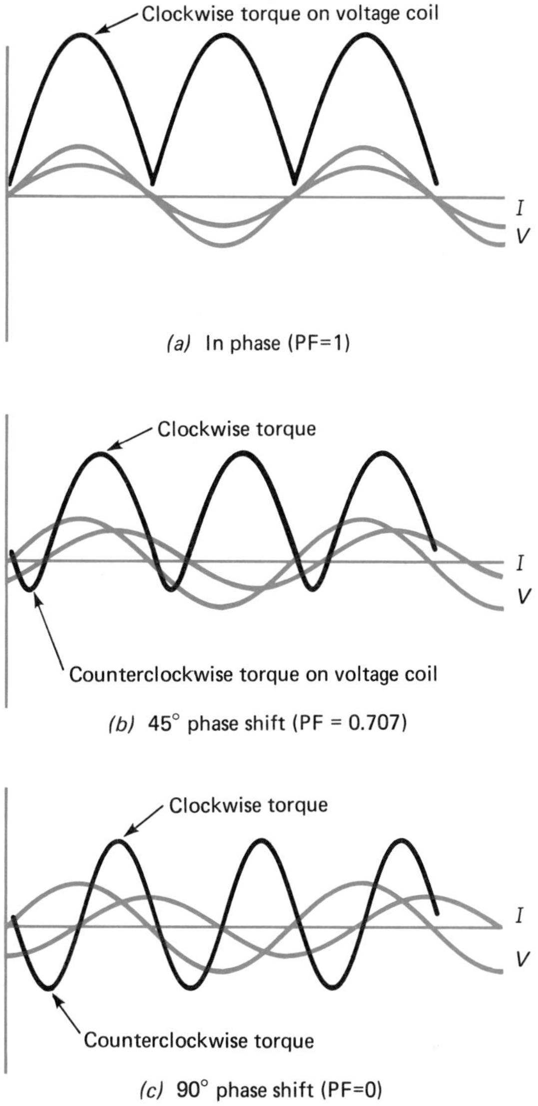

Fig. 14-11 How a wattmeter compensates for power factor.

Measuring Three-Phase Power: Balanced Load

With a balanced three-phase load, the power in each of the three phases must be equal. Therefore, the total power of the three-phase system will be equal to three times the power of any one phase. Consequently, a single wattmeter can be used to measure the power in any one of the phase loads (Fig. 14-12). The measured power is multiplied by three to give the total power of the system.

Notice in Fig. 14-12(*a*) that one must be able to connect the voltage coil of the wattmeter to the neutral or star point. Likewise, in Fig. 14-12(*b*) one has to break the junction of two of the phase loads to measure the current in a phase leg. In many three-phase loads these junction points are not readily accessible. In such cases, the power can be measured using two wattmeters as described in the next section on unbalanced loads.

Unbalanced load

Universal method

Algebraic sum

Voltage rating

Current rating

Power rating

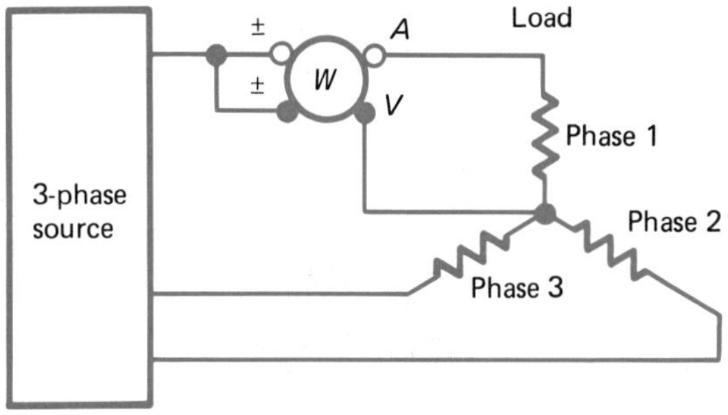

(a) Wye-connected load

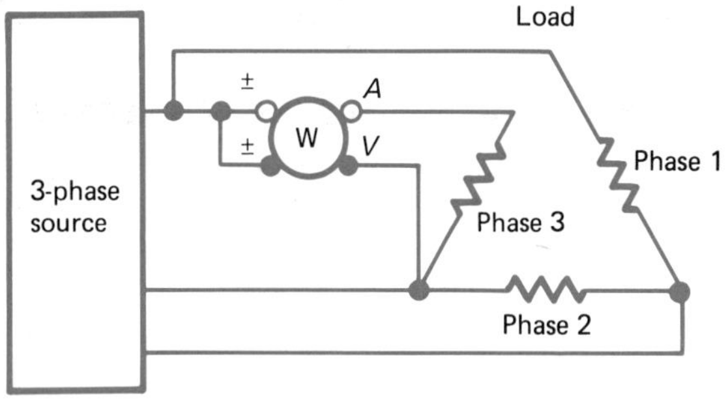

(b) Delta-connected load

Fig. 14-12 Measuring power with balanced loads. Only one wattmeter is needed when the phase loads are equal.

Measuring Three-Phase Power: Unbalanced Load

The two-wattmeter method of measuring three-phase power, as shown in Fig. 14-13, is a universal method. It works equally well with wye or delta loads and with balanced or unbalanced loads.

Notice in Fig. 14-13 that the wattmeters are monitoring line currents and line voltages. Yet, the algebraic sum of the two meter readings equals the total power of the system. (Proof of this statement requires phasor analysis beyond the scope of this text.) The phrase "algebraic sum of the two meter readings" needs some explanation. Normally, the meters are connected with the polarities shown in Fig. 14-13(*a*). This connection will usually result in clockwise deflection of both meters. If it does, the algebraic sum of the two readings is obtained by simply adding the two readings. For example, if one meter indicated 2800 W and the other indicated 1600 W, the power of the three-phase system would be 4400 W (2800W + 1600 W).

Sometimes, when the loads are extremely unbalanced, the polarities in Fig. 14-13(*a*) will result in a reverse deflection of one of the meters. In this case, the polarity of the reverse deflecting meter must be changed. This can be done by reversing the connections to the voltage coil as shown in Fig. 14-13(*b*). Now the algebraic sum of the two meter readings is found by subtracting the smaller reading from the larger reading. Let us look at an example. Suppose we connect the meters as shown in Fig. 14-13(*a*) and W_1 deflects counterclockwise. Therefore, we reverse the voltage coil connection on W_1. This yields the connections shown in Fig. 14-13(*b*). Now suppose W_1 indicates 300 W and W_2 indicates 2600 W. The total power of the system would be 2300 W (2600 W − 300 W = 2300 W).

Wattmeter Ratings

Wattmeters are rated for maximum current and voltage as well as maximum power. These maximum ratings are necessary to protect the coils in the electrodynamometer movement. If the voltage rating is exceeded, the movable coil (and the multiplier resistor) will get too hot. Exceeding the current rating of the meter overheats the stationary coils.

It is easy to exceed the current rating without exceeding the maximum power rating of the meter. For example, the meter may be operating at rated voltage, but connected to a

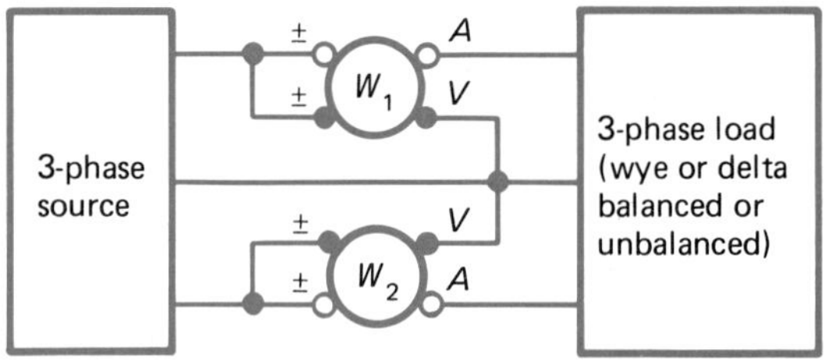

(a) Add the meter readings

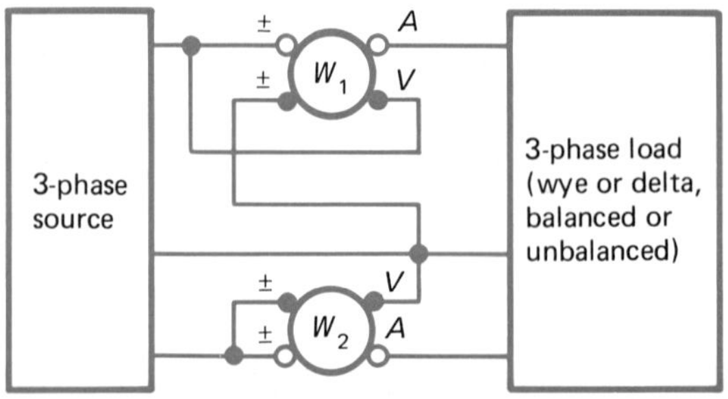

(b) Subtract the meter readings

Fig. 14-13 Power measurement with unbalanced or balanced loads. The connections shown in (*b*) are needed with some unbalanced loads.

very low power-factor load. In this case, the current rating would be exceeded long before the meter indicated full-scale power. For another example, suppose the meter was rated at 300 V maximum but was being used on a 120 V circuit. Again, the current rating would be exceeded before full-scale power was reached even if the load had a power factor of one.

Review Questions

21. What type of meter movement is used in a wattmeter?
22. Which coil of a wattmeter is wound with the larger diameter wire?
23. Why does an ac wattmeter require polarity markings?
24. Does a wattmeter indicate true or apparent power?
25. A wattmeter is connected to a 440-V circuit that draws 10 A. The power factor is 0. How much power does the meter indicate?
26. A wattmeter in a balanced three-phase circuit indicates 1380 W. What is the total power of the circuit?
27. What are the limitations of the single-wattmeter method of measuring three-phase power?
28. Two wattmeters are connected to a three-phase system. Both are connected for normal polarity. One meter indicates 4800 W, and the other meter indicates 1400 W. What is the power of the three-phase system?
29. If the 1400-W meter in the above question required reverse polarity connections, what would the power be?
30. How is it possible to exceed the current rating of a wattmeter without causing the pointer to deflect beyond full scale?

14-3 FREQUENCY METERS

Frequency can be measured with a variety of both electrical and electronic devices. Electronically, frequency can be measured with such devices as digital frequency counters and

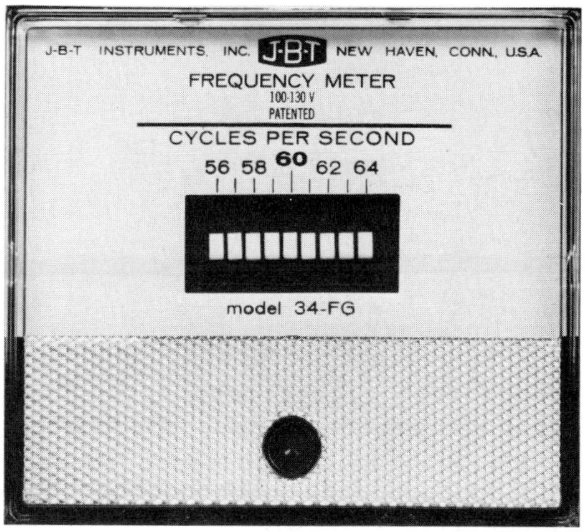

Fig. 14-14 Vibrating-reed frequency meter. (Courtesy of J-B-T Instruments, Inc.)

heterodyne frequency meters. These devices are capable of measuring a wide range of frequencies extending into the hundreds of megahertz.

Electrical frequency meters operate over a very restricted range of frequencies in the power frequency range. We will concern ourselves with the operation of two types of electrical frequency meters.

Frequency meter

Power frequency range

Vibrating-Reed Meter

A typical vibrating-reed meter is pictured in Fig. 14-14. It is designed to monitor and measure the frequency of a 60-Hz power source. The small white rectangles in the center of Fig. 14-14 are the ends of the vibrating reeds. In operation, the end of the reed which is tuned to the frequency of the voltage applied to the meter will vibrate. The end of the reed which is vibrating will appear to be two or three times its actual height. For example, the meter in Fig. 14-15 indicates 60 Hz. It is connected to a 60-Hz power line.

Mechanical resonant frequency

The principle of operation of a vibrating-reed meter is illustrated in Fig. 14-16. Each reed, due to its weight, thickness, width, etc., has a mechanical resonant frequency at which it vibrates when bent and then released. The coil in Fig. 14-16 lightly attracts (bends) and releases each reed twice during each cycle of the applied frequency. If the resonant frequencies of the reed and the applied voltage are the same, the reed's vibrations will be reinforced by the coil's attraction. The magnitude of the reed's vibration will increase.

Two fixed coils

Measuring impedance

Indirect technique

Current-voltage method

Fig. 14-15 Typical response of a vibrating-reed meter. This meter indicates 60 Hz. (Courtesy of J-B-T Instruments, Inc.)

If the applied frequency and the reed's natural frequency are not the same, the reed's vibrations will be opposed by the coil. When the reed is trying to bend up to do its natural vibration, the coil will be attracting it. Thus, the magnitude of vibration is severely limited for those reeds with a frequency different than the frequency received by the coil.

Vibrating-reed meters are rugged, dependable units. They can be calibrated to within 0.1 percent of specified frequency, and they are very stable.

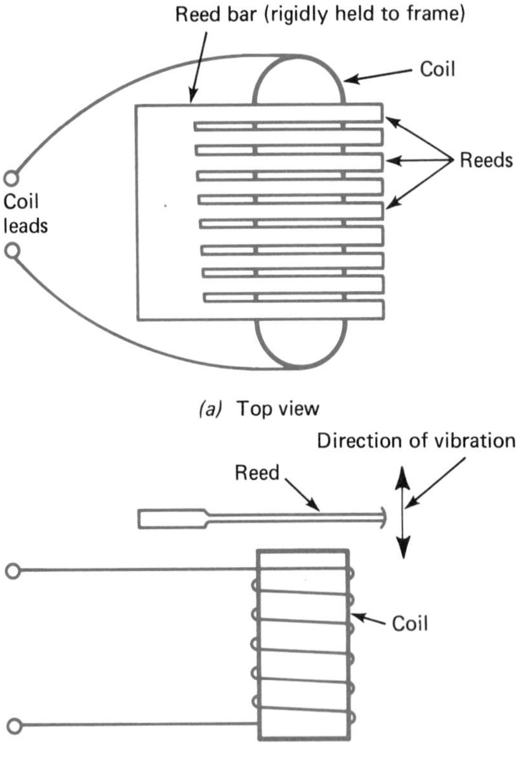

Fig. 14-16 Principle of the vibrating reed meter. Each reed is tuned to a different frequency. The reed which is tuned to the frequency of the coil current will vibrate.

Iron-Vane Frequency Meter

The iron-vane type meter (see Fig. 14-17) uses two fixed coils and a movable iron vane. The magnetic fields of both coils act on the iron vane. Each coil tries to align the vane with its field. Thus, the position of the vane (and therefore the pointer) will depend on the relative strength of the two magnetic fields. Of course, the strength of a magnetic field depends upon the amount of coil current.

As drawn in Fig. 14-17, the vane is closely aligned with the field of coil 1. Therefore, coil 1 is drawing more current than coil 2. Notice that coil 1 current flows through an external inductor. Coil 2 current flows through a capacitor. If the frequency of the applied voltage is increased, X_L will increase and X_C will decrease. This will cause more current to flow through coil 2 and less current to flow through coil 1. Therefore, the vane will rotate in a clockwise direction and indicate a higher frequency.

Review Questions

31. List two types of electrical frequency meters.
32. Do iron-vane or digital-frequency meters usually respond to a higher range of frequencies?
33. Does a vibrating-reed meter use one or two stationary coils?

14-4 MEASURING IMPEDANCE

Impedance and reactance must be measured by some indirect technique. Of course, both impedance and reactance are functions of frequency so they are measured and specified at some stated frequency. In this section we will concern ourselves only with measuring impedance (and reactance) at power and audio frequencies.

Current-Voltage Method

One of the simplest ways to indirectly measure impedance is to measure the current and volt-

Equivalent-resistance

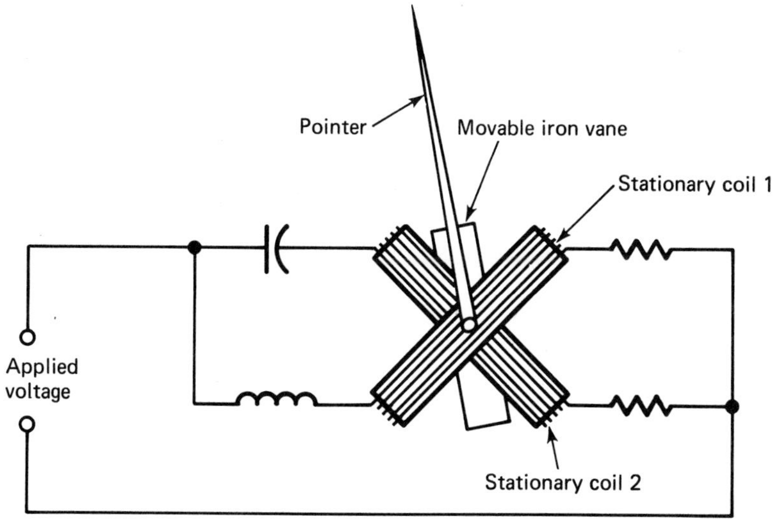

Fig. 14.17 Iron-vane frequency meter. The vane is attracted toward the coil conducting the most current.

age and then use Ohm's law. This method is illustrated in Fig. 14-18. It works well at any low frequency, but it is especially well suited to measuring impedance at 60 Hz. Most ac ammeters are not accurate at frequencies above several thousand hertz. Thus, unless high-frequency ammeters are available, the method is not useful at the higher audio frequencies.

Also, low-current ranges (less than 1 mA) on ac meters are not common. Therefore, the ammeter-voltmeter method of Fig. 14-18 is not too useful at very high impedances. In using the method of Fig. 14-18, it is best to use the rated voltage of the load being measured.

Equivalent-Resistance Method

A more versatile technique of measuring impedance is shown in Fig. 14-19(*a*). In using this technique, the source (generator) is first set to the frequency at which the impedance is to be measured. Any convenient output voltage level (within the rating of the load) is selected. Next, the voltage across the load is noted. Then the switch (*S*) is switched to its upper position and the voltage across the resistor (*R*) is measured. If the voltage across *R* is less than the voltage across the load, the value

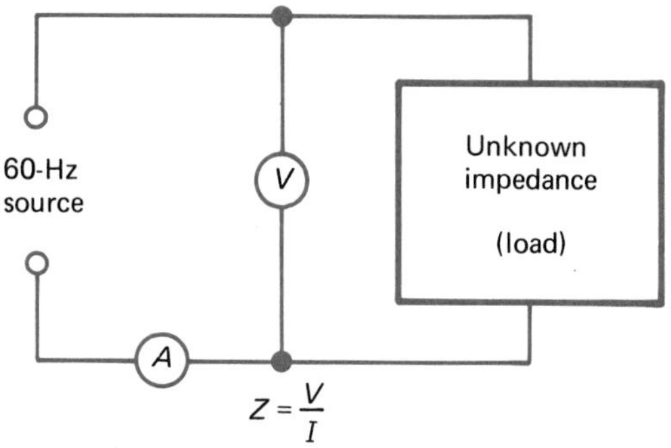

Fig. 14-18 Determining an unknown impedance. After the current and voltage are measured, Ohn.'s law is used to calculate impedance.

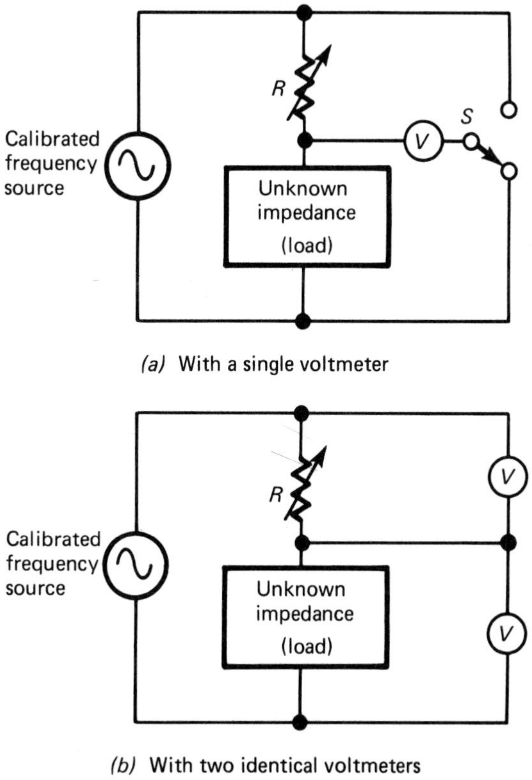

Fig. 14-19 Impedance-measuring technique. When the voltmeter readings in (*b*) are equal, the resistance of *R* will be the same value as the unknown impedance.

Voltmeter dip

of R is increased. If the reverse is true, the value of R is decreased. The switch is then returned to its original position to measure the new value of the load voltage. Then, the switch is again thrown to the up position and the resistor and load voltages are again compared. If they are not equal, R is again adjusted. This process is repeated until the measured voltages across R and the load are equal. When these voltages are equal, the impedance and resistance in this series circuit must also be equal. If R is a calibrated resistor, the value of the impedance of the load can be read from the dial. If not, then the source voltage is removed and the value of R (which equals Z) can be measured.

When two identical voltmeters are available, the circuit in Fig. 14-19(*b*) is easier and faster to use. Once the generator's output voltage and frequency have been set, R is adjusted until the two meters indicate equal voltages. Then the value of R is determined.

Review Questions

34. List two limitations of the current-voltage method of measuring impedance.
35. What is the advantage of using two voltmeters when determining Z by the equivalent resistance method?
36. A speaker and a 10-Ω resistor are connected in series to a generator. They each develop 2 V at 1400 Hz. What is the impedance of the speaker at 1400 Hz?

14-5 MEASURING INDUCTANCE AND CAPACITANCE

Inductance and capacitance can be measured in a variety of ways. Two common methods are presented below.

Resonant-Circuit Method

One of the easiest methods of determining values of L or C is by the resonant-circuit method. This method is shown in Fig. 14-20 for measuring either an unknown capacitance or an unknown inductance. Notice in Fig. 14-20 that the circuit for both cases is the same; only the formula has been rearranged.

The formulas in Fig. 14-20 are derived from the formula for finding the resonant frequency of an LC circuit. Values of L, C, and f in these formulas are in henries, farad, and hertz respectively.

The circuits in Fig. 14-20 are at resonance when the voltmeter "dips" to its lowest indication. *Voltmeter dip* means that the meter pointer suddenly moves toward zero and then returns to its previous indication. This dip occurs when the generator's frequency control is moved through the resonant frequency of the circuit. The resonant frequency of the circuit is that frequency at which the voltmeter reaches the bottom of its dip. The reason this dip occurs is because the series LC circuit's impedance approaches zero at resonance. Thus, most of the generator's available voltage is dropped across its own internal impedance.

Example 14-1

Suppose the circuit in Fig. 14-20(*b*) uses a 0.001-μF capacitor. Maximum voltmeter dip occurs at 180 kHz. What is the value of the inductance?

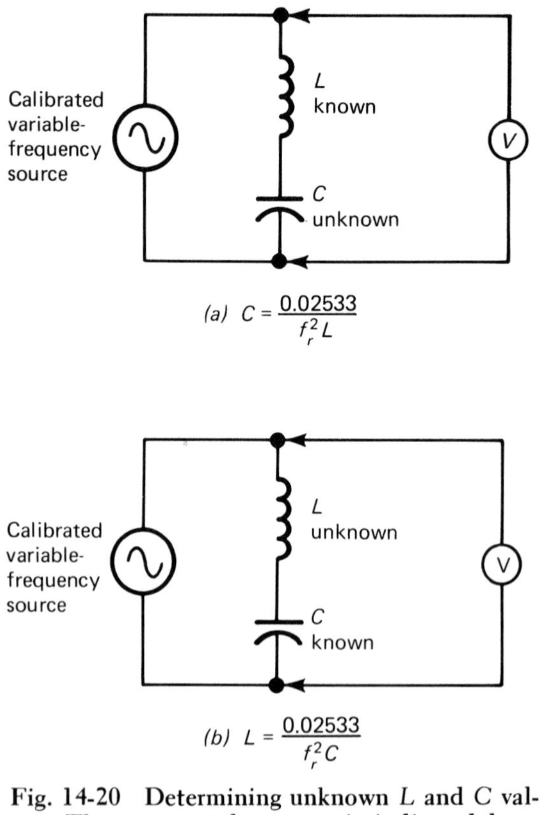

Fig. 14-20 Determining unknown L and C values. The resonant frequency is indicated by a sharp dip in the voltmeter reading.

Given: $C = 0.001\ \mu F$
$f_r = 180$ kHz

Find: L

Known: $L = \dfrac{0.02533}{f_r^2 C}$

Solution: $L = \dfrac{0.02533}{(180 \times 10^3)^2 \times 0.001 \times 10^{-6}}$
$= 0.000782$ H or 0.782 mH

Answer: The inductance is 0.782 millihenry.

AC-Bridge Method

The value of a capacitor (or an inductor) can also be determined with an ac bridge. The ac-bridge circuit in Fig. 14-21 is used to measure the value of the unknown capacitor (C_u). To use the bridge, R_1 and R_3 are adjusted for the lowest possible indication on the voltmeter. There will be some interaction between R_1 and R_3, so that one and then the other resistor is adjusted several times to achieve minimum voltmeter reading. When properly adjusted, the meter should indicate essentially zero volts.

The purpose of R_3 is to make the resistive losses in C_1 equal to the resistive losses in C_u. If C_u has as high a Q (quality) as C_1, then, R_3 will be approximately 0 Ω. In fact, R_3 is only needed in the bridge when low-quality capacitors are to be measured. For our analysis of the capacitor bridge we will assume high-Q capacitors are being measured. Thus, R_3 will be 0 Ω.

When R_1 and R_3 are adjusted for minimum voltmeter indication, the bridge is balanced. Under balanced conditions, the ratio of R_1 to X_{C_1} will be equal to the ratio of R_2 to X_{Cu}. Mathematically, we can write:

$$\frac{R_1}{X_{C_1}} = \frac{R_2}{X_{Cu}}$$

Since capacitive reactance is inversely proportional to capacitance

$$R_1 C_1 = R_2 C_u$$

or

$$C_u = \frac{C_1}{R_2} \times R_1$$

In the above formula, capacitance is given in farads and resistance in ohms.

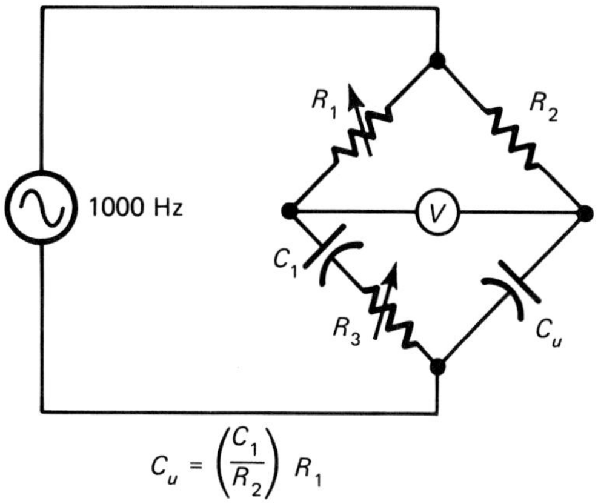

Capacitor bridge

Fig. 14-21 Measuring C with a bridge. The variable resistors are adjusted to obtain the smallest possible voltmeter reading.

A dial plate can be attached to R_1 in Fig. 14-21. Then the plate can be calibrated in units of capacitance for a given value of C_1 and R_2. Once this is done, a wide range of capacitance can be measured by using different combinations of C_1 and R_2. The dial-plate reading is then multiplied by a number (multiplier) which represents the change in the ratio C_1/R_2.

Example 14-2

As shown in Fig. 14-21, $R_2 = 2000\ \Omega$ and $C_1 = 0.002\ \mu F$. If balance occurs when R_1 is adjusted to 1700 Ω, what is the capacitance of C_u?

Given: $R_2 = 2000\ \Omega$
$C_1 = 0.002\ \mu F$
$R_1 = 1700\ \Omega$

Find: C_u

Known: $C_u = \dfrac{C_1}{R_2} \times R_1$

Solution: $C_u = \dfrac{0.002 \times 10^{-6}\ F}{2000\ \Omega} \times 1700\ \Omega$
$= 0.0017 \times 10^{-6}$ F or 0.0017 μF

Answer: The unknown capacitance is 0.0017 microfarads.

In the above example the 1700-Ω position of R_1 could be marked 0.0017 μF on the dial plate. Then the dial would be direct reading for the specified value of R_2 and C_1. Now suppose R_2 were changed to 200 Ω. The new C_1/R_2 would be ten times greater than the old value. If the bridge again balanced at the

1700-Ω setting of R_1, the unknown capacitance would be 0.017 μF.

The inductance bridge in Fig. 14-22 works on the same principles as the capacitance bridge of Fig. 14-21. The formula $L_u = CR_2R_1$ can be derived in the same way that the capacitance bridge formula was derived.

Notice in Fig. 14-22 that R_3 is in parallel with C. (In some inductance bridges it is in series.) Again, R_3 is adjusted to provide the same amount of resistance loss in the capacitor as there is in the inductor. Since R_3 is in parallel with C, the resistive losses will increase as R_3 is decreased. If the unknown inductance has a high Q, then the value of R_3 will be very high.

The resistance of the inductor (R_{ind}) can also be determined from the bridge circuit of Fig. 14-22 by the formula

$$R_{ind} = \frac{R_2R_1}{R_3}$$

Once the resistance and inductance of the inductor are known, it is easy to determine the Q of the inductor. (The generator frequency is known and $Q = X_L/R = 6.28fL/R$.) Many inductance bridges have the dial attached to R_3 calibrated directly in Q values.

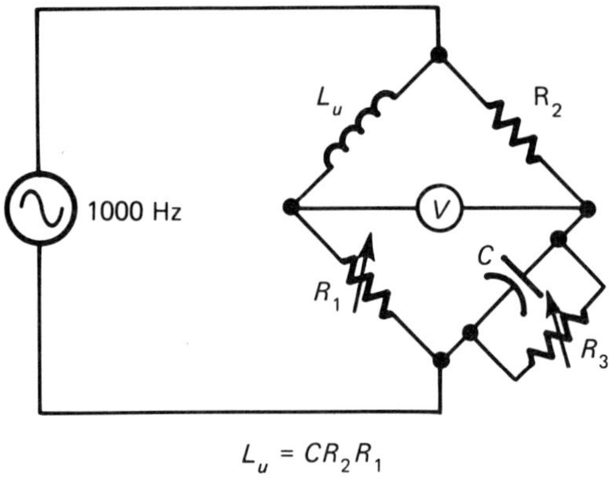

Fig. 14-22 Measuring L with a bridge. The bridge is balanced when further adjustment of R_1 and R_3 cannot reduce the voltage indicated by the voltmeter.

Example 14-3

Suppose the bridge in Fig. 14-22 balances when $R_1 = 2$ kΩ, $R_2 = 5$ kΩ, $R_3 = 500$ kΩ and $C = 0.01$ μF. What is the inductance L and the quality Q of the inductor?

Given: $R_1 = 2$ kΩ
$R_2 = 5$ kΩ
$R_3 = 500$ kΩ
$C = 0.01$ μF
$f = 1000$ Hz

Find: L and Q

Known: $L_u = CR_2R_1$
$$Q = \frac{X_L}{R}$$
$X_L = 6.28fL$
$$R_{ind} = \frac{R_2R_1}{R_3}$$

Solution: $L_u = 0.01 \times 10^{-6} \times 5 \times 10^3 \times 2 \times 10^3$
$= 0.1 \text{ H} = 100 \text{ mH}$

$$R_{ind} = \frac{5 \times 10^3 \times 2 \times 10^3}{500 \times 10^3} = 20\ \Omega$$

$X_L = 6.28 \times 1000 \times 0.1 = 628\ \Omega$

$$Q = \frac{628}{20} = 31.4$$

Answer: The inductance is 100 millihenries, and the quality is 31.4.

Review Questions

37. List two methods of measuring capacitance or inductance.
38. How can you tell when a signal generator is adjusted to the resonant frequency of a series LC circuit?
39. Refer to Fig. 14-21 and find C_u when $R_2 = 1$ kΩ, $R_1 = 3$ kΩ and $C = 1000$ pF.
40. Refer to Fig. 14-22 and find L_u when $R_2 = 1$ kΩ, $R_1 = 3$ kΩ, and $C = 1000$ pF.

Summary

1. Rectifier-type meters use a d'Arsonval meter movement and a rectifier.
2. A rectifier converts ac to pulsating dc.
3. A d'Arsonval meter movement responds only to dc.
4. A rectifier allows current to flow in only one direction.
5. Shunts and multipliers can be used with rectifier-type, iron-vane, and electrodynamometer meters.

6. The rectifier-type meter is more sensitive than the iron-vane meter.
7. The iron-vane meter has no moving coil or permanent magnet.
8. The iron-vane meter responds to both ac and dc. It has a nonlinear scale and low sensitivity.
9. Thermocouple meters are used to measure high frequency currents. They use a d'Arsonval meter movement.
10. Electrodynamometer meter movements use stationary and moving coils to develop interacting magnetic fields. They respond to ac or dc and are used in wattmeters. Electrodynamometer meters have low sensitivity and high accuracy.
11. Current transformers are used to extend the range of an ac ammeter.
12. Clamp-on ammeters can measure current in an ac circuit without physically interrupting the circuit.
13. In a wattmeter the moving coil responds to voltage and the stationary coil responds to current.
14. Wattmeters are polarized instruments. They indicate true power. They have current, voltage, and power ratings.
15. A single wattmeter can measure power in a balanced three-phase load.
16. Two wattmeters are required to measure power in an unbalanced three-phase load. The wattmeter readings are algebraically added.
17. Frequency can be measured with a vibrating-reed meter or an iron-vane frequency meter. These meters measure only a narrow range of frequencies.
18. Impedance can be measured by the current-voltage method and the equivalent-resistance method. The equivalent-resistance method is the more versatile method.
19. Inductance and capacitance can be measured by the resonant circuit method or the ac-bridge method. The bridge method also can measure the Q of the component.

Chapter Review Questions

For questions 14-1 to 14-12, determine whether each statement is true or false.

14-1. Iron-vane meter movements are used in rectifier-type meters.

14-2. The stationary coils in a wattmeter respond to current.

14-3. The iron-vane movement is often used to measure high-frequency current.

14-4. The clamp-on ammeter uses the current transformer principle.

14-5. The vibrating-reed meter is used to measure impedance.

14-6. The current in an ac circuit can be measured without physically interrupting the circuit.

14-7. A current transformer is used to measure high values of high-frequency ac voltages.

14-8. An ac wattmeter is a nonpolarized instrument.

14-9. Two wattmeters are needed to measure ac power in any three-phase circuit.

14-10. Bridge balance is obtained when the null indicator (voltmeter) indicates its lowest voltage.

14-11. An iron-vane meter movement has a linear scale.

14-12. Meter loading in ac circuits depends on frequency.

For questions 14-13 to 14-17, choose the letter that best answers each statement.

14-13. Which of these meter movements is the most sensitive?
(A) D'Arsonval (B) Electrodynamometer (C) Iron-vane

14-14. Which of these meter movements cannot measure alternating current?
(A) D'Arsonval (B) Electrodynamometer (C) Iron-vane

14-15. Which of these meter movements is used to measure power?
(A) D'Arsonval (B) Electrodynamometer (C) Iron-vane

14-16. Which of these meter movements uses a permanent magnet?
(A) D'Arsonval (B) Electrodynamometer (C) Iron-vane

14-17. Which of these meter movements does not use a moving coil?
(A) D'Arsonval (B) Electrodynamometer (C) Iron-vane

Answer the following questions.

14-18. List two methods of measuring impedance.

14-19. List two methods of measuring inductance.

14-20. List two methods of measuring capacitance.

14-21. Which of the two methods of measuring inductance can also be used to measure the quality of the inductor?

14-22. What is a rectifier?

Answers to Review Questions

1. D'Arsonval movement
2. A device that allows current to flow only in one direction
3. Dc
4. Both ac and dc
5. The d'Arsonval
6. The d'Arsonval
7. No
8. D'Arsonval
9. Measuring high frequency currents
10. Both ac and dc
11. The interaction of the magnetic fields of the stationary coils and the movable coil
12. Two
13. One
14. D'Arsonval, electrodynamometer, and iron vane
15. A transformer that is used, in conjunction with an ammeter, to measure large values of current
16. Yes
17. No
18. The current transformer principle
19. Yes
20. Yes
21. Electrodynamometer movement
22. The stationary or current coil
23. So that the current in the stationary coils will be in the correct direction relative to the current in the movable coil
24. True power
25. Zero
26. 4140 W
27. The load must be balanced and the phase junctions of the load must be accessible
28. 6200 W
29. 3400 W
30. The voltage may be

below rated value, or the load may have a low power factor

31. Vibrating-reed and iron-vane
32. Digital
33. One coil
34. It is difficult to measure current at higher frequencies and it is difficult to measure high values of impedance
35. It is much faster. No switching of meter leads is required
36. 10 Ω
37. Resonant circuit method and ac-bridge method
38. By the sudden drop in the generator's output voltage
39. 0.003 μF
40. 0.003 H or 3 mH

Appendix 1

GLOSSARY OF TERMS AND SYMBOLS

Term	Definition	Symbol or Abbreviation
Alternation	One-half of a cycle	
Ammeter	Device used to measure current	
Ampere	Base unit of current (coulomb per second)	A
Ampere-hour	Unit used to show energy storage capacity of cell or battery	Ah
Ampere-turn	Base unit of magnetomotive force	A·t
Ampere-turn per meter	Base unit of magnetic field strength	
Apparent power	The product of current times voltage in a circuit containing reactance	
Atom	Building block of all matter	
Bandwidth	The space, expressed in hertz, between the lowest and highest frequencies an *LC* circuit will respond to with at least 70.7% of its maximum response	BW
Battery	Two or more cells electrically connected together	
Bemf	Back electromotive force. Another name for cemf	
Bimetallic strip	Sandwich of two metals with different coefficients of expansion	
Capacitance	Ability to store energy in the form of electrical energy	C
Capacitive reactance	The opposition capacitance offers to ac	X_C
Capacitor	An electrical component that possesses capacitance	
Cell	Chemical system that produces dc voltage	
Cemf	Counter electromotive force. The voltage produced by self-inductance	
Charge	Electrical property of electrons and protons	Q
Choke	Another name for an inductor	
Circuit breaker	Device which protects a circuit from excessive currents	
Clamp-on meter	An ammeter used to measure current without physically interrupting the circuit	

GLOSSARY OF TERMS AND SYMBOLS

Term	Definition	Symbol or Abbreviation
Coefficient of coupling	Denotes the portion of flux from one coil that links with another coil	
Coil	Another name for an inductor	
Compound	Matter composed of two or more elements	
Condenser	Outdated name for capacitor	
Conductance	Ability to conduct current	G
Conductors	Materials that have very low resistivity	
Continuity	Continuous path for current	
Copper loss	Refers to the conversion of electrical energy to heat energy in the windings of magnetic devices	
Core loss	Refers to the conversion of electrical energy to heat energy in the core material of magnetic devices	
Coulomb	Base unit of charge (6.25×10^{18} electrons)	C
Current	Movement of charge in specified direction	*I*
Current carrier	Charged particle (electron or ion)	
Current transformer	A transformer used to extend the range of an ac ammeter	
Cycle	That part of a periodic waveform that does not repeat itself	
Delta connection	A method of connecting a three-phase system so that the line and phase voltage are equal	
Dielectric	Insulation used between the plates of a capacitor	
Dielectric constant	A number which compares a material's ability to store energy to the ability of air	
Dissipation factor	A number obtained by dividing resistance by reactance. The reciprocal of Q. Used to indicate relative energy loss in a capacitor	DF
Eddy current	Current induced into the core of a magnetic device. Causes part of the core losses	
Electric field	Invisible field of force that exists between electric charges	
Electrical degree	One three-hundred-sixtieth ($\frac{1}{360}$) of an ac cycle	
Electron	Negatively charged particle of the atom	
Element	Matter composed entirely of one type of atom	
Energy	Ability to do work	W
Energizing current	The primary current in an unloaded transformer	

GLOSSARY OF TERMS AND SYMBOLS

Term	Definition	Symbol or Abbreviation
Farad	Base unit of capacitance. Equal to a coulomb per volt	F
Filter	A circuit designed to separate one frequency, or group of frequencies, from all other frequencies	
Fluctuating dc	A dc that varies in amplitude but does not periodically drop to zero	
Flux	Lines of force around a magnet	ϕ
Flux density	Amount of flux per unit cross-sectional area	B
Free electrons	Electrons that are not attached (held) to any atom	
Frequency	Rapidity with which a periodic waveform repeats itself	f
Fuse	Device which protects a circuit from excessive currents	
Henry	Base unit of inductance. One volt of cemf for a current change of one ampere per second	H
Hertz	The base unit of frequency. One cycle per second	Hz
Horsepower	Unit of power (1 hp = 746 watts)	hp
Hydrometer	Device used to measure specific gravity	
Hysteresis	Magnetic effect caused by residual magnetism in ac operated magnetic devices. Causes part of the core losses	
Impedance	The total opposition of a circuit consisting of resistance and reactance. The base unit is ohms	Z
Induced voltage	Voltage created in a conductor when the conductor interacts with a magnetic field	
Inductance	Electrical property which opposes changes in current	L
Inductive kick	The high cemf that is generated when an inductive circuit is opened	
Inductive reactance	The opposition inductance offers to ac	X_L
Inductor	An electrical component that possesses inductance	
Insulators	Materials that have very high resistivity	
Internal resistance	Resistance contained within a power or energy source	

GLOSSARY OF TERMS AND SYMBOLS

Term	Definition	Symbol or Abbreviation
Ion	Atom which has an excess or deficiency of electrons	
Iron-vane meter	A type of meter which uses a moving iron vane and a stationary coil. Commonly used to measure alternating current	
Joule	Base unit of energy (newton-meter)	J
Kilowatt-hour	Unit of energy (1 kWh = 3,600,000 joules)	kWh
Lenz's law	States that the cemf will always oppose the force which created it.	
Litz wire	A special wire used to reduce the skin effect	
Load	Device that converts electric energy to some other form of energy	
Magnetic field strength	Amount of magnetomotive force per unit length. Other terms for magnetic field strength are magnetic field intensity and magnetizing force	H
Magnetomotive force	Force that creates a magnetic field	mmf
Meter loading	Changes in circuit current or voltage caused by putting a meter in circuit	
Multimeter	Electric instrument designed to measure two or more electrical quantities	
Multiplier	Precision resistor used to extend voltage range of a meter movement	
Mutual inductance	Inductance caused by the flux from one circuit inducing a voltage into another circuit	
Neutron	Electrically neutral portion of atom	
Nucleus	Center of atom which contains protons and neutrons	
Ohm	Base unit of resistance (volt per ampere)	Ω
Ohmic resistance	The dc resistance of an inductor	
Ohmmeter	Device used to measure resistance	
Period	Time required to complete one cycle	T
Permeability	Ease with which flux is created in a material	μ
Phase	A time relationship between two electrical quantities	ϕ
Phase shift	The result of two waveforms being out of step with each other	

GLOSSARY OF TERMS AND SYMBOLS

Term	Definition	Symbol or Abbreviation
Phasing	Interconnecting transformer, generator, or motor windings so that they have the correct time (phase) relationships between them	
Phasor	A line representing alternating current or voltage at some instant of time	
Photoconductive	Changing conductance, or resistance, by changing the light energy level	
Photovoltaic	Producing voltage from light energy	
Piezoelectric	Producing voltage by applying pressure to a crystal	
Polarity	Electrical characteristic (negative or positive) of a charge	
Polarization	Accumulation of gas ions around electrode of a cell	
Potentiometer	Three-terminal variable resistor	
Power	Rate of doing work or using energy	P
Power factor	The cosine of theta. The ratio of the true power over the apparent power. Also equal to resistance divided by impedance	PF
Primary cell	Cell that is not meant to be recharged	
Proton	Positively charged particle of the atom	
Pulsating dc	A dc which periodically returns to zero	
Quality	A number which indicated the ratio of reactance to resistance	Q
Reactance	Opposition to current which converts no energy (uses no power). A property of both inductance and capacitance	X
Reactor	Another name for an inductor	
Rectification	The process of converting ac to pulsating dc	
Relative permeability	Permeability of a material compared with permeability of air	μ_r
Reluctance	Opposition to creation of magnetic flux	$\mathcal{R}$
Residual magnetism	Magnetic flux left in temporary magnet after magnetizing force has been removed	
Resistance	Opposition to current which converts electric energy into heat energy	R
Resistivity	Characteristic resistance of a material (resistance of a cubic meter of the material)	

GLOSSARY OF TERMS AND SYMBOLS

Term	Definition	Symbol or Abbreviation
Resonance	A circuit condition in which $X_L = X_C$	
Resonant frequency	That frequency at which $X_L = X_C$ for a given value of L and C	f_r
Rheostat	Two-terminal variable resistor	
Secondary cell	Cell that can be recharged	
Selectivity	The ability of a circuit to separate signals (voltages) that are at different frequencies	
Self-inductance	The process of a conductor inducing a cemf in itself	
Semiconductor	Elements with four valence electrons	
Shunt	A parallel branch or component. Precision resistor used to extend current range of a meter movement	
Sine wave	A symmetrical waveform whose instantaneous value is related to the trigonometric sine function	
Skin effect	Phenomenon that causes the resistance of a conductor to increase with increased frequency	
Specific gravity	Weight of a substance compared with weight of equal volume of water	
Tank circuit	A parallel LC circuit	
Temperature coefficient	Number of units change per degree Celsius change from a specified temperature	
Tesla	Base unit of flux density	T
Theta	The phase angle between phasors	θ
Three-phase	Three voltages or currents displaced from each other by 120 electrical degrees	3-ϕ
Time constant	The time a capacitor requires to charge or discharge (through a resistor) 63.2% of the available voltage	T
Transformer losses	Power losses in a transformer. Caused by hysteresis, eddy current, and winding resistance	
Trigonometric functions	The relationships between the angles and sides of a right triangle. The three common functions are sine, cosine, and tangent	
Turns-per-volt ratio	The number of turns for each volt in a transformer winding	
Valence electrons	Electrons in outermost shell of an atom	

GLOSSARY OF TERMS AND SYMBOLS

Term	Definition	Symbol or Abbreviation
Vector	A line that represents both the direction and the magnitude of a quantity. Vectors of electrical quantities are called phasors	
Vibrating-reed meter	A meter used to measure frequency—especially the low frequencies used in power systems	
Volt	Base unit of voltage (joule per coulomb)	V
Voltage	Potential energy difference (electrical pressure)	V
Voltmeter	Device used to measure voltage	
Voltmeter sensitivity	Ohms-per-volt rating of voltmeter. Numerically equal to reciprocal of full-scale current of the meter movement used in voltmeter	
Watt	Base unit of power (joule per second)	W
Watt-hour	Unit of energy (3600 joules)	
Watt-second	Unit of energy (1 joule)	
Wattmeter	An electrical meter that measures true power	
Weber	Base unit of flux	Wb
Wheatstone bridge	Circuit configuration used to measure electrical qualities such as resistance	
Wye connection	Connecting the phases of a three-phase system at a common point so that the line and phase currents are equal	

Appendix 2

FORMULAS AND CONVERSIONS

Amplitude Conversions

$V_{p\text{-}p} = 2\,V_p$ and $V_p = \frac{V_{p\text{-}p}}{2}$

$V_p = 1.414\,V_{rms}$ and $V_{rms} = 0.707\,V_p$

$V_p = 1.57\,V_{av}$ and $V_{av} = 0.637\,V_p$

$V_{av} = 0.9\,V_{rms}$ and $V_{rms} = 1.11\,V_{av}$

V_{rms} = effective value

These formulas can also apply to current by substituting *I* for V.

Bandwidth

$$BW \simeq \frac{f_r}{Q}$$

Capacitance

$$C = \frac{Q}{V}$$

In Parallel

$$C_T = C_1 + C_2 + C_3 + \text{etc.}$$

In Series

$$C_T = \frac{1}{\frac{1}{C_1} + \frac{1}{C_2} + \frac{1}{C_3} + \text{etc.}}$$

$$C_T = \frac{C_1 \times C_2}{C_1 + C_2}$$

Energy Storage

$$W = 0.5\,CV^2$$

For Resonance

$$C = \frac{0.02533}{f_r^2\,L}$$

Capacitive Reactance

$$X_C = \frac{1}{6.28fC}$$

In Parallel

$$X_{CT} = \frac{1}{\frac{1}{X_{C1}} + \frac{1}{X_{C2}} + \frac{1}{X_{C3}} + \text{etc.}}$$

$$X_{CT} = \frac{X_{C1} \times X_{C2}}{X_{C1} + X_{C2}}$$

In Series

$$X_{CT} = X_{C1} + X_{C2} + X_{C3} + \text{etc.}$$

Conductance

$$G = \frac{1}{R}$$

Current (Definition)

$$I = \frac{Q}{t}$$

Current (In Impedance Circuits)

For Any Circuit

$$I_T = \frac{V_T}{Z}$$

For Parallel Circuits

$$I_T = \sqrt{I_R^2 + I_C^2}$$

$$I_T = \sqrt{I_R^2 + I_L^2}$$

$$I_T = \sqrt{I_R^2 + (L_L - I_C)^2}$$

Efficiency

$$\% \text{ eff.} = \frac{W_{out}}{W_{in}} \times 100$$

$$\% \text{ eff.} = \frac{P_{out}}{P_{in}} \times 100$$

Frequency–Period Conversions

$$T = \frac{1}{f}$$

$$f = \frac{1}{T}$$

Generator Frequency

$$f = \frac{\text{r/min}}{60} \times \text{pairs of poles}$$

Impedance

For Any Circuit

$$Z = \frac{V_T}{I_T}$$

For Series Circuits

$$Z = \sqrt{R^2 + X_L^2}$$

$$Z = \sqrt{R^2 + X_C^2}$$

$$Z = \sqrt{(X_L - X_C)^2 + R^2}$$

For Parallel Circuits

$$Z = \frac{RX_L}{\sqrt{R^2 + X_L^2}}$$

$$Z = \frac{RX_C}{\sqrt{R^2 + X_C^2}}$$

$$Z = \frac{RX_L X_C}{\sqrt{(RX_L - RX_C)^2 + (X_L^2 X_C^2)}}$$

Inductance

Quality

$$Q = \frac{X_L}{R}$$

In Parallel

$$L_T = \frac{1}{\frac{1}{L_1} + \frac{1}{L_2} + \frac{1}{L_3} + \text{etc.}}$$

$$L_T = \frac{L_1 \times L_2}{L_1 + L_2}$$

In Series

$$L_T = L_1 + L_2 + L_3 + \text{etc.}$$

For Resonance

$$L = \frac{0.02533}{f_r^2 C}$$

Inductive Reactance

$$X_L = 6.28fL$$

In Parallel

$$X_{LT} = \frac{1}{\frac{1}{X_{L1}} + \frac{1}{X_{L2}} + \frac{1}{X_{L3}} + \text{etc.}}$$

$$X_{LT} = \frac{X_{L_1} \times X_{L_2}}{X_{L_1} + X_{L_2}}$$

In Series

$$X_{LT} = X_{L1} + X_{L2} + X_{L3} + \text{etc.}$$

Ohm's Law

$$V = IR$$

$$I = \frac{V}{R}$$

$$R = \frac{V}{I}$$

Parallel Circuits

$$P_T = P_{R1} + P_{R2} + P_{R3} + \text{etc.}$$

$$R_T = \frac{1}{\frac{1}{R_1} + \frac{1}{R_2} + \frac{1}{R_3} + \text{etc.}}$$

$$R_T = \frac{R_1 \times R_2}{R_1 + R_2} \quad \text{for two resistances}$$

$R_T = \frac{R}{n}$ for n equal resistances

$V_T = V_{R1} = V_{R2} = V_{R3} =$ etc.

$I_T = I_{R_1} + I_{R_2} + I_{R_3} +$ etc.

Power

$P = IV \cos\theta$

$P_{app} = IV$

$P = I_R^2 R$

$P = V_R I_R$

$P = \frac{W}{t}$

Power Factor

$PF = \cos\theta$

$PF = \frac{P}{P_{app}}$

Quality

$Q = \frac{X}{R}$

Resistance

$R = \frac{\text{resistivity} \times \text{length}}{\text{area}}$

$= \frac{Kl}{A}$

where K = resistivity
l = length
A = area

Resonant Frequency

$f_r = \frac{1}{6.28\sqrt{LC}}$

Series Circuits

$P_T = P_{R1} + P_{R2} + P_{R3} +$ etc.

$R_T = R_1 + R_2 + R_3 +$ etc.

$V_T = V_{R1} + V_{R2} + V_{R3} +$ etc.

$I_T = I_{R1} = I_{R2} = I_{R3} =$ etc.

Time Constant

$T = RC$

$T = \frac{L}{R}$

Transformers

$\frac{N_{pri}}{N_{sec}} = \frac{V_{pri}}{V_{sec}}$

$\left(\frac{N_{pri}}{N_{sec}}\right)^2 = \frac{Z_{pri}}{Z_{sec}}$

$\%\ \text{eff.} = \frac{P_{sec}}{P_{pri}} \times 100$

Trigonometric Functions

$\cos\theta = \frac{\text{adjacent}}{\text{hypotenuse}}$

$\sin\theta = \frac{\text{opposite}}{\text{hypotenuse}}$

$\tan\theta = \frac{\text{opposite}}{\text{adjacent}}$

For Any Circuit

$\cos\theta = \frac{P}{P_{app}} = PF$

For Series Circuits

$\cos\theta = \frac{V_R}{V_T} = \frac{R}{Z}$

$\sin\theta = \frac{V_X}{V_T} = \frac{X}{Z}$

$\tan\theta = \frac{V_X}{V_R} = \frac{X}{R}$

For Parallel Circuits

$\cos\theta = \frac{I_R}{I_T}$

$\sin\theta = \frac{I_X}{I_T}$

$\tan\theta = \frac{I_X}{I_R}$

Voltage (Definition)

$V = \frac{W}{Q}$

Voltage (In Impedance Circuits)

For Any Circuit

$V_T = I_T Z$

For Series Circuits

$V_T = \sqrt{V_R^2 + V_C^2}$

$V_T = \sqrt{V_R^2 + V_L^2}$

$V_T = \sqrt{V_R^2 + (V_L - V_C)^2}$

Appendix 3

COPPER WIRE TABLE

AWG (B & S) gage	Standard metric size (mm)	Diameter in mils	Cross-sectional area: Circular mils	Cross-sectional area: Square inches	Ohms per 1,000 ft at 20°C (68°F)	Lbs per 1,000 ft	Ft per lb
0000	11.8	460.0	211,600	0.1662	0.04901	640.5	1.561
000	11.0	409.6	167,800	0.1318	0.06180	507.9	1.968
00	9.0	364.8	133,100	0.1045	0.07793	402.8	2.482
0	8.0	324.9	105,500	0.08289	0.09827	319.5	3.130
1	7.1	289.3	83,690	0.06573	0.1239	253.3	3.947
2	6.3	257.6	66,370	0.05213	0.1563	200.9	4.977
3	5.6	229.4	52,640	0.04134	0.1970	159.3	6.276
4	5.0	204.3	41,740	0.03278	0.2485	126.4	7.914
5	4.5	181.9	33,100	0.02600	0.3133	100.2	9.980
6	4.0	162.0	26,250	0.02062	0.3951	79.46	12.58
7	3.55	144.3	20,820	0.01635	0.4982	63.02	15.87
8	3.15	128.5	16,510	0.01297	0.6282	49.98	20.01
9	2.80	114.4	13,090	0.01028	0.7921	39.63	25.23
10	2.50	101.9	10,380	0.008155	0.9989	31.43	31.82
11	2.24	90.74	8,234	0.006467	1.260	24.92	40.12
12	2.00	80.81	6,530	0.005129	1.588	19.77	50.59
13	1.80	71.96	5,178	0.004067	2.003	15.68	63.80
14	1.60	64.08	4,107	0.003225	2.525	12.43	80.44
15	1.40	57.07	3,257	0.002558	3.184	9.858	101.4
16	1.25	50.82	2,583	0.002028	4.016	7.818	127.9
17	1.12	45.26	2,048	0.001609	5.064	6.200	161.3
18	1.00	40.30	1,624	0.001276	6.385	4.917	203.4
19	0.90	35.89	1,288	0.001012	8.051	3.899	256.5
20	0.80	31.96	1,022	0.0008023	10.15	3.092	323.4
21	0.71	28.46	810.1	0.0006363	12.80	2.452	407.8
22	0.63	25.35	642.4	0.0005046	16.14	1.945	514.2
23	0.56	22.57	509.5	0.0004002	20.36	1.542	648.4
24	0.50	20.10	404.0	0.0003173	25.67	1.223	817.7
25	0.45	17.90	320.4	0.0002517	32.37	0.9699	1,031.0
26	0.40	15.94	254.1	0.0001996	40.81	0.7692	1,300
27	0.355	14.20	201.5	0.0001583	51.47	0.6100	1,639
28	0.315	12.64	159.8	0.0001255	64.90	0.4837	2,067
29	0.280	11.26	126.7	0.00009953	81.83	0.3836	2,607
30	0.250	10.03	100.5	0.00007894	103.2	0.3042	3,287
31	0.224	8.928	79.70	0.00006260	130.1	0.2413	4,145
32	0.200	7.950	63.21	0.00004964	164.1	0.1913	5,227
33	0.180	7.080	50.13	0.00003937	206.9	0.1517	6,591
34	0.160	6.305	39.75	0.00003122	260.9	0.1203	8,310
35	0.140	5.615	31.52	0.00002476	329.0	0.09542	10,480
36	0.125	5.000	25.00	0.00001964	414.8	0.07568	13,210
37	0.112	4.453	19.83	0.00001557	523.1	0.06001	16,660
38	0.100	3.965	15.72	0.00001235	659.6	0.04759	21,010
39	0.090	3.531	12.47	0.000009793	831.8	0.03774	26,500
40	0.080	3.145	9.888	0.000007766	1049.0	0.02993	33,410

Appendix 4

RESISTIVITY OF METALS AND ALLOYS

Material	Resistivity (Ω)
Aluminum	0.00000262
Brass (66% Cu, 34% Zn)	0.0000039
Carbon (graphite form)	0.0014
Constantan (55% Cu, 45% Ni)	0.0000442
Copper (commercial annealed)	0.0000017241
German silver (18% Ni)	0.000033
Iron	0.00000971
Lead	0.0000219
Mercury	0.0000958
Monel metal (67% Ni, 30% Cu, 1.4% Fe, 1% Mn)	0.000042
Nichrome (65% Ni, 12% Cr, 23% Fe)	0.000100
Nickel	0.0000069
Phosphor bronze (4% Sn, 0.5% P, 95.5% Cu)	0.0000094
Silver	0.00000162
Steel (0.4–0.5% C)	0.000013–0.000022
Tantalum	0.0000131
Tin	0.0000114
Tungsten	0.00000548
Zinc	0.000006

Source: *Reference Data for Radio Engineers*, 6th ed., 1975, Howard W. Sams & Co., Inc.

Appendix 5

TEMPERATURE COEFFICIENTS OF RESISTANCE

Material	Temperature Coefficient at 20°C (per °C)
Aluminum	0.0039
Brass (66% Cu, 34% Zn)	0.002
Carbon (graphite form)	−0.0005
Copper (commercial annealed)	0.0039
German silver (18% Ni)	0.0004
Iron	0.0052–0.0062
Lead	0.004
Mercury	0.00089
Monel metal (67% Ni, 30% Cu, 1.4% Fe, 1% Mn)	0.002
Nichrome (65% Ni, 12% Cr, 23% Fe)	0.00017
Nickel	0.0047
Phosphor bronze (4% Sn, 0.5% P, 95.5% Cu)	0.003
Silver	0.0038
Steel (0.4–0.5% C)	0.003
Tantalum	0.003
Tin	0.0042
Tungsten	0.0045
Zinc	0.0037

Source: *Reference Data for Radio Engineers*, 6th ed., 1975, Howard W. Sams & Co., Inc.

Appendix 6

TRIGONOMETRIC FUNCTIONS

Angle (degrees)	Sin	Cos	Tan
0	0.000	1.000	0.000
1	.018	1.000	.018
2	.035	.999	.035
3	.052	.999	.052
4	.070	.998	.070
5	.087	.996	.088
6	.105	.995	.105
7	.122	.993	.123
8	.139	.990	.141
9	.156	.988	.158
10	.174	.985	.176
11	.191	.982	.194
12	.208	.978	.213
13	.225	.974	.231
14	.242	.970	.249
15	.259	.966	.268
16	.276	.961	.287
17	.292	.956	.306
18	.309	.951	.325
19	.326	.946	.344
20	.342	.940	.364
21	.358	.934	.384
22	.375	.927	.404
23	.391	.921	.425
24	.407	.914	.445
25	.423	.906	.466
26	.438	.899	.488
27	.454	.891	.510
28	.470	.883	.532
29	.485	.875	.554
30	.500	.866	.577
31	.515	.857	.601
32	.530	.848	.625
33	.545	.839	.649
34	.559	.829	.675
35	.574	.819	.700
36	.588	.809	.727
37	.602	.799	.754
38	.616	.788	.781
39	.629	.777	.810
40	.643	.766	.839
41	.656	.755	.869
42	.669	.743	.900
43	.682	.731	.933
44	.695	.719	.966
45	.707	.707	1.000
46	.719	.695	1.036
47	.731	.682	1.072
48	.743	.669	1.111
49	.755	.656	1.150
50	.766	.643	1.192
51	.777	.629	1.235
52	.788	.616	1.280
53	.799	.602	1.327
54	.809	.588	1.376
55	.819	.574	1.428
56	.829	.559	1.483
57	.839	.545	1.540
58	.848	.530	1.600
59	.857	.515	1.664
60	.866	.500	1.732
61	.875	.485	1.804
62	.883	.470	1.881
63	.891	.454	1.963
64	.899	.438	2.050
65	.906	.423	2.145
66	.914	.407	2.246
67	.921	.391	2.356
68	.927	.375	2.475
69	.934	.358	2.605
70	.940	.342	2.748
71	.946	.326	2.904
72	.951	.309	3.078
73	.956	.292	3.271
74	.961	.276	3.487
75	.966	.259	3.732
76	.970	.242	4.011
77	.974	.225	4.332
78	.978	.208	4.705
79	.982	.191	5.145
80	.985	.174	5.671
81	.988	.156	6.314
82	.990	.139	7.115
83	.993	.1229	8.144
84	.995	.105	9.514
85	.996	.087	11.43
86	.998	.070	14.30
87	.999	.052	19.08
88	.999	.035	28.64
89	1.000	.018	57.29
90	1.000	.000	—

Index